Chemical and Biological Aspects of Steroid Conjugation

Chemical and Biological Aspects of Steroid Conjugation

Edited by

SEYMOUR BERNSTEIN

Lederle Laboratories
A Division of American Cyanamid Company

and

SAMUEL SOLOMON

McGill University and Royal Victoria Hospital

SPRINGER-VERLAG BERLIN · HEIDELBERG · NEW YORK 1970

ISBN-13: 978-3-642-95179-4 e-ISBN-13: 978-3-642-95177-0
DOI: 10.1007/978-3-642-95177-0

 Library of Congress Catalog Number 70-98420.

Softcover reprint of the hardcover 1st edition 1970

Title Number 1635

To Ruth and Sheila

PREFACE

This book represents a collaborative endeavor by a group of investigators to bring together in a single volume a critical discussion of the major facets of our knowledge, ranging from chemical to clinical aspects, of steroid conjugation. However, the important field of bile acid and bile alcohol conjugation has been discussed only superficially since it was decided arbitrarily to be outside the projected scope of the book.

The reader is referred to the companion volume of this book, namely *Physical Properties of Steroid Conjugates* (by Bernstein, Dusza, and Joseph, Springer-Verlag New York 1968), for complemental information on individual conjugates.

The Editors wish to thank Miss Elise Kramer for typing assistance. Also we wish to express our deep appreciation and gratitude to Mrs. Dorothy Budd, our copy-editor, and Mr. J. Joseph for their editorial assistance in making this book possible.

Pearl River, N.Y.
January, 1970

SEYMOUR BERNSTEIN
SAMUEL SOLOMON

CONTENTS

NOMENCLATURE

a) Discussion of Nomenclature System

Countless problems on steroid nomenclature have arisen in editing a book of this type which represents the joint efforts of many contributors. No uniformly accepted standards of nomenclature have been reached by steroid scientists, regardless of IUPAC recommendations. The Editors have, consequently, taken on the unenviable task of standardizing trivial and systematic nomenclature throughout this book, and have edited all chapters accordingly. The nomenclature system chosen may or may not represent the choice of the individual contributors. The principal aspects of this nomenclature system may be suitably illustrated by examples:

estrone (trivial name) ≅ 3–hydroxyestra–1,3,5(10)–trien–17–one (systematic name)

dehydroisoandrosterone glucuronide ≅ 17–oxoandrost–5–en–3β–yl–β–D–glucopyranosiduronic acid

dehydroisoandrosterone sulfate ≅ 3β–sulfooxyandrost–5–en–17–one

estriol–3–sulfate–16–glucuronide ≅ 17β–hydroxy–3–sulfooxyestra–1,3,5(10)–trien–16α–yl–β–D–glucopyranosiduronic acid

17α–estradiol–3–glucuronide–17–β–N–acetylglucosaminide ≅ estra–1,3,5(10)–triene–3,17α–di–yl–3–β–D–glucopyranosiduronic acid–17–2′–acetamido–2′–deoxy–β–D–glucopyranoside

pregn–5–ene–3β,20α–diol–3–sulfate–20–N–acetylglucosaminide ≅ 3β–sulfooxypregn–5–en–20α–yl–2′–acetamido–2′–deoxy–β–D–glucopyranoside.

Abbreviations for compounds, such as *F* ≅ cortisol (hydrocortisone) or *DS* ≅ dehydroisoandrosterone sulfate, generally have not been used in this book. Only in a few special instances (*e.g.*, Chapter 6, Fig. 2) has it been deemed advantageous to employ them.

b) Trivial and Systematic Names

Each chapter is self–explanatory in regard to the equivalent trivial and systematic names of compounds discussed therein. The first time (subsequent to the Introduction) a trivial name for a compound is cited in a chapter its systematic name is shown. Thereafter only the trivial name is employed in the text.

LIST OF CONTRIBUTORS

Seymour Bernstein, Ph. D.

Senior Research Chemist and Group Leader, Lederle Laboratories, A Division of American Cyanamid Company, Pearl River, New York

Bhagu R. Bhavnani, Ph.D.

Lecturer, Dept. of Experimental Medicine, McGill University and Associated Scientist, Royal Victoria Hospital Montreal, Canada

Alfred M. Bongiovanni, M.D.

Physician-in-Chief, The Children's Hospital of Philadelphia, Professor and Chairman, Department of Pediatrics, School of Medicine, University of Pennsylvania, Philadelphia, Pa.

H. Leon Bradlow, Ph.D.

Senior Investigator, Institute for Steroid Research, Montefiore Hospital and Medical Center, Bronx, New York

Robert M. Cohn, M.D.

Fellow in Metabolic Diseases, The Children's Hospital of Philadelphia, Instructor of Pediatrics, School of Medicine, University of Pennsylvania Philadelphia, Pa.

Egon Diczfalusy, M.D.

Professor and Director, Swedish Medical Research Council Reproductive Endocrinology Research Unit, Karolinska Sjukhuset, Stockholm, Sweden

John P. Dusza, Ph.D.

Senior Research Chemist, Lederle Laboratories, A Division of American Cyanamid Company, Pearl River, New York

Francis Herr, M.D., D.Sc.

Associate Director of Research, Ayerst Laboratories, St. Laurent, Montreal, Canada

John B. Jewell, M.D.

Medical Director, Ayerst Laboratories, St. Laurent, Montreal, Canada

Joseph P. Joseph, B.S.

Research Chemist, Lederle Laboratories, A Division of American Cyanamid Company, Pearl River, New York

Mortimer Levitz, Ph.D.

Professor, Dept. of Obstetrics and Gynecology, New York University Medical Center, New York, N.Y.

Seymour Lieberman, Ph.D.

Professor of Biochemistry, Dept. of Biochemistry, College of Physicians and Surgeons, Columbia University, New York, N.Y.

A. John Manson, Ph.D.

Assistant to the Research Director, Ayerst Laboratories, St. Laurent, Montreal, Canada

Clara Revesz, M.D.

Head of Endocrinology, Ayerst Laboratories, St. Laurent, Montreal, Canada

Kenneth D. Roberts, Ph.D.

Assistant Professor of Biochemistry, University of Montreal, and Department of Laboratories, Maisonneuve Hospital, Montreal, Canada

Alexander B. Roy, Ph.D., D.Sc.

Senior Fellow, Dept. of Physical Biochemistry, John Curtin School of Medical Research, Australian National University, Canberra, Australia

Pentti K. Siiteri, Ph.D.

Associate Professor, Dept. of Obstetrics and Gynecology and Biochemistry, The University of Texas, Southwestern Medical School, Dallas, Texas

Samuel Solomon, Ph.D.

Professor, Departments of Biochemistry and Experimental Medicine, McGill University, and Director, Steroid Laboratory, Royal Victoria Hospital, Montreal, Canada

Chemical and Biological Aspects of Steroid Conjugation

Chapter 1

Chemistry: Synthesis and Characterization

SEYMOUR BERNSTEIN, JOHN P. DUSZA, and JOSEPH P. JOSEPH

Contents

I. Introduction and Scope

A large number and wide variety of steroid conjugates isolated from animal and human sources have been described in the literature. Many of these conjugates represent novel and interesting steroid conjugation mechanisms, but for some, the evidence for their existence is either preliminary and/or inadequate.

The following list records with extensive documentation the various conjugation mechanisms reported to date:

(a) *Acylation* [1–5].
(b) *O-Methylation* [6–11].
(c) *Glucuronidation* [12].
(d) *N-Acetylglucosaminidation* [13–18].
(e) *Sulfation* [12].
(f) *Sulfatidation* [19–26].
(g) *Phosphatidation* [21, 25, 27–29].
(h) *Nucleosidation* [30].
(i) *Nucleotidation* [31].
(j) *Conjugation with glutathione* [32–35], *glycine* [36], *and taurine* [36].

The scope of this chapter will be limited to a discussion of selected chemical and physical aspects of steroid conjugation, and will deal with glucuronidation, sulfation, and the recently established novel mode of conjugation, N-acetylglucosaminidation. For these purposes the literature as of September 1968 has been surveyed.

The reader is referred to the companion volume to this book, namely, *Physical Properties of Steroid Conjugates* [by Bernstein, Dusza, and Joseph, Springer-Verlag, New York, New York 1968] for additional chemical and physical information on individual conjugates.

The following abbreviations are employed in this chapter:

St ≅ steroidal
G ≅ glucuronide ≅

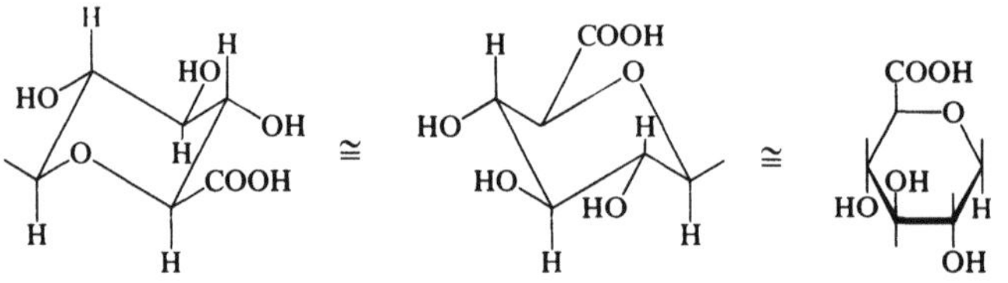

GAc_3Me ≅ glucuronide triacetate methyl ester

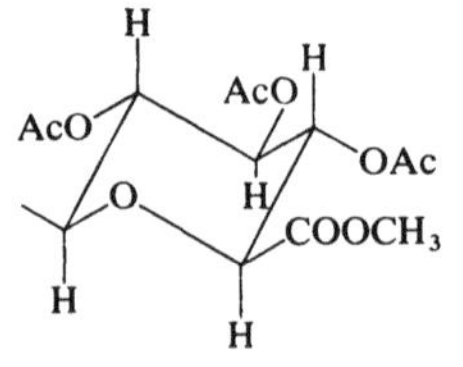

1α-$BrGAc_3Me$ ≅ methyl [1α-bromo-1-deoxy-2,3,4-tri-*O*-acetyl-D-glucopyranosid]uronate

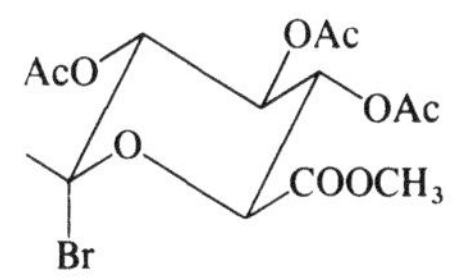

GlNAc ≅ N–acetylglucosaminide

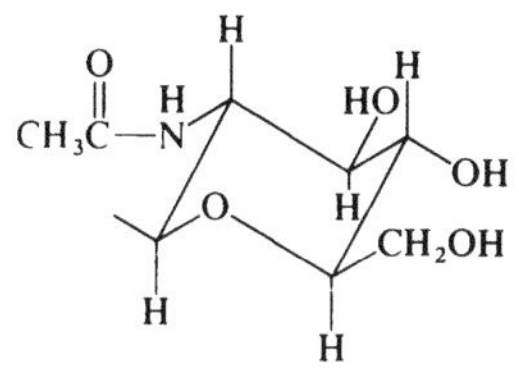

GMe ≅ glucuronide methyl ester
GMe_3Me ≅ glucuronide trimethyl ether methyl ester
GM ≅ glucuronide metal salt (*e.g.*, M = K)
GNH_2 ≅ glucuronamide

II. Glucuronides

A. Types

A variety of steroid glucuronides has been isolated from natural sources, and their structures may be classified chemically as follows:

1. Alkyl Glucuronides

This glucuronide type is probably the most commonly occurring in nature, and is formed during the metabolism of practically all steroids, estrogens, androgens, corticoids, *etc*. Representative examples are androsterone glucuronide (I) [17–oxo–5α–androstan–3α–yl–β–D–glucopyranosiduronic acid], estriol–16–glucuronide (II) [3,17β–dihydroxyestra–1,3, 5(10)–trien–16α–yl–β–D–glucopyranosiduronic acid], and tetrahydrocortisol–3–glucuronide (III) [11β,17α,21–trihydroxy–20–oxo–5β–pregnan–3α–yl–β–D–glucopyranosiduronic acid]:

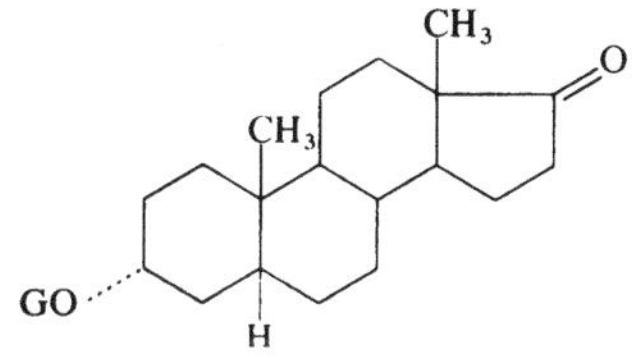

I

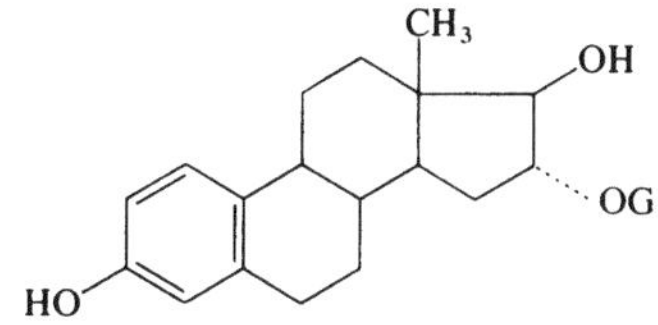

II

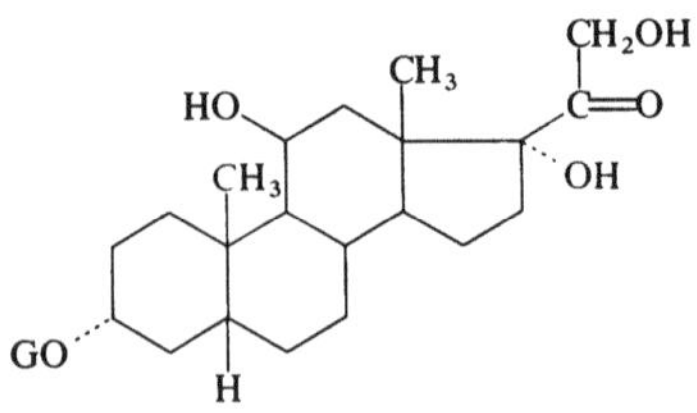

III

A variant of this type is illustrated by the enol–glucuronide (IV) of testosterone [17β–hydroxyandrosta–3,5–dien–3–yl–β–D–glucopyranosid-uronic acid]:

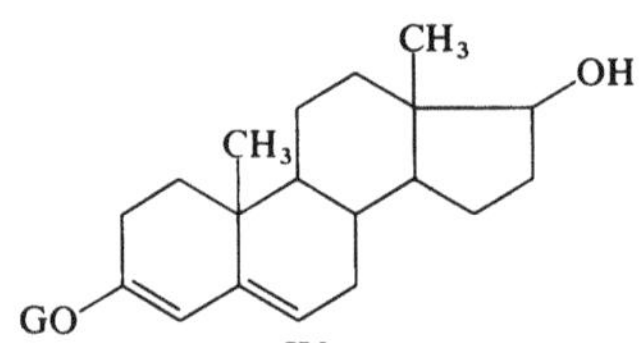

IV

2. *Aryl Glucuronides*

Estrogen metabolites have been obviously the only source of this type of glucuronide in the steroid field. This may be illustrated by estriol–3–glucuronide (V) [16α,17β–dihydroxyestra–1,3,5(10)–trien–3–yl–β–D–gluco-pyranosiduronic acid] found in human pregnancy urine.

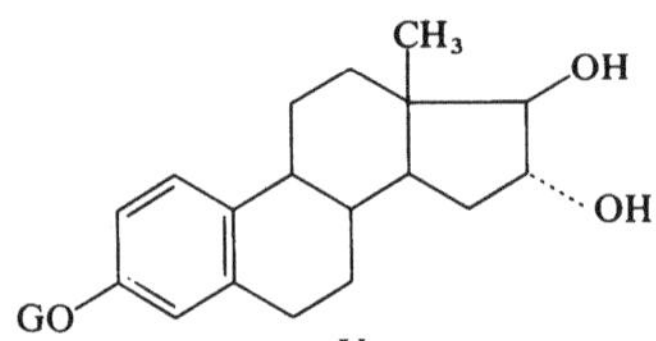

V

3. *Acyl Glucuronides*

Although acyl glucuronidation represents a common metabolic pathway [37] only one example of this type has been reported in the steroid hormone literature [38], namely, the glucuronide (VI) of 17β–carboxy-androst–4–en–3–one. Related to this is the isolation of the glucuronide (VII) of the steroid antibiotic, fusidic acid [39], and the synthesis of glycyrrhetic acid glucuronide in the form of its tri–*O*–acetyl methyl ester (VIII) [40].

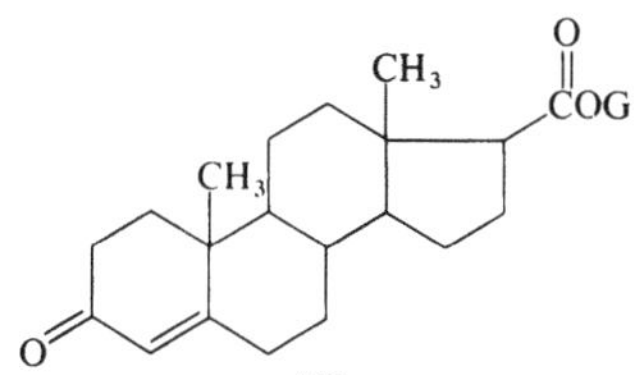

VI

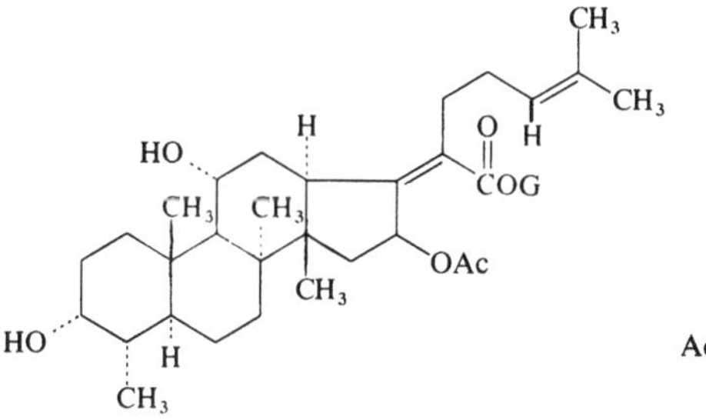

VII

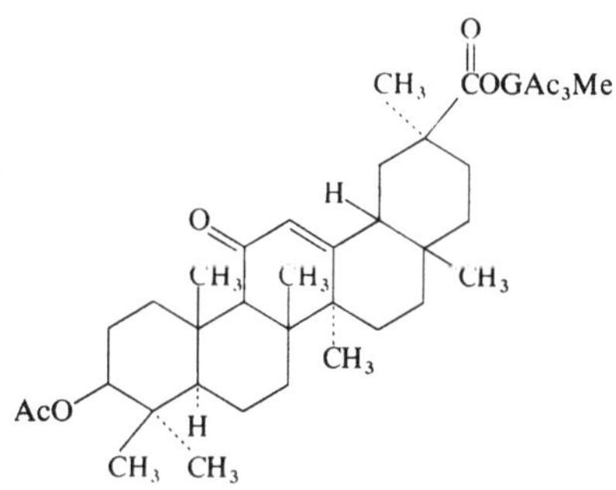

VIII

4. "*Hemiacetal*" *Glucuronides*

The "oxo–conjugate" (IX) of aldosterone [11β,21–dihydroxy–3,20–dioxopregn–4–en–18–al] presumably is the outstanding example of this type [41–51].

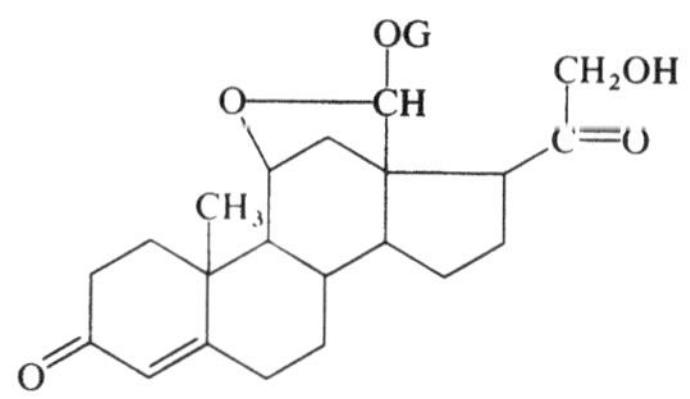

IX

B. Chemical Synthesis

The most commonly employed method for the synthesis of steroid glucuronides involves the condensation of a halosugar with a steroid aglycon under Koenigs–Knorr conditions [52]. This may be accomplished directly by using methyl 1α–bromo–1–deoxy–2,3,4–tri–*O*–acetyl–β–D–glucopyranosiduronate (I) (1α–BrGAc$_3$Me) as the condensing halosugar or indirectly *via* 1α–halo–1–deoxy–2,3,4,6–tetra–*O*–acetyl–β–D–glucopyranose with subsequent removal of the sugar acetoxy groups and oxidation of the C6–hydroxyl group to a carboxylic acid function.

The Koenigs–Knorr reaction may be described as the condensation of a blocked sugar halide with an alcohol or phenol, generally under anhydrous conditions, in the presence of an acid acceptor such as silver oxide or silver carbonate. This reaction has been the subject of many review articles [53–60], and will not be discussed extensively in this chapter.

It would appear that this condensation reaction proceeds mechanistically *via* Walden inversion; hence an α-halosugar on reaction with a steroidal aglycon produces a β-glycoside. Most of the theoretical and practical considerations of this reaction have dealt with glycoside rather than with glucuronide formation. Nevertheless, these studies and conclusions are applicable to glucuronidation.

In an excellent review on the chemistry of D–glucuronic acid, Marsh [37] states that all glucuronides isolated from animal sources, whose structures have been confirmed by chemical synthesis, occur as β–D–glucopyranosiduronic acids.

The major part of this section will be devoted to a discussion of the direct glucuronidation of steroids, since this is the method employed by the vast majority of investigators. The pertinent literature will be reviewed on the basis of (1) reactivity of the halosugar, (2) suitable conditions for reaction and the acid acceptor, and (3) the synthesis of aryl– and alkyl–glucuronides.

1. Reactivity of the Halosugar

Since the condensation proceeds *via* Walden inversion, the formation of a β–glucuronide is favored by a *cis*–relationship between the C2–acetoxy group and the α–halogen at C1 (anchimeric assist), and the reactivity of the latter:

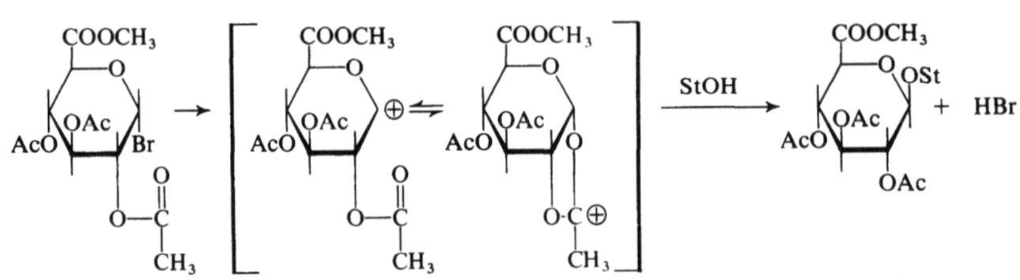

I II

Although methyl 1α–chloro–triacetyl glucuronide was first used by Goebel and Babers [61] it was the subsequent preparation of the more reactive methyl 1α–bromo–1–deoxy–2,3,4–tri–*O*–acetyl–β–D–glucopyranosiduronate (I) by these investigators [62] which facilitated the preparation of steroid glucuronides. The methyl 1α–iodo–triacetyl–glucuronide [63] has also been prepared and assayed in this reaction, but has not been found superior to the 1α–bromoderivative.

The methyl 1α–bromo–triacetyl–glucuronide (I) may be most conveniently prepared from glucuronolactone (III) by base–catalyzed esterification in methanol to give the methyl glucuronide (IV). Acetylation and treatment with 30–32% hydrobromic acid in glacial acetic acid afford the 1α–bromo-sugar (I) [62, 64].

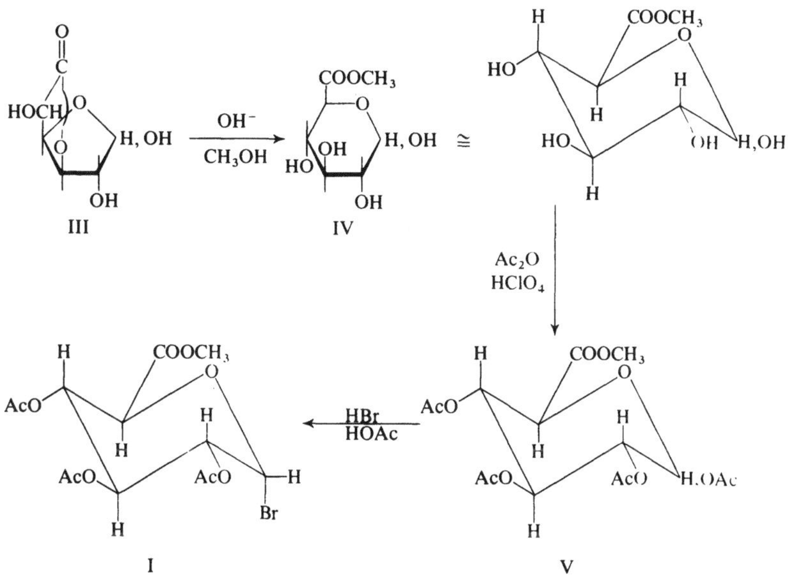

Marsh [37] reviews the preparation of ^{14}C–labeled glucuronic acid from the catalytic oxidation of 1,2–isopropylidene–D–glucose. The latter is obtained from glucose which has been randomly labeled with ^{14}C by photosynthesis from $^{14}CO_2$. Other techniques employed involve the use of the Kiliani reaction with a xylose derivative and $Na^{14}CN$ to build the next higher sugar homologue. The separation of stereoisomers and necessary oxidation and/or reduction will give glucuronic acid labeled at $^{14}C1$ or $^{14}C6$. In this connection no reference has been found for the chemical synthesis of steroidal glucuronides using labeled glucuronic acid.

In an unsuccessful attempt to prepare a crystalline intermediate in the synthesis of hydrocortisone–21–glucuronide [11β,17α–dihydroxy–3,20–dioxopregn–4–en–21–yl–β–D–glucopyranosiduronic acid], Zorbach and Valiaveedan [80] utilized the modified bromosugar, methyl 1α–bromo–1–deoxy–2,3,4–tri–*O*–benzoyl–β–D–glucopyranosiduronate. Pravdic and Keglevic [65], in their studies on the preparation of simple acyl–glucuronides *via* the carbodiimide method, made benzyl 1β–hydroxy–2,3,4–tri–*O*–benzyl–β–D–glucopyranosiduronate and its 1β–chloro derivative [66]. The removal of the benzyl groups by hydrogenolysis should make this a useful starting material for the formation of steroid glucuronides which may be sensitive to the usual conditions required for the removal of acyl blocking groups.

2. Conditions for Reaction and Acid Acceptors

There are two standard modifications of the Koenigs–Knorr reaction as applied to steroids:

(a) the Schapiro [67] modification which involves agitation of the reactants, *i.e.*, the halosugar, the silver salt as acid acceptor, and steroid, in an anhydrous solvent at room temperature for 18 or more hours; and

(b) the Meystre and Mischer [68] modification, which uses an azeotropic distillation of solvent and water during the addition of the halosugar in order to maintain anhydrous conditions.

It is generally recognized that water plays a detrimental role in the Koenigs–Knorr reaction; anhydrous solvents and completely dried reaction vessels are necessary for successful reaction. Goldschmid and Perlin [69] in a detailed investigation of factors affecting the reaction have stated that water from such sources as solvent, acid acceptor, and the atmosphere, may yield an orthoacetate, a 1–hydroxylated sugar or a polymeric sugar. Water may also be formed by the reaction between the bromosugar, the aglycon and silver oxide (Eq. (a)):

$$2RX + 2StOH + Ag_2O \rightarrow 2StOR + 2AgX + H_2O. \qquad (a)$$

The slow addition of the halosugar to the steroid under azeotropic distillation conditions in some cases increased the yield presumably due to the constant removal of water. However, Wotiz *et al.* [70] in a comparison of the Schapiro (room temperature) *vs* Meystre and Mischer (azeotropic distillation) procedures have reported inconsistent results.

Another technique used to maintain anhydrous conditions has been the employment of desiccants such as calcium chloride, calcium sulfate, calcium hydride, *etc.*, directly in the reaction mixture [71–74]. The use of a desiccant (calcium sulfate) has been noted to reduce the rate of reaction [75], and to compensate for this, elemental iodine has been added as a catalyst. Iodine has been suggested as affecting the polarity of the solvent which should increase the reaction rate in an SN1–type mechanism as is proposed for the synthesis involving a glycosyl halide. Iodine has also been shown to suppress side reactions between glycosyl halides and silver oxide [69, 71].

It is generally agreed that the silver salts used as catalysts and acid acceptors in the Koenigs–Knorr reaction should be freshly prepared. Wolfrom *et al.* [76, 77] have outlined a preparation of silver carbonate used in sugar condensations. In addition, there are two preparations of silver oxide described by Cowdrey [78] and co-workers, and by Helferich and Klein [79]. Other agents used with limited success in this condensation have been mercuric cyanide and silver perchlorate [80, 81].

Although anhydrous benzene has been the most common of the inert solvents to be used as a medium for the Koenigs–Knorr reaction, several other solvents have also been employed successfully; namely, chloroform [82], carbon tetrachloride [80], dioxane [80], 1,2–dichloroethane [81], and nitromethane [80, 83].

Only one detailed study of the reaction time necessary for complete condensation of the bromosugar (I) (1α–$BrGAc_3Me$) with a steroid has been reported [84, 85]. Testosterone–17–glucuronide triacetate methyl ester [methyl(3–oxoandrost–4–en–17β–yl–2',3',4'–tri–*O*–acetyl–β–D–glucopyranosid)uronate] was formed in 18% yield when the bromosugar (1α–$BrGAc_3Me$) was added over a period of two hours at 25 to 37°. No further reaction was noted after this time although the reaction was allowed to proceed for 42 hours. Generally in the azeotropic distillation procedure the reaction has been terminated when all the halosugar had been added, while in the Schapiro method agitation of reaction mixture was provided for periods of 18 to 24 hours.

3. The Synthesis of Aryl and Alkyl Glucuronides

a) Aryl Glucuronides. An analysis of the literature has revealed that aromatic hydroxyl groups on steroids are considerably less reactive toward glucuronide formation than are alkyl hydroxyl groups. Condensation of estrone [3–hydroxyestra–1,3,5(10)–trien–17–one] or estriol–16,17–diacetate [16α,17β–diacetoxyestra–1,3,5(10)–trien–3–ol] with 1α–$BrGAc_3Me$ has resulted in yields of less than 10% [67, 70, 86]. Parenthetically, the literature shows that 17β–estradiol–3–glucuronide has been prepared only by the metal borohydride reduction of estrone–glucuronide triacetate methyl ester [methyl(17–oxoestra–1,3,5(10)–trien–3–yl–2',3',4'–triacetyl–β–D–glucopyranosid)uronate] [86, 87] although a suitably blocked 17–derivative of 17β–estradiol should undergo direct condensation. No explanation has yet been offered to account for the low yields encountered in the preparation of aryl steroidal glucuronides, despite the fact that simple phenol glucuronides have been obtained in respectable yields [64].

b) Alkyl Glucuronides. Historically, Schapiro [67, 88], recognizing the importance of the discoveries of Cohen, Marrian, and Odell [89, 90], and Venning and Browne [91] that water-soluble sex hormones could be found in pregnancy urine, carried out the first chemical synthesis of steroid glucuronides; namely, dehydroisoandrosterone–3–glucuronide [17–oxoandrost–5–en–3β–yl–β–D–glucopyranosiduronic acid] and 17β–estradiol–17–glucuronide. The structure of the water–soluble material isolated from human pregnancy urine, by Venning and Browne in 1936, was determined to be sodium pregnanediol glucuronide [20α–hydroxy–5β–pregnan–3α–yl–β–D–glucopyranosiduronic acid sodium salt] by Heard, Hoffman, and Mack [92]. Its chemical synthesis was carried out by Huebner, Overman, and Link [93] who employed the Schapiro modification of the Koenigs–Knorr reaction, thus confirming the structure of this important metabolite. Subsequently numerous syntheses of steroidal alkyl glucuronides have been reported [67, 70, 80–85, 93–109], for which a detailed discussion appears unwarranted. For a more complete coverage the reader is referred to the companion volume, "Physical Properties of Steroid Conjugates".

The synthesis of alkyl glucuronides may be influenced by steroidal conformational factors [84]. The spatial arrangement of the axial 3α–hydroxyl group of androsterone with regard to 1,3 diaxial interactions presents a greater restriction to the incoming bulky bromosugar (1α–BrGAc$_3$Me) than the isomeric etiocholanolone with a 3α–equatorial hydroxyl. The findings of Staib and Dönges [95] appear to support this observation. These investigators reported the synthesis of androsterone and etiocholanolone glucuronides [17–oxo–5β–androstan–3α–yl–β–D–glucopyranosiduronic acid] in 8 and 31% yields, respectively. Emiliozzi [105], however, reports the preparation of 3 out of the 4 (except 3β,5β) possible C3,5 conformational forms of "androsterone glucuronide" in yields of 15–20%. The interference of the C18–CH$_3$ on C17–β–glucuronide formation may have some effect on yields (18 to 29%) [70] in the case of testosterone. The validity of this effect is in doubt since 17β–estradiol [estra–1,3,5(10)–triene–3,17β–diol] blocked glucuronide (E$_2$–17–GAc$_3$Me) has been prepared in yields up to 70% [86]. The equatorial 3β–hydroxyl group of dehydroisoandrosterone [3β–hydroxyandrost–5–en–17–one] offers little stereochemical interference to the approaching bromosugar, and condensation occurs in 60% yield [94].

Estriol–16–glucuronide (XI), first isolated from human pregnancy urine by Cohen and Marrian [89], has been the subject of considerable scientific investigation; its structure has been determined rigorously by biochemical and chemical methods [110–116]. The synthesis of this and related compounds presenting an interesting exercise in selective condensation *via* a Koenigs–Knorr reaction, has been reported by two groups of investigators [86, 87, 117, 118]. The synthetic approach by both groups was similar and was based on two key reactions, the use of a benzyl ether [119, 120] as a selective blocking group for the aromatic hydroxyl function, and the conversion of estrone–3–benzyl ether (VI) [3–benzyloxyestra–1,3,5(10)–

trien–17–one] into 16α–hydroxyestrone–3–benzyl ether (VII) [3–benzyl-oxy–16α–hydroxyestra–1,3,5(10)–trien–17–one] according to the procedure of Leeds, Fukushima, and Gallagher [121]. The synthesis of the 16–glucuronide can be illustrated as follows:

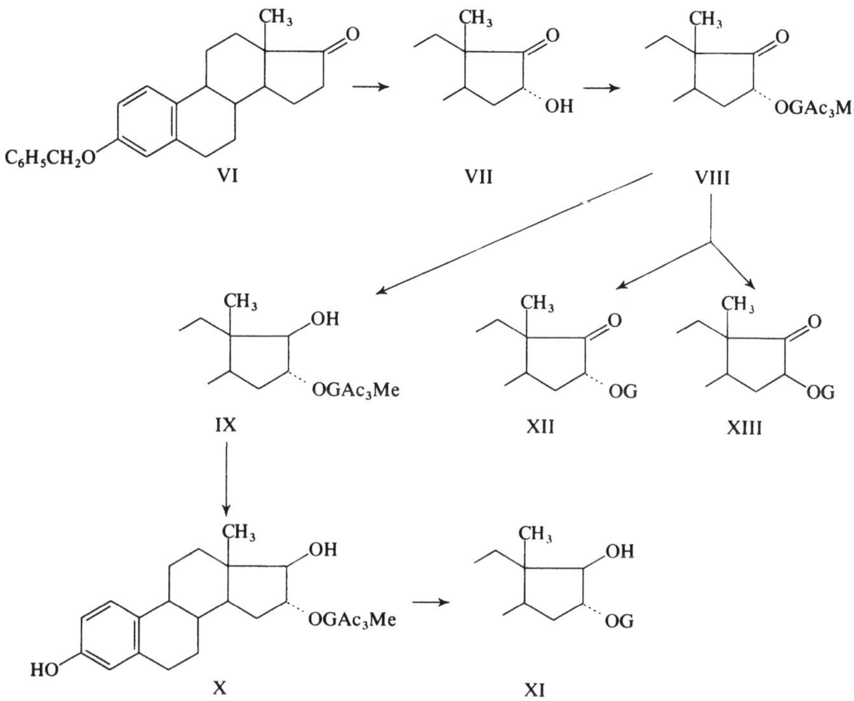

The reduction of the 17–ketone (VIII) was accomplished by Elce and co-workers [86] with sodium borohydride and the 17β–hydroxy compound (IX) was primarily obtained. Nambara and Imai [87] have reported the use of potassium borohydride for this reduction having found that the carbomethoxy group of the sugar moiety is reduced under sodium borohydride conditions. Supporting evidence for this may be found in the reduction of methyl (methyl–α–D–glucopyranosid)uronate to methyl α–D–glucopyranoside, by Wolfrom and Anno [122]. Removal of the benzyl group from IX followed by basic hydrolysis of acyl blocking groups readily gave estriol–16–glucuronide (XI). The base hydrolysis of the protective groups on the sugar moiety of VIII gave predictably a 1:1 equilibration mixture of XII and XIII. Employing compound VII, estriol–17–glucuronide (XVII) [3,16α–dihydroxyestra–1,3,5(10)–trien–17β–yl–β–D–glucopyranosiduronic acid] was also prepared:

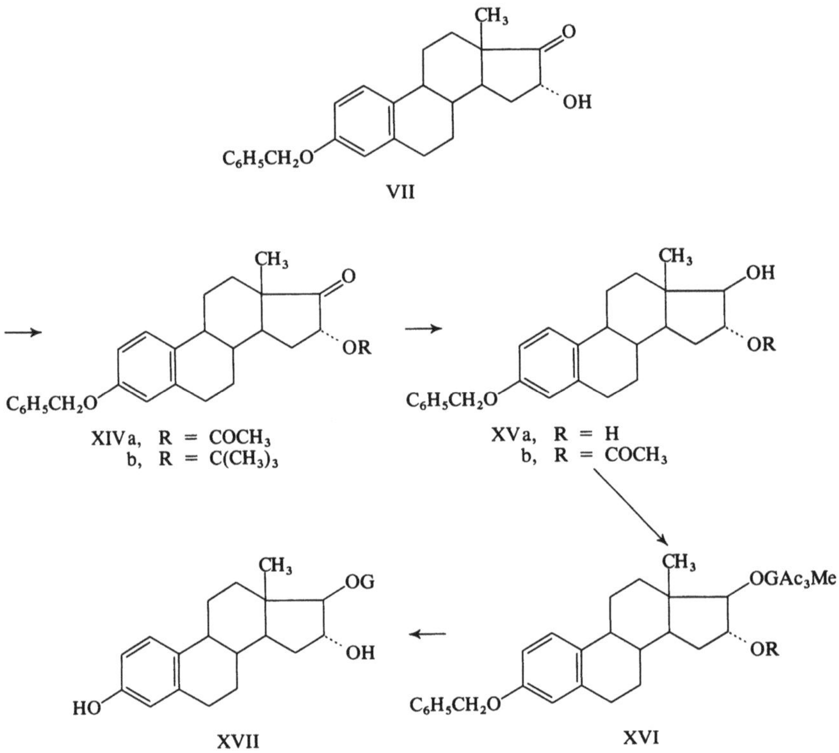

Whereas Elce and co-workers used an acetyl group (XIVa) to protect the 16α–hydroxyl group, Nambara and Imai successfully employed the tertiary butoxy group (XIVb) at this position to avoid possible acyl migration.

Elce *et al.* also observed a variation in yield of glucuronide formation dependent upon the position of the hydroxyl group, and the steric and electronic influence of the neighboring groups. For 17β–estradiol–3–benzyl ether [3–benzyloxyestra–1,3,5(10)–trien–17β–ol] a 60–70% yield of glucuronide was achieved. However, when a 16α–acetoxy group (XVb) was introduced, the yield was lowered to 40% presumably due to hydrogen bonding. With 16β–acetoxy–17β–hydroxy, however, the yield of glucuronide was only 15%, this difference being ascribed to even stronger hydrogen bonding in this case. Condensation of the bromosugar with the 16α–hydroxy–17–ketone gave only a 23% yield of glucuronide regardless of the spatial availability of the hydroxyl group. This low yield was attributed to the withdrawal of electrons by the adjacent carbonyl group resulting in the 16α–hydroxyl group being less nucleophilic than an ordinary secondary hydroxyl group. The condensation of the isomeric α–ketol, 17β–hydroxy–

16–one, appeared to be influenced by both steric (C18–CH_3) and electronic effects, since only a 2% yield of the C17–glucuronide was obtained in this instance.

Joseph, Dusza, and Bernstein [123] have condensed estriol–3–benzyl ether (XVa) [3–benzyloxyestra–1,3,5(10)–triene–16α,17β–diol] directly with methyl 1α–bromo–1–deoxy–2,3,4–tri–*O*–acetyl–β–D–glucopyranosiduronate (I) to obtain IX, the 16–glucuronide. This finding neatly illustrated the difference in reactivity between the C16α– and C17β–hydroxyl groups, thereby obviating the need of a protective grouping at C17.

Steroidal enol–glucuronides may be regarded as a special type of alkyl glucuronide. Schubert [124], Wotiz and Fishman [125], and Hadd [126] have presented evidence for their existence as metabolites. The synthesis of the enol glucuronides of androstenedione [androst–4–ene–3,17–dione] and progesterone [pregn–4–ene–3,20–dione] by the Koenigs–Knorr reaction has been described by Wotiz *et al.* [70, 127]. The compounds were characterized by spectral data as well as by enzymatic hydrolysis with β–glucuronidase [128]. The spectral evidence for these compounds is presented elsewhere in this chapter.

An additional preparative procedure for glucuronides involves the oxidation of glycopyranosides, a technique known for many years. A review of this oxidation reaction has been written by Heynes and Paulsen [129]. In this regard, Winguth [98] has described the conversion of the primary hydroxyl group of cholesterol, ergosterol, androsterone, and dehydroisoandrosterone–β–D–glucopyranosides with molecular oxygen and platinum on carbon to the corresponding glucuronides. The crude products were isolated in the form of their chinconine or sodium salts in yields of approximately 50%.

4. Other Methods of Glucuronidation

In an effort to improve the yield of steroid glucuronide formation Siiteri [103], employing a method of Osman *et al.* [130] and Cadotte *et al.* [131], condensed glucuronolactone (III) with several steroids in the presence of a cationic exchange resin. The products isolated however were α– and β–glucofuranosiduronates, which could not be converted to the desired glucopyranosiduronates.

Zorbach [80] prepared deoxycorticosterone–21–glucuronide [3,20–dioxopregn–4–en–21–yl–β–D–glucopyranosiduronic acid] in low yield by reacting 21–trityl–deoxycorticosterone [21–triphenylmethoxypregn–4–ene–3,20–dione] with the bromosugar with silver perchlorate, but was completely unsuccessful in an attempt to prepare hydrocortisone–21–glucuronide by this procedure.

Recently Kochetkov, Khorlin, and Bochkov [132, 133] have claimed a new synthesis of β–glucosides based on the reaction of the acetylated ortho-esters of glucopyranose with mercuric bromide. In this manner, cholesterol β–D–glucopyranoside was prepared in 45% yield. There has been no equivalent steroid glucuronide preparation using this procedure.

C. Chemical Properties

A summary of the chemical reactions of steroid glucuronides will be presented here. Prime emphasis will be directed toward those reactions which are performed at various sites on the steroid molecule and/or involve the glucuronide part of the molecule. Information concerning the rupture of the steroid sugar linkage can be found in Chapter 3 which deals with the hydrolysis of steroid conjugates.

The number of glucuronide reactions is limited, and does not approach, in either quantity or interest, that of steroid sulfate conjugates which will be covered later in this chapter.

The reactions to be discussed have been grouped under reaction types. The discussion under each type will not include all examples published in the literature pertinent to the reaction. Only significant studies will be cited.

1. Reduction Studies

For a discussion of reduction studies of glucuronides, initial consideration will be given to catalytic methods. An illustrative example is the reduction of sodium 6–dehydro–17β–estradiol–17–glucuronide (I) [3–hydroxyestra–1,3,5(10),6–tetraen–17β–yl–β–D–glucopyranosiduronic acid] [134, 135] with hydrogen–10%–palladium on charcoal in acetic acid to yield the 17β–estradiol derivative. The primary interest of this reaction was for the development of a procedure for the introduction of tritium labeling into the steroid part of the conjugate, and thus the preparation of the 6,7–^{3}H–17β–estradiol–17–glucuronide (II) [3–hydroxyestra–1,3,5(10)–trien–17β–yl–β–D–glucopyranosiduronic acid] was conveniently accomplished. The authors preferred in the reduction step to use a mixture of tritium and hydrogen to avoid instability of the final product due to autoradiolysis.

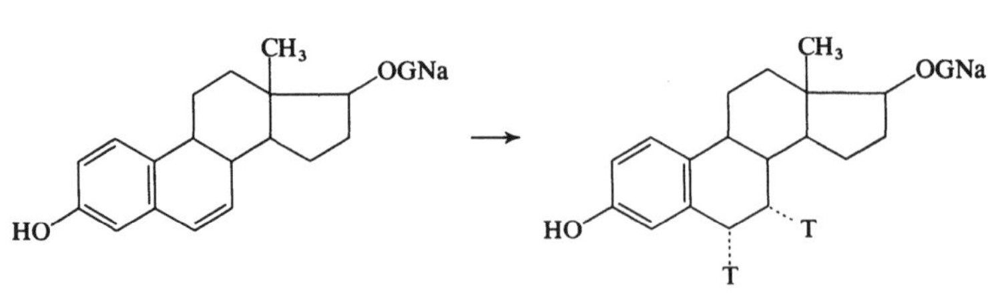

I II

Another example of catalytic reduction includes the conversion of an oxime to an amino group [97]. Reduction of the 20–oxime of pregnenolone–3–glucuronide triacetate methyl ester [methyl(20–oxopregn–5–en–3*β*–yl–2′,3′,4′–tri–*O*–acetyl–*β*–D–glucopyranosid)uronate] in acetic acid using platinum oxide as catalyst afforded the 20–amino derivative.

Glucuronides have been found stable to hydrogenolysis conditions, namely the removal of a benzyl protective group. Illustrated in the equation below [86, 152], the procedure has found extensive utility in the synthesis of glucuronides and double sulfate–glucuronides. A more detailed discussion of this reaction can be found in other sections of this chapter devoted to methods of synthesis (Sections II.B and IV.B).

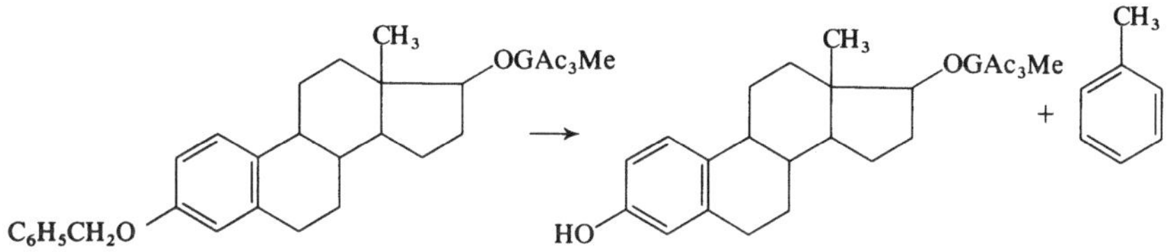

III IV

The synthesis of 17*β*–estradiol–3–glucuronide (VI) [17*β*–hydroxyestra–1,3,5(10)–trien–17*β*–yl–*β*–D–glucopyranosiduronic acid sodium salt] has been accomplished by sodium borohydride reduction at low temperature of either sodium estrone glucuronide (V) [17–oxoestra–1,3,5(10)–trien–3–yl–*β*–D–glucopyranosiduronic acid sodium salt] or its triacetate methyl ester [86]. In the case of the sodium salt, the reduction was carried out in methanol

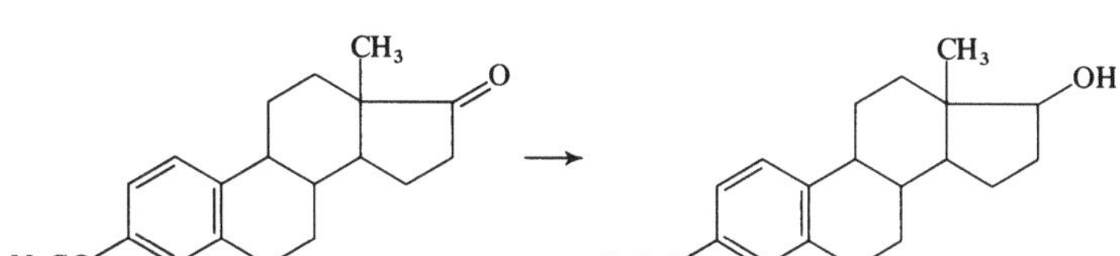

V VI

as the solvent, while a mixture of methanol–benzene (2:1) was employed for the blocked derivative. The problems involved with the use of hydride reducing agents in the presence of the glucuronide triacetate methyl ester grouping is amplified in the section devoted to the synthesis of estriol glucuronides. However the importance of low temperature and solvent choice in minimizing the reduction of the carbomethoxy grouping to the primary alcohol function should be noted. Testosterone–3–enol–glucuronide (VIII) [276] has been prepared by sodium borohydride reduction of androstenedione enol glucuronide (VII) [17–oxoandrosta–3,5–dien–3–yl–β–D–glucopyranosiduronic acid].

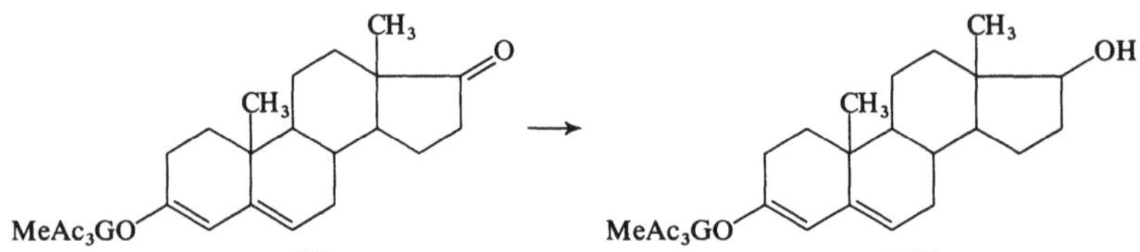

VII VIII

Borohydride reduction has been used in the determination of the structure of glucuronide metabolites isolated from urine [136]. In this procedure the crude glucuronide fraction was esterified with diazomethane and acetylated to yield the completely blocked glucuronide. For example, it was necessary to locate the glucuronide moiety in tetrahydro–Compound S glucuronide [17α,21–dihydroxy–20–oxo–5β–pregnan–3α–yl–β–D–glucopyranosiduronic acid]. This was easily accomplished since sodium borohydride reduction of the completely blocked metabolite (IX) followed by sodium bismuthate cleavage of X yielded etiocholanolone–3–glucuronide triacetate methyl ester (XI) [methyl(17–oxo–5β–androstan–3α–yl–2',3',4'–tri–*O*–acetyl–β–D–glucopyranosid)uronate], which the authors had prepared independently. Had the glucuronide moiety been situated at C21, oxidative cleavage would have separated the sugar and steroid residues.

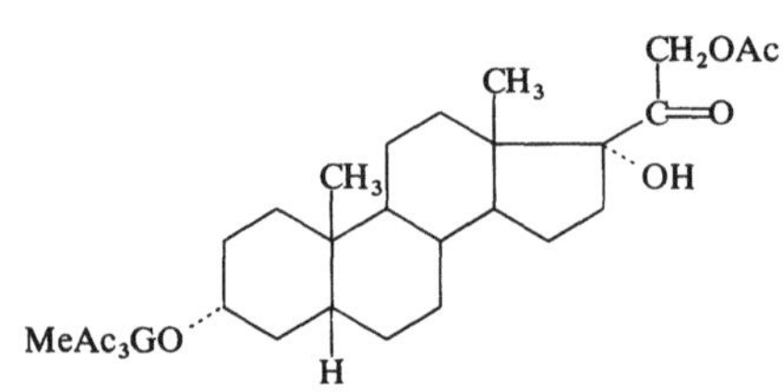

IX

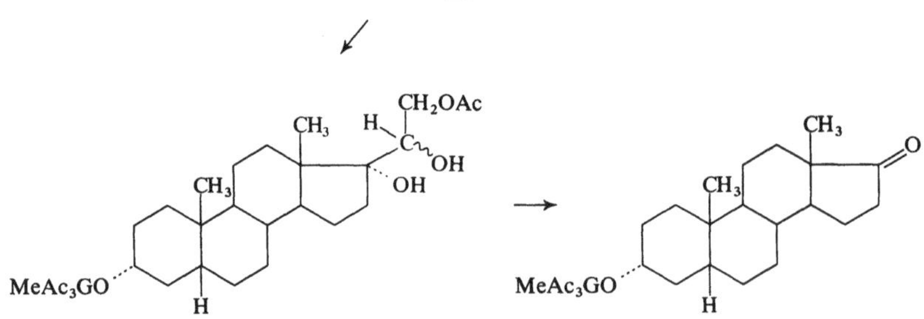

X XI

2. Oxidation Studies

In the preceding example in which a glucuronide conjugate was the substrate for an oxidation reaction, bismuthate cleavage of the molecule between the vicinal hydroxyl groups presented no problems.

To establish conclusively the position of Koenigs–Knorr glucuronidation on estriol–3–benzyl ether (XII), Joseph, Dusza, and Bernstein [123] were able to relate this product to a known 17–keto–compound (XIV) by oxidation with chromic acid in acetone. Thus, exclusive glucuronidation at the 16–position (XIII) was demonstrated by compound comparison.

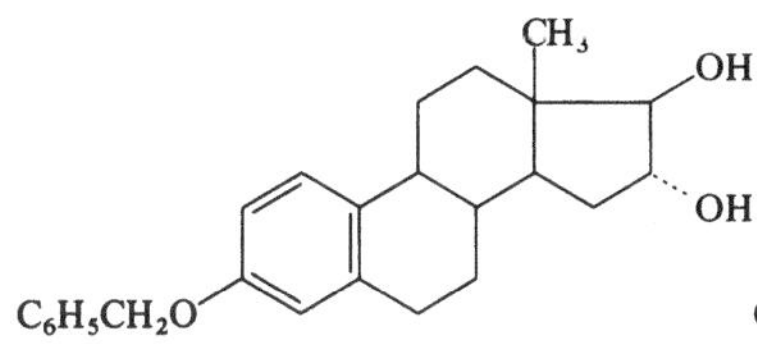

XII

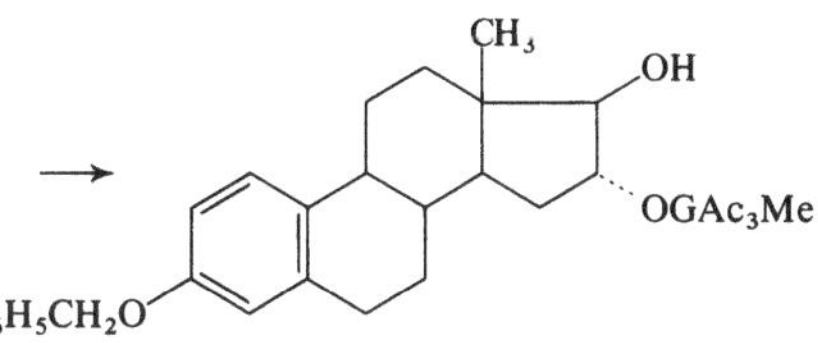

XIII

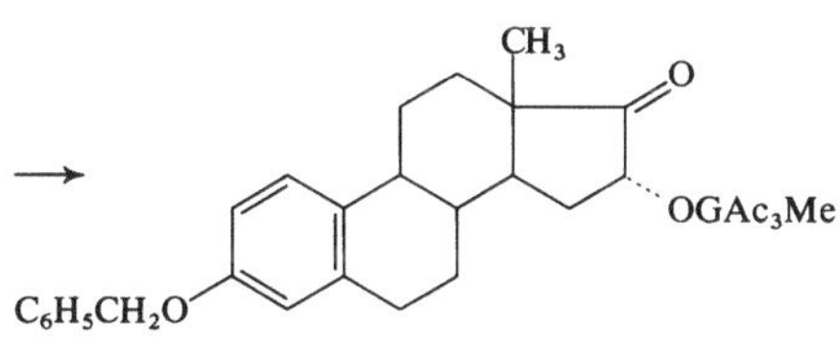

XIV

Hydrocortisone–21–glucuronide–triacetate methyl ester [methyl(11β,17α–dihydroxy–3,20–dioxopregn–4–en–21–yl–2′,3′,4′–tri–*O*–acetyl–β–D–glucopyranosid)uronate] has also been oxidized with chromium trioxide in pyridine to the corresponding cortisone derivative [137]. It is obvious that these oxidation conditions can be applied to a wide variety of steroid glucuronides.

3. Hydrolysis Studies

For a general discussion of the hydrolysis of glucuronides the reader is referred to Chapter 3. In this section only certain aspects of the behavior of glucuronides to hydrolytic conditions will be discussed. Strong acidic conditions are required to rupture the normal sugar–steroid acetal linkage. The hydrolytic conditions employed to retain the glucuronide linkage are usually 1 *N* aqueous base (sodium or potassium hydroxide) or barium hydroxide. Stronger base treatment may lead to a number of side reactions which involve dehydration in the glucuronide part of the conjugate [37].

If the glucuronide conjugate (XV) is hydrolyzed in a methanol–ammonia mixture, the acetate groups are readily removed and a glucuronamide (XVI) is formed [104].

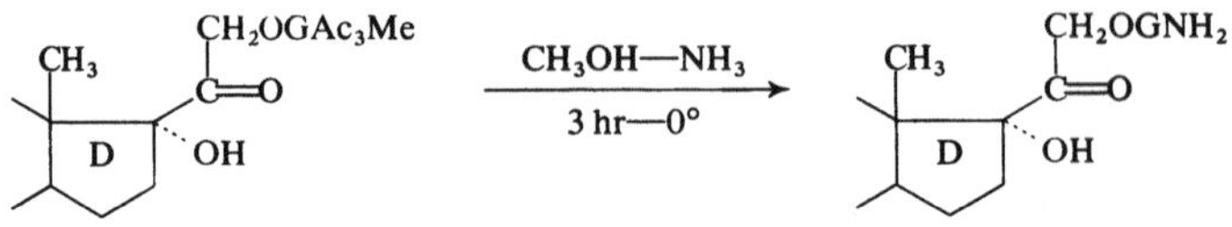

XV XVI

In an analogous ester interchange reaction, treatment of a completely blocked glucuronide (XVII) with methanol–sodium methoxide results in complete acetate hydrolysis and an interchange to the methyl ester (XVIII) [104]. The partial deblocked glucuronide methyl ester can also be prepared by diazomethane esterification of the free acid (XIX) [113].

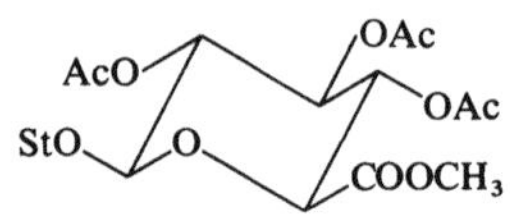

XVII

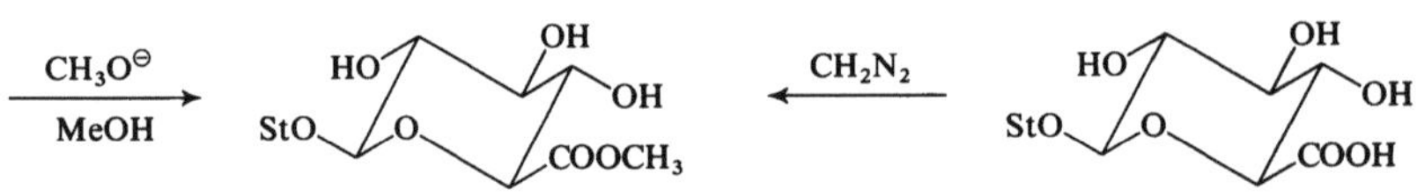

XVIII XIX

An acid hydrolysis procedure which does not effect the blocked glucuronide unit was described by Nambara and Imai [87] in their synthesis of the monoglucuronides of estriol [estra-1,3,5(10)-triene-3,16α,17β-triol]. In partic-

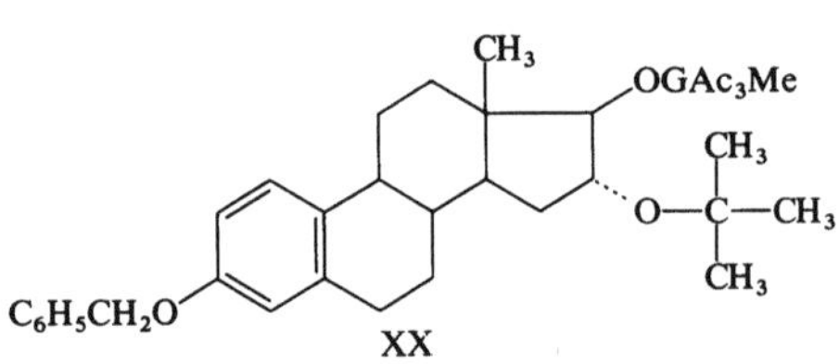

XX

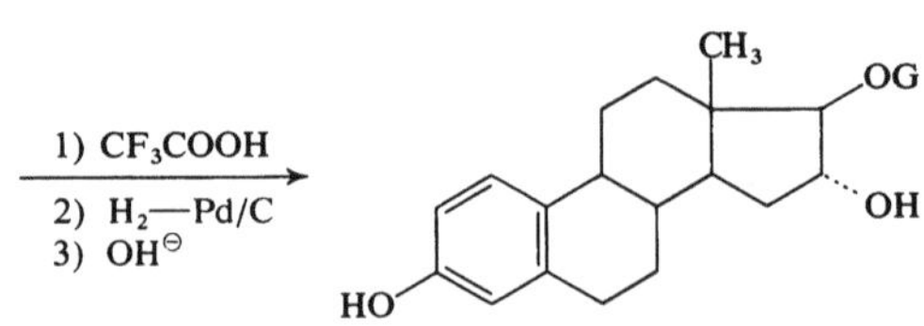

XXI

ular, in their synthesis of estriol–17–glucuronide, these workers prepared the key blocked estriol intermediate 3–benzyloxy–16α–*t*–butyloxyestra–1,3,5(10)–trien–17β–ol. Coupling of this compound with the bromo sugar gave the blocked 17–glucuronide (XX). Removal of the unique 16–*t*–butyl ether was accomplished by brief treatment (1 hr–0°) with trifluoroacetic acid which did not affect the acetal linkage at C17$\rightarrow$1′. Debenzylation by catalytic hydrogenation followed by alkaline hydrolysis afforded estriol–17–glucuronide (XXI).

4. Diazomethane Studies

As a general procedure diazomethane has been used extensively for the esterification of the C6′–carboxyl group of steroid glucuronides. To establish unequivocally the structure of urinary estriol glucuronide (XXII), Nieman and Hashimoto [113, 114] made use of diazomethane in a rather unique fashion. The treatment of estriol glucuronide with diazomethane gave the 3-methyl ether–6′–methyl ester (XXIII). The further treatment of this material with diazomethane and a strong acid catalyst, boron trifluoride etherate, gave the fully methylated methyl ester (XXIV) having methoxyl groups at 3,17,2′,3′ and 4′–carbon positions. In order to affect cleavage of the methylated glucuronide a number of hydrolytic conditions were tried and proved unsuccessful. Cleavage of the sugar was finally achieved by refluxing 4*N* hydrochloric acid in 10% aqueous ethanol which afforded 3,17β–dimethoxyestra–1,3,5(10)–trien–16α–ol (XXV) in 81% yield. The structure of this material was confirmed by synthesis, thus establishing the position of the glucuronide group at C16.

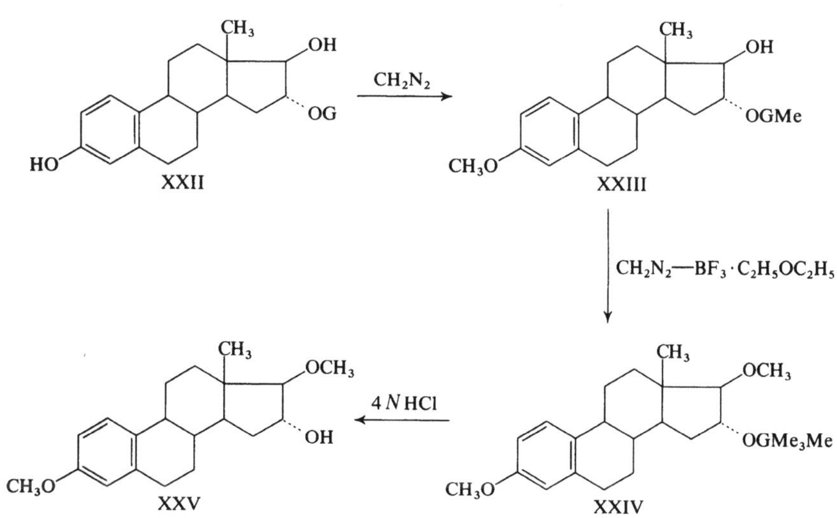

5. Acylation

From the preceding remarks it is obvious that the glucuronide linkage is stable to acylation conditions. Extensive use has been made of acylation on both the steroid and sugar portions of the steroid glucuronides.

6. Miscellaneous Studies

Trimethylsilylation [138, 140] of glucuronides has proved to be suitable for vapor phase chromatography. The initial step in derivatization of the glucuronide is esterification with diazomethane, and then silylation usually with hexamethyldisilazane, trimethylchlorosilane, or *bis*–trimethylsilyl–acetamide in pyridine. Suitable stationary phases and column packing applicable to these separations have been worked out.

An interesting reaction applied to a N–acetyl–β–glucosaminide is that of the epimerization to an α–sugar [141]. Steroidal N–acetyl–β–glucosaminides as the tri–*O*–acetyl derivatives have been epimerized in anhydrous chloroform with titanium tetrachloride at room temperature for 2 days in fair yields. Hydrolysis of the *O*–acetyl groups was achieved by transesterification in good yield.

D. Physical Properties

In this section selected physical properties of steroid glucuronides will be discussed. The reader is referred to Chapter 4 where other physical properties such as electrophoretic and chromatographic behavior are discussed. Also the reader is referred to the companion volume "Physical Properties of Steroid Conjugates" for further details and references on the physical properties of glucuronides.

1. Solvation and Stability

Generally, glucuronides in the free or salt forms crystallize with water or solvent of crystallization which are not removed by prolonged heating *in vacuo* [70, 127]. Unlike steroid sulfates (see Section III-B) this property should have essentially no effect on the stability of glucuronides.

2. Ultraviolet Spectroscopy

No ultraviolet absorption maximum is associated with the incorporation of the glucuronide moiety into the steroid nucleus. If the parent steroid possesses an independent ultraviolet chromophore, the corresponding glucuronide will obviously exhibit this chromophore. Two general exceptions are noted among the alkyl glucuronides. An ultraviolet absorption maximum is observed for the completely blocked $\Delta^{3,5}$–3–enol glucuronides (I) at 237.5 mμ, displaced from the maximum at 240 mμ observed for the parent

Δ^4–3–ones. It is interesting to note that methanolysis of the completely blocked glucuronide group giving rise to the deacetylated methyl ester leads to a bathochromic effect, and a return to λ_{max} 240 mμ. If the substantiation of the Δ^2–1–one 3–enol glucuronides (II) [142] is forthcoming, these will represent another series with an ultraviolet absorption maximum near 248–250 mμ.

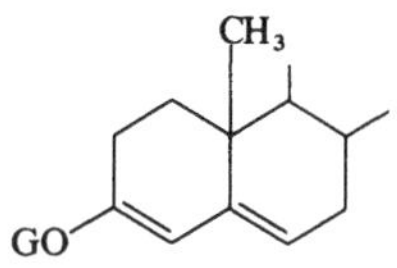

I

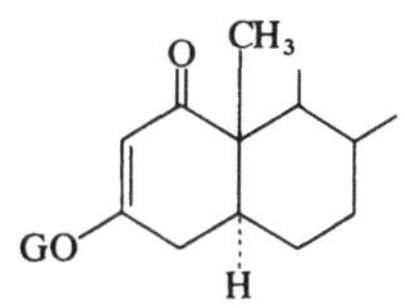

II

Aryl glucuronides do not have an ultraviolet absorption spectrum different from the parent phenolic steroid. Both estriol–3–glucuronide and estriol–16–glucuronide have absorption maxima at approximately 280 mμ with nearly equal extinction coefficients. The two compounds can however be easily distinguished through their ultraviolet spectra after the addition of 0.5 *N* sodium hydroxide solution. The typical bathochromic displacement to 298 mμ for the phenolate ion clearly sets apart the spectrum of estriol–16–glucuronide [113] from the unchanged spectrum of estriol–3–glucuronide [87].

3. Optical Rotatory Properties

Becker [107] has used the theory of molecular rotation differences for supporting the structures of four synthetic β–D–glucopyranosiduronic acids of the 3–hydroxyandrostan–17–ones. The observed optical rotations of the compounds were compared with the calculated values for the α–D– and β–D–configurations. The observed figures agreed closely with the β–D–structure in all four cases. The calculated figures were derived from the Klyne rule for steroid glycosides [143] which was modified by the replacement of the glycosyl by glucuronyl term. The validity of this modified Klyne rule was checked against the rotation of three acetylated derivatives of 3–hydroxysteroids recorded in the literature. The three examples, acetylated glucuronide derivatives of isoandrosterone [3β–hydroxy–5α–androstan–17–one], dehydroisoandrosterone, and estrone indicated that there was good agreement between the observed and calculated values for the β–D–form.

Optical rotatory dispersion (ORD) has been used by Hadd [84, 85] for additional evidence for the purity and structure of a synthetic testosterone glucuronide. Testosterone, testosterone glucuronide, and the triacetate methyl ester derivative of the latter all gave similar ORD curves with a negative Cotton effect.

Emiliozzi [105] has observed some anomalous results in the rotatory dispersion of the acetylated methyl ester glucuronide derivatives in the hydroxyandrostane series.

4. Infrared Absorption

The companion volume of this work lists glucuronide conjugates for which infrared absorption data have been published by indicating such data to be available in the reference cited. In most cases this is confined to the listing of certain maximal absorption frequencies. This section cannot include references to these many citations, but will generally summarize the results. In addition a list of publications which give graphic representations of glucuronide infrared spectra will be cited.

The infrared spectra of steroid glucuronides will be discussed in regard to the completely blocked triacetate methyl esters, the deblocked glucuronide salt, and the free glucuronic acid. The completely blocked glucuronide spectrum is dominated by a very intense carbonyl ester stretching peak near 1750 cm^{-1}. The peak represents the three acetates plus methyl ester functions. Two other intense vibrational peaks are located at 1220 cm^{-1} and 1040 cm^{-1}. In addition, it has been stated by Barker and co-workers [144], and also by Tipson and Isbell [145] that bands at 872 cm^{-1} and 914 cm^{-1} are representative of the β-anomer of methyl glucopyranosides. The α-anomer has a peak at 940 cm^{-1}. These bands are generally weak and in most steroid glucuronides, especially aromatic derivatives, a plethora of bonds is observed in this region making assignments based on these peaks very tenuous.

In the case of the glucuronide salts a strong carboxylate band is observed at 1610 cm^{-1}, and a characteristic strong, broad and diffuse band(s) is located near 1050 cm^{-1}. The free acid exhibits carboxylic acid carbonyl absorption at 1760 cm^{-1}, but in these derivatives there are few features in the 1000 to 1100 cm^{-1} region which can be used for characterization due to the broad absorption in this part of the spectrum. In the case of the glucuronide salts and the free acids, the polyhydroxyl absorption bands at 3300–3600 cm^{-1} are quite prominent. The free acids also show weak hydroxyl absorption in the region 3000–2500 cm^{-1} which is associated with the hydrogen bonding of the carboxylate groups.

The infrared spectra of several $\Delta^{3,5}$-3-enol glucuronides have been recorded by Wotiz *et al.* [70]. These spectra are characterized by the appearance of two weak but distinct peaks at 1655 and 1631 cm^{-1}, in addition to the other bands associated with the glucuronide moiety which we have indicated above. Absent from these spectra is the characteristic Δ^4–3–one absorption at $1670\,cm^{-1}$ present in the starting material.

The largest number of graphically depicted steroid glucuronide infrared spectra has been published by Smakula *et al.* [127]. This compilation is useful since many different types of compounds are recorded; *e.g.*, mono– and dialkyl–glucuronides, aryl–glucuronides, and $\Delta^{3,5}$–3–enol–3–glucuronides. A series of infrared spectra of steroid 21–glucopyranosiduronates and uronamides have been published by Nitta, Shindo and Takamura [104]. The spectra of several derivatives of estriol–16–glucuronide have been published by Hashimoto and Neeman [115]; and in the androgen series, the infrared spectra of the four 3–glucuronide–17–ones have been depicted [107], as well as testosterone–17–glucuronide [3–oxoandrost–4–en–17β–yl–β–D–glucopyranosiduronic acid] and several derivatives [84, 146].

5. Nuclear Magnetic Resonance (nmr)

Among the various spectroscopic measurements generally employed for structural characterization studies, nuclear magnetic resonance must be considered among the most useful. Since this analytical tool is extremely sensitive to structural changes occurring in the near vicinity to an appropriately resonating atomic nucleus, its application in structural elucidation studies is enormous. For a complete discussion of the utility of this technique, any one of the standard texts [147–149] covering this subject matter should be consulted. In addition, specialty texts [150, 151] provide additional very valuable references for the reader.

There have been a considerable number of papers devoted to the nmr analysis of sugars, including glucopyranoses, glucofuranoses, glucopyranosiduronic acids, and anomer studies of all the preceding. The nmr analysis of steroid glucuronides was initially reported by Neeman and Hashimoto [114] in their structural studies of estriol–16–glucuronide. They were able to conclude by using published data and model nonsteroidal glucuronides that the sugar derivative involved in the conjugate was definitely glucuronic acid. From coupling constant analysis of the C1′ proton, they also concluded the sugar to be in the β–series. These nmr studies were carried out on the triacetate methyl ester and the fully methylated methyl ester glucuronides (carbon tetrachloride as solvent). Testosterone–17–glucuronide isolated and also synthesized by Hadd [84] was analyzed by nmr as its triacetyl methyl ester in deuteriochloroform as solvent. The β–sugar series was also confirmed by the splitting constant of the C1′ proton. Layne, Sheth and Kirdani [13] noted in the nmr spectrum of an estrone metabolite isolated from rabbit urine that the conjugate exhibited a methyl signal at 120 Hz. Along with other considerations they concluded that the structure of the substance was 17α–estradiol–17–N–acetylglucosaminide [3–hydroxyestra–1,3,5(10)–trien–17α–yl–2′–acetamido–2′–deoxy–β–D–glucopyranoside]. The methyl of the N–acetyl function was responsible for the peak at 120 Hz. The nmr spectra of several 17β–estradiol–17–glucuronide derivatives are mentioned by Cantrall, McGrath, and Bernstein [152]. A number of sodium estrogen glucuronidates [153] have been prepared and their nmr determined in d_6–DMSO. Their spectra are not as informative as are their triacetyl methyl esters.

6. Mass Spectrometry

Although thus far not routinely employed, mass spectrometry in the study of steroid conjugates will no doubt become an extremely important analytical method. Not only can this technique indicate the molecular weight of the compound in question, but analysis of the fragmentation pattern can lead to identification of both the steroid and the conjugating moiety along with an indication of the bonding between the two.

Detailed discussion of mass spectrometry can be found in a number of excellent reference works [154, 155]. For the subject matter outlined in this chapter, steroid glucuronides, a volume [156] containing practical aspects of the mass spectrometry of steroids and of carbohydrates is available which contains separate chapters devoted to each subject.

Certainly one important goal in the mass spectrometry of steroid glucuronides would be to obtain the molecular ion of the conjugate. The conjugate in the form of a derivative appears to be the only species to give a molecular ion signal. In the study of estriol–16–glucuronide a number of derivatives were prepared and submitted for mass spectral analysis [157]. Molecular ions were obtained for estriol–17–acetate–16–glucuronide triacetate methyl ester [methyl(17β–acetoxy–3–hydroxyestra–1,3,5(10)–trien–16α–yl–2′,3′,4′–tri–*O*–acetyl–β–D–glucopyranosid)uronate] and the 3–methyl ether of the latter. No molecular ion could be detected when the 3–triethylammonium sulfate was studied. Molecular ions have been observed for androsterone–3–2′–amino–2′–deoxyglucopyranose tetraacetate [17–oxo–5α–androstan–3α–yl–2′–acetamido–3′,4′,6′–tri–*O*–acetyl–2′–deoxy–β–D–glucopyranoside] [158], 17α–estradiol–3–acetate–17–2′–amino–2′–deoxyglucopyranose tetraacetate [3–acetoxyestra–1,3,5(10)–trien–17α–yl–2′–acetamido–3′,4′,6′–tri–*O*–acetyl–2′–deoxy–β–D–glucopyranoside] [159], and a weak molecular ion for 15α–hydroxyestrone 3–acetate–15–2′–amino–2′–deoxyglucopyranose tetraacetate [3–acetoxy–17–oxoestra–1,3,5(10)–trien–15α–yl–2′–acetamido–3′,4′,6′–tri–*O*–acetyl–2′–deoxy–β–D–glucopyranoside] indicated extensive decomposition at the temperature required for instrument response.

The combination of gas–liquid chromatography (glc) and mass spectrometry in the study of steroid glucuronides has recently received intensive study. Since glc is capable of high theoretical resolution, and mass spectrometry requires a small sample, the combination of the two procedures is excellent for the analysis of steroid metabolites. The problem of underivatized glucuronides and glc has been discussed by Jaakonmaki [160]. Column hydrolysis of steroid glucuronides was observed and thus the mass spectrum records the molecular ion of the parent steroid. Successful resolution of this problem has been reported by several groups of workers [138–140]. The approach is essentially the same in that the glucuronide fraction or pure compound is esterified with diazomethane. Etherification of the remaining hydroxyl groups with hexamethyldisilazane and trimethylchlorosilane affords a derivative which appears to be stable to glc conditions. When subsequently introduced into the mass spectrometer, a weak but discernible molecular ion peak can be observed.

III. Sulfates

A. Types

There are two broad classes of steroid sulfates. These are alkyl (I) and aryl (III) sulfates. A variant of an alkyl sulfate is the enol–sulfate (II). The following structure (in ionic form) are typical of the types encountered in the steroid field:

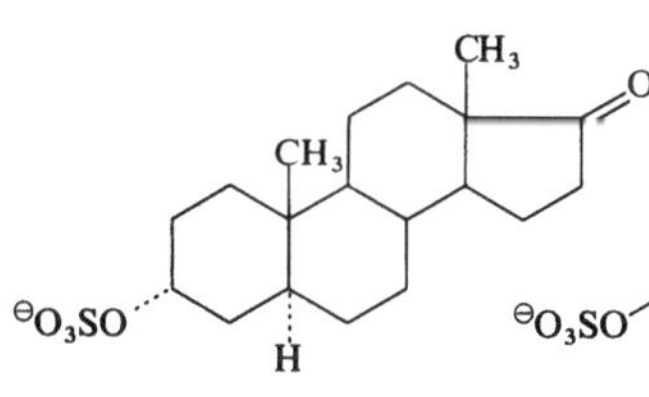

Androsterone sulfate
[3α–sulfooxy–5α–androstan–17–one]
I

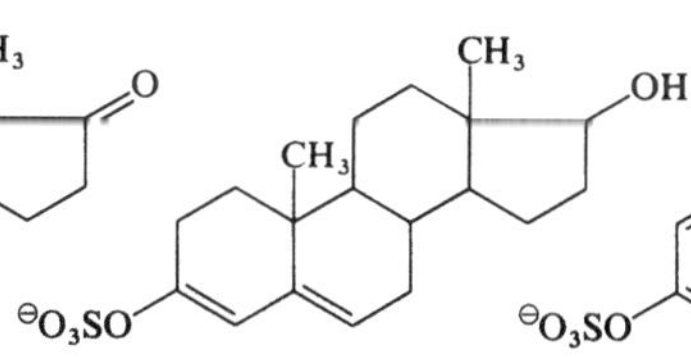

Testosterone enol-sulfate
[3–sulfooxyandrosta–3,5–dien–17β–ol]
II

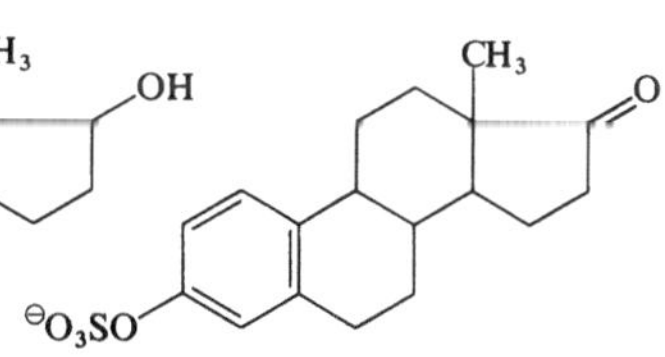

Estrone sulfate
[3–sulfooxyestra–1,3,5(10)–trien–17–one]
III

B. Chemical Synthesis

Steroid sulfates have been synthesized by a variety of methods which have utilized the following sulfating agents:

(1) sulfur trioxide complexes, (2) sulfuric acid, (3) sulfamic acid.

1. Sulfur Trioxide Complexes [161]

The most widely used method for the sulfation of steroids has been through the employment of sulfur trioxide complexes. These may be generated *in situ*, or added as such to the reaction mixture. The most commonly used reagents for this purpose are: chlorosulfonic acid, methyl chlorosulfonate, ethyl chlorosulfonate, sulfuric acid, pyridine–sulfur trioxide, trimethylamine–sulfur trioxide, and triethylamine–sulfur trioxide.

Sulfur trioxide, a Lewis acid, combines with bases to form coordination compounds (complexes) (Eq. (a)). In the steroid field, the bases employed have been the fairly strong tertiary amines such as trimethyl– or triethyl–amine, or the considerably weaker base, pyridine; picolines have also been used successfully [162]. The intermediate amine or base sulfate salt (Eq. (b))

$$\begin{array}{c}\mathrm{R}\\ \ddot{}\\ \mathrm{R{:}N{:}}\\ \ddot{}\\ \mathrm{R}\end{array} + \begin{array}{c}\mathrm{O}\\ \|\\ \mathrm{O{:}S}\\ \|\\ \mathrm{O}\end{array} \longrightarrow \begin{array}{c}\mathrm{R\ O}\\ \ddot{}\ \|\\ \mathrm{R{:}N{:}S{:}O}\\ \ddot{}\ \|\\ \mathrm{R\ O}\end{array} \qquad \text{(a)}$$

$$\mathrm{StOH} + \mathrm{R_3N{\cdot}SO_3} \longrightarrow \mathrm{StOSO_3^{\ominus}\,{}^{\oplus}NHR_3} \qquad \text{(b)}$$

$$\downarrow \mathrm{MOH}$$

$$\mathrm{StOSO_3^{\ominus}M^{\oplus}} \qquad \text{(c)}$$

$$\mathrm{M^{\oplus}} = \mathrm{Na,\ K,\ NH_4}\ \textit{etc.}$$

may or may not be isolated, and on treatment with alkali metal or ammonium hydroxide, is converted into the corresponding alkali metal or ammonium salts (Eq. (c)). Sulfur trioxide complexes such as dioxane–sulfur trioxide, dimethylformamide–sulfur trioxide, *etc.*, have not been investigated as yet for the preparation of steroid sulfates.

a) Chlorosulfonic Acid. The use of chlorosulfonic acid in pyridine may be considered the classical method for the preparation of steroid sulfates. Mandel and Neuberg [163] in 1915 employed this reagent for the first preparation of a steroid sulfate, namely, cholesterol sulfate [3β–sulfooxy-cholest–5–ene]. The procedure was based on the method of Verley [164] for the preparation of phenol sulfate. Cholesterol (I) in chloroform solution was treated (15 min at room temperature, followed by 15 min at 50°) with chlorosulfonic acid in pyridine and chloroform. The intermediate, pyridinium salt (II) on treatment with base gives the corresponding metallic salt. Potassium hydroxide gave potassium cholesterol sulfate (III) in an 89% over–all yield from cholesterol.

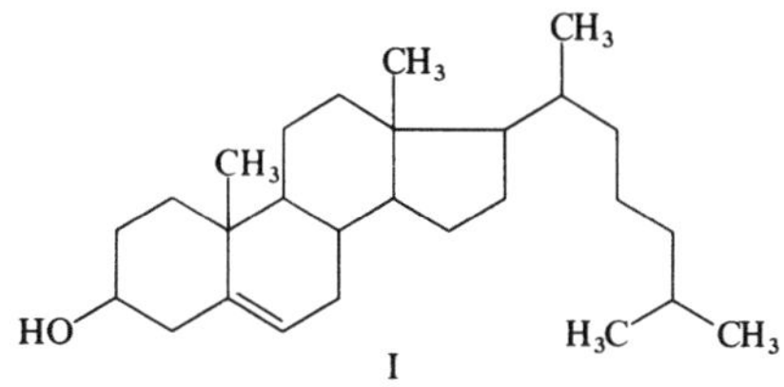

I

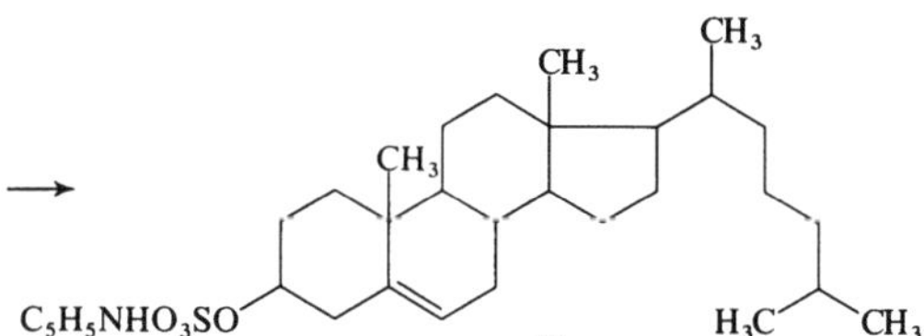

II

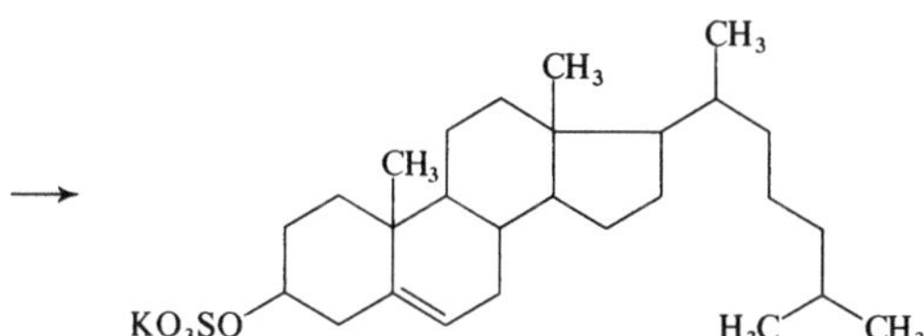

III

The active sulfating species generated from chlorosulfonic acid in pyridine is pyridine–sulfur trioxide complex [165] (Eq. (d)):

$$2C_5H_5N + ClSO_3H \rightarrow C_5H_5N \cdot SO_3 + C_5H_5NHCl. \qquad \text{(d)}$$

Grant *et al.* [169, 172] in a very thorough study of the chemical properties of estrogen sulfates found it advantageous to add the estrogen solution to the chlorosulfonic acid mixture, and to adjust carefully to pH 7.8–8.0 during the subsequent neutralization with N–methanolic sodium hydroxide. After the removal of inorganic salts, the sodium salt of the particular estrogen sulfate was isolated as usual from the methanol solution by fractional precipitation with ether. Subsequently the chlorosulfonic acid procedure was extensively employed by numerous investigators for the preparation in fair to excellent yields of unlabeled and labeled sulfates [162, 166–178].

b) Methyl and Ethyl Chlorosulfonate. Methyl and ethyl chlorosulfonate have found only limited use as sulfating agents in the steroid field. These reagents, in the presence of a base such as pyridine, give rise to the active sulfating species, pyridine–sulfur trioxide complex [165] (Eq. (e)):

$$C_5H_5N + ClSO_3R \rightarrow C_5H_5N \cdot SO_3 + RCl \qquad R = CH_3,\ C_2H_5 \tag{e}$$

The Lederle group [179] has successfully employed ethyl chlorosulfonate in pyridine for the preparation of potassium estrone sulfate.

The 3,5–dienol–sulfate (sodium salt) of androstenedione has been prepared by the method of Roy [180] with methyl chlorosulfonate (as well as with chlorosulfonic acid) in pyridine [181].

c) Pyridine Sulfate. Sobel *et al.* [182] found that pyridine sulfate could be used for the sulfation of cholesterol; thus pyridine sulfate does not react at

$$StOH + C_5H_5N \cdot H_2SO_4 \xrightarrow{Ac_2O} StOSO_3NHC_5H_5 \tag{f}$$

all unless an internal dehydrating agent such as acetic anhydride is also used in the reaction mixture (Eq. (f)). It is apparent that here also the sulfating species is pyridine–sulfur trioxide. Sobel *et al.* noted that the reaction in the presence of acetic anhydride goes to completion even in the presence of small traces of water or alcohol which might otherwise seriously interfere in the procedure on a micro–scale. Moreover, no acetylation of cholesterol was observed under the conditions employed. Pyridine sulfate was prepared by the addition of 1 mole of sulfuric acid to 3 moles of pyridine in chloroform. The precipitated pyridine sulfate was then dried in a desiccator.

Levitz [183] has reported on the synthesis of 6,7–^{3}H–estrone sulfate and estrone–^{35}S–sulfate by use of the pyridine sulfate method of Sobel, Drekter, and Natelson [182] and noted that the sulfuric acid containing ^{35}S is inexpensive and safe to handle. In the presence of a 10% molar excess of acetic anhydride, pyridine sulfate reacted with estrone overnight at room temperature to produce pyridinium estrone sulfate in good yield. Without isolation, the latter was converted into sodium estrone sulfate by the addition of sodium hydroxide. The amount of acetic anhydride was found to be critical, when it exceeded that of pyridine sulfate by more than two–fold, the yield of estrone sulfate was negligible, estrone acetate being the major product. Similarly sodium 6,7–^{3}H–estrone sulfate was prepared from 6,7–^{3}H–estrone.

Although the Levitz method was designed for the synthesis of radioactive estrone sulfate, it is apparently applicable for the routine preparation of steroid sulfates on any laboratory scale [184]. Sodium 17β–estradiol–

3,17–disulfate [3,17β–disulfooxyestra–1,3,5(10)–triene disodium salt] has been prepared by Kirdani [176] according to the Levitz method.

Sodium 5α–pregnane–3β,17α,20β–triol–20–sulfate [20β–sulfooxy–5α–pregnane–3β,17α–diol sodium salt] [185] and sodium 2–hydroxyestrone–2,3–disulfate [2,3–disulfooxyestra–1,3,5(10)–trien–17–one disodium salt] [186] have both been prepared by the use of pyridine sulfate in the presence of acetic anhydride.

d) Pyridine–Sulfur Trioxide. Using pyridine–sulfur trioxide, Sobel and colleagues [182, 187] first prepared pyridinium cholesterol sulfate (II) as a microanalytic method for the determination of cholesterol (I). The reagent may be readily prepared by the addition of chlorosulfonic acid to dry pyridine in chloroform [165, 188, 189].

A review by Gilbert [161] includes numerous methods of preparing this complex. Pyridine–sulfur trioxide complex [available commercially from General Chemical Division, Allied Chemical, New York City, and Eastman Organic Chemicals Dept., Distillation Products Industries, Inc.], is a white crystalline solid, which is quite stable to, and insoluble in cold water and cold aqueous alkali, but decomposes rapidly and completely upon warming in either medium [165]. It is hydrolyzed quantitatively to pyridine and sulfuric acid by hot water [188], and is insoluble (*i.e.*, less than 1% by weight) in pyridine, nitrobenzene, cyclohexane, methycyclohexane, n–hexane, chloroform, carbon tetrachloride, dioxane, diethylether, and acetone at 25° [190]. It is soluble in dimethylformamide [191].

Pyridine–sulfur trioxide complex for sulfating steroids is generally applied in an organic solvent such as chloroform, benzene and pyridine. The reaction is generally conducted at from room temperature to about 60°. Occasionally the reaction is performed in the presence of acetic anhydride to ensure anhydrous conditions. Apparently no definitive study of the reaction conditions for the diverse types of steroid sulfates has been reported. The reaction times have varied from 2 minutes to 68 hours. As with the chlorosulfonic acid–pyridine method, the intermediate pyridinium salt may or may not be isolated as such, but on treatment with base gives the corresponding cationic sulfate salt.

Sobel and Spoerri [187] carried out the pioneer study of the application of this complex for preparing steroid sulfates. These investigators prepared in quantitative yield with this reagent the sulfate salts of cholesterol, ergosterol [ergosta–5,7–dien–3β–ol], lanosterol [lanosta–8,24–dien–3β–ol], and dibromocholesterol [5α,6β–dibromocholest–5–en–3β–ol]. They also found that the reaction mixture on treatment with petroleum ether (b.p. 30–60°) precipitated the intermediate, *e.g.*, pyridinium cholesterol sulfate, which could be purified by recrystallization from cold chloroform with petroleum ether. In this manner, the pyridinium salt was freed of pyridine–sulfur trioxide, as the latter was insoluble in cold chloroform. The ester linkage in pyridinium cholesterol sulfate was found to be stable in strong acid at room temperature, and in strong alkali even at refluxing temperatures.

Subsequent to the Sobel and Spoerri publication, a very large number of steroid sulfates have been prepared by this method, both in unlabeled and labeled forms [180, 192–210].

e) Trimethylamine–Sulfur Trioxide. Trimethylamine–sulfur trioxide complex has only recently been employed as a sulfating agent in the steroid field. References to its preparation may be found in the Gilbert review article [161]. The complex is available commercially (General Chemical Division, Allied Chemical).

The trimethylamine–sulfur trioxide complex is a stable solid melting with decomposition at 239° [211]. It has generally low solubility in organic solvents with which it does not react (*cf.*, f, Triethylamine–Sulfur Trioxide). It is the most stable of the sulfur trioxide complexes studied to date, and is considerably more so than pyridine–sulfur trioxide complex [161]. It is believed that this stability may be ascribed to the greater basic strength of trimethylamine compared with that of pyridine. As a result of its stability, this complex may be used without regard for anhydrous conditions. Moreover, it may be used as a sulfating agent in aqueous solutions. Usually, loss of this reagent by hydrolysis is minor since the rate of sulfation is much faster.

Hydorn, Lerner, and Schwartz [212] were the first to report the use of trimethylamine–sulfur trioxide as a sulfating agent in the steroid field. The 21–sulfate of triamcinolone acetonide [9α–fluoro–11β,21–dihydroxy–16α,17α–isopropylidenedioxypregn–1,4–diene–3,20–dione] and its Ring A–saturated analog were prepared with the reagent in pyridine at room temperature. The sulfates were isolated as their sodium salts.

Fex *et al.* [213] have utilized trimethylamine–sulfur trioxide complex for the preparation of sodium estriol–3–sulfate (VI) [3–sulfooxyestra–1,3,5(10)–triene–16α,17β–diol sodium salt] in 31% conversion yield from estriol (IV). The latter in acetone and aqueous 0.3 *N* sodium hydroxide was heated at 55° for 7 hr with trimethylamine–sulfur trioxide complex. The intermediate trimethylammonium sulfate (V) was not characterized but treated with sodium acetate to give the desired compound. This work established an interesting and important procedure for the selective sulfation of an aromatic hydroxyl over an aliphatic one in the estrogen class of steroids.

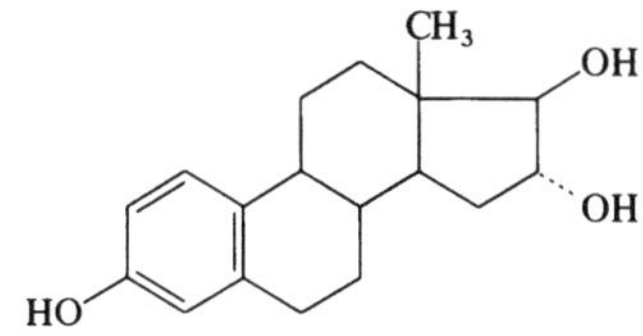

IV

Acetone
aq. 0.3 N
NaOH
$Me_3N \cdot SO_3$
55°—7 hr

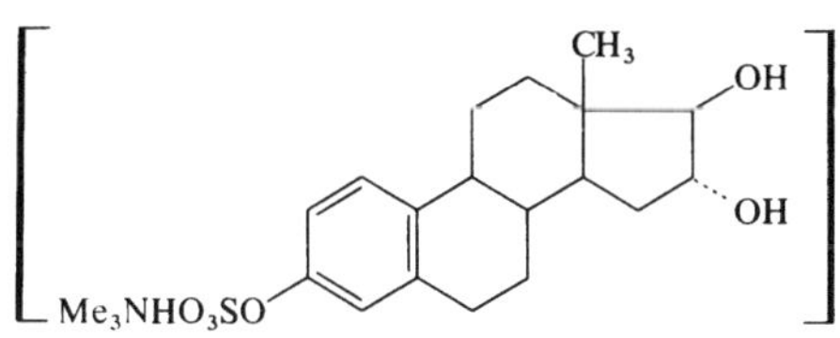

V

NaOAc

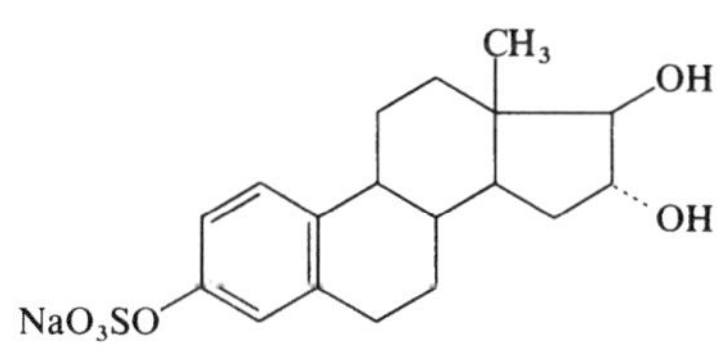

VI

Miyazaki and Fishman [186] have recently reported on the synthesis by a multi–stage procedure of sodium estra–1,3,5(10)–triene–2,3,17β–triol–2–sulfate (VII) [2–sulfooxyestra–1,3,5(10)–triene–3,17β–diol sodium salt] and sodium estra–1,3,5(10)–triene–2,3,17β–triol–3–sulfate (VIII) [3–sulfo-oxyestra–1,3,5(10)–triene–2,17β–diol sodium salt]. The sulfation step involved the use of trimethylamine–sulfur trioxide complex in pyridine at room temperature for 18 hr to give the intermediate trimethylammonium salts.

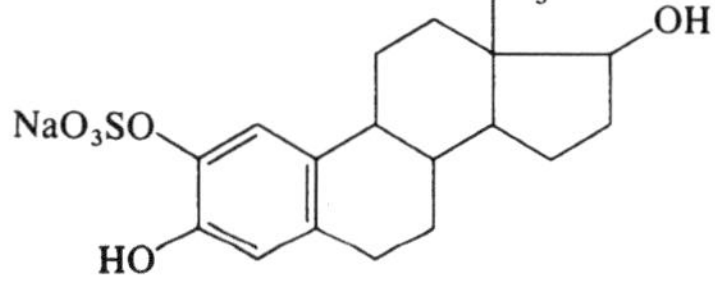

VII

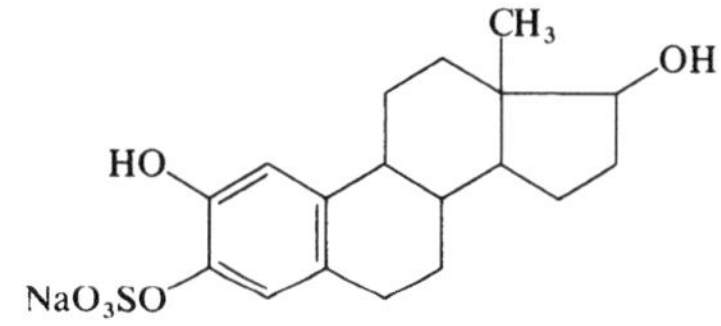

VIII

Similarly these investigators [214] have synthesized sodium estra-1,3,5(10)-triene-2,17β-diol-2-sulfate (X) [2-sulfooxyestra-1,3,5(10)-trien-17β-ol sodium salt] from the free phenol 17-acetate (IX).

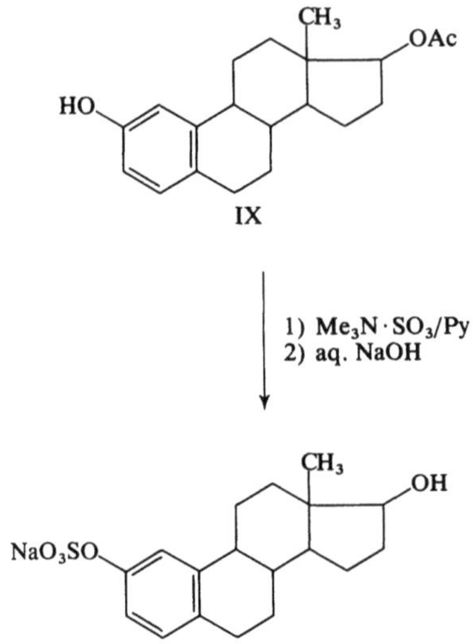

f) Triethylamine-Sulfur Trioxide. As with trimethylamine-sulfur trioxide complex, the use of triethylamine-sulfur trioxide complex as a steroid sulfating agent has only recently been developed. References to its preparation are cited in Gilbert's review article [161] and the complex is available commercially from General Chemical Division, Allied Chemical. Commercial grade triethylamine-sulfur trioxide complex may be purified by recrystallization from methylene chloride-diethyl ether after passage of the methylene chloride solution through Magnesol® [215]. The triethylamine-complex melts at 93°, and it is recommended that it be stored under refrigeration [216]. Although the bases, trimethyl- and triethyl-amine, have nearly the same basic strength, the triethylamine complex is less stable and more reactive. Triethylamine-sulfur trioxide has a greater solubility at room temperature in pyridine than does its trimethyl analog, which is a distinct advantage in the steroid field. Also it is fairly soluble in acetone and 1,2-dichloroethane, therein differing from the trimethylamine complex. At 25°, 2.7 g of triethylamine complex dissolves in 100 ml of water, which is double the solubility of the trimethylamine complex. Triethylamine-sulfur trioxide complex is sufficiently stable to be used in aqueous medium.

Fex *et al.* [213] have prepared sodium 17β-estradiol-3-sulfate (XIII) in 67% conversion yield from 17β-estradiol (XI). The latter in acetone was treated with 0.4 *N* sodium hydroxide, and the acetone was removed at 55°. Triethylamine–sulfur trioxide complex was added, and the mixture was heated at 55° for 7 hr. The intermediate triethylammonium salt (XII) was not characterized but was reacted with sodium acetate to give the desired sodium salt.

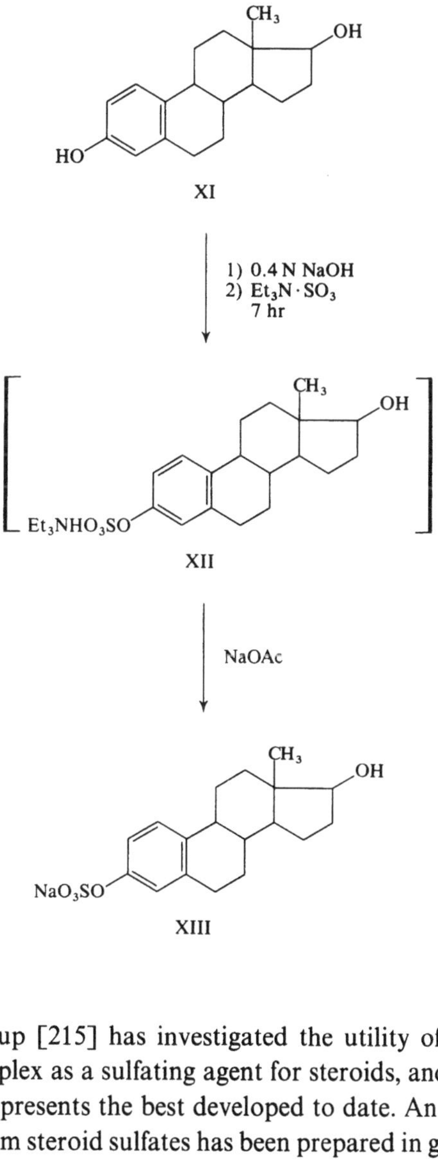

The Lederle group [215] has investigated the utility of triethylamine–sulfur trioxide complex as a sulfating agent for steroids, and has concluded that this method represents the best developed to date. An extensive series of triethylammonium steroid sulfates has been prepared in good to excellent yields. In addition, selective sulfation was observed in those cases where reaction rates greatly differentiated between the various hydroxyl groups. Modifications of the general reaction conditions, such as heating, have allowed the sulfation of hindered hydroxyl groups, and the preparation of several disulfate salts.

The triethylammonium sulfates were found to be, generally, highly crystalline compounds which could be readily recrystallized. Analytical data indicated the salts to be generally nonhydrated, nonhygroscopic, and stable under normal storage conditions. An advantage of the triethylammonium salts is their solubility in methylene chloride which conveniently allows their recrystallization from methylene chloride–anhydrous ether. This solubility also aids in their purification, since passage of a methylene chloride solution of the salt through a pad of an adsorbent such as Magnesol®, a hydrous magnesium silicate (Food Machinery Chemical Corp.), removes colored impurities and, in some cases, unreacted starting material. The adsorbent also removes traces of water which might hamper recrystallization. The triethylammonium salts can easily be converted into their corresponding ammonium salts by passage of a stream of ammonia into a methylene chloride or methylene chloride–anhydrous ether (2:1) solution of the former. The precipitated ammonium salt can be recovered by filtration.

A wide variety of sulfates have been prepared with this reagent: *e.g.*, 3β–sulfooxyandrost–5–en–17–one triethylammonium and ammonium salts, 17β–sulfooxyandrost–4–en–3–one triethylammonium and ammonium salts, 3β,17α–disulfooxypregn–5–en–20–one diammonium salts, 6β–sulfooxypregn–4–ene–3,20–dione triethylammonium salt, 17α–hydroxy–21–sulfooxypregn–4–ene–3,20–dione triethylammonium salt, 17α,21–disulfooxypregn–4–ene–3,20–dione triethylammonium salt, 17α–hydroxy–21–sulfooxypregn–4–ene–3,11,20–trione triethylammonium and ammonium salts, 11β,17α–dihydroxy–21–sulfooxypregn–4–ene–3,20–dione triethylammonium salt, and 3–sulfooxyestra–1,3,5(10)–trien–17–one triethylammonium salt.

2. Sulfuric Acid

Mumma [217, 218] has recently described a novel method for the preparation of steroid sulfates in good yield under mild conditions. It is claimed that there are practically no side reactions, and the products can be easily isolated. The method is based on the reaction of sulfuric acid with an alcohol and dicyclohexylcarbodiimide in a polar solvent such as dimethylformamide (Eq. (g)). Accordingly, potassium cholesterol sulfate and potassium testosterone sulfate (^{35}S) [17β–sulfooxyandrost–4–en–3–one potassium salt] were prepared in over 70% yields. The author has pointed out

$$\mathrm{StOH + H_2SO_4 + C_6H_5N{=}C{=}NC_6H_5 \xrightarrow[30\,min-25^\circ]{Dimethylformamide} StOSO_3H + C_6H_5NHCONHC_6H_5} \qquad (g)$$

that the method of synthesis is well adapted for radiolabeled syntheses since $^{35}SO_4$ can be used directly. Also further studies are reported to be in progress on the mechanism and versatility of the reaction.

Related to the Mumma synthesis is the recently announced [219] synthesis of dehydroisoandrosterone sulfate [3β–sulfooxyandrost–5–en–17–one] in high yield from dehydroisoandrosterone with L–ascorbic acid 3–sulfate in dimethylformamide with dicyclohexylcarbodiimide, followed by mild acid hydrolysis. No experimental details were presented, and the nature of the salt prepared was not specified. The enol–sulfate reagent was prepared from ascorbic acid 5,6–*O*–benzylidene derivative with the pyridine–sulfur trioxide. The complex was used as such and with ^{35}S.

Nikolaev and Krastelevska [220], in a study of the Salkowski color test for cholesterol, have concluded that the appearance of a red coloration on agitating a cholesterol solution in chloroform with concentrated sulfuric acid is due to the formation of cholesterol sulfate.

Brieskorn and Hofmann [221], in an investigation of the Liebermann–Burchard color reaction of Δ^5–sterols, have postulated sterol sulfate as intermediates in the development of the chromophore.

3. *Sulfamic Acid*

Sulfamic acid has been employed by a number of investigators for the sulfation of steroids. The steroid in pyridine is heated with sulfamic acid for 1–2 hr to give directly the sulfate ammonium salt (Eq. (h)) which, on treatment with a methanolic metal hydroxide, is converted into the metal salt (Eq. (i)).

$$StOH + NH_2SO_3OH \xrightarrow{\text{Pyridine}} StOSO_3NH_4 \qquad \text{(h)}$$

$$\downarrow \text{MeOH} \mid \text{MOH}$$

$$StOSO_3M \qquad \text{(i)}$$

$$M = K,\ Na,\ \textit{etc.}$$

Anhydrous reaction conditions, and careful control of temperature and pH are not required. Sulfamic acid has been used in the preparation of synthetic estrogen sulfates [222], estrone sulfate salts [223, 224], ammonium dehydroisoandrosterone sulfate [194], and vitamin D_2 sulfate [225].

The Lederle group [226] has carried out the most extensive study of the utility of sulfamic acid in pyridine as a sulfating agent in the steroid field. The sulfates of the following were prepared and characterized as hydrates: estrone, 2–methoxyestrone [3–hydroxy–2–methoxyestra–1,3,5(10)–trien–17–one], ethynylestradiol [17α–ethynylestra–1,3,5(10)–triene–3,17β–diol], ethynylestradiol–3–methyl ether [17α–ethynyl–3–methoxyestra–1,3,5(10)–trien–17β–ol], dehydroisoandrosterone, testosterone [17β–hydroxyandrost–

4-en-3-one], pregnenolone [3β-hydroxypregn-5-en-20-one], and cholesterol. It was noted that this method was not applicable to compounds possessing a dihydroxyacetone side chain, as this grouping underwent extensive decomposition under the reaction conditions. The procedure so developed was subsequently employed for the preparation of the mixed conjugate, 17β-estradiol-3-sulfate-17-glucuronide [3-sulfooxyestra-1,3,5(10)-trien-17β-yl-β-D-glucopyranosiduronic acid] [152].

C. Chemical Properties

In the voluminous literature concerning steroid sulfate conjugates, a number of chemical transformations have been reported. The primary intent of these studies has been the elucidation of structure and/or synthesis of the naturally occurring conjugates. Although little has been described concerning the chemistry apart from the above obvious purposes, some interesting observations can be made from the examples cited. The literature has been surveyed and the reactions categorized. In each of the listed categories, after indicating the illustrative reaction, any broad implication suggested by the reaction will be discussed.

1. Reduction Studies

The examples in this area will illustrate reduction studies of double bonds and carbonyl groups located on the steroid conjugate molecule. At least two examples of double-bond reduction have been reported. In one study the catalytic reduction of 1-methylpyridinium cholesterol sulfate (I) [227] was investigated. In this instance the use of hydrogen-platinum oxide (atmospheric pressure) afforded reduction of the 5,6-double bond to yield cholestanol sulfate (II) [3-β-sulfooxycholestane] and 1-methylpiperidine. 1-Methylpyridinium cholestanol sulfate was also reduced under the identical conditions and yielded the same products.

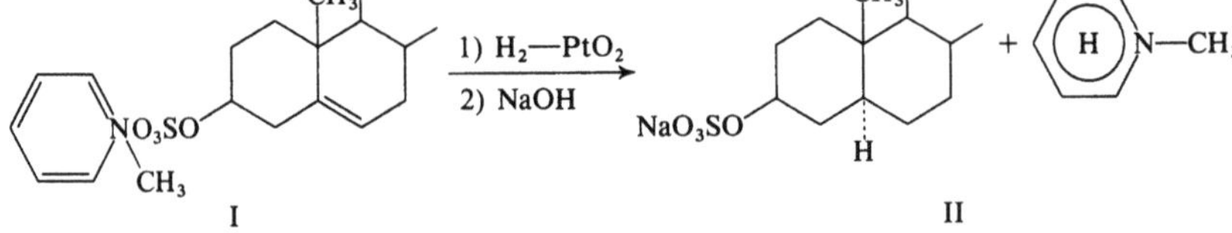

I II

The ring A reduction of triamcinolone acetonide 21-sulfate (III) [9α-fluoro-11β-hydroxy-16α,17α-isopropylidenedioxy-21-sulfooxypregna-1,4-diene-3,20-dione] to yield IV has been reported by Hydorn, Lerner, and Schwartz [212]. These authors state that in their experience, the reduction of the sulfate ester in aqueous medium [hydrogen (5 psig)-5%

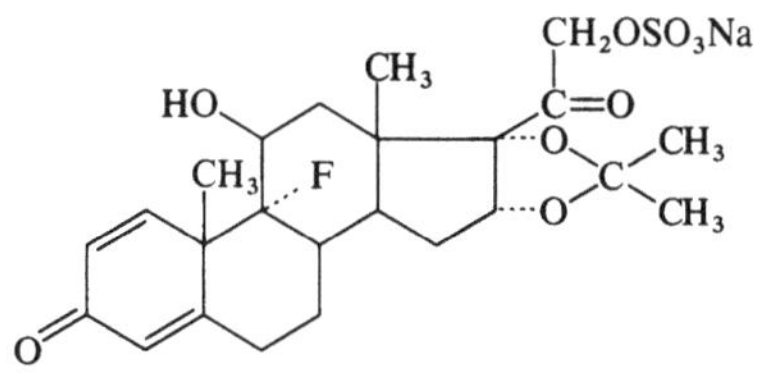

III

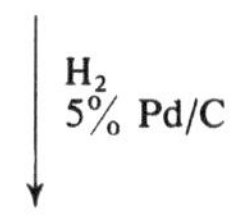

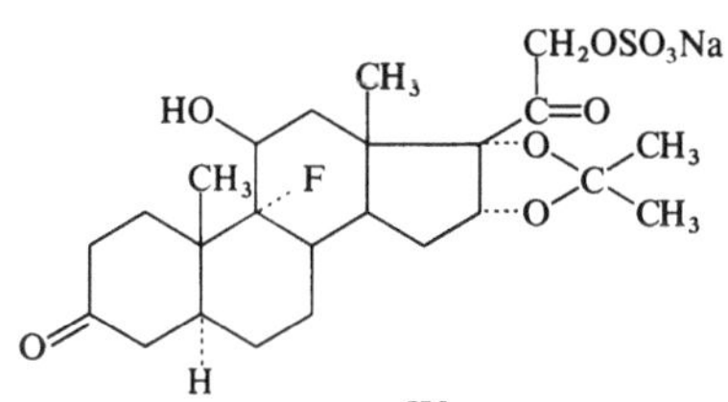

IV

palladium on charcoal–3 hr] is much more readily accomplished than the related reduction of triamcinolone [9α–fluoro–11β,16α,17α,21–tetrahydroxy-pregna–1,4–diene–3,20–dione] in ethyl acetate. It was suggested that the stronger adsorption of the sulfate ester to the catalyst surface aids this reduction considerably.

The initial study of the reduction of a carbonyl group in sulfate conjugates was reported by Grant and Beall [172] in their preparation of 17β–estradiol–3–sulfate (VI) [3–sulfooxyestra–1,3,5(10)–trien–17β–ol]. A catalytic reduction of sodium estrone sulfate (V) in an aqueous solution (0.2 M phosphate buffer, pH 7.5) employing hydrogen and a platinum oxide catalyst afforded reduction of the 17–carbonyl grouping. A more convenient reduction technique for this transformation was later employed by Kirdani [176] who carried out the reduction in methanol with sodium borohydride. The use of

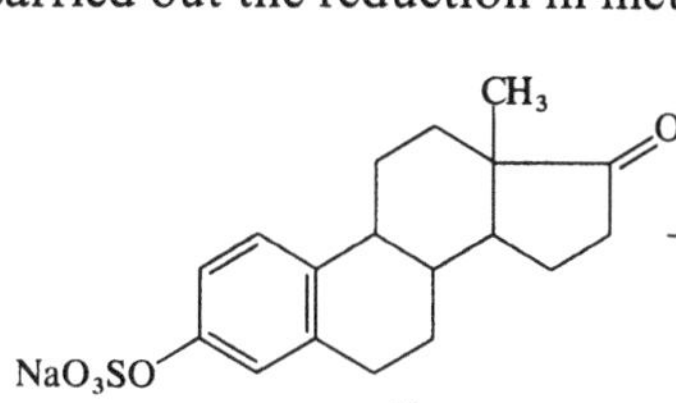

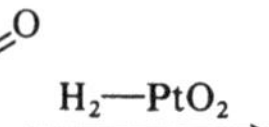

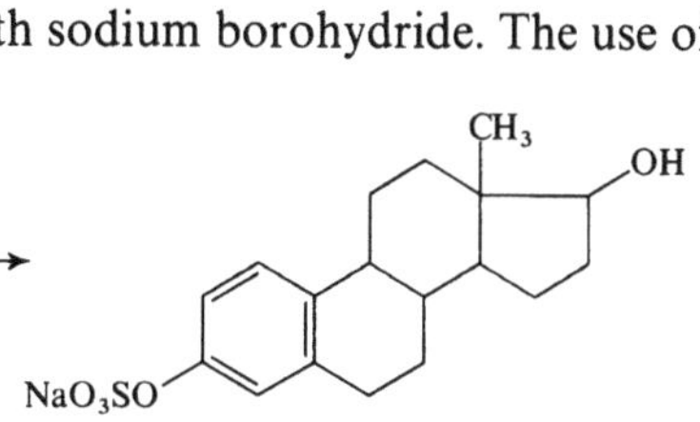

V VI

borohydride ion for the reduction of carbonyl groups in steroid sulfate conjugates was initially demonstrated by Baulieu and Emiliozzi [202] in their synthetic studies on androgen sulfates. For their purposes potassium borohydride reduction of potassium dehydroisoandrosterone sulfate (VII) in methanol gave potassium androstenediol-3-monosulfate (VIII) [3β-sulfooxyandrost-5-en-17β-ol potassium salt]. Roberts *et al.* [228] have reported on the sodium borohydride reduction of sodium dehydroisoandrosterone sulfate.

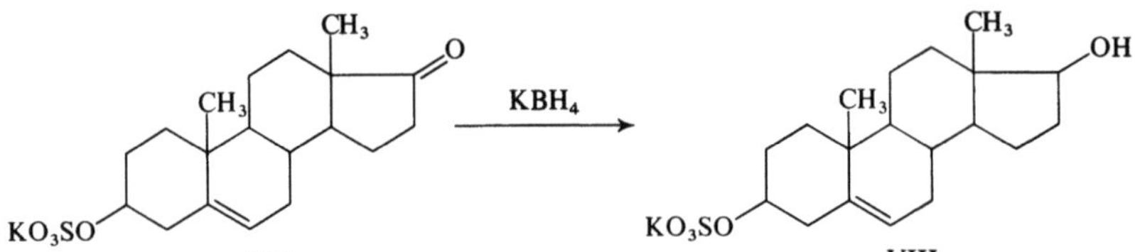

VII VIII

The identity of 17α-hydroxypregnenolone sulfate (IX) [3β-sulfooxy-17α-hydroxypregn-5-en-20-one] as a metabolite arising from the direct 17α-hydroxylation of a sulfate substrate employing a homogenate of hyperplastic adrenal tissue was established by Calvin and Lieberman [229] through potassium borohydride reduction to yield the 20-hydroxyl derivative (X) which in turn was cleaved with periodate to yield dehydroisoandrosterone sulfate (XI).

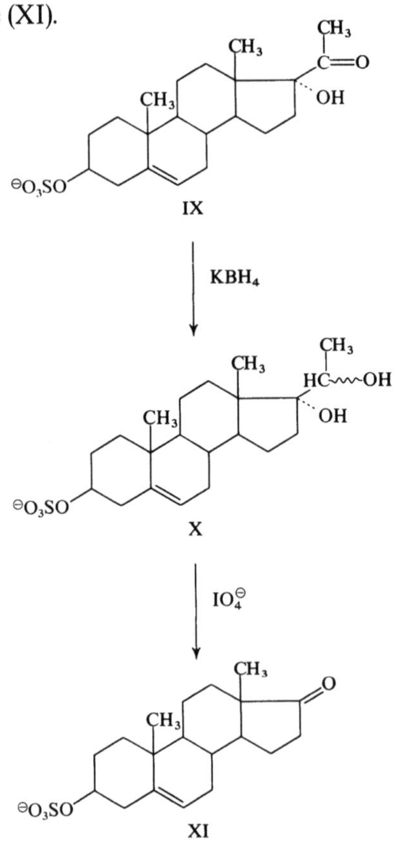

IX X XI

The solubility of steroid sulfate conjugates in aqueous and alcoholic solvents, due to the polarity of the sulfated hydroxyl groups, affords reduction of carbonyl groups in the molecules through the use of the various borohydride reducing agents. The variety of reducing agents allows for considerable variation in both solvent conditions and reaction temperature and hence in the stereochemistry of reduction processes. The use of a polar grouping such as sulfate to achieve favorable solubility properties for subsequent chemical transformation steps is not widely used in preparative organic chemistry.

2. Oxidation Studies

Although the number of examples of oxidation reactions applied to steroid sulfate conjugates is very limited, those reported are quite interesting and illustrate several different reagents, reaction conditions, and product possibilities. One oxidation study reported helped to elucidate the structure of scymnol [230] in the following manner. As isolated from shark bile, scymnol sulfate is a polyhydroxyl monosulfated bile alcohol with the structure, 26–sulfooxy–5β–cholestane–3α,7α,12α,24ξ,27–pentol (XII). Oxidation of the sulfate in acetic acid with chromic acid gave dehydrocholic acid (XIII) [3,7,12–trioxo–5β–cholanic acid] with the sulfate portion of the molecule being lost in the glycol fragmentation. The hydroxyl functions at positions 3, 7, and 12 were simultaneously oxidized to carbonyl functions during the reaction.

An interesting example in which sulfate groups protect hydroxyl functions during oxidation reactions has been outlined in the study of the structure

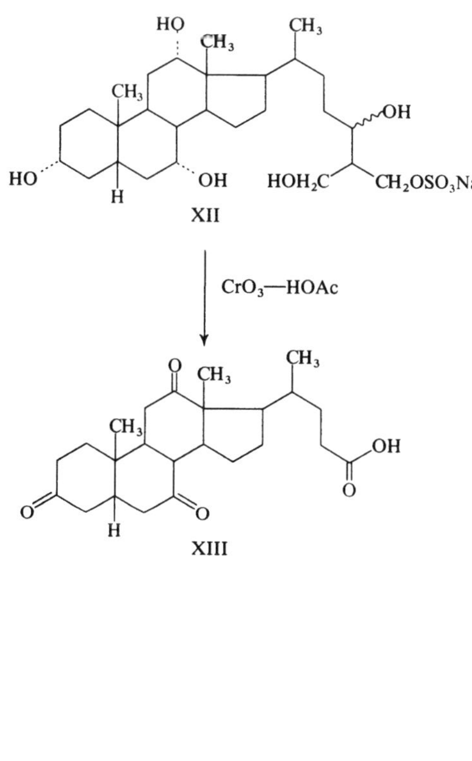

of myxinol [231]. The sulfated bile alcohol isolated from hagfish has the structure 3β,26–disulfooxy–5α–cholestane–7α,16α–diol (XIV) [232]. When disodium myxinol disulfate in acetic acid was oxidized with chromic acid and then hydrolyzed with 0.25 *N* hydrochloric acid, 3β,26–dihydroxy–5α–cholestane–7,16–dione (XV) was isolated. Thus the sulfate groupings served as protective groups for the hydroxyl functions at positions 3 and 26

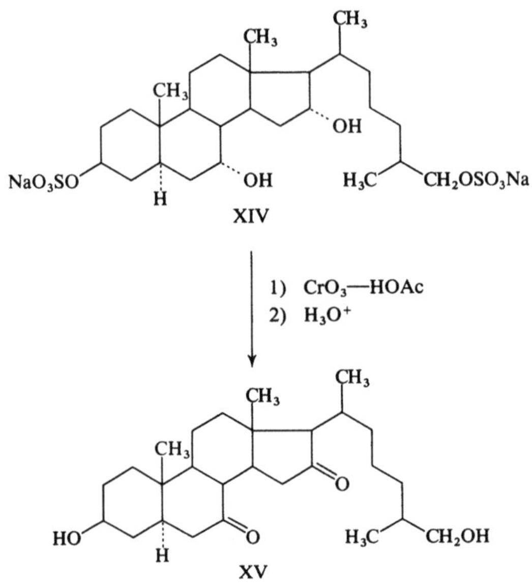

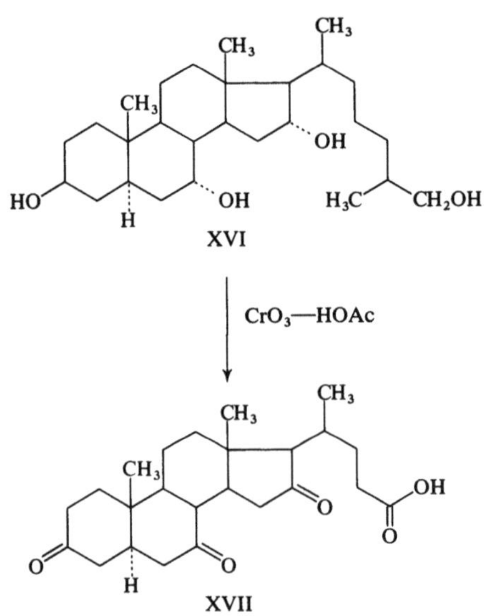

preventing their oxidations under the reaction conditions. A similar oxidation of myxinol (XVI) itself gave 3,7,16–trioxo–5α–cholanic acid (XVII) in which the unprotected 3–hydroxyl group was oxidized to a carbonyl group and the side chain has undergone the typical oxidation to the C_{24} carboxylic acid. The use of the oxidation reagent, chromium trioxide in pyridine, has been described by Starka *et al.* [233, 234] in their investigation of the metabolite, 7α–hydroxydehydroisoandrosterone [3β,7α–dihydroxy-androst–5–en–17–one]. As isolated, the compound was a mixture of mono– and disulfated derivatives. When the monosulfate (XVIII) was oxidized, 7–ketodehydroisoandrosterone–3–sulfate (XIX) [3β–sulfooxyandrost–5–ene–7,17–dione] was obtained, convincingly establishing the position of monosulfation at position 3.

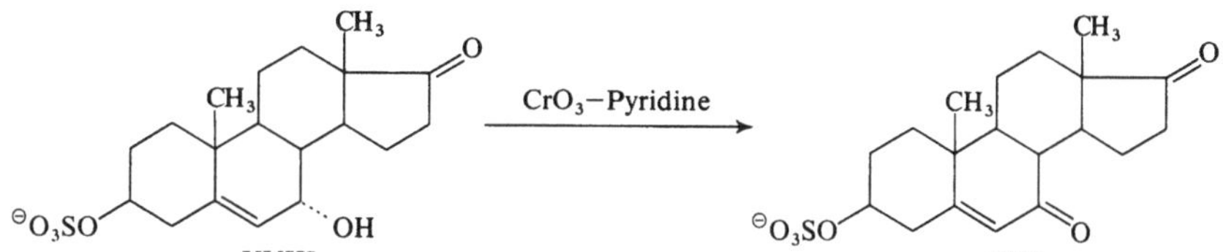

XVIII XIX

In an unpublished observation [157], chromic acid–pyridine oxidation of triethylammonium corticosterone–21–sulfate (XX) [11β–hydroxy–21–sulfooxypregn–4–ene–3,20–dione triethylammonium salt] proceeds quite smoothly to yield triethylammonium dehydrocorticosterone–21–sulfate (XXI) [21–sulfooxypregn–4–ene–3,11,20–trione triethylammonium salt]. The use of triethylamine–sulfur trioxide for selective sulfation of hydroxyl

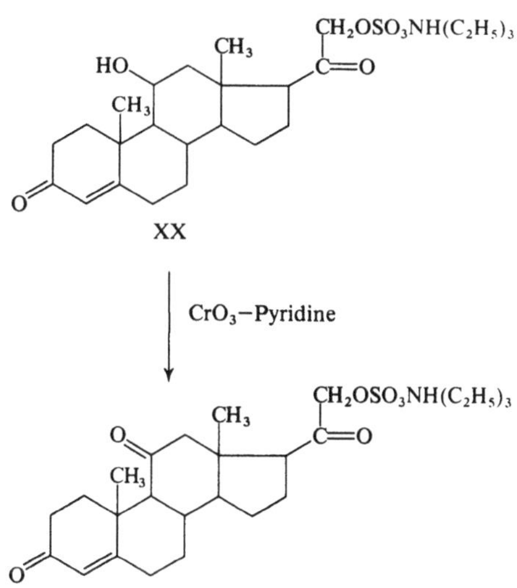

XXI

functions followed by chromic acid–pyridine oxidation may prove of some synthetic importance in steroid transformation studies since solvolytic removal of the sulfate grouping takes place under such mild reaction conditions.

Oxidation of sulfate conjugates having vicinal hydroxyl groups already has been mentioned in the work of Calvin and Liebermann [229]. Periodate oxidation in aqueous media of sulfate conjugates is favored by the polar nature of the sulfate grouping. An interesting oxidation of an aromatic sulfate in aqueous solution, though not a steroid sulfate, is the conversion of potassium isoeugenol sulfate (XXII) to vanillin (XXIV) [164] using ozone and hydrolysis of the intermediate XXIII.

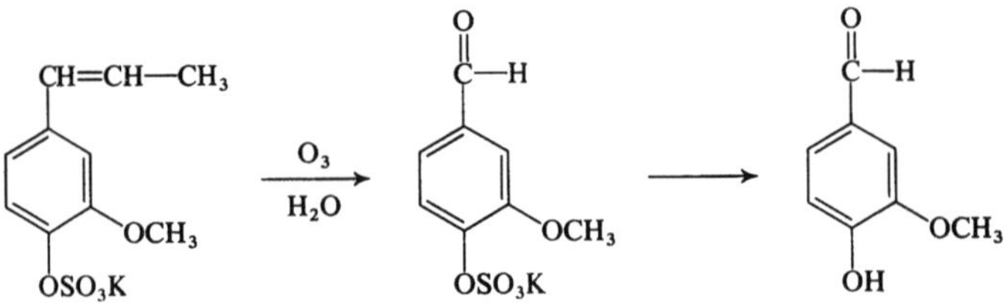

XXII XXIII XXIV

3. Elimination Studies

Elimination along with hydrolysis are certainly among the earliest reactions reported for steroid sulfates. Natelson and Gottfried studied the pyrolytic elimination reaction of potassium cholesterol sulfate (XXV) [235], which, heated at 100° in *vacuo*, gave cholesterilene (XXVI), an impure cholesta-3,5-diene. When this work was repeated by Sobel and Rosen [236] at various temperatures up to and including 180°, levorotatory solid cholesterilene was obtained. The latter investigators found that pure cholesta-3,5-diene (XXVII) could be obtained by heating potassium cholesterol sulfate (XXV) at 177° in 2-octanol in the presence of sodium octylate. The use of sodium n-butylate in n-butanol at 120° gave a solid mixture of potassium and sodium cholesterol sulfates corresponding to the formula $C_{27}H_{45}NaO_4S \cdot 2C_{27}H_{45}KO_4S$. When the elimination reactions were studied in the cholestanol series, the results paralleled those above such that the thermal decomposition of potassium cholestanol sulfate (XXVIII) in octanol in the presence of sodium octylate at 180° for 6 hours gave cholest-2-ene (XXIX). If the reaction was performed at 169° for one hour, a mixture of sodium and potassium cholestanol sulfates was obtained corresponding to a 1:1 ratio.

Cholesterilene

XXVI

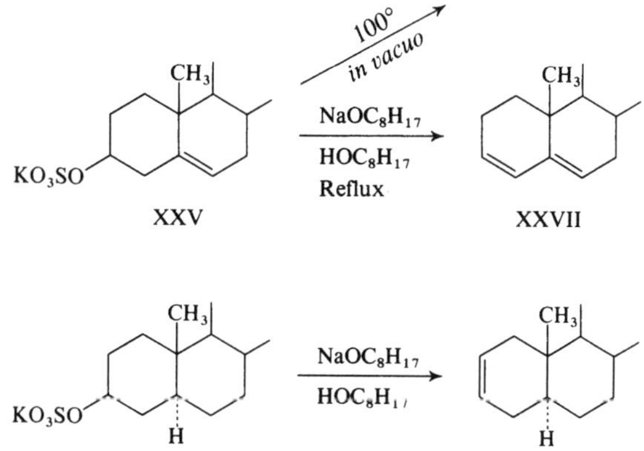

XXV XXVII

XXVIII XXIX

The decomposition of calcium cholesterol sulfate (XXX) was studied by Sobel and Spoerri [237]. Under the relatively mild conditions of heating the dry salt at 100° for 40 min, the elements of calcium sulfate were removed and dicholesteryl ether (XXXI) was identified as the product. When refluxed

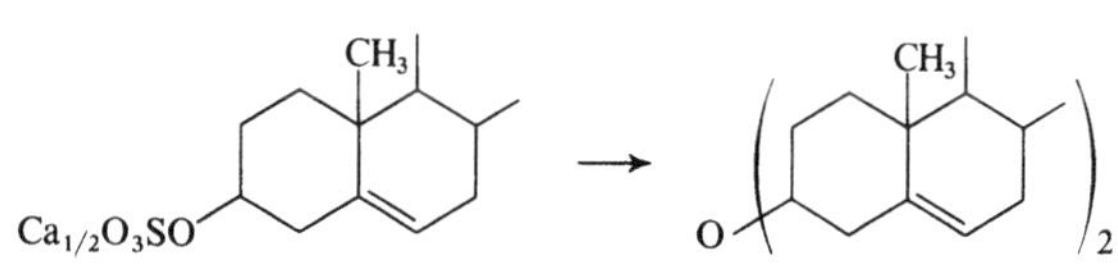

XXX XXXI

in water, however, calcium cholesterol sulfate gave 7% of cholesterol and largely unchanged starting material. Dicholesteryl ether has also been observed during the decomposition of other cholesterol sulfates at varying temperatures. The ferrous and ferric salts [238] decompose at approximately 100–115° while the aluminum salt *on standing at room temperature* slowly decomposes to yield dicholesteryl ether. The same series of conversions was reported by Khaletskii [203] in the sitosterol series with the decomposition of the calcium and aluminum salts giving rise to disitosteryl ether.

A synthesis of 7-dehydrocholesterol (XXXIII) [cholesta–5,7–dien–3β–ol] [238] has been based on sulfate elimination in the following manner. 7β–Hydroxycholesteryl phthalate was converted to the 7–sulfate ester (XXXII) with chlorosulfonic acid in pyridine. Decomposition of the dry salt at 138°, or reaction of the salt with sodium propylate in propanol at reflux, gave 7–dehydrocholesterol (XXXIII).

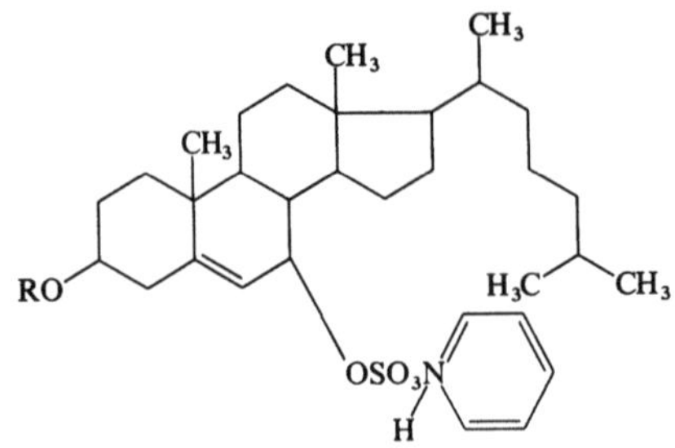

XXXII

↓

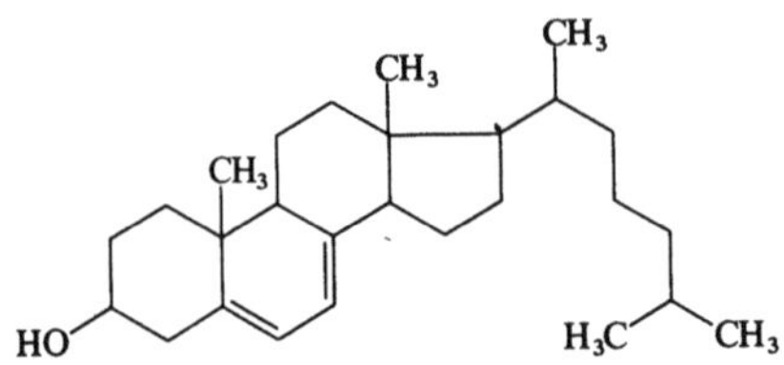

XXXIII

In a sparsely documented paper, the two previously mentioned modes of decomposition of sulfated cholesteryl derivatives [239] were combined such that 3–barium–7α–pyridinium cholesteryl disulfate (XXXIV) on heating was converted to what was assumed to be di–7–dehydrocholesteryl ether

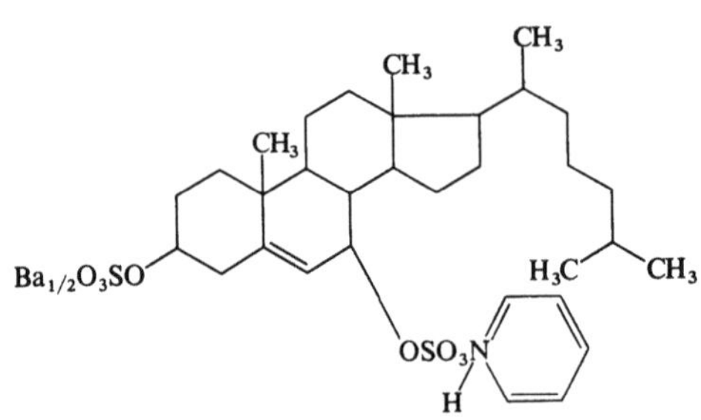

XXXIV

↓

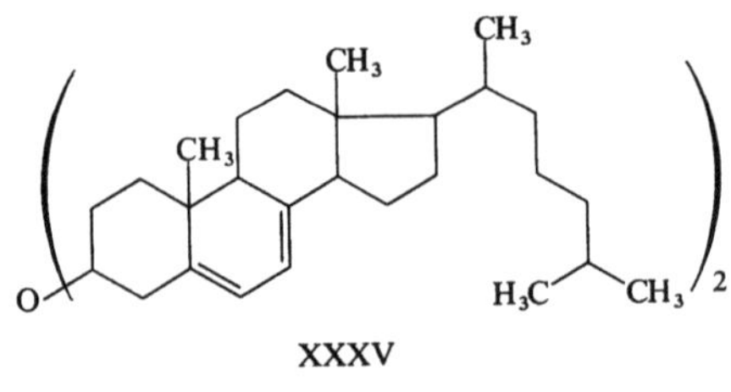

XXXV

(XXXV) on the basis of the ultraviolet absorption maximum at 282 mμ which is characteristic of the 5,7–diene chromophore.

In order to establish the structure of 5α–pregn–2–en–20–one as one of the numerous steroidal constituents of human urine (which may be an artifact introduced through processing procedures), Lieberman *et al.* [240] prepared the above–mentioned compound from allopregnanolone [3β–hydroxy–5α–pregnan–20–one] by (1) sulfation of the 3–hydroxyl group with chlorosulfonic acid in pyridine and (2) pyrolysis of the potassium sulfate salt *in vacuo* (10^{-5} mm) at 250°.

4. Diazomethane Studies

In point of time, the reaction of diazomethane and a steroid sulfate represents one of the earliest reactions of a steroid conjugate. Among the reactions reported for "estrone sulfuric acid" (XXXVI) by Butenandt and Hofstetter [166] was that with diazomethane to give a "*dimethyl ester*". Undoubtedly the compound is the mono–methyl ester (XXXVII) although the analytical data do not support this contention. Since the methylated

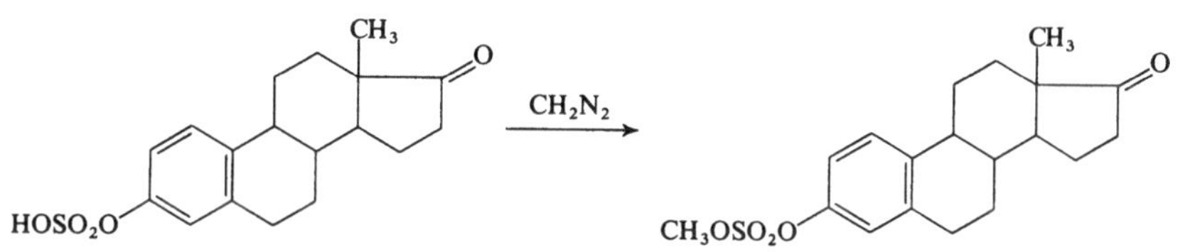

XXXVI XXXVII

derivative was recrystallized several times from methanol, the sample may well have become contaminated with some solvolytic products of the sulfate ester.

A more convenient synthesis of methyl steroid sulfate esters was described by McKenna and Norymberski [227]. In their procedure the steroid was first sulfated with pyridine–sulfur trioxide reagent, and the pyridinium salt then treated with ethereal diazomethane to yield the methyl esters of cholestanol and cholesterol sulfate. With the compounds thus prepared the authors report their extensive study of the solvolysis reactions of these esters.

Emiliozzi [241] prepared a large number of steroid sulfate esters in the androgen series, using pyridine–sulfur trioxide as the sulfating agent. The compounds were eventually isolated as their sodium salts. Adjustment of the hydrogen ion concentration to pH 7 to 8 with dilute hydrochloric acid, and subsequent treatment with a benzene solution of diazomethane gave the methyl esters. Essentially the same procedure was used to prepare the methyl ester of β–sitosterol sulfate [3β–sulfooxystigmast–5–ene] [200].

5. Hydrolysis Studies

Since the purpose of this section is to discuss the reactions of sulfate conjugates, minimal reference will be made to these studies which involve hydrolytic removal of the sulfate grouping from the steroid substrate. These reactions are discussed in Chapter 3, where the general topic of the solvolysis of sulfate esters can be found.

An early synthesis of 17β–estradiol–17–sulfate (XL) [17β–sulfooxyestra–1,3,5(10)–trien–3–ol] by Holden and Bromley [171] was based on the sulfation of 17β–estradiol–3–benzoate (XXXVIII), and subsequent hydrolysis of the intermediate XXXIX with methanolic sodium hydroxide. The authors state that 17β–estradiol–17–sulfate is not hydrolyzed by a barium chloride–sodium acetate mixture; however, employment of strong acid conditions, aqueous hydrochloric acid in benzene, brings about hydrolysis to 17β–estradiol in 55% yield.

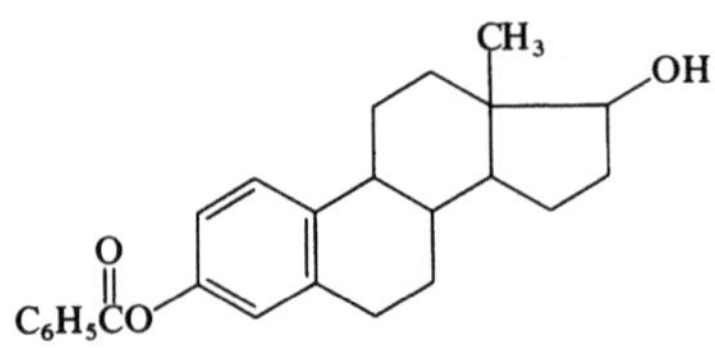

XXXVIII

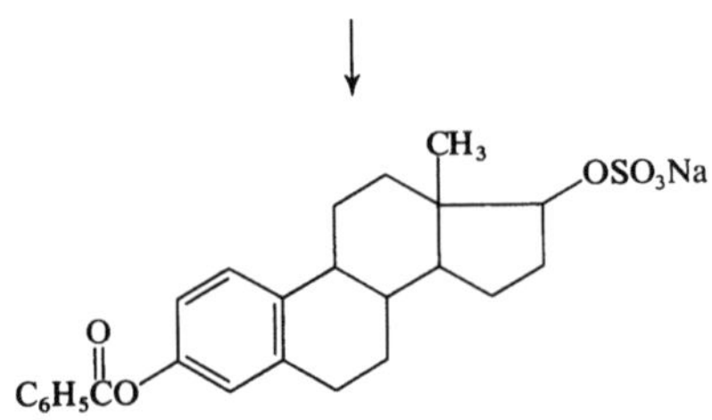

XXXIX

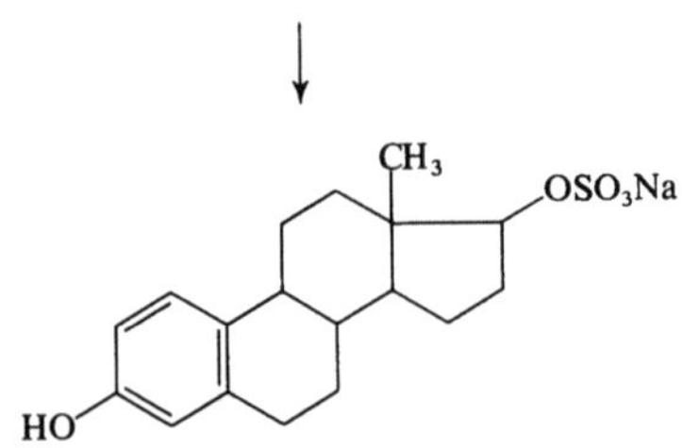

XL

Beall and Grant [242] in the preparation of 17β–estradiol–3–sulfate, and Fex and coauthors [213] for 17β–estradiol–3–sulfate and various estriol sulfates made use of basic hydrolysis conditions for the removal of the acetate projective groups at C 17, and C 16 and C 17 (XLI → XLII), respectively, without hydrolysis of the aromatic sulfate grouping. In general, aryl

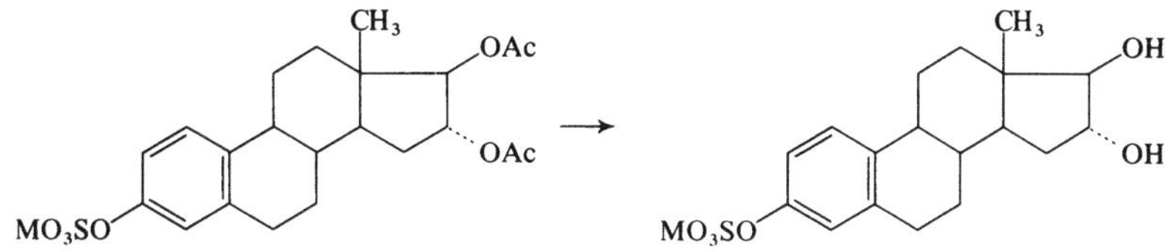

XLI XLII

sulfates will withstand strongly basic conditions but are conveniently hydrolyzed by acid or such solvents as dioxane, tetrahydrofuran, *etc.*, at reflux. For achieving the synthesis of several double estrogen conjugates [123, 152] the synthesis entailed the removal of the protective groupings on the glucuronide moiety under such conditions which would retain the sulfate grouping at C 3 (XLIII → XLIV; XLV → XLVI).

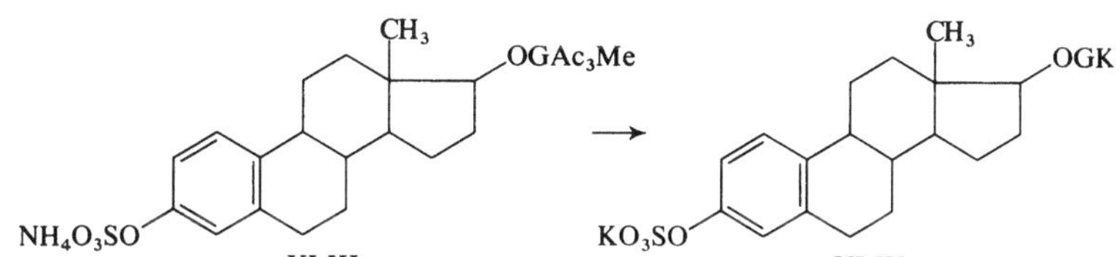

XLIII XLIV

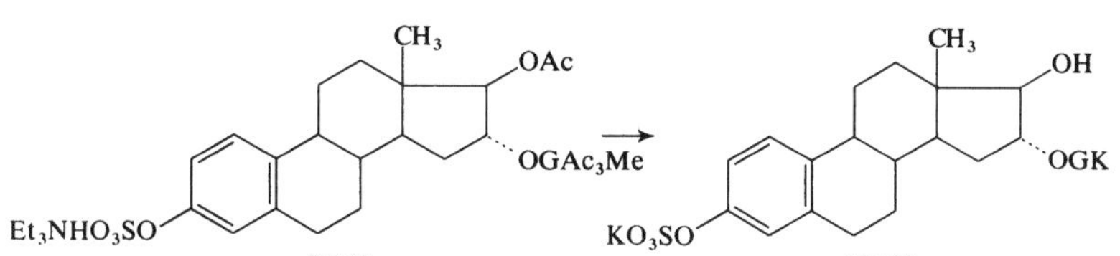

XLV XLVI

In assessing the synthesis of the various 17β–estradiol and estriol sulfates, other than those we have discussed thus far, hydrolysis conditions have been uniquely applied in each case. Both 17β–estradiol–17–sulfate and estriol–16,17–disulfate [16α,17β–disulfooxyestra–1,3,5(10)–trien–3–ol] were prepared by hydrolysis of the completely sulfated derivatives (*e.g.*, XLVII) employing acid hydrolysis conditions, such as hydrochloric acid in aqueous methanol. These conditions were considered preferable to 70% aqueous dioxane (60°). The extended hydrolysis of estriol–16,17–disulfate (XLVIII) resulted in the isolation of a monosulfate which was identified as estriol–17–sulfate (XLIX) [17β–sulfooxyestra–1,3,5(10)–triene–3,16α–diol].

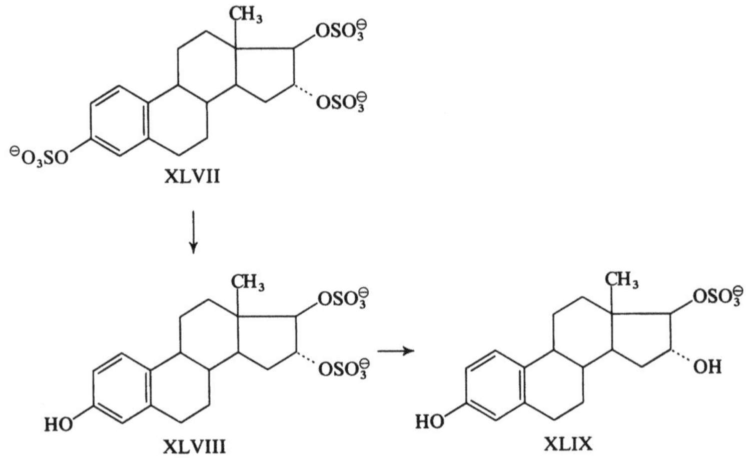

A discussion of the hydrolysis reaction of sulfate conjugates would not be complete without mention of uranediol [243]. The structure of this hydrolysis product was established by Klyne [244] to be 17α–methyl–D–homo–5α–androstane–3β,17aβ–diol (LI). The manner in which this arose was conclusively demonstrated by Hirschmann and Williams [245]. 5α–Pregnane–3β,20β–diol–20–sulfate (L) [20β–sulfooxy–5α–pregnan–3β–ol] isolated from pregnant mare's urine was subjected to the original hydrolytic conditions of Marker, and a compound was indeed isolated that proved identical to uranediol. The compound was also characterized as its diacetate.

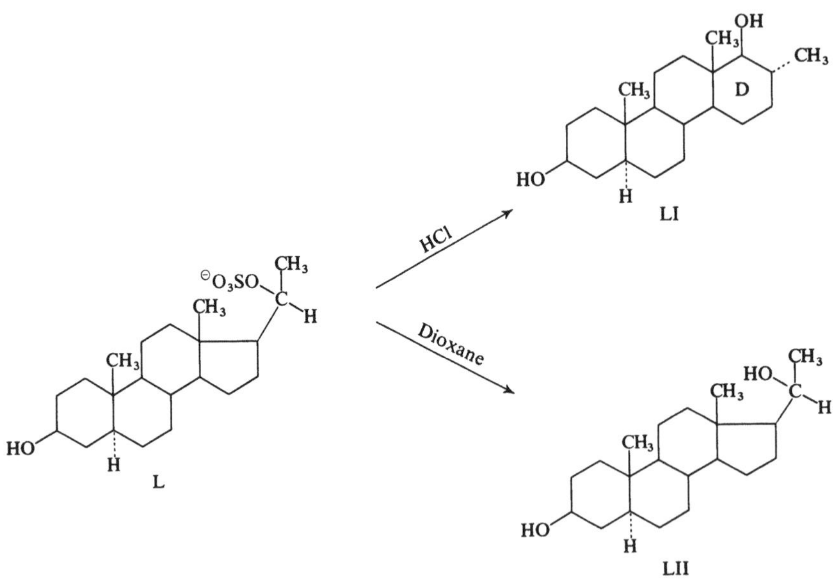

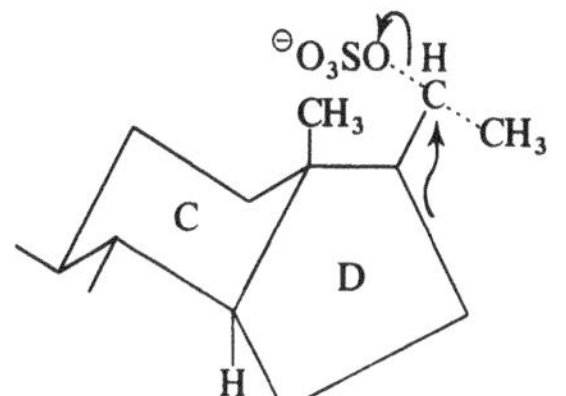

La

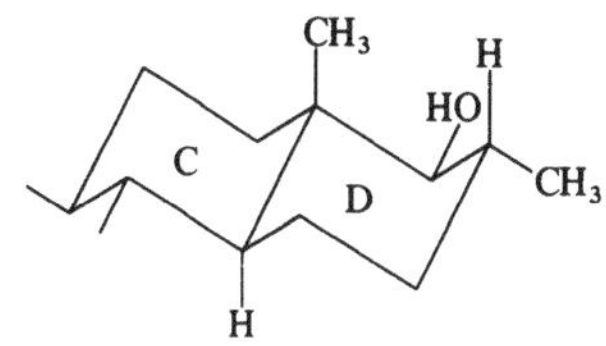

LIa

Milder hydrolysis conditions (refluxing dioxane) yielded 5α–pregnane–3β,20β–diol (LII) which supported the postulate that acid hydrolytic conditions were responsible for the D–homo–rearrangement (La → LIa).

A similar D–homoannulation reaction was observed by Williams, Smulowitz and Fukushima [185] in the pregnane–17,20–glycol series. Thus when 5α–pregnane–3β,17α,20β–triol–20–sulfate (LIII) [20β–sulfooxy–5α–pregnane–3β,17α–diol] was heated on a steam bath for 1 hr with 1 *N* hydrochloric acid and the crude reaction mixture acetylated (acetic anhydride–pyridine), there was isolated two compounds (LIV, LV). The less polar material was identified as the D–homoannulated product, 3β–acetoxy–17α–methyl–D–homo–5α–androstan–17a–one (LV). Also encountered in the reaction mixture was the simple hydrolysis product 3β,20β–diacetoxy–5α–pregnan–17α–ol (LIV) (acetylated after hydrolysis). Unrearranged product could also be obtained by acidic hydrolysis with sulfuric acid (0.1 *N*) in moist ethyl acetate. Base treatment of the 20–sulfate gave the 17α,20α–epoxide (LVI).

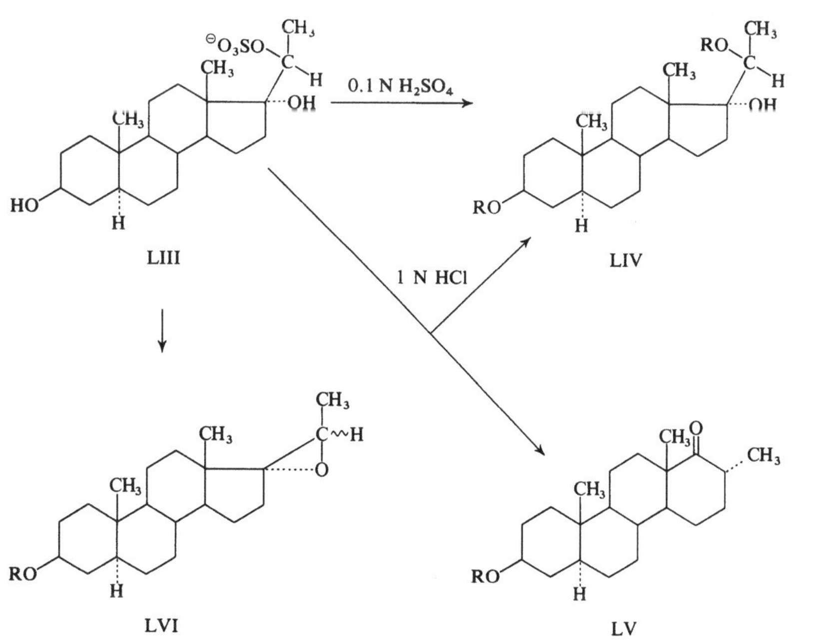

The appearance of epoxides (mainly substituted trimethylene epoxides) has been recognized as a structural feature in the side chain of certain bile alcohols, and arises during alkaline hydrolysis of the naturally occurring bile alcohol sulfates. Through displacement of a sulfate moiety by an appropriately positioned hydroxyl group, the "anhydro–bile alcohols" have been implicated in the alkaline hydrolysis of the sulfates of scymnol (XII → LVII), chimaerol, cyprinol, latimenol and ranol [36]. In the case of scymnol, the side chain structure has been completely elaborated through the use of nuclear magnetic spectroscopy [246]. This study has confirmed the originally postulated side chain structure proposed by Fieser and Fieser [247].

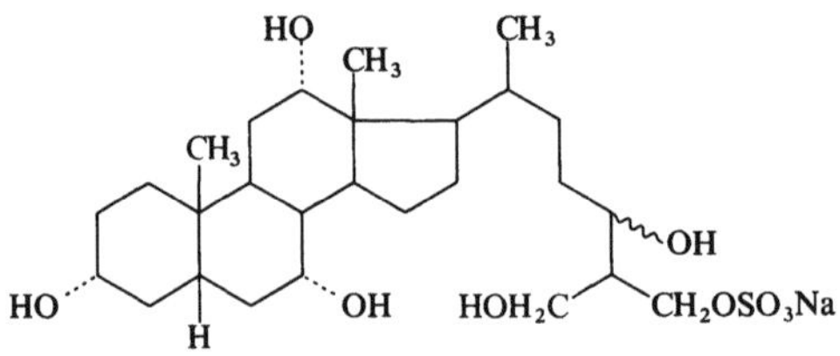

XII

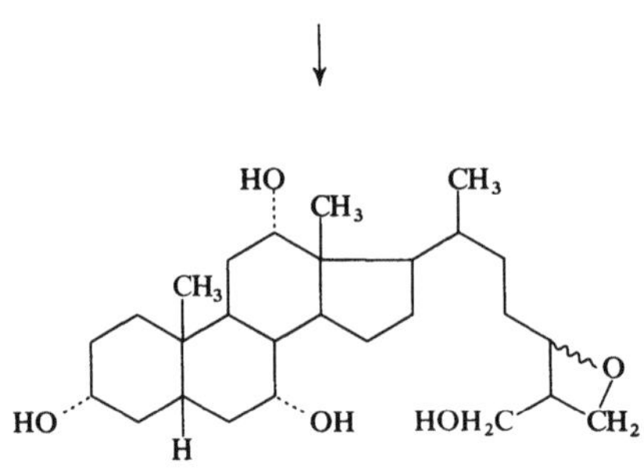

LVII

Mention must be made at this time of the classical reaction encountered in the hydrolysis of dehydroisoandrosterone sulfate under near neutral conditions employing a heterogeneous reaction mixture. Under these conditions a very high yield of 6*β*–hydroxy–3α,5α–*cyclo*androstan–17–one is encountered [248]. The appearance of 6*β*–hydroxy–3α,5α–*cyclo*steroids (LX) can be easily rationalized by concluding that hydrolysis of the sulfate (LVIII) proceeds with neighboring group participation of the 5,6–double bond and concomitant generation of the homoallylic cation (LIX). Attack by solvent (water) at approximately neutral pH conditions gives the 6*β*–hydroxy*cyclo*steroid.

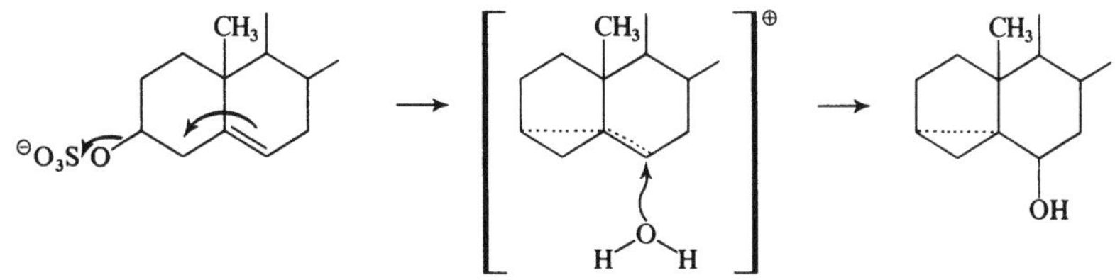

LVIII LIX LX

The extent to which refluxing aqueous $1N$ hydrochloric acid hydrolysis of steroid sulfate conjugates can complicate the general study of sulfate solvolysis, and thus steroid metabolic studies has been demonstrated by Ramseyer, Williams and Hirschmann [249]. Potassium androsterone sulfate (LXI) is cleaved by refluxing dioxane to yield only androsterone (LXII) [3α–hydroxy–5α–androstan–17–one] as the product. Hydrolysis conditions employing refluxing aqueous $1N$ hydrochloric acid gave a very complicated reaction mixture which consisted of an olefin fraction (73%) and a hydroxyl fraction (16%). The olefin fraction contains a mixture of Δ^2– and Δ^3–isomers

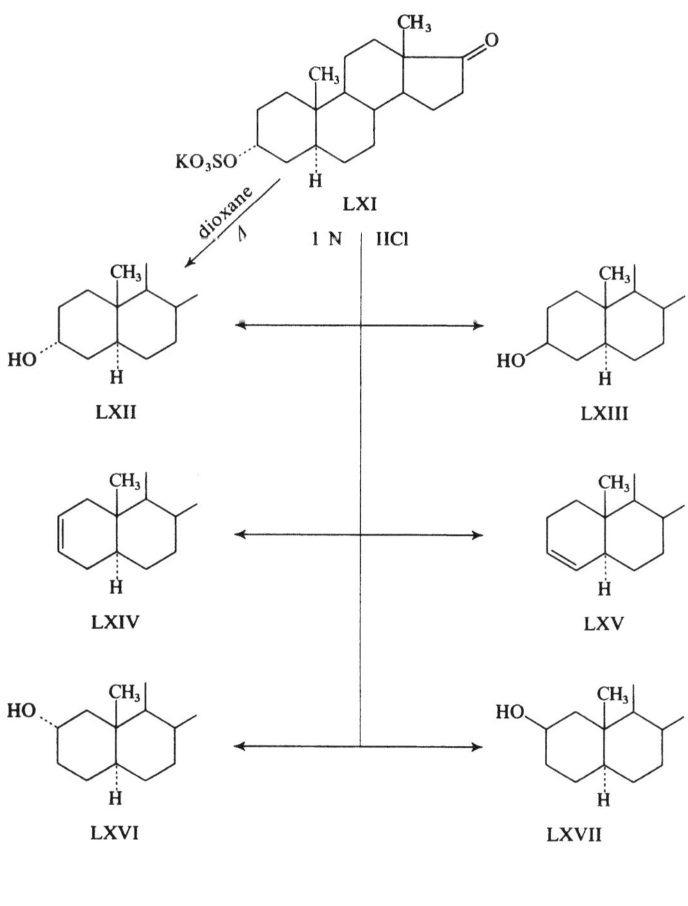

LXI LXII LXIII LXIV LXV LXVI LXVII

(LXIV, LXV resp.) of 5α–androstan–17–one. The hydroxyl fraction was shown to contain androsterone (LXII) (66%), 2β–hydroxy–5α–androstan–17–one (LXVII) (5%), isoandrosterone (XLIII) (21%) and 2α–hydroxy–5α–androstan–17–one (LXVI) (7%). The reaction products are those of C—O bond cleavage with the generation of the secondary carbonium ion. The many reaction modes thus open to the carbonium ion invariably leads to a complex reaction mixture.

In summarizing the general field of acid catalyzed hydrolysis of alkyl and aryl sulfates it must be concluded after careful analysis of kinetic studies that the mechanism of hydrolysis for both categories is of the A–1 [250] type illustrated by the following equation (Eq. (a))

$$ROSO_3^{\ominus} + H^{\oplus} \xrightleftharpoons{\text{fast}} RO^{\oplus}H - SO_3^{\ominus} \xrightarrow{\text{slow}} ROH + SO_3 \xrightarrow[\text{fast}]{H_2O} H_2SO_4 \quad (a)$$

6. *Acylation Studies*

In those cases where a steroid sulfate conjugate contains a free hydroxyl function and provided the hydroxyl group is normally reactive, acylation (and/or aroylation) reactions proceed with no difficulties. In the cases studied [157], the triethylammonium steroid sulfate salts were soluble in pyridine. The acylating agent (acid chloride or anhydride) in excess was added, and the reaction mixture kept at room temperature usually 2 to 24 hrs (LXVIII → LXIX; LXX → LXXI). The product could be isolated by ether precipitation and purified by recrystallization.

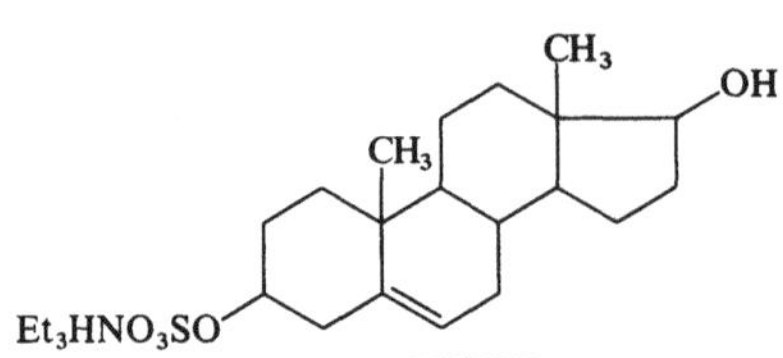

LXVIII

Ac_2O
Pyridine

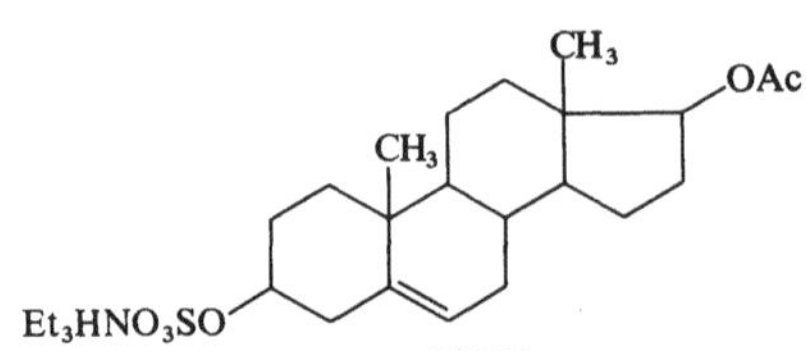

LXIX

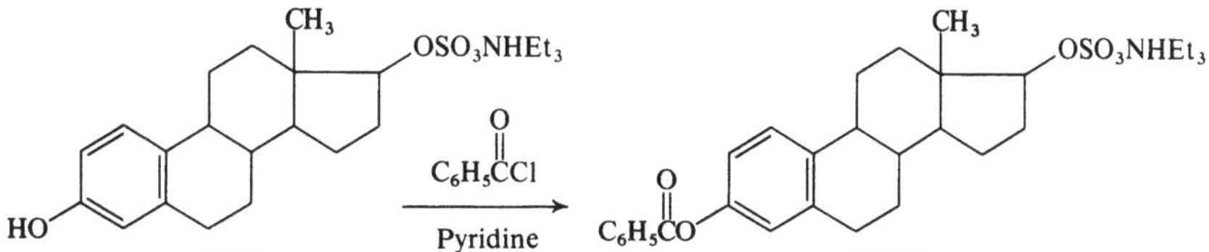

LXX LXXI

7. *Microbiological Hydroxylation Studies*

The synthesis of hydroxylated steroids by microbiological means represents an important process for the preparation of compounds difficult to obtain through chemical synthesis. The same procedure has been applied to steroid sulfate conjugates in the preparation of 16α–hydroxydehydroisoandrosterone–3–sulfate [3–sulfooxy–16α–hydroxyandrost–5–en–17–one] by YoungLai and Solomon [251]. In their studies, the incubation of dehydroisoandrosterone sulfate (XI) with a strain of *Streptomyces roseochromogenus* (ATCC 3347) produced the 16α–hydroxylated derivative in 44% yield. This was processed in such a manner that the 3–sulfate grouping was retained and the compound isolated as the ammonium salt (LXXII).

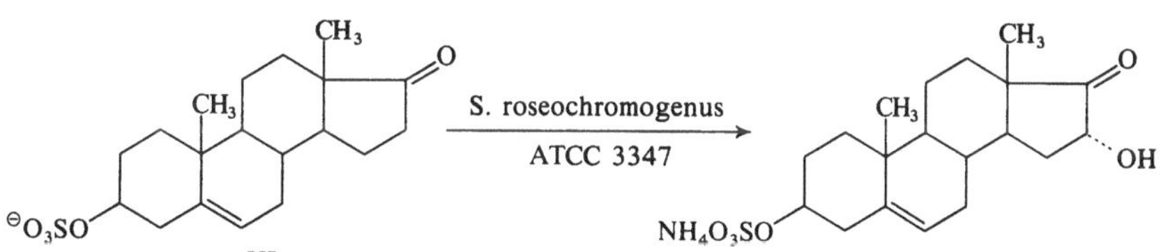

XI LXXII

One might classify the work of Lieberman and associates [252] in the same general category since they were able to show through double labeling (^{35}S, 7α–^{3}H) cholesterol sulfate that the sulfate group remained intact as the substrate underwent various biological hydroxylation and desmolase reactions to yield finally dehydroisoandrosterone sulfate. A biological system was also employed to effect the transformation of pregnenolone sulfate [229] to 17α–hydroxypregnenolone sulfate.

D. Physical Properties

In this section selected physical properties of steroid sulfates will be discussed. The reader is referred to Chapter 4 where other physical properties such as electrophoretic and chromatographic behavior are reviewed. Also the reader is referred to the companion volume "Physical Properties of Steroid Conjugates" for further details and references on the physical properties of sulfates.

1. Solvation and Stability

A common characteristic of steroid sulfates is the incorporation of water or solvent of crystallization. One consequence of this has been the secondary significance of melting point (generally, decomposition point) determination for the characterization of the sulfate. Also the retention of water or solvent of crystallization has a profound deleterious effect on the stability of the sulfate. No systematic study on stability has been made, but some apparently conflicting observations have been recorded in the literature. Butenandt and Hofstetter [166] have reported that sodium estrone sulfate is unstable during storage. However Fex *et al.* [213] found that a crude preparation of this compound containing a small amount of sodium acetate was stable for more than a year at $-20°$. At 35° it remained unchanged for some 2 days, but was decomposed within 10 days. The other sulfates such as those of 17β–estradiol and estriol were found to be stable when stored at room temperature as their alkali salts.

The stability of cholesterol sulfate salts has been studied by Sobel and Spoerri [187] who found that most of the salts were stable when kept in a desiccator. However the barium, cupric and ferric salts were unstable under these conditions. At 100° the potassium salt was found to be unusually stable, whereas under similar conditions the calcium and magnesium salts were unstable. Sobel and Spoerri also noted that pyridinium dibromocholesterol sulfate was stable for at least 4 months when kept in a stoppered bottle. In contrast under similar conditions, dibromocholesterol always decomposed in 10 days, with the evolution of hydrogen bromide even when kept in a dark bottle.

D'Alo [194] has observed that dehydroisoandrosterone sulfate sodium salt was stable for a long time in a dry state at room temperature, whereas at 100° only for a few minutes. Sodium estra–1,3,5(10)–triene–2,17β–diol–2–sulfate and the corresponding 3–sulfate sodium salt have been found to be at least as stable as phenol sulfates, but are hygroscopic and retain moisture with tenacity [186].

Dusza, Joseph, and Bernstein [215] during the course of studying the use of triethylamine–sulfur trioxide complex for the preparation of steroidal sulfates noted that the triethylammonium salts of the latter were obtained in an anhydrous form. Consequently these sulfate salts displayed a unique stability. In fact, periodic examination of various samples stored at room temperature in capped vials showed virtually no sign of decomposition from several months to a year when examined by thin layer chromatography.

Levitz [253] has utilized an effective way of preserving the enzymatically prepared mixed conjugate, estriol–3–sulfate–16–glucuronide [3–sulfooxy–

17β–hydroxyestra–1,3,5(10)–trien–16α–yl–β–D–glucopyranosiduronic acid], by simply storing a methanolic solution containing the compound at −20°.

2. Ultraviolet Spectroscopy

The utility of ultraviolet spectroscopy in the study of sulfate conjugation is definitely limited in scope. Since steroid alkyl sulfates do not absorb in the region covered by instruments currently in use, this analytical tool cannot aid in characterization studies. The only example which might be expected to have usable absorption values is the $\Delta^{3,5}$–3–enol sulfate prepared by Oertel and co–workers [181]. The authors report the ultraviolet absorption maximum to be at 238 mμ without specifying the extinction coefficient. Steroid alkyl sulfates whose parent steroid exhibits ultraviolet absorption will, of course, exhibit the same absorption characteristics.

Aryl sulfates differ significantly in the ultraviolet spectra from the corresponding phenol such that considerable use can be made of ultraviolet spectra in locating the sulfate grouping in, for example, estrogen sulfates.

Table 1–1 summarizes the ultraviolet spectra of estrone, 17β–estradiol, and estriol, and several of their sulfate conjugates. It can be seen that the sulfate positioning in an unidentified estriol monosulfate can be partially resolved by examination of its ultraviolet spectrum, while a definite assignment can be made in distinguishing between the two 17β–estradiol monosulfates.

The isomeric 2– and 3–monosulfates of 2–hydroxy–17β–estradiol [estra–1,3,5(10)–triene–2,3,17β triol] cannot be differentiated by their ultraviolet spectra since both compounds absorb in the 281–282 mμ region with nearly equal extinction coefficients [214]. The 3–sulfate–2–methyl ether also absorbs in the same region, while the presence of sulfate groupings at both 2 and 3 gives a compound in the estrone series with a maximum at 247 mμ (ε 1740).

Sulfation of the 6β–hydroxyl group in Δ^4–3–one steroids does not impose any additional effect on the Δ^4–3–one chromophore from that of the normal hypsochromic effect of the 6β–hydroxyl group and decrease in extinction coefficient associated with that grouping [215]. The ultraviolet spectra of several alkyl and aryl steroid sulfates have been recorded in concentrated sulfuric acid [254]. When compared to their unsulfated counterparts no difference could be detected. The time study curves of each pair were also indistinguishable. Data were obtained on the parent steroid and its sulfate for androsterone, dehydroisoandrosterone, estrone and 2–methoxyestrone.

3. Optical Rotatory Properties

Teich *et al.* [248] have employed the Method of Molecular Rotation Differences to support the structure of potassium dehydroisoandrosterone

TABLE 1–1. *Ultraviolet absorption maxima of estrogen sulfates*[a]

Sulfate derivative \ Parent steroid	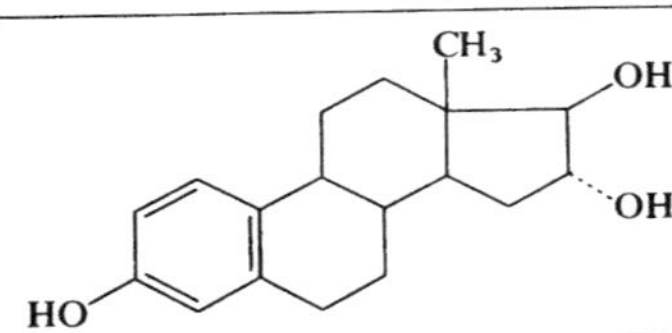		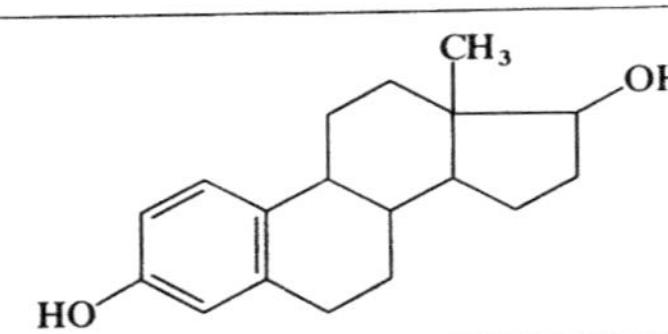		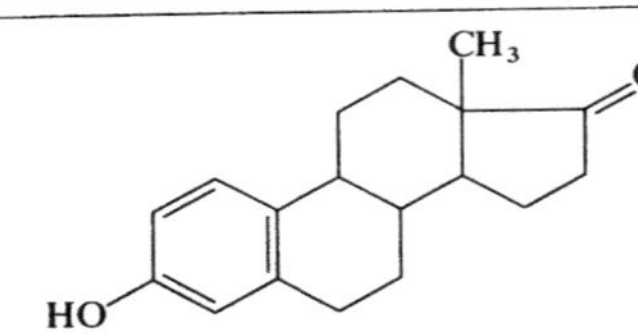	
	λ_{max} mμ	ε	λ_{max} mμ	ε	λ_{max} mμ	ε
3–Sulfate	269, 275	790, 737	269, 275	783, 768	269, 275	780, 756
17–Sulfate	279	2005	279	1978		
3,17–Disulfate	269, 275	775, 717	269, 275	825, 762		
16,17–Disulfate	279	1965				
3,16,17–Trisulfate	269, 275	813, 776				

[a] Fex, H., Lundvall, K.-E., Olsson, O.: Acta Chem. Scand. **22**, 254 (1968).

sulfate. A C3–sulfate group contributes a small negative molecular rotation to the parent steroid. Zelnik *et al.* [199] have studied the optical rotatory contribution of sulfate formation in several 20–oxo–pregnane and 20–oxo-pregnene compounds. They have concluded that, in chloroform solution, both pyridinium sulfate and methyl sulfate formation imparts a positive molecular rotation contribution on the parent steroid.

4. Infrared Spectroscopy

For a general discussion of the infrared absorption of alkyl and aryl sulfates, any one of the standard reference texts on infrared spectroscopy [255, 256] is recommended. Cited in these texts are numerous other references to original research articles that report on the systematic study of the vibrational modes of the sulfate grouping. To generalize, it can be stated that the sulfate grouping gives rise to strong absorption bands at 1240 and 1125 cm^{-1} which in some cases may be poorly resolved because of their intensity. Another strong band appears in the 1070–1060 cm^{-1} region. The spectra themselves are conveniently recorded in potassium bromide pellets; however solution spectra in a variety of solvents are possible by varying the cationic component of the salt and thereby its solubility. It should be pointed out, although quite obvious, that organic bases of steroid sulfates contribute their own characteristic peaks to the infrared spectrum of the salt; this can be very helpful on occasion.

Although citation of the absorption frequency of various sulfate bands could prove useful on certain occasions in identification of a steroid sulfate conjugate it was not expedient in this chapter to list the numerous citations found in the literature. For a complete citation of the published data, the companion volume of this text should be consulted.

A number of curves of representative infrared absorption spectra have been published, and are extremely useful for consultation on the relative intensities of various peaks in the spectrum. In their work on the physical and chemical characterization of androgen sulfate conjugates, Baulieu and Emiliozzi [202] have given a discussion of the infrared properties of these conjugates along with graphic presentation of the spectra of a number of salts. These curves are also included in the series on analysis of steroid hormone edited by Jayle [257].

Another collection of spectra has been compiled by Neudert and Röpke [258] in their general steroid atlas. Representative compounds included in this collection are estrogen, androgen and glucocorticoid sulfate salts with considerable variation being encountered in the cations. Many of these compounds do not appear elsewhere in the conjugate literature. In addition to the previously mentioned catalogs, several papers have presented extensive pictorial infrared presentations. The infrared curves of the various 17β–estradiol mono– and di–sulfates have been reported by Kirdani [176]. An extensive series of estriol sulfate conjugates were synthesized by Fex and co-workers [213] and infrared curves for these derivatives have appeared in this paper. The spectra of sulfate methyl esters in the androgen series are depicted by Emiliozzi [241].

5. Nuclear Magnetic Resonance (nmr)

One of the limiting factors in the utility of nuclear magnetic resonance spectroscopy for the study of steroid sulfate conjugates has been the solubility of the conjugates in an appropriate (nmr acceptable) solvent. The use of carbon tetrachloride or carbon disulfide, while ideal in not having proton absorption to obscure part of the spectrum, have limited utility due to their solution capabilities. On the other hand deuteriochloroform has been used extensively and is an excellent solvent for sulfate conjugates especially those associated with amine bases. With the advent of the powerful solvent, d_6–dimethylsulfoxide (DMSO), many more compounds have been studied. Although d_6–DMSO exhibits residual water peaks in its spectrum, it has the ability to sharpen bands associated with hydroxyl groups in the molecules; these may appear somewhat broad and diffuse with deuteriochloroform as solvent. The use of CD_3OD in spectra determination is limited since this solvent brings about the exchange of X–D for X–H in active exchangeable hydrogens, *e.g.*, OH, NH, etc. This proves useful however, when CD_3OD is added to the measurement solution after an initial curve has been traced. In this respect, a measure of the exchangeable hydrogens can prove quite useful. Undoubtedly additional solvents and solvent systems will be devised to handle the wide range of solubility problems encountered with sulfate conjugates.

Aside from indicating gross structural features of sulfate conjugates, *e.g.*, methyl groups, olefinic protons, aliphatic and aromatic protons, acetyl groups, hydroxyl and methoxy groups, etc., the displacement of the characteristic peak of any of the above or closely related groups from that in the parent steroid structure is the basis for structural study of sulfate conjugation.

A large number of nmr spectra have been reported for triethylammonium steroid sulfate salts [215]. The spectra are not graphically presented; however, salient features are designated. In the synthesis of 2– and 3–monosulfates of 17β–estradiol, Miyazaki and Fishman [214] were able to assign chemical shift values to the C1 and C4 protons. For the 3–monosulfate, the signals appeared at 407 and 411 Hz. The 2–monosulfate gave signals at 450 and 438 Hz. Those authors attempted the monosulfation of 2–hydroxyestrone directly using one equivalent of sulfating agent. A solid was isolated that showed peaks at 395, 407, 411 and 424 Hz, thus indicating that sulfation gave a 1:1 mixture of the two monosulfates with no peaks evident for the disulfated derivative, which had also previously been prepared and had signals at 450 and 438 Hz. As more studies are reported on the nmr spectra of steroid sulfates, the usefulness of this technique will be self evident. The assignment of axial and equatorial protons can be conveniently handled by nmr analysis, and this can greatly assist in structural assignments which are not handled as conveniently by the other analytical techniques.

6. Mass Spectrometry

The utility of mass spectrometry for the characterization of steroid sulfate conjugates is limited since the temperature required to obtain adequate signal response leads to decomposition of the salts presently used. The thermally induced salt decomposition therefore precludes the appearance of the molecular ion and its molecular weight information. The general description of the fragmentation exhibited by ammonium dehydroisoandrosterone sulfate (I) and potassium 2–methoxyestrone sulfate [2–methoxy–3–sulfooxyestra–1,3,5(10)–trien–17–one potassium salt] has been indicated [226]. The base peak for the dehydroisoandrosterone derivative was at m/e 270 (M–115) indicating that the elements of ammonium sulfate were removed and the mass ion of the $\Delta^{3,5}$–derivative (II) was formed.

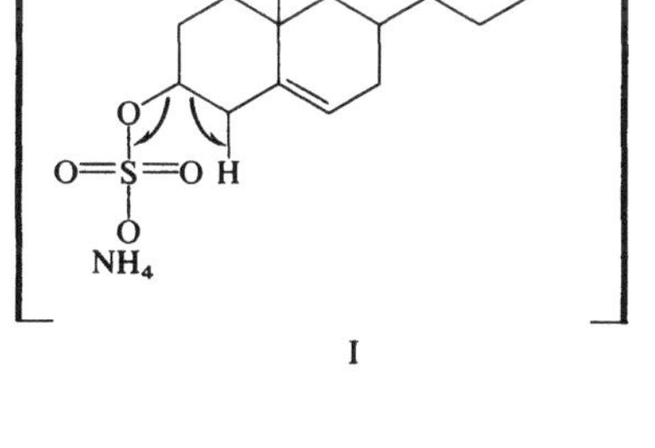

I

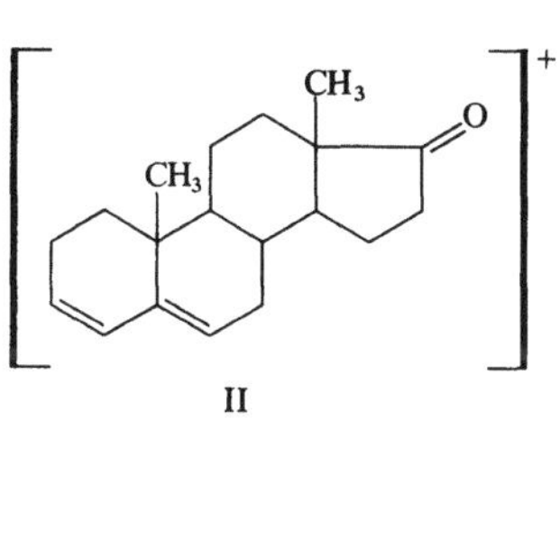

II

Prominent signals for SO^+ (m/e 48) and SO_2^+ (m/e 64) were also observed from the decomposition of the "sulfate" fragment. The base signal for the estrogen metabolite was m/e 300 (m−119), the molecular ion of 2–methoxyestrone indicating in this example S—O bond cleavage. No signals were observed for SO^+ or SO_2^+ in this case since the temperature employed was not deemed sufficient to cause decomposition of the resultant potassium salt. A number of triethylammonium steroid sulfate salts have been subjected to mass spectral analysis [157]. In these no molecular ion has been observed. The presence of SO^+ and SO_2^+ is indicated along with the fragment $(C_2H_5)_3N^+$.

A preliminary report [160] of the combination of gas–liquid chromatography and mass spectrometry of steroid sulfate conjugates has been made. The combination of these two powerful analytical tools allows analyses of complicated steroid mixtures to be attempted. The sulfates in these mixtures undergo dehydration in the gas–liquid chromatographic separation and thus when measured by the mass spectrograph they give rise to a molecular ion 18 mass units lower than the corresponding free steroid. We have described this previously in case of ammonium dehydroisoandrosterone sulfate.

IV. Double Conjugates

A. Types

To date, the following double steroid conjugates have either been isolated from a natural source and/or synthesized:

(1) 2–methoxyestrone sulfate (I) [226, 259].
(2) 17β–estradiol–3–sulfate–17–glucuronide (II) [152].
(3) estriol–3–sulfate–16–glucuronide (III) [123, 260–270].
(4) 15α–hydroxyestrone–3–sulfate–15–ξ–N–acetylglucosaminide (IV) [18] [17–oxo–3–sulfooxyestra–1,3,5(10)–trien–15α–yl–2′–acetamido–2′–deoxy–ξ–D–glucopyranoside].
(5) 15α–hydroxy–17β–estradiol–3–sulfate–15–ξ–N–acetylglucosaminide (V) [18] [17β–hydroxy–3–sulfooxyestra–1,3,5(10)–trien–15α–yl–2′–acetamido–2′–deoxy–ξ–D–glucopyranoside].
(6) 17α–estradiol–3–glucuronide–17–β–N–acetylglucosaminide (VI) [13–15] [estra–1,3,5(10)–triene–3,17α–di–yl–3–β–D–glucopyranosiduronic acid–17–2′–acetamido–2′–deoxy–β–D–glucopyranoside].
(7) pregn–5–ene–3β,20α–diol–3–sulfate–20β–N–acetylglucosaminide (VII) [16, 278].
(8) glycolithocholic acid sulfate (VIII) [210].
(9) taurolithocholic acid sulfate (IX) [210].

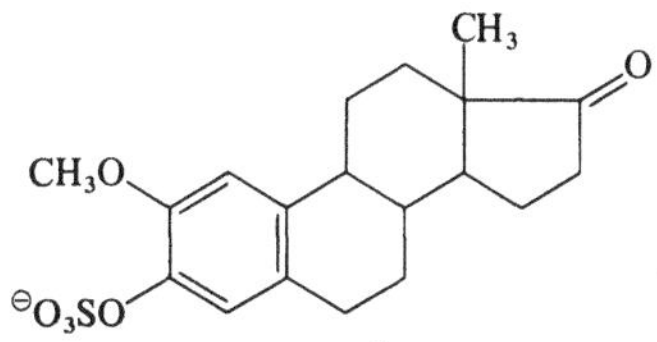

I

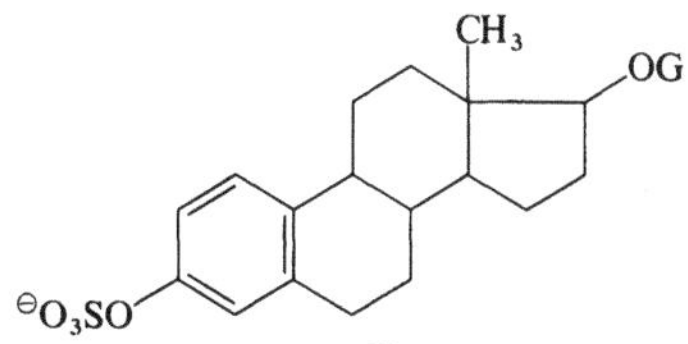

II

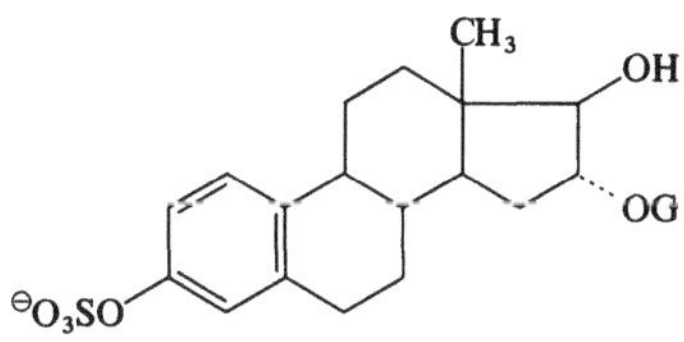

III

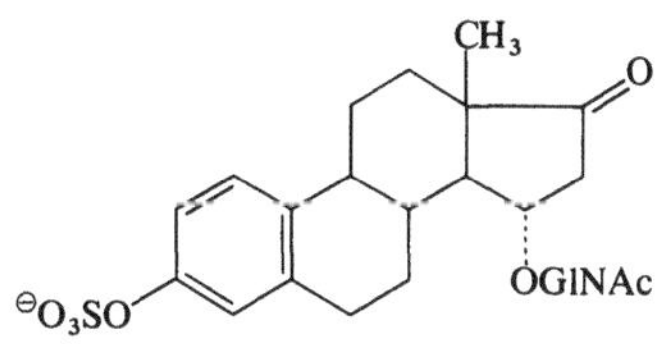

IV

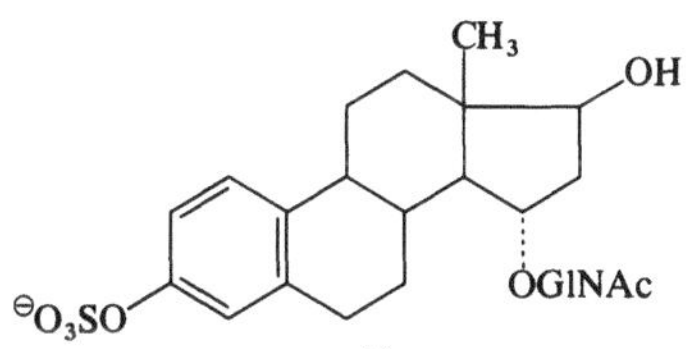

V

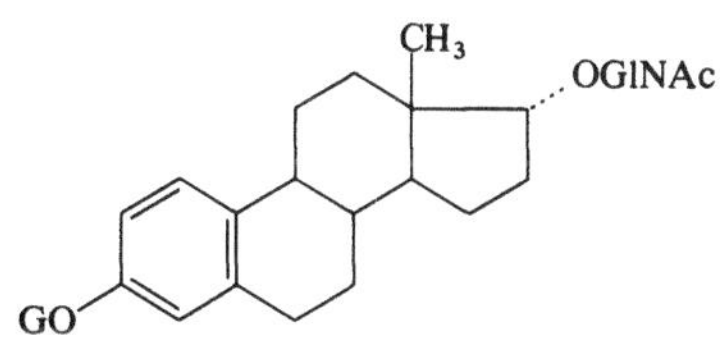

VI

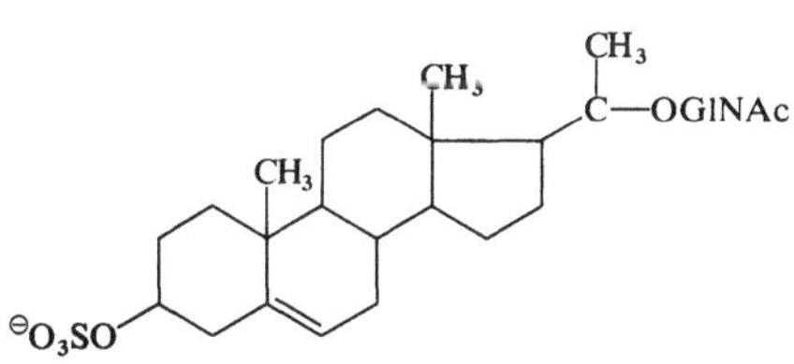

VII

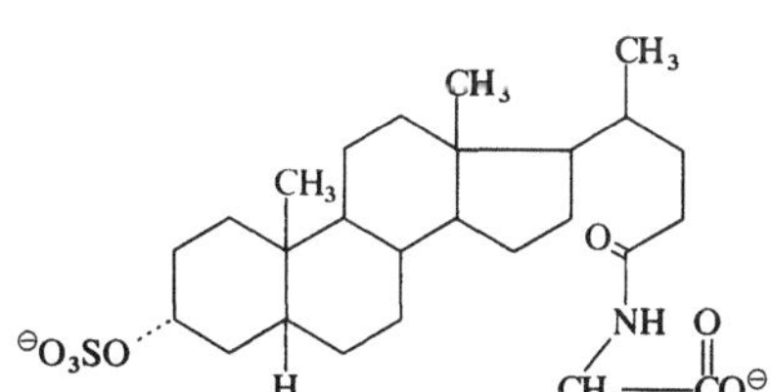

VIII

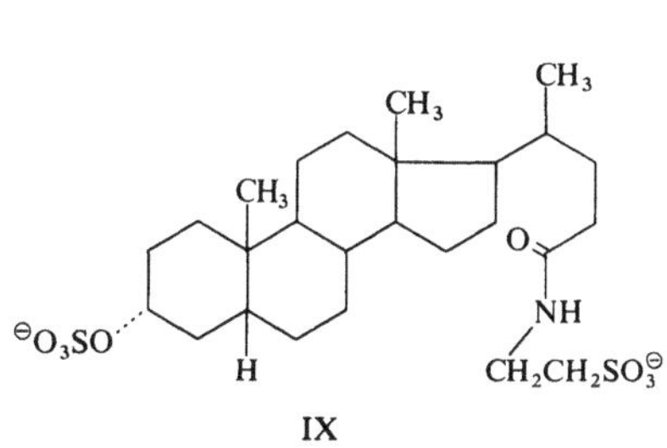

IX

B. Chemical Synthesis

1. 2–Methoxyestrone Sulfate

2–Methoxyestrone sulfate (I) is a double conjugate which has not as yet been isolated from a natural source. The Lederle group has synthesized this compound from 2–methoxyestrone [6, 271] employing sulfamic acid–pyridine [226] and triethylamine–sulfur trioxide complex in pyridine (triethylammonium salt) [215].

2. 17β–Estradiol–3–sulfate–17–glucuronide

17β–Estradiol–3–sulfate–17–glucuronide (II) has not been isolated from a natural source, but has been prepared in the form of its dipotassium salt XV by a five–stage synthesis from 17β–estradiol (X) [152] outlined as follows:

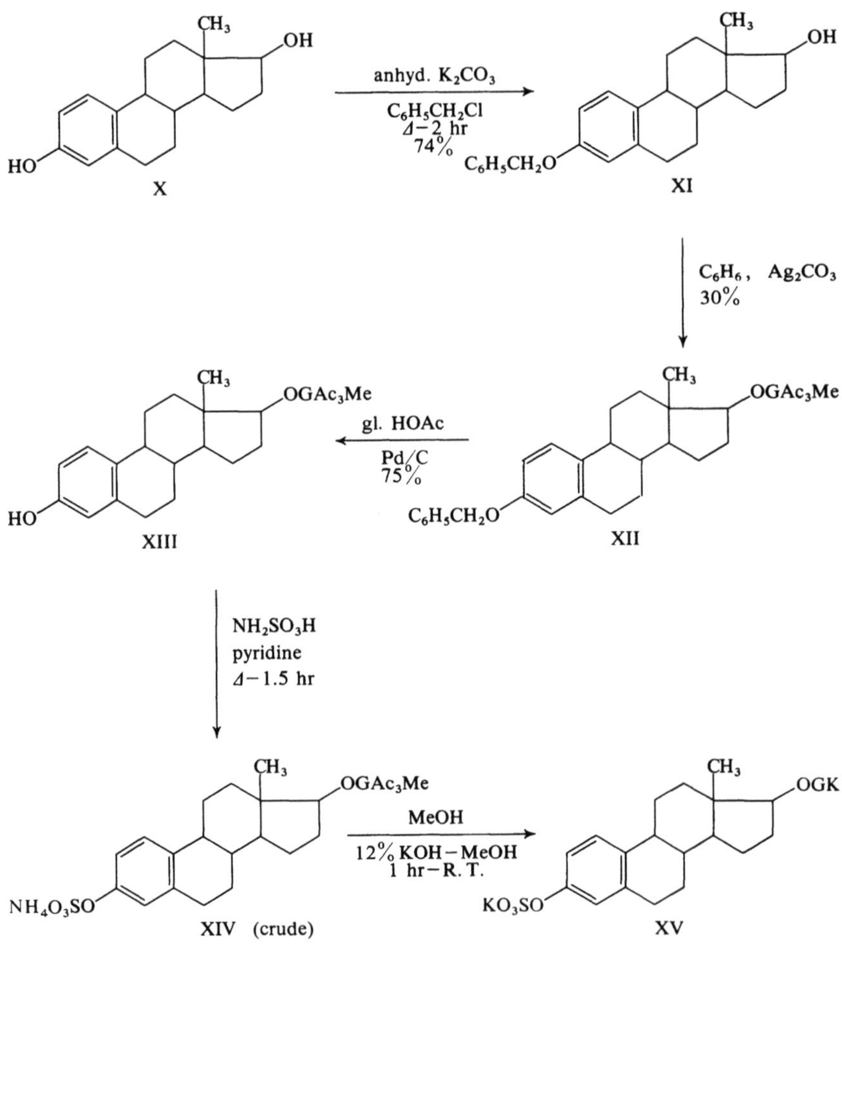

3. *Estriol-3-sulfate-16-glucuronide*

Levitz, Katz, and Twombly [272] have presented compelling evidence for the successful biosynthesis of this conjugate by incubation of estriol-16-glucuronide in a 100,000 g supernatant of homogenized guinea pig liver in the presence of adenosine triphosphate and magnesium sulfate. The Lederle group has reported on the first chemical synthesis of this biologically important conjugate [123]. The synthesis based on the starting material, estriol (XVI), follows:

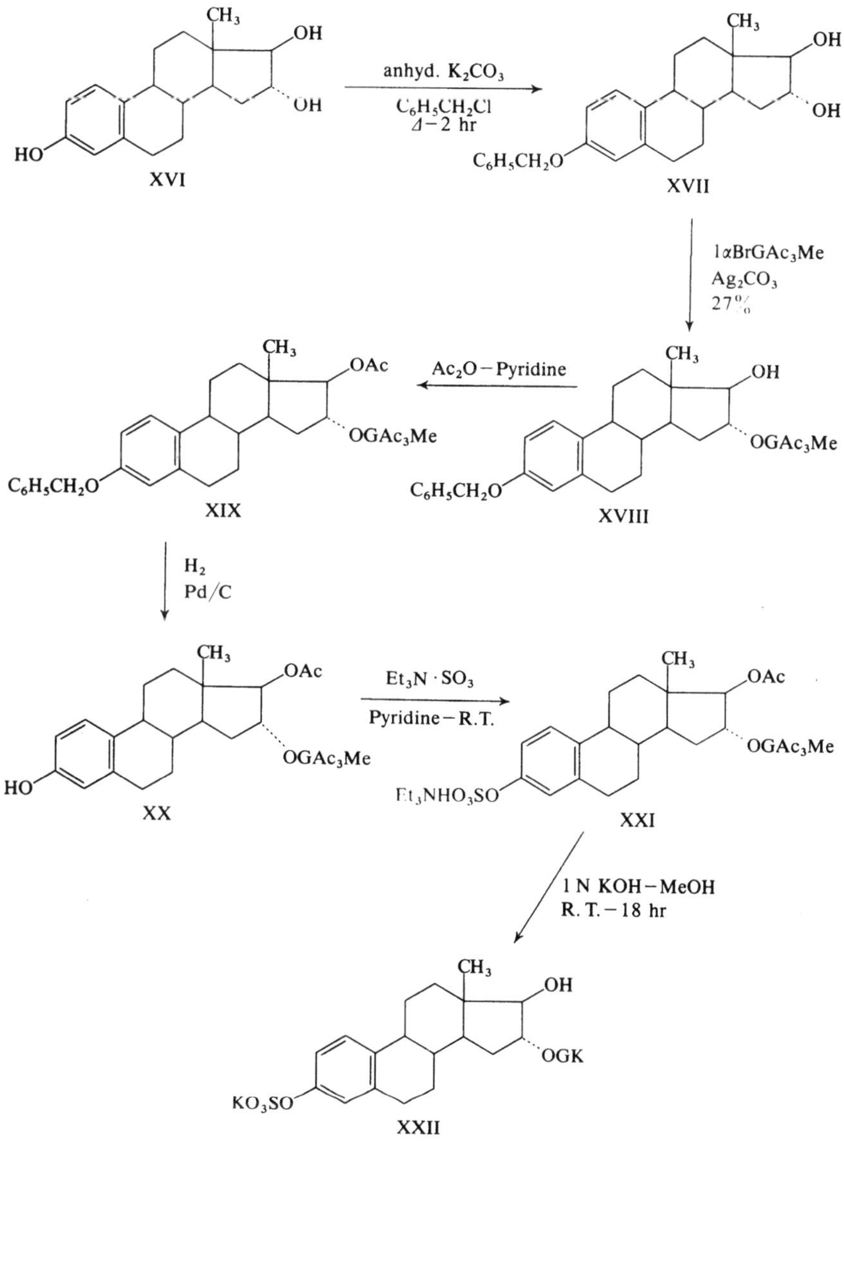

The conjugate (XXII) did not exhibit the diagnostic methylene blue test for sulfates [273]; this has also been observed for estradiol–3–sulfate–17–glucuronide (XV). However the blocked conjugate (XXI) gave the typical color test with this reagent.

4. N–Acetylglucosaminide Conjugates

None of the four double conjugates containing a N–acetylglucosaminide grouping has been chemically synthesized. However Jirku and Layne [15] have reported on the biochemical synthesis of estradiol–3–glucuronide–17α–N–acetylglucosaminide (VI). When 6,7–^{3}H–17β–estradiol was incubated with rabbit liver homogenates at pH 7.6 with uridine diphosphate N–acetylglucosamine in the presence of uridine diphosphate glucosiduronic acid, 17α–estradiol–3–glucuronide–17–β–N–acetylglucosaminide was formed in amounts equivalent to about 10% of the total conjugated radioactivity. In this work no evidence was found for the enzymatic transfer of N–acetylglucosamine to the 17–hydroxyl group of the starting material, 17β–estradiol.

Collins and Layne [277] have recently reported on an interesting aspect of the metabolism of 6,7–^{3}H–estriol in the rabbit. There was no evidence for the conjugation of estriol with N–acetylglucosamine. However, a conjugate was obtained which was identified as the 17–N–acetylglucosaminide of 17–epiestriol [3,16α–dihydroxyestra–1,3,5(10)–trien–17α–yl–2′–acetamido–2′–deoxy–β–D–glucopyranoside]. Indications were also found for the presence of 16,17–epiestriol–N–acetylglucosaminide [3,16β–dihydroxyestra–1,3,5(10)–trien–17α–yl–2′–acetamido–2′–deoxy–β–D–glucopyranoside].

Two publications have appeared on the preparation of steroid N–acetylglucosaminides. Hirschmann *et al.* [274] have synthesized 11β,17α–dihydroxy–3,20–dioxy–1,4–pregnadien–21–yl–2′–acetamido–2′–deoxy–β–D–glucopyranoside by Koenigs–Knorr condensation of prednisolone with 2–acetamido–3,4,6–tri–*O*–acetyl–2–deoxy–α–D–glucopyranosyl chloride followed by methanolysis.

Recently Fukushima and Sauer [141, 275] have reported on the synthesis of the N–acetyl–β–D–glucosaminides of androsterone, isoandrosterone, etiocholanolone [3α–hydroxy–5β–androstan–17–one], dehydroisoandrosterone, 17β–hydroxy–5α–androstan–3–one, and androstenediol [androst–5–ene–3β,17β–diol] (at C17). These investigators found that the reaction of 1α–chloro–2–acetamido–3,4,6–triacetyl–2–deoxy–D–glucopyranose and the steroid with silver carbonate proceeded in rather poor yields (*ca.* 5%). With a mixture of mercuric chloride and mercuric cyanide the yields improved and ranged from 10–25%. Hydrolysis of the groups was achieved by transesterification with sodium methoxide in methanol. Moreover the steroidal N–acetyl–β–glucosaminides (as the tetraacetyl derivatives) were epimerized in anhydrous chloroform with Lewis acids, *e.g.*, titanium tetrachloride, at room temperature for two days in fair yields. Hydrolysis of the *O*–acetyl groups was again achieved by transesterification in good yields. The conjugates are soluble in water, and polar organic solvents. They are readily hydrolyzed with hot aqueous acids, and partially with Ketodase. The glycosidic linkage is stable under the solvolytic conditions of Burstein and Lieberman [195] for sulfates.

5. Sulfates of Bile Acid Conjugates

Palmer [210] has described an interesting and perhaps important new pattern of bile acid metabolism. ^{14}C–Labeled glycolithocholic acid sulfate (VIII) and taurolithocholic acid sulfate (IX) were recovered from bile following the oral administration of lithocholic acid–24–^{14}C [3α–hydroxy–5β–cholanic acid] to a human. These sulfate esters of glycolithocholic acid and taurolithocholic acid were prepared using pyridine–sulfur trioxide complex, and crystallized as the ammonium salts. Details of the syntheses were not presented.

V. Concluding Remarks

Recent developments in the biological aspects of steroid conjugation have generated an increasing interest in related chemical aspects.

The synthesis of steroid glucuronides by the modified Koenigs–Knorr procedure continues to require considerable research, notably in regard to the synthesis of aryl–glucuronides where the yields are generally below 10%.

Significant advances have been made in the development of synthetic methods for both alkyl– and aryl–sulfates particularly with novel sulfur trioxide complexes, such as trimethylamine—and triethylamine—sulfur trioxide.

Gratifying progress has been made in the synthesis of double conjugates especially those of the sulfoglucuronide type, and some inroads have been reported in steroid N–acetylglucosaminide chemistry. However, the synthesis of double conjugates containing a N–acetylglucosaminide group remains an interesting unsolved problem for the steroid chemist.

A considerable number of steroid conjugates of all types have been synthesized to date, and many in labeled form are now commercially available.

An important advance in the characterization of steroid sulfates, glucuronides and N–acetylglucosaminides has been made with use of nuclear magnetic resonance and mass spectrometry.

VI. References

1. Weichselbaum, T. E., Margraf, H.W.: J.Clin. Endocrinol. Metab. **20**, 1341 (1960).
2. Margraf, H.W., Margraf, C. O., Weichselbaum, T. E.: Steroids **2**, 155 (1963).
3. Grosser, B. I., Axelrod, L. R.: Steroids **11**, 827 (1968).
4. Purdy, R. H., Grosser, B. I., Axelrod, L. R.: Steroids **11**, 837 (1968).
5. — Axelrod, L. R.: Steroids **11**, 851 (1968).
6. Kraychy, S., Gallagher, T. F.: J. Am. Chem. Soc. **79**, 754 (1957).
7. — — J. Biol. Chem. **229**, 519 (1957).
8. Engel, L. L.: Cancer **10**, 711 (1957).
9. Fishman, J., Gallagher, T. F.: Arch. Biochem. Biophys. **77**, 511 (1958).
10. Loke, K. H., Marrian, G. F.: Biochim. Biophys. Acta **27**, 213 (1958).
11. Frandsen, V. A.: Acta Endocrinol. **31**, 603 (1959).
12. Bernstein, S., Cantrall, E.W., Dusza, J. P., Joseph, J. P.: Steroid conjugates. Chemical Abstract Services, American Chemical Society, Washington, D. C., 1966.
13. Layne, D. S., Sheth, N. A., Kirdani, R. Y.: J. Biol. Chem. **239**, 3221 (1964).
14. — Endocrinology **76**, 600 (1965).
15. Jirku, H., Layne, D. S.: Biochemistry **4**, 2126 (1965).
16. Arcos, M., Lieberman, S.: Biochemistry **6**, 2032 (1967).
17. Collins, D. C., Williams, K. I. H., Layne, D. S.: Arch. Biochem. Biophys. **121**, 609 (1967).
18. Jirku, H., Levitz, M.: Abstr. 389, Third Internat. Congr. of Endocrinology, Mexico, D.F., July 1968, p. 156.
19. Oertel, G.W.: Biochem. Z. **334**, 431 (1961).
20. — Naturwissenschaften **48**, 621 (1961).
21. — Biochem. Z. **339**, 125 (1963).
22. — Groot, K.: Z. Physiol. Chem. **341**, 1 (1965).
23. Weinand, K., Oertel, G.W.: Endokrinologie **50**, 145 (1966).
24. Oertel, G.W., Rindt, W., in: Androgens in normal and pathological conditions (A. Vermeulen, ed.), p. 192. New York: Exerpta Medica Foundation 1966.
25. — Z. Physiol. Chem. **343**, 276 (1966).
26. — Knapstein, P., Treiber, L.: Z. Physiol. Chem. **345**, 221 (1966).
27. — Naturwissenschaften **49**, 394 (1962).
28. — Biochem. Z. **339**, 135 (1963).
29. — Wurie, A.: J. Endocrinol. **33**, 337 (1965).
30. Kornel, L., Fujiki, T.: Abstr. 39, 48th Meeting of the Endocrine Soc. Chicago, Ill., June 1966, p. 46.
31. Oertel, G.W.: Arch. Biochem. Biophys. **85**, 564 (1959).
32. Jellinck, P. H., Lazier, C.: Proc. Soc. Exptl. Biol. Med. **118**, 519 (1965).
33. — — Copp, M. L.: Can. J. Biochem. **43**, 1774 (1965).
34. — Lewis, J., Boston, F.: Steroids **10**, 329 (1967).
35. Kuss, E.: Z. Physiol. Chem. **348**, 1707 (1967).
36. Haslewood, G. A. D.: Bile salts. London, England: Methuen & Co., Ltd. 1967.
37. Marsh, C. A.: Glucuronic acid, free and combined (G. I. Dutton, ed.). New York, N.Y.: Academic Press 1966.

38. Schneider, J. J., in: Hormonal steroids–biochemistry, pharmacology and therapeutics, vol. 1, (L. Martini and A. Pecile, eds.), p. 132–133. New York, N.Y.: Academic Press 1964.
39. Godtfredsen, W. O., Vangedal, S.: Acta Chem. Scand. **20**, 1599 (1966).
40. Biorex Labs. Ltd., Brit. Pat. 1,112,805 (May 8, 1968).
41. Simpson, S. A., Tait, J. F.: Mem. Soc. Endocrinol. (Lond.) **2**, 9 (1953).
42. — — Wettstein, A., Neher, R., Euw, J. v., Reichstein, T.: Experientia **9**, 333 (1953).
43. Luetscher, J. A., Jr., Axelrad, B. J.: J. Clin. Endocrinol. Metab. **14**, 1086 (1954).
44. — Neher, R., Wettstein, A.: Experientia **10**, 456 (1954).
45. Simpson, S. A., Tait, J. F., Wettstein, A., Neher, R., Euw, J. v., Schindler, O., Reichstein, T.: Helv. Chim. Acta **37**, 1163 (1954).
46. Axelrad, B. J., Cates, J. E., Johnson, B. B., Luetscher, J. A., Jr.: Brit. med. J. **1955I**, 196.
47. Pasqualini, J. R.: Compt. Rend. **259**, 934 (1964).
48. — in: Structure and metabolism of corticosteroids, p. 77. London and New York: Academic Press 1964.
49. Underwood, R. H., Tait, J. F.: J. Clin Endrocrinol. Metab. **24**, 1110 (1964).
50. Pasqualini, J. R., Uhrich, E., Jayle, M.-F.: Biochim. Biophys. Acta **104**, 515 (1965).
51. Möhring, J.: Klin. Wochschr. **46**, 18 (1968).
52. Koenigs, W., Knorr, E.: Chem. Ber. **34**, 957 (1901).
53. Frush, H. L., Isbell, H. S.: J. Res. Natl. Bur. Std. **27**, 413 (1941).
54. Isbell, H. S., Frush, H. L.: J. Res. Natl. Bur. Std. **43**, 161 (1949).
55. Ness, R. K., Fletcher, H. G., Jr., Hudson, C. S.: J. Am. Chem. Soc. **73**, 959 (1951).
56. Evans, W. L., Reynolds, D. D., Talley, E. A.: Advan. Carbohydrate Chem. **6**, 27 (1951).
57. Lemieux, R. U.: Advan. Carbohydrate Chem. **9**, 1 (1954).
58. Haynes, L. J., Newth, F. H.: Advan. Carbohydrate Chem. **10**, 207 (1955).
59. Mattack, G. L., Phillips, G. O.: J. Chem. Soc. 1836 (1956).
60. Conchie, J., Levvy, G. A., Marsh, C. A.: Advan. Carbohydrate Chem. **12**, 157 (1957).
61. Goebel, W. F., Babers, F. H.: J. Biol. Chem. **106**, 63 (1934).
62. — — J. Biol. Chem. **111**, 347 (1935).
63. Anderson, F. B., Leaback, D. H.: Chem. Ind. (London) 967 (1960).
64. Bollenback, G. N., Long, J. W., Benjamin, D. G., Lindquist, J. A.: J. Am. Chem. Soc. **77**, 3310 (1955).
65. Pravdic, N., Keglevic, D.: Tetrahedron **21**, 1897 (1965).
66. — — Carbohydrate Res. **7**, 167 (1968).
67. Schapiro, E.: Biochem. J. **33**, 385 (1939).
68. Meystre, C., Mischer, K.: Helv. Chim. Acta **27**, 231 (1944).
69. Goldschmid, H. R., Perlin, A. S.: Can. J. Chem. **39**, 2025 (1961).
70. Wotiz, H. H., Smakula, E., Lichtin, N. N., Leftin, J. H.: J. Am. Chem. Soc. **81**, 1704 (1959).
71. Helferich, B., Bohn, E., Winkler, S.: Chem. Ber. **63**, 989 (1930).
72. Reynolds, D. D., Evans, W. L.: J. Am. Chem. Soc. **60**, 2559 (1938).
73. Helferich, B., Goerdeler, J.: Chem. Ber. **73**, 532 (1940).

74. Lettre, H., Hagedorn, A.: Z. Physiol. Chem. **242**, 210 (1936).
75. Coulon-Morelec, M. J.: Bull Soc. Chem. Biol. **49**, 825 (1967).
76. Wolfrom, M. L., Pittet, A. O., Gillam, J. C.: Proc. Natl. Natl. Acad. Sci. U.S. **47**, 700 (1961).
77. — Lineback, D. R.: Methods Carbohydrate Chem. **2**, 342 (1963).
78. Cowdrey, W. A., Hughes, E. D., Ingold, C. K.: J. Chem. Soc. 1243 (1937).
79. Helferich, B., Klein, W.: Ann. Chem. **450**, 219 (1926).
80. Zorbach, W. W., Valiaveedan, G. D.: J. Org. Chem. **29**, 2462 (1964).
81. — — Kasheliker, D. V.: J. Org. Chem. **27**, 1766 (1962).
82. Schneider, J. J., Lewbart, M. L., Levitan, P., Lieberman, S.: J. Am. Chem. Soc. **77**, 4184 (1955).
83. Zorbach, W. W.: J. Org. Chem. **23**, 1797 (1958).
84. Hadd, H. E.: Dissertation, Indiana Univ. (1964).
85. — Blickenstaff, R. T.: Steroids **4**, 503 (1964).
86. Elce, J. S., Carpenter, J. G. D., Kellie, A. E.: J. Chem. Soc. (C) 542 (1967).
87. Nambara, T., Imai, K.: Chem. Pharm. Bull. (Tokyo) **15**, 1232 (1967).
88. Schapiro, E.: Nature **142**, 1036 (1938).
89. Cohen, S. L., Marrian, G. F.: Biochem. J. **30**, 57 (1936).
90. — — Odell, A. D.: Biochem. J. **30**, 2250 (1936).
91. Venning, E. M., Browne, J. S. L.: Proc. Soc. Exptl. Biol. Med. **34**, 792 (1936).
92. Heard, R. D. H., Hoffman, M. M., Mack, G. E.: J. Biol. Chem. **155**, 607 (1944).
93. Huebner, C. E., Overman, R. S., Link, K. P.: J. Biol. Chem. **155**, 615 (1944).
94. Pelzer, H.: Z. Physiol. Chem. **314**, 234 (1959).
95. Staib, W., Dönges, K.: Z. Physiol. Chem. **319**, 233 (1960).
96. Sheps, M. C., Purdy, R. H., Engel, L. L., Oncley, J. L.: J. Biol. Chem. **235**, 3042 (1960).
97. Lucas, R. A., Deckel, D. F., Dziemian, R. L., Ceglowski, M. J., Hensle, B. L., MacPhillamy, H. B.: J. Am. Chem. Soc. **82**, 5688 (1960).
98. Winguth, K.: Dissertation, Univ. Hamburg (1961).
99. Brooksbank, B. W. L., Haslewood, G. A. D.: Biochem. J. **80**, 488 (1961).
100. Shindo, M.: Japan Pat. 62/4527, June 14, 1962.
101. Siiteri, P. K., Lieberman, S.: Biochemistry **2**, 1171 (1963).
102. — Vande Wiele, R. L., Lieberman, S.: J. Clin. Endocrinol. Metab. **23**, 588 (1963).
103. — Dissertation, Columbia Univ. (1963).
104. Nitta, Y., Shindo, M., Takamura, K.: Chem. Pharm. Bull. (Tokyo) **12**, 450 (1964).
105. Emiliozzi, R.: Compt. Rend. **258**, 3875 (1964).
106. Robel, P., Emiliozzi, R., Baulieu, E.-E.: Compt. Rend. **258**, 1331 (1964).
107. Becker, J. F.: Biochem. Biophys. Acta **100**, 574 (1965).
108. Nagayama, F., Saito, A., Idler, D. R.: Can. J. Biochem. **44**, 1109 (1966).
109. Robel, P., Emiliozzi, R., Baulieu, E.-E.: J. Biol. Chem. **241**, 20 (1966).
110. Carpenter, J. G. D.: Dissertation, Univ. London (1961).
111. — Kellie, A. E.: Biochem. J. **78**, 1 P (1961).
112. — — Biochem. J. **84**, 303 (1962).
113. Neeman, M., Hashimoto, Y.: Tetrahedron Letters 183 (1961).
114. — — J. Am. Chem. Soc. **84**, 2972 (1962).

115. Hashimoto, Y., Neeman, M.: J. Biol. Chem. **238**, 1273 (1963).
116. Neeman, M., Hashimoto, Y.: Biochem. Prep. **12**, 81 (1968).
117. Elce, J. S., Carpenter, J. G. D., Kellie, A. E.: Biochem. J. **91**, 30 P (1964).
118. Elce, J. S.: Dissertation, Univ. London (1966).
119. Weisz, R.: U.S. Patent No 2,208,915 (1940); Chem. Abstr. **45**, 137 (1941).
120. Kerestzky, Wolf: Austrian Patent No 160,891 (1943); Chem. Abstr. **47**, 7558 (1943).
121. Leeds, N. S., Fukushima, D. K., Gallagher, T. F.: J. Am. Chem. Soc. **76**, 2943 (1954).
122. Wolfrom, M. L., Anno, K.: J. Am. Chem. Soc. **74**, 5583 (1952).
123. Joseph, J. P., Dusza, J. P., Bernstein, S.: J. Am. Chem. Soc. **89**, 5078 (1967).
124. Schubert, K.: Acta Endocrinol. **27**, 36 (1958).
125. Wotiz, H. H., Fishman, W. H.: Steroids **1**, 211 (1963).
126. Hadd, H. E.: Abstr. 46, 47th Meeting of the Endocrine Soc., New York, N.Y., June, 1965.
127. Smakula, E., Leftin, J. H., Wotiz, H. H.: J. Am. Chem. Soc. **81**, 1708 (1959).
128. Wakabayashi, J., Wotiz, H. H., Fishman, W. H.: Biochim. Biophys. Acta **48**, 198 (1961).
129. Heynes, K., Paulsen, H.: Angew. Chem. **69**, 600 (1957).
130. Osman, E. M., Hobbs, K. C., Walston, W. E.: J. Am. Chem. Soc. **73**, 2726 (1951).
131. Cadotte, J. E., Smith, F., Spriesterbach, D.: J. Am. Chem. Soc. **74**, 1501 (1952).
132. Kochetkov, N. K., Khorlin, A. J., Bochkov, A. F.: Tetrahedron Letters 289 (1964).
133. Khorlin, A. Y., Bochkov, A. F., Kochetkov, N. K.: Chem. Natur. Prod. **2**, 4 (1966).
134. Corbellini, A., Torti, G., Botti, C., Ferrara, G.: Rend. Ist. Lombardo Sci. Lettere A **100**, 273 (1966).
135. Ferrara, G., Botti, C., Torti, G., Corbellini, A.: Steroids **8**, 111 (1966).
136. Foggitt, F., Kellie, A. E.: Biochem. J. **91**, 209 (1964).
137. Mattox, V. R., Vrieze, W., Goodrich, J.: Federation Proc. **24**, 415 (1965).
138. Horning, E. C., Horning, M. G., Ikekawa, N., Chambaz, E. M., Jaakonmaki, P. I., Brooks, C. J. W.: J. Gas Chromatog. **5**, 283 (1967).
139. Vanden Heuvel, W. J. A.: J. Chromatog. **28**, 406 (1967).
140. Jaakonmaki, P. I., Yarger, K. A., Horning, E. C.: Biochim. Biophys. Acta **137**, 216 (1967).
141. Fukushima, D. K., Sauer, G.: Federation Proc. **27**, 623 (1968).
142. Edwards, R. W. H., Trafford, D. J. H.: Biochem. J. **108**, 185 (1968).
143. Klyne, W.: Biochem. J. **47**, 12 P (1950).
144. Barker, S. A., Bourne, E. J., Stephans, R., Whiffen, D. H.: J. Chem. Soc. 3468 1954).
145. Tipson, R. S., Isbell, H. S.: J. Res. Natl. Bur. Std. **64** A, 405 (1960).
146. Wotiz, H. H., Sie, H.-G., Fishman, W. H.: J. Biol. Chem. **232**, 723 (1958).
147. Jackman, L. M.: Application of nuclear magnetic resonance spectroscopy in organic chemistry. New York: Pergamon Press 1959.
148. Roberts, J. D.: Nuclear magnetic resonance application in organic chemistry. New York: McGraw-Hill Book Co. 1959.
149. Pople, J. A., Schneider, W. G., Bernstein, H. J.: High resolution nuclear magnetic resonance. New York: McGraw Hill Book Co. 1959.

150. Bhacca, N. S.: Application of NMR spectroscopy in organic chemistry: Illustrations from the steroid field. San Francisco: Holden-Day, Inc. 1964.
151. Howell, M. G., Kende, A. S., Webb, J. S.: Formula index to NMR Literature data, vol. I, New York: Plenum Press 1965; vol. II, 1966.
152. Cantrall, E.W., McGrath, M. G., Bernstein, S.: Steroids **8**, 967 (1966).
153. Conrow, R. B.: Private Communication.
154. McLafferty, F.W.: Mass spectroscopy of organic ions. New York: Academic Press 1963; – Interpretation of mass spectra. New York: W. A. Benjamin, Inc. 1966.
155. Budzikiewicz, H., Djerassi, C., Williams, D. H.: Mass spectroscopy of organic compounds. San Francisco: Holden–Day, Inc. 1967.
156. — — — Structure elucidation of natural products by mass spectrometry, vol. II. San Francisco: Holden-Day, Inc. 1964.
157. Joseph, J. P., Dusza, J. P.: Unpublished data.
158. Unpublished results from this Laboratory. We wish to thank Dr. D. Fukushima for supplying the compound.
159. — We wish to thank Dr. M. Levitz for supplying the compound.
160. Jaakonmaki, P. I.: Acta Endocrinol. **56**, Suppl. **119**, 118 (1967).
161. Gilbert, E. E.: Chem. Rev. **62**, 549 (1962).
162. Natelson, S., Sobel, A. E., Kramer, B.: J. Biol. Chem. **105**, 761 (1934).
163. Mandel, J. A., Neuberg, C.: Biochem. Z. **71**, 186 (1915).
164. Verley, A.: Bull Soc. Chim. (Paris) **3**, **25**, 46; Chem. Zentr. **1**, 313 (1901).
165. Baumgarten, P.: Chem. Ber. **69**, 1166 (1926).
166. Butenandt, A., Hofstetter, H.: Z. Physiol. Chem. **259**, 222 (1939).
167. Grant, G. A., Souch, J. K.: Abstr., 106th Meeting, Amer. Chem. Soc., p. 19 B (1943).
168. — Seemann, C. von: Abstr., 106th Meeting, Amer. Chem. Soc., p. 18 B (1943).
169. — Glen, W. L.: J. Am. Chem. Soc. **71**, 2255 (1949).
170. Holden, G.W., Len, I., Bromley, R.: J. Am. Chem. Soc. **71**, 3844 (1949).
171. — Bromley, R.: J. Am. Chem. Soc. **72**, 3807 (1950).
172. Grant, G. A., Beall, D.: Recent Progr. Hormone Res. **5**, 307 (1950).
173. Cavina, G.: Rend. Ist. Super Sanita **20**, 923 (1957).
174. Siiteri, P. K., MacDonald, P.: Steroids **2**, 713 (1963).
175. Payne, A. H., Mason, M.: Steroids **6**, 323 (1965).
176. Kirdani, R. Y.: Steroids **6**, 845 (1965).
177. Calvin, H. I., Lieberman, S.: J. Clin. Endocrinol. Metab. **26**, 402 (1966).
178. Cantrall, E.W., Littell, R., Bernstein, S.: U.S. Pat. 3,274,056 (Sept. 20, 1966).
179. — — — Unpublished data.
180. Roy, A. B.: Biochem. J. **62**, 41 (1956).
181. Oertel, G.W., Treiber, L., Rindt, W.: Experientia **23**, 97 (1967).
182. Sobel, A. E., Drekter, I. J., Natelson, S.: J. Biol. Chem. **115**, 381 (1936).
183. Levitz, M.: Steroids **1**, 117 (1963).
184. Bradlow, H. L.: Private communication.
185. Williams, K. I. H., Smulowitz, M., Fukushima, D. K.: J. Org. Chem. **30**, 1447 (1965).

186. Miyazaki, M., Fishman, J.: J. Org. Chem. **33**, 662 (1968).
187. Sobel, A. E., Spoerri, P. E.: J. Am. Chem. Soc. **63**, 1259 (1941).
188. Sisler, H. H., Audrieth, L. F.: Inorg. Syn. **2**, 173 (1946).
189. Wagner, J.: Chem. Ber. **19**, 1157 (1886).
190. Ratcliff, G. A.: Ph. D. Dissertation, Cornell University, Dissertation Abstr. **14**, 2018 (1954), as cited by E. E. Gilbert, Chem. Rev. **62**, 549 (1962).
191. Allied Chemical Corp., General Chemical Division, unpublished research data; as cited by E. E. Gilbert, Chem. Rev. **62**, 549 (1962).
192. Birchenough, M. J., Burton, H.: J. Chem. Soc. 2443 (1952).
193. Salkin, R.: U.S. Pat. 2,636,042 (Apr. 21,1953).
194. D'Alo, F.: Farmaco, Ed. Sci. **11**, 477 (1956).
195. Burstein, S., Lieberman, S.: J. Am. Chem. Soc. **80**, 5235 (1958).
196. Griebsch, E., Garn, W.: German Pat. 1,047,780 (Dec. 31, 1958).
197. — — German Pat. 1,066,581 (Oct. 8, 1959).
198. Schering, A. G.: British Pat. 848,515.
199. Zelnik, R., Desfosses, B., Emiliozzi, R.: Compt. Rend. **250**, 1671 (1960).
200. Khaletskii, A. M., Vasil'eva, M. V.: Zh. Obshch. Khim. **31**, 2996 (1961).
201. Pasqualini, J. R., Zelnik, R., Jayle, M.-F.: Bull. Soc. Chim. France 1171 (1962).
202. Baulieu, E.-E., Emiliozzi, R.: Bull. Soc. Chim. Biol. **44**, 823 (1962).
203. Khaletskii, A. M., Vasil'eva, M. V.: Zh. Obshch. Khim. **33**, 1104 (1963).
204. Pasqualini, J. R., Dutter, F., Jayle, M.-F.: Biochim. Biophys. Acta **69**, 331 (1963).
205. Baulieu, E.-E., Corpéchot, C., Emiliozzi, R.: Steroids **2**, 429 (1963).
206. Engelfried, O., Schargan, K., Wiechert, R.: German Pat. 1,170,949 (May 27, 1964).
207. Boström, H., Franksson, C., Wengle, B.: Acta Endocrinol. **47**, 633 (1964).
208. Klein, G. P., Giroud, C. J. P.: Steroids **5**, 765 (1965).
209. Baulieu, E.-E., Corpéchot, C.: Bull. Soc. Chim. Biol. **47**, 443 (1965).
210. Palmer, R. H.: Proc. Natl. Acad. Sci. U.S. **58**, 1047 (1967).
211. Burg, A. B.: J. Am. Chem. Soc. **65**, 1629 (1943).
212. Hydorn, A. E., Lerner, L. J., Schwartz, J.: Steroids **6**, 247 (1965).
213. Fex, H., Lundvall, K.-E., Olsson, O.: Acta Chem. Scand. **22**, 254 (1968).
214. Miyazaki, M., Fishman, J.: J. Org. Chem., in process of publication. We wish to thank Dr. Fishman for a preprint of this mss.
215. Dusza, J. P., Joseph, J. P., Bernstein, S.: Steroids **12**, 49 (1968).
216. American Cyanamid Company: Trialkylamine-sulfur trioxide compounds. New York: New York 1955.
217. Mumma, R. O.: Lipids **1**, 221 (1966).
218. Hoiberg, C. P., Mumma, R. O.: Abstr. 27 Division of Biological Chem., 156th Meeting Amer. Chem. Soc., Atlantic City, N.J., Sept. 8-13, 1968.
219. Ming Chu, T., Slaunwhite, W. R.: Federation Proc. **27**, 820 (1968), Abstr. No. 3416. See also, R. O. Mumma, Biochem. Biophys. Acta **165**, 571 (1968).
220. Nikolaev, V. I., Krastelevska, S. A.: J. Russ. Phys.-Chem. Soc. **60**, 1211 (1928); Chem. Abstr. **23**, 2996 (1929).
221. Brieskorn, C. H., Hofmann, H.: Arch. Pharm. **297**, 577 (1964).
222. Short, W. F., Oxley, P.: Brit. Patent 577,666 (May 27, 1946); Chem. Abstr. **41**, 2084c (1947).

223. Hasbrouck, R. B.: U.S. Patent 2,642,427 (June 16, 1953); Chem. Abstr. **48**, 6474i (1954).
224. Price, W. H.: U.S. Patent 2,917,522 (Dec. 15, 1959); Chem. Abstr. **54**, 6823f (1960).
225. Higaki, M., Takahashi, M., Suzuki, T., Sahashi, Y.: J. Vitaminol. (Kyoto) **11**, 261 (1965).
226. Joseph, J. P., Dusza, J. P., Bernstein, S.: Steroids **7**, 577 (1966).
227. McKenna, J., Norymberski, J.: J. Chem. Soc. 3893 (1957).
228. Roberts, K. D., Bandi, L., Calvin, H. I., Drucker, W. D., Lieberman, S.: Biochemistry **3**, 1983 (1964).
229. Calvin, H. I., Lieberman, S.: Biochemistry **3**, 259 (1964).
230. Bridgwater, R. J., Briggs, T., Haslewood, G. A. D.: Biochem. J. **82**, 285 (1962).
231. Haslewood, G. A. D.: Biochem. J. **100**, 233 (1966).
232. Anderson, I. G., Haslewood, G. A. D., Cross, A. D., Tokes, L.: Biochem. J. **104**, 1061 (1967).
233. Starka, L., Sukova, J., Silink, K.: Clin Chim. Acta **7**, 309 (1962).
234. — Hampl, R.: Naturwissenschaften **51**, 164 (1964).
235. Natelson, S., Gottfried, S. P.: J. Am. Chem. Soc. **61**, 971 (1939).
236. Sobel, A. E., Rosen, M. J.: J. Am. Chem. Soc. **63**, 3536 (1941).
237. — Spoerri, P. E.: J. Am. Chem. Soc. **64**, 482 (1942).
238. — Owades, P. S., Owades, J. L.: J. Am. Chem. Soc. **71**, 1487 (1947).
239. Apotheker, D., Owades, J. L., Sobel, A. E.: J. Am. Chem. Soc. **76**, 3684 (1954).
240. Lieberman, S., Dobriner, K., Hill, B. R., Fieser, L. F., Rhoads, C. P.: J. Biol. Chem. **172**, 263 (1948).
241. Emiliozzi, R.: Bull. Soc. Chim. France 911 (1960).
242. Beall, D., Grant, G. A.: Canada Pat. 482,631, April 22, 1952.
243. Marker, R. E., Rohrmann, E., Wittle, E. L.: J. Am. Chem. Soc. **60**, 1561 (1938).
244. Klyne, W.: Nature **166**, 559 (1948).
245. Hirschmann, H., Williams, J. S.: J. Biol. Chem. **238**, 2305 (1963).
246. Cross, A. D.: J. Chem. Soc. 2817 (1961).
247. Fieser, L. F., Fieser, M.: Natural products related to phenanthrene, 3rd ed. New York: Reinhold 1949; Steroids, p. 431–432. New York: Reinhold 1959.
248. Teich, S., Rogers, J., Lieberman, S., Engel, L. L., Davis, J. W.: J. Am. Chem. Soc. **75**, 2523 (1953).
249. Ramseyer, J., Williams, J. S., Hirschmann, H.: Steroids **9**, 347 (1967).
250. Kice, J. L., Anderson, J. M.: J. Am. Chem. Soc. **88**, 5242 (1966).
251. YoungLai, E., Solomon, S.: Endocrinology **80**, 177 (1967).
252. Roberts, K. D., Bandi, L., Calvin, H. I., Drucker, W. D., Lieberman, S.: Am. Chem. Soc. **86**, 958 (1964).
253. Levitz, M.: Private communication.
254. Bernstein, S.: Unpublished data.
255. Colthrup, N. B., Daly, L. H., Wiberley, S. E.: Introduction to infrared and raman spectroscopy. New York: Academic Press 1964.
256. Bellamy, L. J.: The infrared spectrum of complex molecules, 2nd ed. New York: John Wiley 1958.

257. Weinman, S., Weinman, J.: Spectres infrarouges régions 1300–450 cm^{-1}. In: Analyse des stéroïdes hormonaux, vol. III, edit. by M. F. Jayle. Paris: Masson & Cie. 1965.
258. Neudert, W., Röpke, H.: Atlas of steroid spectra. Berlin-Heidelberg-New York: Springer 1965.
259. Fishman, J., Miyazaki, M., Yoshizawa, I.: J. Am. Chem. Soc. **89**, 7147 (1967).
260. Straw, R. F., Katzman, P. A., Doisy, E. A.: Endocrinology **57**, 87 (1955).
261. Troen, P., Nilsson, B., Wiqvist, N., Diczfalusy, E.: Acta Endocrinol. **38**, 361 (1961).
262. Touchstone, J. C., Greene, J. W., Jr., McElroy, R. C., Murawec, T.: Biochemistry **2**, 653 (1963).
263. Wilson, R., Ericksson, G., Diczfalusy, E.: Acta Endocrinol. **46**, 525 (1964).
264. Diczfalusy, E., Barr, M., Lind, J.: Acta Endocrinol. **46**, 511 (1964).
265. Levitz, M., Shotsky, S., Twombley, G. H.: Estrogen assays in clinical medicine (C. A. Paulsen, ed.), p. 155. Seattle, Wash.: University of Washington Press 1965.
266. Smith, O. W., Hagerman, D. D.: J. Clin Endocrinol. Metab. **25**, 732 (1965).
267. Levitz, M.: J. Clin. Endocrinol. Metab. **26**, 773 (1966).
268. Goebelsmann, U., Eriksson, G., Diczfalusy, E., Levitz, M., Condon, G. P.: Acta Endocrinol. **53**, 391 (1966).
269. Emerman, S., Twombly, G. H., Levitz, M.: J. Clin Endocrinol. Metab. **27**, 539 (1967).
270. Smith, E. R., Kellie, A. E.: Biochem. J. **104**, 83 (1967).
271. Fishman, J.: J. Am. Chem. Soc. **80**, 1213 (1958).
272. Levitz, M., Katz, J., Twombly, G. H.: Steroids **6**, 553 (1965).
273. Crépy, O., Rulleau-Meslin, F.: Rev. Franc. Etudes Clin. Biol. **5**, 283 (1960).
274. Hirschmann, R., Strachan, R. G., Buchschacher, P., Sarett, L. H., Steelman, S. L., Silber, R.: J. Am. Chem. Soc. **80**, 3903 (1964).
275. Fukushima, D. K., Sauer, G.: Private communication.
276. Okerholm, R. A., Smakula, E. E., Wotiz, H. H.: Federation Proc. **27**, 823 (1968).
277. Collins, D. C., Layne, D. S.: Can. J. Biochem. **46**, 1089 (1968).
278. Matsui, M., Fukushima, D. K.: Biochemistry **8**, 2997 (1969).

Chapter 2

Enzymological Aspects of Steroid Conjugation

ALEXANDER B. ROY

Contents

I. Introduction

The subject matter of this chapter could, if taken in its widest sense, encompass a very large part of modern enzymology because of the very close relationships between the enzymes of steroid conjugation and those of the general metabolic pathways of an animal. Here, however, attention will be restricted to those enzymes directly related to the metabolism of steroid conjugates and, furthermore, of only those steroid conjugates whose metabolism has been investigated in detail; that is, of the steroid glucuronides and the steroid sulfates.

Unfortunately limitations of space make it impossible to consider the lesser known, but not necessarily less important, forms of conjugated steroids. We cannot deal with, for instance, the early report of the presence of dehydroisoandrosterone phosphate in plasma [1] nor with the steroid sulfatides which have recently been studied in considerable detail by Oertel *et al.* It is stated that the steroid sulfatides can be synthesized *in vitro* by mitochondria from guinea pig liver [2] but by what route this occurs is unknown. Certainly the mechanism of the formation of the doubly esterified sulfate group would be most interesting. Another recently discovered conjugate is 17α-estradiol-17-*N*-acetylglucosaminide [3-hydroxyestra-1,3,5(10)-trien-17α-yl-2′-acetamido-2′-deoxy-β-D-glucopyranoside] found in rabbit urine by Layne *et al.* [3] and subsequently shown to be synthesized in rabbit liver homogenates fortified with uridine diphosphate glucuronate and uridine diphosphate *N*-acetylglucosamine, probably by the transfer of the amino sugar to 17α-estradiol-3-glucuronide [4].

Even more interesting is the characterization [5] of pregnenediol-3-sulfate-20α-*N*-acetylglucosaminide [3β-sulfooxypregn-5-en-20α-yl-2′-acetamido-2′-deoxy-α-D-glucopyranoside] from human urine because this is the first naturally occurring steroid conjugate which has been shown to contain an α-glycosidic link*. As Arcos and Lieberman [5] point out, however, few such compounds have been chemically characterized with respect to the stereochemistry of the glycoside bond, the evidence for their structures usually being based simply on their susceptibility to hydrolysis by rather crude enzyme preparations.

The discovery of conjugates of this type makes it important to stress once again the dangers of using crude preparations of β-glucuronidase—especially from mollusks—for the hydrolysis of conjugated steroids. Molluscan tissues are known to be rich sources of many different carbohydrases, in-

* Matsui and Fukushima [Biochemistry **8**, 2997 (1969)] have shown that this compound is in fact the β-anomer.

cluding α– and β–N–acetylglucosaminidases [6], and the errors which could arise from the uncritical use of such enzyme preparations should be obvious.

Other conjugates may of course arise by nonenzymatic reactions. Such is probably the origin of the ureasterone [3α–ureido–11β–hydroxyandrost–4–en–17–one] isolated from human urine by Noguchi and Fukushima [7], which can be formed by the reaction at pH 5 of urea with 3α,11β–dihydroxyandrost–4–en–17–one, a normal metabolite of androstenedione [androst–4–ene–3,17–dione] [8]. One wonders if similar reactions might not lead to the formation of the corticosteroid nucleosides which have recently been reported [9] to occur in human urine, although it is claimed that here the steroid is linked to the sugar. Perhaps yet another example of a nonenzymatically formed conjugate could be that given by the combination of steroids with SH–containing compounds [10].

However, despite the attractions of those more exotic conjugates, attention in the rest of this chapter will be devoted mainly to the steroid glucuronides and steroid sulfates with only a brief consideration of steroid *N*–acetyl–glucosaminides.

II. The Biosynthesis of Steroid Glucuronides

The fact that isolated tissue preparations are able to synthesize steroid glucuronides has been known since Crépy [11] first demonstrated the formation of estrone glucuronide [17–oxoestra–1,3,5(10)–trien–3–yl–β–D–glucopyranosiduronic acid] by slices of guinea pig liver. However, only with the studies of Dutton [12] did it become obvious that the mechanism of this synthesis was similar to that involved in the formation of, for instance, *o*-aminophenyl glucuronide, and that uridine diphosphate glucuronate was the glucuronyl donor. The synthesis of steroid glucuronides therefore involves three consecutive stages: the formation of uridine diphosphate glucose, the conversion of this into uridine diphosphate glucuronate, and the transfer of the glucuronyl residue from the latter to the appropriate acceptor. These three reactions can be represented as follows:

$$\mathrm{UTP + G{-}1{-}P \rightleftharpoons UDPG + PP_i}, \tag{1}$$

$$\mathrm{UDPG + 2\,NAD \rightarrow UDPGa + 2\,NADH_2}, \tag{2}$$

$$\mathrm{UDPGa + ROH \rightarrow UDP + RO \cdot Ga}. \tag{3}$$

The first two of these reactions are obviously common to all pathways of glucuronide metabolism, and only in the last must enzymes specific to steroids be considered.

A very detailed review of the biosynthesis of glucuronides has recently appeared [13].

A. The Formation of Uridine Diphosphate Glucose

The enzyme catalyzing reaction (1) above is of obvious importance because it links the pathway leading to the formation of glucuronides with the glycolytic sequence of reactions through the common intermediate, glucose 1–phosphate. This enzyme is uridine diphosphate glucose pyrophosphorylase, more commonly known as UDPG pyrophosphorylase (UTP: α–D–glucose–1–phosphate uridylyltransferase, EC 2.7.7.9). It was first detected in yeast [14] but is now known to be widely distributed in plants, in microorganisms, and in mammalian tissues [15–17].

The most highly purified preparation of UDPG pyrophosphorylase so far described is that obtained from mung beans by Ginsburg [18], a purification of about 800–fold having been achieved. The rather unstable enzyme was assayed by following the pyrophosphorolysis of uridine diphosphate glucose and measuring spectrophotometrically the rate of production of glucose 1–phosphate through the action of phosphoglucomutase and glucose 6–phosphate dehydrogenase in the presence of NADP. The pH optimum of the reaction was close to 8, the activity falling off steeply on the alkaline side, and the K_m values for the two substrates were 0.11 mM uridine diphosphate glucose and 0.23 mM pyrophosphate. Divalent metal ions (Mg^{2+}, Mn^{2+} or Co^{2+}) were required as cofactors and the enzyme was apparently quite specific with respect to uridine diphosphate glucose.

The UDPG pyrophosphorylase from yeast has also been quite highly purified [19], again to give a preparation which is rather unstable in dilute solution. The pH optimum was about 7, the K_m for uridine diphosphate glucose was 0.07 mM, and Mg^{2+} ions were required in concentrations of about 2 mM. Rather surprisingly, EDTA did not inhibit the enzyme. Once again the reaction studied was the pyrophosphorolysis of uridine diphosphate glucose, and the equilibrium constant of the reaction was shown to be close to unity. According to Turner and Turner [20] the reaction catalyzed by UDPG pyrophosphorylase from pea seedlings has an apparent equilibrium constant of about 7, and this, as would be expected, varies with the pH and with the concentration of Mg^{2+} ions. It must be stressed that this equilibrium constant is only apparent, because no account has been taken of the ionic state of the various reactants, nor indeed are these even known. Likewise, it should be remembered that the values for K_m reported for the enzyme may have little absolute significance, because in no case has the system been studied as a two-substrate reaction; simple Michaelis–type kinetics have been assumed to occur and it is unlikely that this assumption is valid.

It was originally claimed that the yeast enzyme catalyzed a phosphate-exchange reaction [19] and that this was consistent with the formation of a uridylyl–enzyme as an intermediate. This exchange was later shown to be an artefact caused by the presence of phosphorus–containing compounds in the enzyme [21].

B. The Formation of Uridine Diphosphate Glucuronate

The reaction represented in Eq. (2), above, is catalyzed by the enzyme, UDPG dehydrogenase [UDP glucose: NAD oxidoreductase, EC 1.1.1.22] which again occurs widely in plant, microbial, and animal tissues. It is the key reaction in the formation of glucuronides leading to the *de novo* synthesis of the glucuronyl residue.

Strominger *et al.* [22] purified several hundred–fold the enzyme from calf liver and followed its activity spectrophotometrically by measuring the reduction of NAD. Other similar preparations have recently been described [23, 24]. The pH optimum for the oxidation was approximately 8.7, and the K_m for uridine diphosphate glucose and for NAD were 0.07 and 0.1 mM, respectively. Once again there was no study of possible interaction between the substrates. Although the reaction involves four electrons and so must occur in two stages, no aldehyde could be detected as an intermediate; presumably, therefore, both stages are catalyzed by the one protein with the intermediate remaining firmly bound to the enzyme. It was not possible to demonstrate the reverse reaction, the reduction of uridine diphosphate glucuronate to uridine diphosphate glucose. This irreversibility is doubtless partly due to the fact that the carboxyl group of the uronic acid derivative is ionized in the usual reaction mixture, whereas the product of the oxidation, and so, presumably, the starting point for the reverse reaction, would be the un-ionized acid. UDPG dehydrogenase is reversibly inhibited by mercuribenzoate and, rather interestingly, by thiosemicarbazide; this latter inhibition suggests perhaps the occurrence of a carbonyl group on the enzyme [25].

It has recently been shown [26] that uridine diphosphate xylose is a very powerful inhibitor of liver UDPG dehydrogenase, causing virtually complete inhibition at concentrations of about 0.05 mM. This inhibition is obviously complex in type, giving nonlinear reciprocal plots in the usual kinetic analysis, but it is stated to be competitive in nature with a K_i of 4 μM uridine diphosphate xylose. Not surprisingly, UDPG dehydrogenase is inhibited by the reaction product, uridine diphosphate glucuronate, with a K_i of 0.05 mM; the inhibition was competitive with respect to uridine diphosphate glucose and noncompetitive with respect to NAD [26]. Uridine diphosphate galactose is also a competitive inhibitor with a K_i of 0.13 mM [27].

The specificity of liver UDPG dehydrogenase has not been fully studied. Silbert and Hughes [28] have shown that uridine diphosphate glucosamine is slowly, and incompletely, oxidized by preparations of UDPG dehydrogenase, but it is, of course, by no means certain that this enzyme is directly responsible for the oxidation. More interesting and informative are the observations of Goldberg *et al.* [29] that the 5–fluorouracil and 6–azauracil analogues of uridine diphosphate glucose are substrates for UDPG dehydrogenase. With those substrates, changes in pH affect all the maximum velocities similarly, but there are striking differences in the variations in K_m with pH; between pH 7.0 and 8.7 the K_m for the 5–fluoro derivative increases forty–fold and that for the 6–aza derivative thirteen–fold, while that for uridine diphosphate glucose itself varies only three–fold. These differences can be correlated with the different ring–associated p*K* values of the uridines, and it seems that only the un-ionized forms of the pyrimidine can combine with the enzyme, presumably through the –CO–NH– system.

Stereochemically, reduction of NAD by UDPG dehydrogenase causes transfer of the hydrogen to the B side of the nicotinamide ring [30].

The only other UDPG dehydrogenase which has been studied in some detail is that from peas [31]. This enzyme is obviously very similar to that from liver, in that the reaction it catalyzed was irreversible, the pH optimum was about 9, and the K_m values for uridine diphosphate glucose and for NAD were 0.02 and 0.12 mM, respectively. It also was inhibited by mercuribenzoate, but apparently irreversibly. The purified enzyme would not oxidize uridine diphosphate acetylglucosamine, uridine diphosphate acetylgalactosamine, nor uridine diphosphate galactose. It also was inhibited by uridine diphosphate xylose [26], an inhibition which could conceivably be of some value as a control mechanism in plants.

C. Uridine Diphosphate Transglucuronylase

The final step in the biosynthesis of steroid glucuronides involves the transfer of the glucuronyl residue from uridine diphosphate glucuronate to the appropriate acceptor, this transfer being catalyzed by a UDP glucuronyltransferase [UDP glucuronate glucuronyltransferase, EC 2.4.1.17). Although the nomenclature used by the Enzyme Commission recognizes the existence of only a single such enzyme — "acceptor unspecific" — it is clear that, in fact, a number of more or less specific UDP transglucuronylases do exist, but only now is much progress being made in their separation.

From the early work of Dutton *et al.* [12, 32] it was obvious that liver contained a rather labile UDP transglucuronylase which occurred in the microsomes and which could not readily be solubilized. Any of a wide range of acceptors could be utilized by the enzyme to yield either glycoside–linked (ether) or acylal–linked (ester) glucuronides. Transglucuronylase activity was also present in the kidney and gastrointestinal tract, although in smaller amounts than in liver [12, 33–35]. In fetal and neonatal animals (man, guinea pig, and mouse) the ability to synthesize glucuronides was generally low, but in such animals the gastrointestinal tract and the kidney were quantitatively more important than the liver [35, 36]. This latter point is, of course, of some importance in view of the work of Diczfalusy on the steroid metabolism of the human fetus. Much of the work on the distribution, and on the assay, of the UDP transglucuronylases has recently been reviewed [13, 37, 38].

It is now clear that there are several UDP transglucuronylases with different specificities, although unfortunately most of the evidence for this view is indirect because of the difficulty of obtaining soluble preparations of the enzymes. For instance, studies in Gunn rats [39, 40] and in cats [40, 41] have shown that these animals must utilize different glucuronyltransferases for the formation of bilirubin glucuronide and of the ether glucuronides more usually studied *in vitro*. More direct confirmation of this came from the work of Isselbacher *et al.* [42] who, by using a heat–treated venom of *Trimeresurus flavoviridis*, were able to solubilize from the microsomes of rabbit liver the transglucuronylase(s) catalyzing the formation of ether and ester glucuronides; the enzyme catalyzing the formation of *N*–glucuronides was not solubilized under these conditions. As will be discussed below, further evidence of this type is now available from the work of Breuer on the steroid–specific transglucuronylases. There is also some indirect evidence for a multiplicity of UDP transglucuronylases which utilize phenols as acceptors. Storey [43], from the results of detailed studies of the action of SH inhibitors on the reaction, has suggested that, in the mouse, *o*–aminophenyl glucuronide and *p*–nitrophenyl glucuronide are formed by different enzymes. Similarly, Dutton and Lawes [44] have suggested that *o*–aminophenyl glucuronide and phenolphthalein glucuronide are formed by different transglucuronylases, their evidence being the characteristic pattern of development of the two activities in fetal and neonatal animals.

In view of this doubt as to the number of UDP transglucuronylases present in crude liver preparations, there is obviously little significance to be attached to any kinetic studies of these enzymes. All that can safely be said is that they are rather labile enzymes which are difficult to solubilize, which have pH optima around neutrality, which require Mg^{2+} ions as cofactors, and which are inhibited by SH reagents. The values of the K_m for both substrates appear to be of the order of 0.1 mM. Goldberg, quoted by Dutton [13], reported that the K_m for uridine diphosphate glucuronate is independent of the concentration of *p*–nitrophenol so that the two substrates must have independent combining sites, a situation comparable to that found in the sulfotransferases discussed later. Further developments must await the availability of stable, soluble preparations of the enzymes. With such preparations one point which will be worthy of detailed investigation is the change of configuration that occurs during the transfer of the glucuronyl residue, this having the α–configuration in uridine diphosphate glucuronate, but the β–configuration in the glucuronide that is formed [45].

In the synthesis of steroid glucuronides by UDP transglucuronylase(s), again there is no clear indication of the number of enzymes which must be involved. Undoubtedly there are several, as might be predicted from the number of different types of steroid glucuronide known to be formed *in vivo*. The early work, using unfractionated preparations of liver microsomes, made it clear that many different steroids could act as acceptors; 17β-estradiol [estra–1,3,5(10)–triene–3,17β–diol], testosterone [17β–hydroxyandrost–4–en–3–one], androsterone [3α–hydroxy–5α–androstan–17–one], pregnanediol [5β–pregnane–3α,20α–diol], allopregnanediol [5α–pregnane–3α,20α–diol], and tetrahydrocortisone [3α,17α,21–trihydroxy–5β–pregnane–11,20–dione] all gave glucuronides, although cortisone [17α,21–dihydroxypregn–4–ene–3,11,20–trione] and dihydrocortisone [17α,21–dihydroxy–5β–pregnane–3,11,20–trione] did not [12, 46, 47]. Most of the recent work on the synthesis of steroid glucuronides has been carried out by Breuer and his group who have accumulated quite definite evidence for a considerable degree of specificity among the transglucuronylases.

Smith and Breuer [48] showed the synthesis of estrone–3–glucuronide by rabbit liver microsomes at an optimum pH of 8.0–8.2: the value of the K_m for estrone was 0.097 mM. 17β–Estradiol–3–methyl–ether [3–methoxyestra–1,3,5(10)–trien–17β–ol] reacted much more slowly than did estrone [3–hydroxyestra–1,3,5(10)–trien–17–one], reflecting some degree of substrate specificity by the enzyme. More recently Breuer and Wessendorff [49], with a similar enzyme preparation, used 17β–estradiol as a substrate and showed once again a preferential reaction with the phenolic hydroxyl group. The major product of the reaction catalyzed by the transglucuronylase was 17β–estradiol–3–glucuronide [17β–hydroxyestra–1,3,5(10)–trien–3–yl–β–D–glucopyranosiduronic acid], 17β–estradiol–17–glucuronide [3–hydroxyestra–1,3,5(10)–trien–17β–yl–β–D–glucopyranosiduronic acid], and 17β–estradiol–3,17–diglucuronide [estra–1,3,5(10)–

trien-3,17β-yl-di-β-D-glucopyranosiduronic acid] were produced only in much smaller amounts, and presumably by a different enzyme. The optimum pH was between 8.7 and 9.2, and the values for K_m were 0.28 mM 17β-estradiol and 0.29 mM uridine diphosphate glucuronate.

Breuer *et al.* also studied glucuronide formation in preparations from the mucosa of human intestine. A microsomal preparation from this tissue rapidly formed testosterone-17β-glucuronide [3-oxoandrost-4-en-17β-yl-β-D-glucopyranosiduronic acid] at a pH optimum of 6.6 and with a K_m of 0.024 mM testosterone [50]. The relative rate of formation of a number of steroid glucuronides by this enzyme preparation are shown in Table 2-1. Unlike the microsomal preparation from rabbit liver, that from the mucosa of the gastrointestinal tract was not inhibited by mercuribenzoate.

A further group of UDP transglucuronylases occurs in the soluble fraction of homogenates of human intestinal mucosa. The component enzymes can be partially separated by ammonium sulfate fractionation [51]; the fractions precipitating at 0 to 30% and 30 to 60% saturated ammonium sulfate contained rather unstable transglucuronylases capable of utilizing the phenolic hydroxyl group of estrogens and the 16α-hydroxyl

Table 2-1. *The relative rates of formation of steroid glucuronides by the UDP transglucuronylases from human intestinal mucosa*

Steroid	Position of glucuronidation	Relative rate with enzyme from	
		microsomal fraction	soluble fraction
3α-Hydroxy-5α-androstan-17-one	3	0.10	0.00
3β-Hydroxy-5α-androstan-17-one	—	0.00	0.00
3α-Hydroxy-5β-androstan-17-one	3	<0.03	0.00
3β-Hydroxyandrost-5-en-17-one	—	0.00	0.00
17β-Hydroxyandrost-4-en-3-one	17	1.00	0.79
3β-Hydroxypregn-5-en-20-one	—	0.00	0.00
3-Hydroxyestra-1,3,5(10)-trien-17-one	3	0.23	0.00
3,16α-Dihydroxyestra-1,3,5(10)-	3	0.11	—
trien-17-one	16	0.07	0.00
3,17β-Dihydroxyestra-1,3,5(10)-	3	0.24	—
trien-17-one	17	<0.03	0.86
Estra-1,3,5(10)-triene-3,16α,17β-	3	0.15	—
triol	16	0.11	—
	17	0.07	1.00

group of estriol [estra-1,3,5(10)-triene-3,16α,17β-triol], as acceptors. Both these fractions were contaminated with a further UDP transglucuronylase reacting with a 17β-hydroxyl group (in estriol or in testosterone), but this enzyme, uncontaminated by the other two, was concentrated by precipitation at 60 to 80% saturated ammonium sulfate. This latter fraction was quite stable and showed, with estriol as substrate, a pH optimum of 6.8 and a K_m of 0.34 mM. The specificity of this preparation toward various steroids is also shown in Table 2–1 and it is clear that the enzyme is quite distinct from the microsomal enzyme forming testosterone-17β-glucuronide.

Nevertheless, it should be noted that, although those three estriol-specific transglucuronylases have readily been obtained in a soluble form, they may well have originated from the microsomes, because Dahm and Breuer [52] have demonstrated the formation of estriol-3-, estriol-16α-, and estriol-17β-glucuronides (16α,17β-dihydroxyestra-1,3,5(10)-trien-3-yl-β-D-glucopyranosiduronic acid, 3,17β-dihydroxyestra-1,3,5(10)-trien-16α-yl-β-D-glucopyranosiduronic acid, and 3,16α-dihydroxyestra-1,3,5(10)-trien-17β-yl-β-D-glucopyranosiduronic acid) by both the light and heavy microsome fractions from human intestine. Further, it has been shown [53] that the latter two glucuronides must arise directly from estriol and not by the metabolism of prior-formed 16α-glucuronide or 17β-estradiol-17-glucuronide, respectively.

The properties of those steroid-specific transglucuronylases are summarized in Table 2–2. It must again be emphasized that no detailed kinetic

TABLE 2–2. *The properties of some steroid-specific UDP transglucuronylases from mammalian tissues*

Source of enzyme	Substrate	Position of glucuronidation	K_m (mM)	pH optimum
Rabbit liver microsomes	Estrone	3	0.097	8.0–8.2
Rabbit liver microsomes	17β-Estradiol	3	0.28	8.7–9.2
Human intestine microsomes	Testosterone	17	0.024	6.6
Human intestine supernatant	Estriol	17	0.34	6.8

studies have been carried out, nor indeed has it even been shown whether those preparations can utilize acceptors other than steroids, although such might well be the case with the enzyme reacting with the phenolic hydroxyl groups of estrogens.

There seem to have been no recent reports of the formation of the glucuronides of corticoid metabolites, nor of pregnanediol, although these certainly occur *in vivo*. Breuer has not indicated whether the enzyme preparations he used will react with 5β–pregnane derivatives, but it seems very likely that they will not [54, 55]. Other aspects which remain to be studied are the formation of diglucuronides and of the even more interesting conjugates containing both glucuronide and sulfate residues. Specificity requirements would, at first sight, seem to make it improbable that the same enzyme can be involved in the formation of single and of double conjugates, but there is no information available on this point.

III. β–Glucuronidase

The enzyme β–glucuronidase [β–D–glucuronide glucuronohydrolase, EC 3.2.1.31] is an almost ubiquitous constituent of mammalian tissues and body fluids. It also occurs in other vertebrates, in invertebrates (especially mollusks), in microorganisms, and in plants [56–58]. The reaction catalyzed by the enzyme is the hydrolytic fission of the β–glucuronoside link in the substrate resulting in the liberation of glucuronic acid and the aglycone.

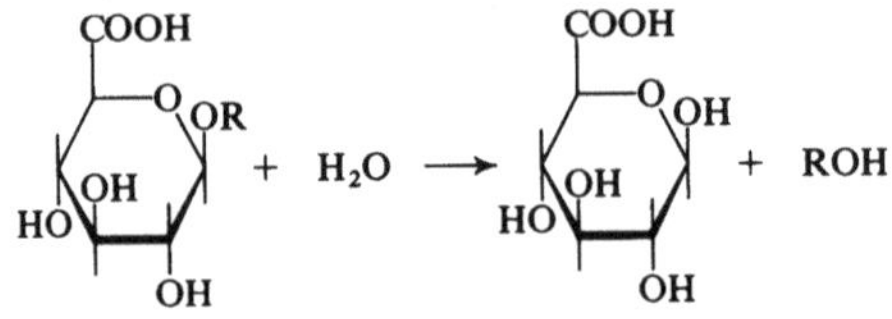

(4)

Most studies of the enzyme have been made with nonsteroid substrates such as menthol glucuronide [59, 60], phenol glucuronide [61], or phenolphthalein glucuronide [62], of which the latter would appear to be the substrate of choice. Standard methods are available for the assay of β–glucuronidase [63]. Only relatively recently have preliminary studies with authentic steroid glucuronides been made, so that much of the information obtained with nonsteroid substrates is very pertinent to the present discussion and must be considered here.

A. Preparation of β–Glucuronidase

Until recently no large–scale methods for obtaining highly purified samples of β–glucuronidase have been described. Bernfeld *et al.* [64] described the preparation of concentrated, although definitely not homogeneous, samples from ox liver, and more recently a claim was made for the crystallization of the enzyme from the same source [65]. Simple methods have now been described which give essentially homogeneous preparations of β–glucuronidase from human liver [66] and from ox liver [67], in yields between 12 and 20%. These proteins have to some extent been characterized and their properties are described below.

For practical purposes the most convenient source of highly purified β–glucuronidase is the preputial gland of the female rat. The preparation of the enzyme from this very rich source has been described [68], and, although the final material has not yet been characterized as a protein, it must, to judge from its very high specific activity, be highly purified indeed. With the ready availability of those essentially homogeneous samples of β–glucuronidase, it is hard to see the justification for the continued use of crude preparations of the enzyme from beef liver as tools in steroid biochemistry.

Insofar as the enzymes of molluscan tissues are concerned, there have been few attempts at purification. Although it is clear that most gastropod mollusks are rich sources of β–glucuronidase, only two species have been widely used as starting material—the European limpet, *Patella vulgata*, and the Roman snail, *Helix pomatia*. Only very crude preparations have been described from the former [69], but Alfsen and Jayle [70] claim to have prepared a crystalline β–glucuronidase from the intestinal juice of *H. pomatia*. This material was by no means fully characterized but it was stated to be homogeneous when examined by zone or by moving boundary electrophoresis. Wakabayashi and Fishman [71] have prepared a more active, but noncrystalline, material from the same source.

In the microbial kingdom only the β–glucuronidase from *E. coli* has been studied to any extent, and no very great purification has been achieved. The majority of workers use the rather crude preparations of this enzyme which are commercially available.

Little need be said about the β–glucuronidases of plants. None have been purified and it is hard to see any circumstance where their use would be required in steroid biochemistry.

Table 2–3 summarizes some information about these preparations of β–glucuronidases, and Table 2–4 gives information about some of the commercially available preparations.

Table 2–3. *A comparison of some preparations of β–glucuronidase from animal tissues*

Source	Yield (units/g tissue)	Specific activity (units/mg protein)
Beef liver [a]	700	60,000
Beef liver [b]	—	150,000
Beef liver [c]	2,800	190,000
Human liver [d]	1,300	57,000
Rat preputial gland [e]	580,000	455,000
Helix pomatia [f]	—	106,000
Helix pomatia [g]	3,500	120,000

[a] Bernfeld, P., Nisselbaum, J. S., Fishman, W. H.: J. Biol. Chem. **202**, 763 (1953).
[b] Bonnichsen, R.: Acta Chem. Scand. **18**, 1302 (1964).
[c] Plapp, B. V., Cole, R. D.: Arch. Biochem. Biophys. **116**, 193 (1966).
[d] Musa, B. U., Doe, R. P., Seal, U. S.: J. Biol. Chem. **240**, 2811 (1965).
[e] Levvy, G. A., McAllan, A., Marsh, C. A.: Biochem. J. **69**, 22 (1958).
[f] Alfsen, A., Jayle, M. F.: Bull. Soc. Chim. Biol. **40**, 2143 (1958).
[g] Wakabayashi, M., Fishman, W. H.: J. Biol. Chem. **236**, 996 (1961).

Table 2–4. *Typical activities of some commercial preparations of β-glucuronidase*

Manufacturer	Source and type		Activity (units/mg)
Sigma	*E. coli*	Type I	50
		VA	100
	Ox liver	B 1	400
		B 10	10,000
	Patella vulgata	L 1	300
	Helix pomatia	H 1	350
Warner–Chilcott Laboratories	Ox liver	"Ketodase"	5,000
Endo Products Inc.	*Helix pomatia*	"Glusulase"	

B. The Chemical and Enzymatic Properties of the β–Glucuronidases

We need not be concerned here with the physiological role of β–glucuronidase, nor with the alterations in its levels in tissues in response to physiological or pathological changes, although such alterations—especially those responsive to the administration of steroids [72]—could well be of considerable importance in the general field of steroid metabolism. Much of this work has been adequately reviewed in recent years (56–58, 73). The intracellular distribution of β-glucuronidase in mammalian tissues has been a matter of some dispute, but it appears now to be generally accepted that this enzyme occurs mainly in the lysosomes, and in the microsomes [74], although Conchie and Levvy [75] have made some pertinent criticisms of this view.

Unfortunately, relatively little information can yet be given on the chemical properties of the β-glucuronidases because, as already pointed out, it is only quite recently that purified preparations of the enzyme have become available in reasonable amounts. Some of the properties of the purified enzymes from human liver and from ox liver are summarized in Table 2–5; the only point that requires comment is the great discrepancy

TABLE 2–5. *Some properties of the β-glucuronidases of ox and of human liver. The kinetic data were obtained using phenolphthalein glucuronide as substrate*

	Ox liver[a]	Human liver[b]
Sedimentation coefficient	14	5.7
Molecular weight	280,000	218,000
$E^{1\%}_{280}$	17	—
Optimum pH	4.8	4.3
K_m (mM)	~0.05	~0.4
Specific activity (μMoles/min/mg protein)	10.0	3.0

[a] Plapp, B. V., Cole, R. D.: Arch. Biochem. Biophys. **116**, 193 (1966).
[b] Musa, B. U., Doe, R. P., Seal, U. S.: J. Biol. Chem. **240**, 2811 (1965).

between the values for the sedimentation coefficients of the two preparations, or more probably, between the sedimentation coefficient and the molecular weight of the enzyme from human liver. Insufficient data have been given to explain this discrepancy, but the rather anomalous relationship between the molecular weight and protein concentration shown in Fig. 8 of reference [66] should be noted. The amino acid compositions of the two preparations have been reported and, rather surprisingly, they are very different [66, 67]. The enzyme from ox liver is a glycoprotein containing about 5% of carbohydrate [67]. Both enzymes are stable over the pH range 4 to 7, and the enzyme from human liver is stable at pH values as high as 9 and at temperatures as high as 70°.

Bernfeld *et al.* [76] have long claimed that the β–glucuronidases of liver and spleen dissociate on dilution to give products which are enzymatically inactive. The evidence for this is simply that the specific activity of the enzyme falls on dilution and that this fall can be prevented, or reversed, by the addition of albumin or DNA. In a later paper [77] it was shown that, in general, polycations could prevent this postulated dissociation. Although there is no doubt that the specific activity of purified preparations of β–glucuronidase falls on dilution, there is, in fact, little to show that this is due to the dissociation of a polymeric form of the enzyme. If it were, it is hard to see why such a dissociation should be reversed by the addition of a second type of protein molecule; reversal should occur only on increasing the total concentration of the β–glucuronidase, and this crucial experiment does not appear to have been performed. Certainly the investigation of the purified β–glucuronidases of liver in the ultracentrifuge have given no indication that these enzymes form a polymerizing system, but it must be pointed out that the concentrations used in the ultracentrifuge are of necessity much higher than those used in enzyme assays, so that it is difficult to relate the results of those two types of study. The only convenient technique for the study of dissociation reactions in such dilute solutions is that of boundary analysis on Sephadex, as was carried out for the arylsulfatase A of ox liver [78]. Whatever may be the interpretation of this dilution effect, it is one of considerable practical importance, and for any study using even partially purified preparations of β–glucuronidase from mammalian tissues, it is essential that the reaction media contain at least 0.01 % of serum albumin to prevent inactivation of the enzyme.

When the kinetic properties of the enzyme are considered, confusion is rampant, even with regard to such an apparently simple property as the position of the optimum pH. In general, β–glucuronidases from mammalian tissues show optima in the region of pH 4.5–5.2, and the position of this optimum apparently varies little with the nature of the aglycone. The situation is complicated by the fact that many such preparations of β–glucuronidase show multiple pH optima. This has perhaps been most clearly demonstrated by Mills *et al.* who obtained preparations of the enzyme from ox spleen and ox liver which showed optima at pH 3.4, 4.5, and 5.2 [79] and at pH 4.5, 5.2, and 6.3 [80], respectively. These authors attributed these optima to the presence of several different β–glucuronidases, but the more commonly held view is that they are due to the combination of a single β–glucuronidase with other proteins or nucleic acids in the enzyme preparations. There is, for instance, no doubt that the addition of DNA to a preparation of β–glucuronidase increases the optimum at pH 5.2 at the expense of that at pH 4.5. The fact that the highly purified preparations of β–glucuronidase now available show only single pH optima (Table 2–5) seemingly provides good evidence that the multiple optima obtained with less purified preparations are caused by the presence of impurities.

The substrate specificity has been studied to only a limited extent and no detailed kinetic investigations have been reported. As far as can be judged at present, all β-D-glucuronides, whether they be glycoside-linked or acylal-linked, are hydrolyzed by β-glucuronidase, but there are, as would be expected, very large differences in the K_m values for the various substrates. These range from 0.057 mM phenolphthalein glucuronide through 0.51 mM phenyl glucuronide (both with the enzyme from rat preputial gland) to 0.03 M *p*-chlorophenyl glucuronide (with the enzyme from ox spleen). Many values of K_m have been tabulated by Levvy and Marsh [57]. In most cases substrate inhibition is pronounced, and there have been several reports that Michaelis-type kinetics are not followed. The specificity for the sugar is not absolute—a fact which should be kept in mind when β-glucuronidase is being used as an analytical tool in steroid chemistry—because the enzyme from mammalian tissues will hydrolyze β-D-galacturonides [81]. It will not, however, hydrolyze α-D-glucuronides nor β-D-glucosides. The affinity of the enzyme for a galacturonide is apparently less than for a glucuronide; the enzyme from rat preputial gland gives K_m values of 0.5 mM phenyl glucuronide and 6.5 mM phenyl galacturonide [68].

A still unsolved problem in the specificity of β-glucuronidase is the requirement regarding the ring structure of the sugar. All naturally occurring substrates of β-glucuronidase are β-D-glucopyranosiduronides, and until recently it was not known whether this pyranose structure was obligatory. The corresponding furanosiduronides were inaccessible, and the fact that the furanosiduronic lactones derived therefrom were inhibitors of β-glucuronidase gave no real information, because, as discussed below, several lactones are inhibitors of the enzyme. Recently, however, Kato *et al.* [82] claimed to have synthesized 2-naphthyl-β-D-glucofuranosiduronic acid and to have studied its hydrolysis by a crude preparation of β-glucuronidase from rabbit liver. Although the pH curve for the hydrolysis of the furanosiduronide differed only slightly from that for the hydrolysis of 2-naphthyl-β-D-glucopyranosiduronide, the value of V for the former was only about 8% of that of the latter and the K_m values for the two compounds were very different: 18 mM 2-naphthyl-β-D-glucofuranosiduronide and 1.3 mM 2-naphthyl-β-D-glucopyranosiduronide [83]. The hydrolysis of the furanose derivative was inhibited by glucarolactone (see below), and it itself inhibited the hydrolysis of the pyranose derivative. This work is very important as it apparently has shown that a pyranose structure is not a prerequisite in a substrate for β-glucuronidase, and therefore casts doubt on the ring structure of any glucuronide characterized only through its ability to be hydrolyzed by this enzyme. Further investigations of this topic are certainly required.

With respect to inhibitors of β-glucuronidase, many organic acids—including citric acid which has been used as a buffer in the assay of the enzyme—are inhibitors [79, 80, 84], as also are heavy metal ions such as Cu^{2+}, Hg^{2+}, and Ag^{+} [57, 58]. The inhibition by copper is interesting because it is apparently dependent upon reduction to the cuprous ion. Ascorbic acid has been claimed to be an inhibitor of β-glucuronidase [79, 85] but this has been disputed [86, 87]. The reason for these discordant findings could be that the inhibition was the result of the reduction by the ascorbic acid of traces of cupric ions which might, or might not, be present in the different enzyme preparations [88].

Undoubtedly the most important inhibitor of β-glucuronidase is the competitive inhibitor glucaro-1,4-lactone [saccharo-1,4-lactone; D-gluco-saccharo-1,4-lactone] [89] which gives a 50% inhibition of the enzyme from rat preputial gland at concentrations about 2.5 μM and has a K_i of about 0.1 μM [68]. The corresponding glucaro-6,3-lactone [L-gulo-saccharo-1,4-lactone] is not an inhibitor. Pure glucaro-1,4-lactone is a rather inaccessible reagent, and from a practical point of view the most important inhibitor of β-glucuronidase is a boiled solution of potassium hydrogen glucarate (saccharate); such a solution, boiled for 30 minutes, contains the equilibrium amount of the 1,4-lactone—about 30% of the total glucarate concentration [89]. Many of the discrepancies in the early studies of the inhibition of β-glucuronidase by glucaric acid (saccharic acid) can be explained by a lack of appreciation of the importance of the spontaneous lactonization which occurred in acid solution. Glucaric acid itself is only a very weak inhibitor of β-glucuronidase, and freshly prepared solutions of the usual salt, potassium hydrogen glucarate, have little inhibitory effect. However, on standing, lactonization occurs spontaneously with a concomitant increase in the inhibitory power.

In view of the multiple specificity of β-glucuronidase, it is not surprising that boiled solutions of mucic acid, which contains galactaro-1,4-lactone, is also an inhibitor of the enzyme. D-Glucuronic acid is only a rather weak inhibitor of β-glucuronidase and D-glucurone [D-glucofuranurono-6,3-lactone] an even weaker one.

It has recently been reported [90] that β-glucuronidases from several sources are rather powerfully inhibited by cholesterol and by retinol—a 70% inhibition at concentrations of 0.1 and 0.08 mM, respectively. Some other related compounds (including carotene, retinoic acid and retinyl acetate) also inhibited, but many others (including several steroids) did not.

The inhibitions by cholesterol and by retinol were not additive; both were probably uncompetitive (the K_i for cholesterol was 0.072 mM) and were reversible by dilution or by the addition of DNA or protein. The mechanism of this inhibition has not been clarified, but it was suggested, on no real evidence, that it might be due to the dissociation of the enzyme into its postulated inactive subunits. Of some interest was the suggestion that the "anti–glucuronidase" of serum which is mentioned below might, in fact, simply be cholesterol. Be this as it may, the further investigation of this inhibition might give useful information on the structure of β–glucuronidase.

In crude tissue preparations complications can arise from the presence of endogenous inhibitors or activators. Human urine, for instance, contains an inhibitor whose potency is greatly increased by boiling under acid conditions. This inhibitor is almost certainly glucaric acid, which in acid conditions forms the much more inhibitory glucaro–1,4–lactone [91]. The β–glucuronidase of rat liver and kidney, but not of mouse tissue, is inhibited by a substance present in rat tissues which has not been identified [92] but is probably similar to the "anti–glucuronidase" of human serum [93]. This endogenous inhibitor is thermostable, nondialyzable, and precipitated at pH 5.2. The inhibition is overcome by the addition of the detergent Triton X–100 to the reaction mixture. This distinguishes the endogenous inhibitor from heparin and chondroitin sulfate, both of which inhibit β–glucuronidase but whose effects are not reversed by Triton. Nevertheless, the endogenous inhibitor of rat tissues is probably also a mucopolysaccharide of some kind. Whatever its nature may be, its presence makes the assay of β–glucuronidase in rat tissues a matter of some difficulty unless they are treated at pH 5.2. A further complication is that rat liver also contains an activator of β–glucuronidase [94]. Once again this is a nondialyzable material which is stable to acid and to alkali. It activates ox liver β–glucuronidase, but not crude preparations of mouse or rat liver β–glucuronidase, presumably because the latter, at least, already contains the activator. With these complications to consider it should seem obvious that all future work on the properties of β–glucuronidase should be carried out only with pure preparations of the enzyme.

One final point is the ability of β–glucuronidase to catalyze the transfer the glucuronyl residue from, say, phenolphthalein glucuronide to an aliphatic alcohol with the formation of the appropriate alkyl glucuronide. This reaction was first investigated by Fishman and Green [95] who used a wide range of donor and acceptor molecules but were unable to show the transfer of the glucuronyl residue to phenolic acceptors. The concentrations of alcohol required are quite high, 0.01 M or higher, but under suitable conditions a 90% transfer of the glucuronyl residue could occur. More recently Yoshida *et al.* [96] have studied the reactions of the more reactive acylal glucuronide, benzoyl glucuronide, and have been able to show transfer of the glucuronyl residue from this to *m*–aminophenol. The reaction product was characterized as a glucuronide but it was not hydrolyzed by β–glucuronidase, and the possibility of the product being an α–glucuronide was suggested. It should be noted that the transfer of the glucuronyl residue from such acylal glucuronides to methanol gave a methyl glucuronide which was hydrolyzed by β–glucuronidase [96]. Apart from the intrinsic interest of such transfer reactions, they could conceivably be of some value in enhancing the hydrolysis of steroid glucuronides, because one of the manifestations of the transfer reaction is that the addition of, say, propylene glycol to the system hydrolyzing phenolphthalein glucuronide causes an increased liberation of the aglycone, although a decreased liberation of glucuronic acid.

The β–glucuronidases from invertebrates have, on the whole, been much less studied and only those occurring in the digestive tracts of mollusks have attracted the attention of most workers. The enzymes from *P. vulgata* and *H. pomatia* seem to be generally similar to those of mammalian sources insofar as they are relatively nonspecific enzymes with respect to the aglycone, in also acting as β–galacturonidases [81], and in having pH optima in the region of 4.0–4.5 [69, 97]. Multiple pH optima do not appear to have been recorded with the molluscan enzymes. Again these β–glucuronidases are powerfully and competitively inhibited by glucaro–1,4–lactone, with a K_i of 0.1 μM, and reversibly inhibited by Ag^+ and Hg^{2+} ions [69].

Rather crude extracts of *P. vulgata* will hydrolyze not only β–D–glucuronides but also α–D–glucuronides [81]. The enzyme responsible for the hydrolysis of the latter was clearly distinct from β–glucuronidase since it had a quite different stability to pH and temperature, and was not inhibited by glucarolactone nor galactarolactone. Nevertheless, a warning must again be sounded against the indiscriminate use of crude preparations of β–glucuronidase as diagnostic tools in steroid biochemistry.

The β–glucuronidases from bacteria have not been investigated in detail. They may be either constitutive or adaptive enzymes. In the most important source, *E. coli*, β–glucuronidase is an adaptive enzyme usually induced by the presence of menthyl glucuronide [98] or bornyl glucuronide [99] in the incubation medium. The enzyme is liberated into the medium on aging of the culture.

The general properties of the β–glucuronidase from *E. coli* are similar to those of the mammalian enzymes. The pH curve shows a single optimum in the region of pH 6, the lack of specificity with respect to the aglycone is similar, and glucarolactone is a competitive inhibitor. It may, however, be a less powerful inhibitor of the microbial enzymes than of the mammalian enzymes; Marsh [100] showed that a β–glucuronidase isolated from sheep–rumen microorganisms had a K_i of 19 μM glucarolactone, a value one hundred times greater than that for most other β–glucuronidases.

One interesting feature of the β–glucuronidase from *E. coli* is that its activity is greatly stimulated, some two–fold or more, by the addition of chloroform or a similar solvent to the reaction mixture [101]. This is in rather sharp contrast to the inhibitory effect of such solvents on the enzyme from ox liver. The mechanism of the activation does not seem to be clearly understood, but it has been shown [102] that the activation is not dependent upon the relative volumes of the organic and aqueous phases but rather on the interfacial area between the two phases. It would be most interesting to have this work extended using highly purified enzymes and using physical methods to detect conformational changes in the protein.

C. The Hydrolysis of Steroid Glucuronides by β–Glucuronidase

Until quite recently there have been few studies on the action of β–glucuronidase on steroid glucuronides, a rather surprising situation in view of the quite early study of the specificity of the steroid sulfatases [103, 104]. Doubtless this situation was partly due to the relative inaccessibility of authentic steroid glucuronides, a situation which has now been ameliorated through the development of reliable synthetic methods [105–107]. One of the first studies of the action of β–glucuronidase on steroid glucuronides was that of Alfsen [97] who used a rather highly purified preparation from *H. pomatia*. Unfortunately the results were not subjected to a detailed kinetic analysis and, although there were obvious differences in the K_m values (shown in Table 2–6), they simply confirmed the general lack of specificity shown by β–glucuronidase toward the aglycone. Until recently little attention has been paid to the possibility that β–glucuronidases from different sources might have different specificities, although Hobkirk and Cohen [108] had suggested that the 3–glucuronides of 5α–androstane derivatives might be hydrolyzed rather less efficiently by liver β–glucuronidase than by bacterial β–glucuronidase.

Recently some studies have been made of the hydrolysis of steroid glucuronides by β–glucuronidases from various sources [109, 110]; Table 2–6 summarizes the few K_m values which have been obtained. Unfortunately it seems impossible to draw any very general conclusion from such scanty data. Determinations of relative values of the maximum

TABLE 2–6. *Values of K_m (mM) for steroid glucuronides with different preparations of β-glucuronidase*

Glucuronide of	Liver		*E. coli*		*Helix pomatia*	*Patella vulgata*
	a	b	a	b		
Estriol	—	—	—	—	0.42	—
3α–Hydroxy–5α–androstan–17–one	—	2.0	—	0.57	1.6	2.2
3β–Hydroxy–5α–androstan–17–one	0.11	1.1	0.17	0.17	—	1.4
3α–Hydroxy–5β–androstan–17–one	0.35	0.61	—	0.35	—	1.0
3β–Hydroxy–5β–androstan–17–one	—	0.55	—	0.15	—	0.74
3β–Hydroxyandrost–5–en–17–one	0.17	—	0.20	—	—	—
5β–Pregnane–3α,20α–diol[c]	0.12	—	0.14	—	8.0	—
3α,17α,21–Trihydroxy–5β–pregnane–11,20–dione[c]	—	—	—	—	3.8	—

[a] Voigt, K. D., Lemmer, M., Tamm, J.: Biochem. Z. **332**, 550 (1960).
[b] Kellie, A. E.: Biochem. J. **100**, 631 (1966).
[c] For the 3–glucuronide.

velocity, *V*, for the hydrolysis of steroid glucuronides by β–glucuronidase are even fewer, although one would have thought that such information should have been readily available from the calculations of K_m, and the only published figures are those of Voigt *et al.* [110]. In general terms, Wakabayashi and Fishman [71] stated that the glucuronides of pregnane derivatives were hydrolyzed more rapidly than the glucuronides of either estrogens or androgens, while Kellie [111] showed that the rate of hydrolysis of steroid glucuronides by the β–glucuronidase of *P. vulgata* decreased in the general order: 3(phenolic)–>3β–>17β–>3α–. These generalizations were based not on kinetic studies but simply on measures of the time taken to achieve complete hydrolysis of given quantities of steroid glucuronide, so that they are not relevant to a discussion of *V* but only to one of the "practical" rates of hydrolysis of such compounds. More detailed studies of the relationship between the rate of hydrolysis of a glucuronide and its structure have been carried out only with the 3–glucuronides of the four isomeric androstanolones which were investigated by Becker [109], who unfortunately did not report values for V, although he did for K_m. Nevertheless, from the published figures [109] it appears that, with the β–glucuronidase from *E. coli*, the values of V for the isomeric androstanolone glucuronides decreased in the order $3\beta 5\beta > 3\beta 5\alpha > 3\alpha 5\beta > 3\alpha 5\alpha$, while with the enzymes from ox liver and from *P. vulgata* the order was $3\beta 5\beta > 3\alpha 5\beta > 3\beta 5\alpha > 3\alpha 5\alpha$. In both cases the decrease in V was about four–fold. The sequence for the latter enzyme is, not surprisingly, rather different from that obtained by Kellie [111] for the "practical" rates of hydrolysis of the same compounds: this was $3\beta 5\alpha$ and $3\beta \Delta^5 > 3\alpha 5\beta > 3\alpha 5\alpha$. Once again there is no immediately obvious simple relationship between the stereochemistry of the substrates and their rates of hydrolysis.

Little need be said of the effect of pH on the hydrolysis of steroid glucuronides. The general picture is similar to that obtained with non-steroid substrates. The most detailed studies have again been those of Wakabayashi and Fishman [71] who showed that for several substrates the β–glucuronidases from ox liver, from *H. pomatia*, and from *P. vulgata* showed fairly symmetrical optima in the region of pH 4 to 5, while the enzyme from *E. coli* showed multiple optima, or a single plateau, extending over a wide pH range between 5 and 8. Essentially similar results were obtained by Becker [109] and by Kellie [111]; the β–glucuronidase of *P. vulgata* had a symmetrical optimum close to pH 4, that of ox liver an optimum close to pH 5, while the enzyme from *E. coli* gave a plateau in the region of pH 6 to 8.

The situation with the enol–glucuronide of androstenedione was interesting [112]; with enzyme preparations from all three of the usual sources, the pH curves showed a double optimum in the region of pH 5 to 6. This behavior was perhaps a reflection of the very facile nonenzymatic hydrolysis of the enol–glucuronide which showed a definite minimum rate at a pH of about 6.

Finally, although it is normally assumed that all steroid glucuronides will be hydrolyzed by β–glucuronidase, it should be realized that this assumption may be invalid. It has recently been claimed [113] that there is excreted in human urine a glucuronide conjugate of aldosterone [11β,21–dihydroxy–3,20–dioxopregn–4–en–18–al] which is not hydrolyzed by liver or bacterial glucuronidase, although it is by the intestinal juice of *H. pomatia*. It was suggested that this β–glucuronidase–resistant conjugate might be an 18–glucuronide. Be this as it may, this work certainly challenges the validity of the usual assumption of the complete nonspecificity of β–glucuronidase.

IV. The Biosynthesis of Steroid Sulfates

Once again the fact that steroid sulfates can be formed *in vitro* by crude preparations from liver has been known since the work of De Meio and Lewycka [114], but it is only recently that details of the reactions have been established. In particular, studies of the transfer of the sulfuryl group from active sulfate—3′-phosphoadenylyl sulfate (adenosine 3′-phosphate 5′-sulfatophosphate)—are only now being actively pursued. As in the synthesis of steroid glucuronides, the over-all pathway for the formation of steroid sulfates involves two essentially discrete stages: first, the formation of 3′-phosphoadenylyl sulfate by reactions common to several pathways of sulfate metabolism; and second, the transfer of the sulfuryl group from phosphoadenylyl sulfate to the appropriate acceptor through the action of steroid-specific enzymes. The reactions involved are:

$$\mathrm{ATP} + \mathrm{SO_4^{2-}} \rightleftharpoons \mathrm{APS} + \mathrm{PP}_i, \tag{5}$$

$$\mathrm{APS} + \mathrm{ATP} \rightarrow \mathrm{PAPS} + \mathrm{ADP}, \tag{6}$$

$$\mathrm{PAPS} + \mathrm{ROH} \rightarrow \mathrm{PAP} + \mathrm{ROSO_3^-}. \tag{7}$$

The first two of these reactions make up the apparently ubiquitous sulfate-activating system which has been clarified primarily through the work of Lipmann.

A. The Formation of 3′-Phosphoadenylyl Sulfate

The most interesting reaction in the formation of sulfate esters is that represented in Eq. (5) above, the production of adenylyl sulfate from ATP and SO_4^{2-} ions through the action of the enzyme, sulfate adenylyltransferase, more commonly known as ATP-sulfurylase [ATP: sulfate adenylyltransferase, EC 2.7.7.4]. The equilibrium of this reaction lies very much in favor of ATP and SO_4^{2-} ions; the apparent equilibrium constant is of the order of 10^{-8} [115, 116] so that the standard free energy change of the reaction is about +11,000 cal. Such a free energy change implies the very high "sulfate group potential" of about 19,000 cal for adenylyl sulfate (117) and means that the assay of ATP-sulfurylase, by the simple production of adenylyl sulfate and pyrophosphate, is impracticable because only negligible quantities of those products will accumulate. Addition of pyrophosphatase will of course drive the reaction in the direction of the synthesis of adenylyl sulfate; but nevertheless, the reaction is most conveniently studied in the reverse direction.

Little is known of the properties, or even of the distribution, of ATP-sulfurylase, and it has been purified only from yeast [115] and from

Desulfotomaculum nigrificans and *Desulfovibrio desulfuricans* [116]. The enzyme from these organisms is a rather stable protein which requires Mg^{2+} ions as a cofactor, which is an SH enzyme, and which has a pH optimum variously reported between 7 and 9.5. The enzyme from mammalian tissues has not been purified, probably because it is a much less stable enzyme. Indeed, the evidence for its occurrence in most animal tissues is simply the ability of those tissues to produce active sulfate. Sundaresan [118] has produced evidence that an acidic derivative of vitamin A is required as a cofactor by the ATP–sulfurylase of mammalian liver. The participation of vitamin A in the activation of sulfate has also been suggested by some studies of the sulfate metabolism of animals deficient in vitamin A [119–121], but unfortunately there are also claims to the contrary [122, 123]. This question of a possible role of vitamin A in the metabolism of SO_4^{2-} ions, in particular in the reaction catalyzed by ATP–sulfurylase, is obviously one of considerable intrinsic interest quite apart from any bearing it may have on the interrelationships of vitamin A and steroid metabolism. It would seem that definitive studies must await the availability of ATP–sulfurylase from mammalian liver in a homogeneous form.

The specificity of ATP sulfurylase has attracted some attention and, insofar as ATP is concerned, the specificity seems to be absolute. A number of anions related to SO_4^{2-} ions can, however, act as the second substrate for the reaction. This aspect has been studied in considerable detail by Wilson and Bandurski [124] who have shown that three similar, but nevertheless distinct, types of reaction are catalyzed by ATP–sulfurylase. In the first type of reaction, shown only with SO_4^{2-} ions as substrate, adenylyl sulfate accumulates and, furthermore, an exchange between pyrophosphate and ATP is catalyzed by the enzyme in the presence of SO_4^{2-} ions. This type of reaction can be represented as:

$$\mathrm{ATP} + \mathrm{SO_4^{2-}} \rightleftharpoons \mathrm{APS} + \mathrm{PP}_i.$$

The second type of reaction occurs only with SeO_4^{2-} ions as substrate; in this, pyrophosphate exchange in the presence of SeO_4^{2-} can be detected, but adenylyl selenate, the selenate-containing analogue of adenylyl sulfate, accumulates only in traces. Nevertheless, the ATP is slowly but completely utilized to appear as AMP. As there is no disappearance of SeO_4^{2-} ions, this type of reaction can be represented as:

$$\mathrm{ATP} + \mathrm{SeO_4^{2-}} \rightleftharpoons \mathrm{APSe} + \mathrm{PP}_i$$

$$\mathrm{AMP} \quad \mathrm{H_2O}.$$

Finally, in the third type of reaction which occurs with SO_3^{2-}, WO_4^{2-}, MoO_4^{2-}, or CrO_4^{2-} ions, no pyrophosphate exchange occurs, no analogues of adenylyl sulfate can be detected, and all the ATP is rapidly and quantitatively converted to AMP and pyrophosphate. The reaction is:

$$ATP + MoO_4^{2-} \rightarrow [APM] + PP_i$$

$$AMP \qquad H_2O.$$

In this latter scheme the postulated anhydride can have only a transitory existence, otherwise pyrophosphate exchange should have been detectable. Reactions of the last type provide what is perhaps the most convenient method of assaying ATP–sulfurylase, particularly if pyrophosphatase be added to the system; under these conditions ATP–sulfurylase activity can be measured simply by the rate of appearance of orthophosphate. This obviously gives a much simpler method of assay than those mentioned previously; nevertheless the theoretical background to the method is far from clear, and rather careful controls are required to take account of any nonenzymatic fission of ATP catalyzed by, say, MoO_4^{2-} ions.

The second stage in the formation of 3′–phosphoadenylyl sulfate, shown in the reaction represented by Eq. (6), above, involves the enzyme, adenylylsulfate kinase, more commonly known as APS–phosphokinase [ATP: adenylyl sulfate 3′–phosphotransferase, EC 2.7.1.25]. This enzyme has been significantly purified only from yeast [125]. It requires Mg^{2+} ions as a cofactor, has a pH optimum between 8.5 and 9, and is completely saturated with its substrate, adenylyl sulfate, at 5 μM which was the lowest concentration tested. By functioning at such low concentrations, and by catalyzing an essentially irreversible reaction, APS–phosphokinase is a most efficient trapping agent for the minute amounts of adenylyl sulfate produced by ATP–sulfurylase, and so allows the accumulation of significant amounts of 3′–phosphoadenylyl sulfate. The distribution of APS–phosphokinase in mammalian tissues is quite unknown, and, in fact, it has never been directly shown to occur therein, although this seems an inescapable conclusion in view of the ability of mammalian tissues to synthesize active sulfate.

The above two enzymes—ATP–sulfurylase and APS–phosphokinase—together make up the sulfate–activating system, and their combined actions allow the production of 3′–phosphoadenylyl sulfate; but it must be stressed again that little indeed is known of the distribution, or properties, of the component enzymes in mammalian tissues. The system as a whole is certainly widely distributed in microorganisms, in plants, and in animals. In mammals it occurs in most tissues, including the endocrine glands.

Wengle [126] and Carroll and Spencer [127] have reported that the activity of the sulfate–activating system is low in the livers of fetal rats, and the latter authors [128] have ascribed this, on rather slender evidence, to a physiological hypovitaminosis A in the fetus. Further work is needed to justify this conclusion, attractive though it is.

The presence of the sulfate–activating system in relatively crude extracts of rat liver allows the ready biosynthesis of substrate amounts of 3′–phosphoadenylyl sulfate. Such methods have been described by several workers [129–131], and they offer some advantages over the chemical methods of synthesis which are also available [132–135].

B. The Steroid Sulfotransferases

The transfer of the sulfuryl group from 3′–phosphoadenylyl sulfate to a steroid (reaction, Eq. (7), above) is catalyzed by the group of enzymes, the steroid sulfotransferases. Insofar as steroids are concerned, the sulfotransferase reaction is irreversible. The nomenclature authorized by the Enzyme Commission recognizes only two such enzymes [3′–phosphoadenylylsulfate: 3β–hydroxysteroid sulfotransferase, EC 2.8.2.2, and 3′–phosphoadenylyl–sulfate: estrone sulfotransferase, EC 2.8.2.4]; but, once again, there are several such enzymes which are often, although incorrectly, known as the steroid sulfokinases. It has been clear since the work of Schneider and Lewbart [136] and of Roy [137] that many different types of steroid could be sulfurylated by crude preparations of rat liver, a conclusion which was perhaps to be expected in view of the many types of steroid sulfate to be found in nature. These early observations gave no indication of the number of different sulfotransferases involved, but certainly implied that, if only one such enzyme did exist, then it was of a much lower specificity than the steroid sulfatases [103, 104]. More recent work by Boström confirmed that many types of steroid could be sulfurylated *in vitro*, but unfortunately no more than an indication of the number of enzymes involved was given.

The first actual separation of steroid sulfotransferases was achieved on an analytical scale by Nose and Lipmann [138] who showed that estrone sulfate [3–sulfooxyestra–1,3,5(10)–trien–17–one] and dehydroisoandrosterone sulfate [3β–sulfooxyandrost–5–en–17–one] were formed by different enzymes. Banerjee and Roy [131] separated the two sulfotransferases responsible for those reactions (estrone sulfotransferase and androstenolone sulfotransferase) from guinea pig liver and produced evidence for the occurrence of two more related enzymes, testosterone sulfotransferase and deoxycorticosterone sulfotransferase. The estrone sulfotransferase from ox adrenal has now been obtained devoid of other steroid sulfotransferase activity [139].

There is also circumstantial evidence for the existence of even more specific sulfotransferases; in particular, of sulfotransferases reacting with the 3–hydroxy groups of 5α– and 5β–androstane derivatives, respectively. This is suggested by the variations in the relative rates of sulfurylation of the various steroids by crude sulfotransferase preparations from different tissues of different species; for instance, in preparations from rat [137] and rabbit [138] liver, 3α– and 3β–hydroxy–5β–androstan–17–ones are sulfurylated much more slowly than dehydroisoandrosterone [3β–hydroxyandrost–5–en–17–one], but in human liver [140] and adrenal [141, 142] the two types of steroid react at similar rates.

With regard to the distribution of the steroid sulfotransferases, including estrone sulfotransferase, in mammals, little need be said. Unlike the ubiquitous phenol sulfotransferase, the steroid sulfotransferases occur in significant amounts only in the liver, the adrenal gland and the jejunal mucosa, at least in the ox [143] and the adult human [144]. Smaller amounts of the steroid sulfotransferases must occur in other tissues, and more sensitive methods have shown their existence in the human ovary [145] and testis [146] where they are undoubtedly of considerable physiological importance. Androstenolone sulfotransferase must also be presumed to occur in boar testes which have been shown to secrete dehydroisoandrosterone sulfate [147].

In human fetal tissues the general level of the steroid sulfotransferases (apart from estrone sulfotransferase) is lower than in adult tissues with the exception of the adrenal gland and the kidney [148]. Estrone sulfotransferase, on the other hand, is very widespread in the fetus and is apparently present in all tissues except the cerebrum, a distribution which suggests that this enzyme is of particular importance during fetal life [148]. This lowered activity of the steroid sulfotransferases in the fetus is also evident in rats [126, 149] in which the liver has no detectable androstenolone sulfotransferase activity (and a very low phenol sulfotransferase activity). Once again, Carroll and Spencer [127, 128] relate this low activity to a physiological deficiency of vitamin A in the fetal rat, a suggestion which is certainly worthy of more detailed investigation. It is perhaps of interest that embryonic chick cartilage is a rich source of estrone sulfotransferase [150]; fetal calf cartilage is devoid of this enzyme.

One point of interest—the significance of which is unknown—is that the steroid sulfotransferases appear to occur only in the soluble fraction of the cell, as do the sulfate–activating enzymes, in striking contrast to the location of the transglucuronylases in the microsomes. It would be interesting indeed to discover the significance of the different distribution of these two conjugating systems. One wonders if more attention should not be paid to this when comparing the metabolism of steroid sulfates and steroid glucuronides, especially when one recalls the complementary difference in the distribution of the two corresponding hydrolases, β–glucuronidase and steroid sulfatase, which occur in the lysosomes and in the microsomes, respectively.

Little can be said of the specificity of the steroid sulfotransferases (except that estrone and dehydroisoandrosterone sulfates are obviously formed by quite independent enzymes) because the only preparation devoid of other sulfotransferase activities is that of the estrone sulfotransferase of ox adrenal [139]. This enzyme can apparently use any estrogen as substrate, the rates of sulfurylation decreasing in the order: estriol, 16–epiestriol [estra–1,3,5(10)–triene–3,16β,17β–triol], 17–epiestriol [estra–1,3,5(10)–triene–3,16α,17α–triol], 16–ketoestradiol [3,17β–dihydroxyestra–1,3,5(10)–trien–16–one], 17β–estradiol, estrone, 17α–estradiol [estra–1,3,5(10)–triene–3,17α–diol], equilenin [3–hydroxyestra–1,3,5(10),6,8–pentaen–17–one] and 17–deoxyestrone [estra–1,3,5(10)–trien–3–ol], from a relative velocity of 1.0 to about 0.3. All the other enzyme preparations which have been used have certainly contained several different steroid sulfotransferases. Some idea of the types of compound sulfurylated by such preparations from various tissues, and the relative amounts of activity in each tissue, is given by the data of Table 2–7, which summarizes a few of the results of Boström and his group [144, 148] who have provided the only accurate quantitative data available. A fairly safe conclusion would be that, in general, hydroxyl groups at positions 3, 17, and 21 of the steroid molecule can readily be sulfurylated, presumably each by its own sulfotransferase. For hydroxyl groups in other positions in the steroid molecule, little, if any, information is available. One ester of particular interest is cholesterol sulfate. Both Schneider and Lewbart [136] and Nose and Lipmann [138] were unable to detect the synthesis of this by rather crude preparations from rat liver. Recently, however, Banerjee and Roy [151] have used a partially purified preparation of androstenolone sulfotransferase from guinea pig liver and have shown that cholesterol sulfate can be produced by this enzyme at about 30% of the rate of formation of dehydroisoandrosterone sulfate. Wengle and Boström [152] have pointed out some relationships between the structures of steroids and their behavior as substrates for the steroid sulfotransferases. For example, a double bond in the 4–5 position inhibits sulfurylation of a 3β–hydroxyl group; a methyl or ethyl, but not an ethinyl, group in the 17α position inhibits sulfurylation of a 17β–hydroxyl group; and a 17α–hydroxyl group inhibits sulfurylation at position 21. Obviously, several of these observations can be correlated with the behavior of similar steroids *in vivo*. Dihydroxy steroids, especially 3,17– and 3,20–

Table 2–7. *The rates of formation of steroid sulfates (in μmMoles/g tissue/hr) under standard conditions by extracts of some human tissues*

		Dehydro-isoandro-sterone	Andro-sterone	Etiochol-anolone	Testo-sterone	3α–Hydroxy–pregnan–20–one	3β–Hydroxy–pregnan–20–one	Deoxy-cortico-sterone	Estrone
Liver	Adult [a]	498	263	483	99	117	615	155	214
	Fetal [b]	120	59	108	15	22	131	34	147
Adrenal	Adult [a]	295	154	326	56	72	383	63	83
	Fetal [b]	3,740	2,268	3,521	553	1,031	4,108	932	661
Jejunum	Adult [a]	52	26	60	8	17	61	5	85
	Fetal [b]	99	64	112	18	16	132	43	141
Kidney	Adult [a]	1	1	1	2	2	1	5	85
	Fetal [b]	21	3	3	2	4	15	25	127

[a] Boström, H., Wengle, B.: Acta Endocrinol. **56**, 691 (1967).
[b] Wengle, B.: Acta Endocrinol. **52**, 607 (1966).

dihydroxy steroids, can form disulfates when incubated with crude sulfotransferase preparations from rat liver [152]; and, rather surprisingly, estriol–16–glucuronide gives estriol–3–sulfate–16–glucuronide [3–sulfooxy–17β–hydroxyestra–1,3,5(10)–trien–16α–yl–β–D–glucopyranosiduronic acid] when incubated with preparations from guinea pig liver [153]. Whether these double conjugates are formed by the same sulfotransferases as yield the single conjugates is not known, but if this is the case, then it would seem to imply a surprising lack of specificity in those enzymes. It might have been thought that the differences in the physical properties of estriol and of estriol–3–glucuronide would have been sufficiently large to make their reaction with a single enzyme most improbable.

There is little information on the kinetics of the steroid sulfotransferases, but they appear to be SH enzymes (or at least to be activated by SH–containing compounds), to require Mg^{2+} ions as cofactors, and to have pH optima in the region of 5.5 to 8, although there is some disagreement in the literature on the position of the latter [131, 139, 142, 144, 150]. The requirement for Mg^{2+} ions makes difficult the interpretation of some results [144, 148] obtained with reaction mixtures containing EDTA added to remove Mg^{2+} ions, and so to inhibit the sulfate–activating system prior to assaying the sulfotransferases; such a procedure is clearly inadmissible if the latter enzymes also require Mg^{2+} ions. The few data that are available are summarized in Table 2–8. Another difficulty in interpreting kinetic studies is the very low solubility of many steroids in aqueous media. Two techniques

TABLE 2–8. *Some kinetic properties of steroid sulfotransferases*

Steroid substrate	Source of enzyme	Optimum pH	K_m(mM)	
			Steroid	PAPS
Estrone	Guinea pig liver	6	0.025	—
	Human liver	7–8	—	—
	Ox adrenal	8	0.014	0.07
	Chick cartilage	7	—	—
Dehydro-isoandrostane	Guinea pig liver	7–8	0.020	0.043
	Human liver	6	—	—
	Human adrenal [a]	6	—	—
	Human adrenal [b]	7.5	—	—
Cholesterol	Guinea pig liver	—	0.020	0.050

[a] Boström, H., Wengle, B.: Acta Endocrinol. **56**, 691 (1967).
[b] Adams, J. B.: Biochim. Biophys. Acta **82**, 572 (1964).

have been adopted to overcome this: either the aqueous reaction mixture is added to the steroid present as a thin film of solid in the reaction vessel, or the steroid is added as a solution in an organic solvent such as propylene glycol. Clearly, neither technique is without its disadvantages and probably no general solution is possible. Certainly the fact that the synthesis of dehydroisoandrosterone sulfate is not inhibited by propylene glycol [151] does not mean that this solvent should be used indiscriminately, as was pointed out some years ago by Cooke and Taylor [154].

So far, no detailed kinetic analysis of the reactions catalyzed by the steroid sulfotransferases, such as has been made with phenol sulfotransferase [155], has been carried out. However, the information that is available [131, 139] would not be inconsistent with the steroid sulfotransferases, like phenol sulfotransferase, catalyzing rapid–equilibrium random bi bi reactions [156] in which there is no interaction between the two substrate–combining sites.

It has not yet been proved that steroid sulfotransferases are devoid of other types of sulfotransferase activity, although Adams has stated [139] that the estrone sulfotransferase of ox adrenal will not utilize simple phenols nor arylamines as substrates. It will, however, sulfurylate diethylstilbestrol [3,4–bis–(4′–hydroxyphenyl)–3–hexene] and hexestrol [3,4–bis–(4′–hydroxyphenyl)–hexane] at about 5% of the rate of estrone. On the other hand, Banerjee and Roy [131] have suggested that both the estrone sulfotransferase and the androstenolone sulfotransferase of guinea pig liver can utilize *p*–nitrophenol and 2–naphthylamine as substrates, and so show phenol sulfotransferase and arylamine sulfotransferase activity. Such a lack of specificity is surprising, but the results with 2–naphthylamine may help to explain the previously observed effects of 17–oxosteroids on arylamine sulfotransferase activity [157–159], particularly since the model 17–oxosteroid used in those investigations, the 3–methyl ether of dehydroisoandrosterone, has now been shown to have a slight inhibitory action on the synthesis of cholesteryl sulfate by androstenolone sulfotransferase [151]. Only further purification of the sulfotransferases can solve those problems, but it seems not improbable that, in the steroid sulfotransferases, we have further examples of enzymes showing allosteric effects [139, 159].

So far only the hormonal steroids or their near relatives have been considered as substrates for the steroid sulfotransferases, but other types of steroid can certainly act as substrates for crude sulfotransferase preparations from mammalian liver. For example, the cardenolide genins, digitoxigenin [3β,14–dihydroxy–5β–card–20(22)–enolide], 3–epidigitoxigenin [3β–hydroxy–14,21–epoxy–5β,20ξ–cardenolide], uzarigenin [3β,14β–dihydroxy–5α–card–20(22)–enolide], and 3–epiuzarigenin [3α,14β–dihydroxy–

5α–card–20(22)–enolide], and sarmentogenin [3β,11α,14β–trihydroxy–5β–card–20(22)–enolide] can all be sulfurylated by enzyme preparations from human or guinea pig liver [160]. These genins are presumably sulfurylated at the 3–hydroxyl groups, although this was not shown. So, perhaps, it is not too unreasonable to assume that one of the known steroid sulfotransferases—perhaps androstenolone sulfotransferase—is responsible for this reaction despite the great differences between the shape of the C and D rings of the genins and those of hormonal steroids. More difficult to explain, perhaps, are the reports of the ability of preparations from rat liver to sulfurylate the *seco*–steroid, vitamin D_2 [161]. Although vitamin D_2 is undoubtedly derived from a 3β–hydroxy steroid, the vitamin itself is so different in shape from the normal substrates of the steroid sulfotransferases that it is hard not to believe that a separate sulfotransferase is involved. Perhaps here we have a function of the uncharacterized sulfotransferase which operates in the formation of alkyl sulfates [162]. The significance of these two types of reaction is, of course, quite unknown; certainly the first, the formation of the cardenolide sulfates, is likely to be of pharmacological rather than physiological interest; but the second, the formation of vitamin D sulfate, could perhaps yet be shown to have some physiological significance. The metabolism of vitamin D in animals is certainly still quite obscure.

Very recently Palmer [163] showed the presence of considerable amounts of the sulfate esters of glycolithocholic acid and of taurolithocholic acid in human bile following the oral administration of 24–[^{14}C] lithocholic acid. Of the label appearing in the bile, about 50% was present in the sulfate fraction. The formation of these compounds is interesting not only in that they once again are double conjugates but that they form an evolutionary link with the bile alcohol sulfates of the lower vertebrates.

A further group of steroids is presumably sulfurylated by rather specific steroid sulfotransferases of quite limited distribution. These are the bile alcohol sulfates found in several lower vertebrates. The structures of those alcohols, and the positions of the sulfate groups in the corresponding bile salts, are shown in Table 2–9. It is quite obvious from their structures that the formation of these most interesting sulfate esters must involve sulfotransferases different from those involved in the synthesis of the sulfate esters of hormonal steroids. Unfortunately no work has been carried out in this field since Bridgwater and Ryan [164] showed that crude preparations from the liver of *Rana temporaria* could sulfurylate the bile alcohol characteristic of this species, ranol, and also scymnol, cholan–24–ol, and 3α,7α,12α,24–tetrahydroxycholane. On the other hand, the rate of formation of dehydroisoandrosterone sulfate by the liver of this species is low [165].

TABLE 2–9. *The structures of the bile alcohols of lower vertebrates*

Alcohol	Source	Structure	Position of sulfurylation
Myxinol[a]	Hagfish	5α–cholestane–3β,7α,16α,26–tetrol	3 and 27
Cyprinol[b]	Carp	5α–cholestane–3α,7α,12α,26,27–pentol	27
Latimerol[c]	*Latimeria*	5α–cholestane–3β,7α,12α,26,27–pentol	27
Ranol[d]	*Rana temporaria*	5α,27–norcholestane–3α,7α,12α,24ξ,26–pentol	24
Chimaerol[e]	*Chimaera monstrosa*	5β–cholestane–3α,7α,12α,24ξ,26–pentol	
Scymnol[f]	Shark	5β–cholestane–3α,7α,12α,24,26,27–hexol	27

[a] Anderson, I. G.. Haslewood, G. A. D., Cross, A. D., Tökés, L.: Biochem. J. **104**, 1061 (1967).
[b] Anderson, I. G., Briggs, T., Haslewood, G. A. D.: Biochem. J. **90**, 303 (1964).
[c] Anderson, I. G., Haslewood, G. A. D.: Biochem. J. **93**, 34 (1964).
[d] Haslewood, G. A. D.: Biochem. J. **90**, 309 (1964).
[e] Bridgwater, R. J., Haslewood, G. A. D., Watt, J. R.: Biochem. J. **87**, 28 (1963).
[f] Bridgwater, R. J., Briggs, T., Haslewood, G. A. D.: Biochem. J. **82**, 285 (1962).

Further work on the sulfotransferases of the livers of the lower vertebrates might well be rewarding both from the standpoint of comparative biochemistry and from specificity of the sulfotransferases.

V. The Enzymatic Hydrolysis of Steroid Sulfates

The enzymatic hydrolysis of the steroid sulfates is a rather more complex problem than the hydrolysis of the steroid glucuronides, because the sulfatases, unlike the glucuronidases, have a rather high degree of specificity toward the steroid part of the molecule. There appear to be, at the very least, separate enzymes responsible for the hydrolysis of the estrogen–3–sulfates, the 3–sulfates of other types of steroid, and the 21–sulfates of corticosteroids. The hydrolysis of those various types of sulfate ester will be considered in turn, but in each case the irreversible reaction can be represented as:

$$R\cdot OSO_3^- + H_2O \rightarrow R\cdot OH + SO_4^{2-} + H^+. \tag{8}$$

A. The Hydrolysis of Estrone Sulfate

Although the possibility of hydrolyzing estrone sulfate by sulfatases has been known since Butenandt and Hofstetter [166] used Takadiastase (a commercial preparation from *Aspergillus oryzae*) for this purpose, there is still remarkably little information on the enzymatic hydrolysis of estrogen sulfates. Nevertheless, the work of Cohen and Bates [167] focused attention on the advantages of using enzymatic methods of hydrolysis in the analysis of urinary estrogens, and, therefore, essentially began the work which culminated in the discovery, and use, of the more specific steroid sulfatases. Interest was further stimulated by the finding by Pulkkinen [168] that human placenta contained large amounts of an enzyme which hydrolyzed estrone sulfate. Recent work on the arylsulfatases [arylsulfate sulfohydrolases, EC 3.1.6.1]—which specifically catalyze the hydrolysis of aryl sulfates, of which estrone sulfate is an example—has shown the existence of many different such enzymes, but for convenience they may be grouped into the type I and type II arylsulfatases [169]. Methods for the assay of those enzymes have been collected [170].

The type I arylsulfatases are characterized by being relatively nonspecific with respect to the structure of the phenolic part of their substrates and by being relatively insensitive to inhibition by sulfate and phosphate ions. They occur in both microorganisms and mammals. Most microbial arylsulfatases seem to be type I enzymes, but the best known example is undoubtedly that of *A. oryzae* which has, unfortunately, not been significantly purified. The only example of this type of sulfatase which has been obtained in a homogeneous form is that from *Aerobacter aerogenes* [171], but the action of this enzyme on estrone sulfate has not been investigated. The type I arylsulfatases of mammalian tissues occur in the microsomes; they have not been obtained in a soluble form, so the study of their properties, and in particular their specificity, is a matter of some difficulty. Although the type I arylsulfatases (sulfatase C) of human [172] and of ox [173] livers have been investigated in some detail, once again their action on estrone sulfate was not studied, and it is only recently that French and Warren [174] have shown the hydrolysis of this ester by the microsomes of human placenta. The pH curve for this hydrolysis of estrone sulfate was bimodal, with optima at pH 6.5 and pH 8.0, and the K_m was approximately 0.1 mM estrone sulfate.

The type II arylsulfatases, in contrast to the type I enzymes, are characterized by being quite powerfully inhibited by sulfate and phosphate ions and, in some cases at least, by showing some specificity toward the phenolic part of their substrates. In particular, the quite unphysiological substrate, nitrocatechol sulfate, is often hydrolyzed much more rapidly than other sulfate esters. Some of the type II arylsulfatases are quite powerfully activated by chloride ions [175, 176]. The type II arylsulfatases are predominantly of animal origin; their most usual sources are the livers of mammals, in which they occur in the lysosomes, and the hepatopancreas (and intestinal juice) of mollusks. Two representatives of this type of sulfatase have been obtained in a homogeneous form—sulfatase A [177] and sulfatase B [178] of ox liver. The properties of these enzymes, together with those of the type I enzyme from *A. aerogenes*, are summarized in Table 2–10. Once again, it is unfortunate that the action of those arylsulfatases from mammalian tissues on

TABLE 2–10. *The physical properties of some arylsulfatases*

	Sulfatase A of ox liver[a]	Sulfatase B of ox liver[b]	Sulfatase of *A. aerogenes*[c]
Molecular weight	109,000[d]	~25,000	40,700
pI	3.6	8.3	—
$E^{1\%}_{280}$	7	14	—
Specific activity (μMoles/min/mg protein)	140	160	73

[a] Nichol, L. W., Roy, A. B.: J. Biochem. (Tokyo) **55**, 643 (1964).
[b] Allen, E., Roy, A. B.: Biochim. Biophys. Acta **168**, 243 (1968).
[c] Fowler, L. R., Rammler, D. H.: Biochemistry **3**, 230 (1964).
[d] At pH 7.5. Below pH 6.0, it forms a tetramer of molecular weight 411,000.

estrone sulfate has not been studied; but it might have been predicted that any action shown by them would be slight compared with that of the type I arylsulfatase of the microsomes. On the other hand, the bimodal pH curve for the hydrolysis of estrone sulfate [174] could well be taken to show the contrary because it is very similar to the curve for the hydrolysis of nitrocatechol sulfate by the microsomes of ox liver, which is due to the activities of both type I and type II arylsulfatases [173]. The hydrolysis of estrone sulfate by the type II arylsulfatases of molluscan tissues is of course a matter of considerable practical interest, because of the presence of these enzymes in preparations of molluscan "β-glucuronidase" and "steroid sulfatase". Jarrige *et al.* [179] showed that such preparations from the intestinal juice of *Helix pomatia* hydrolyzed estrone sulfate at an optimum pH of 5.2 and with a K_m of 0.1 mM estrone sulfate.

Therefore, it is clear that at least some preparations of arylsulfatase can hydrolyze estrone sulfate; and it appears that, in mammalian tissues, the microsomes are the dominant site of this hydrolysis, which suggests that the type I enzyme, sulfatase C, is the enzyme responsible [180], although the possible action of the type II enzymes cannot be ignored. It should be noted, however, that there is some evidence for the existence of a separate estrone sulfatase in animal tissues. This was first suggested by the results of Pulkkinen [168] who showed that, although enzyme preparations from the livers of fetal rats could hydrolyze *p*–nitrophenyl sulfate, a typical substrate for a type I arylsulfatase, they could not hydrolyze estrone sulfate. The recent kinetic studies of French and Warren [174] have similarly led to the suggestion of the existence of an estrone sulfatase distinct from sulfatase C. Clarification of this most important point will come only when—or if—soluble preparations of the microsomal sulfatases are obtained. A further feature which must be stressed is that human fetal liver, unlike rat fetal liver, has considerable estrone sulfatase activity [168], a feature which may be of considerable physiological importance.

It should be quite obvious from this discussion that present knowledge of the enzymes responsible for the hydrolysis of estrone sulfate, either *in vivo* or *in vitro*, is in a rather unsatisfactory state, and more detailed investigations are urgently required.

B. The Hydrolysis of Steroid Alcohol 3–Sulfates

The enzymatic hydrolysis of dehydroisoandrosterone sulfate was first reported by Henry *et al.* [181] who found the long–sought steroid sulfatase in the intestinal juices of *H. pomatia.* This discovery was followed by other reports of similar steroid sulfatases in a number of molluscan species [182–184], and preparations from these animals were soon finding an increasing use in the hydrolysis of urinary steroid sulfates. More recently, quite detailed studies have been made of the distribution of steroid sulfatases in invertebrates, and it is clear that the enzyme is restricted to certain gastropod mollusks [185–187]; the rather sporadic occurrence in these species cannot be related to any obvious biological characteristic. Meanwhile, an apparently similar steroid sulfatase was found in rat liver [188]; it was soon apparent that this enzyme was fairly widely distributed in the livers of mammals although it did not seem to occur in the livers of the lower vertebrates [165, 185, 189]. As the steroid sulfatase from mammalian liver is insoluble, it has not proved as useful a tool as has the molluscan enzyme.

These sulfatases have been detected mainly through the use of dehydroisoandrosterone sulfate as a substrate, and they are commonly called steroid sulfatases, although a better name might be androstenolone sulfatases. The name given by the Enzyme Commission, sterol sulfate sulfohydrolase [EC 3.1.6.2], is unfortunate; not only does the term "sterol" have a significance to steroid chemists that it apparently does not have to enzymologists, but also there are undoubtedly several steroid sulfatases of differing specificity grouped under this one name.

The first androstenolone sulfatase to be purified was the enzyme from the visceral hump of the limpet, *Patella vulgata* [103], but only a fifty-fold concentration was achieved. No separation of the androstenolone sulfatase from the much larger amounts of arylsulfatase was possible, but the separate identities of the two enzymes were clearly demonstrated. The specificity of the androstenolone sulfatase was surprisingly high; it would hydrolyze only the 3β-sulfates of 5α- or Δ^5-steroids of the androstane, pregnane, and cholestane series. In other words, although the androstenolone sulfatase of *P. vulgata* could hydrolyze dehydroisoandrosterone sulfate and isoandrosterone sulfate [3β-sulfooxy-5α-androstan-17-one], it could not hydrolyze androsterone sulfate [3α- and 3β-sulfooxy-5α-androstan-17-one] nor the etiocholanolone sulfate [3α- and 3β-sulfooxy-5β-androstan-17-one]. This immediately put rather stringent limitations on the usefulness of this steroid sulfatase as a general reagent for the hydrolysis of urinary steroid sulfates, limitations which were not lessened by the fact that steroid 17-sulfates and steroid 20-sulfates were likewise resistant to hydrolysis. Rather surprisingly, cortisone-21-sulfate [17α-hydroxy-21-sulfooxypregn-4-ene-3,11,20-trione] was hydrolyzed by the enzyme preparation, but it seemed clear from structural considerations alone that this reaction was catalyzed by a separate enzyme. This rather limited specificity of the steroid sulfatase of *P. vulgata* has since been confirmed by other workers [186].

During their survey of British mollusks, Leon *et al.* [187] noted that three species, the terrestrial *H. pomatia*, and the marine stenoglossids, *Buccinum undatum* and *Nassarius reticulatus*, could hydrolyze etiocholanolone sulfate as well as dehydroisoandrosterone sulfate, although only at a much lower rate. Jarrige [190, 191] has since partially purified the steroid sulfatase(s) of *H. pomatia* and confirmed that this species can indeed hydrolyze the 5β-androstan-17-one sulfates. The specificity of this preparation from *H. pomatia* is shown by the results summarized in Table 2-11. The values of k the relative velocity constants for the breakdown of the enzyme-substrate complex, shown in Table 2-11, might be taken to indicate a very rapid hydrolysis of the 5β-androstan-17-one sulfates; however, these esters have very high K_m values so that, at the substrate concentrations normally used, the relative rate of hydrolysis of the 5β-androstan-17-one sulfates is in fact quite low.

Although the values for the kinetic constants for the hydrolysis of dehydroisoandrosterone sulfate and of the etiocholanolone sulfate are quite different (Table 2-11), these data cannot tell whether the two types of substrate are

TABLE 2–11. *The values of K_m (mM) and the relative values of k (the velocity constant for the breakdown of the enzyme-substrate complex) for the hydrolysis of some steroid sulfates by an enzyme preparation from the intestinal juice of Helix pomatia*

Substrate—the sulfate of	K_m	k_{rel}
3α–Hydroxy–5α–androstan–17–one	—	0.00
3β–Hydroxy–5α–androstan–17–one	0.032	0.74
3α–Hydroxy–5β–androstan–17–one	3.07	0.46
3β–Hydroxy–5β–androstan–17–one	6.35	0.44
3β–Hydroxyandrost–5–en–17–one	0.028	0.57
17β–Hydroxyandrost–4–en–3–one	—	0.00
3α–Hydroxy–5α–pregnan–20–one	—	0.00
3β–Hydroxy–5α–pregnan–20–one	0.023	1.00
3α–Hydroxy–5β–pregnan–20–one	—	[a]
3β–Hydroxy–5β–pregnan–20–one	—	[a]
3β–Hydroxypregn–5–en–20–one	0.020	0.89

[a] Slow hydrolysis.

hydrolyzed by the one enzyme. Consideration of the very different shapes of the 5α–androstan–17–one and 5β–androstan–17–one derivatives alone would suggest the participation of more than one enzyme, and this view is certainly supported by the results of comparative studies of the distribution of the two activities [186, 187]. On the other hand, Jarrige *et. al* [179] imply that only one sulfatase is involved in the hydrolysis of both types of substrate, although some of their own kinetic data do not support this. In particular, etiocholanolone sulfate was shown to be a competitive inhibitor of the hydrolysis of dehydroisoandrosterone sulfate with a K_i of 0.09 mM while the K_m for the hydrolysis of etiocholanolone sulfate by the same enzyme preparations was 3.1 mM. It is hard to avoid the conclusion that these two constants must refer to different enzymes.

The most likely interpretation of the results available at present would seem to be that there must exist in some mollusks two steroid sulfatases: androstenolone sulfatase, hydrolyzing the 3β–sulfates of 5α– and Δ^5–steroids, and etiocholanolone sulfatase, hydrolyzing the 3α– and 3β–sulfates of 5β–androstan–17–one and perhaps of 5β–pregnane. The former enzyme is rather sporadically distributed among the mollusca, but the latter has been found in only three species, of which *H. pomatia* is by far the most important. Once again, only further work can resolve this problem, and it is to be hoped that renewed attempts to separate the steroid sulfatases of *H. pomatia* will be made; it is obvious, however, from the work of Jarrige

[190] that the separation of the various enzymes will not be easy. Whatever the outcome of such further studies may be, it should be obvious that different preparations of molluscan steroid sulfatases must not be used indiscriminately; quite different results could be obtained depending upon whether the enzyme were of the *Patella* type or the *Helix* type.

The steroid sulfatase of mammalian tissues is localized in the microsomes from which it has not been possible to obtain soluble preparations [104, 165]; therefore the question of its separate identity is somewhat problematical, but indirect evidence certainly seems to show that the steroid sulfatase is quite distinct from the associated arylsulfatase [104, 174]. Although originally detected only in liver, steroid sulfatase activity has since been found in the adrenals, ovaries, and testes of several species [192, 193] and also in human placenta [168]. The latter tissue is by far the richest source of steroid sulfatase known, the normal placenta being about four times as active as liver, while the placentas of anencephalic fetuses are several times more active still [193]. In rather sharp contrast to the placenta, both human and rat fetal tissues are devoid of steroid sulfatase activity [168, 194] although such activity develops soon after birth [168]. With respect to the specificity of the steroid sulfatase of mammalian tissues, little need be said; it is an androstenolone sulfatase quite analogous to that of *P. vulgata* in that only the 3β–sulfates of the 5α– and Δ^5–series of steroids are hydrolyzed [104]. No hydrolysis of the 5β–androstan–17–one sulfates, steroid 16α–sulfates, steroid 17–sulfates, nor steroid 20–sulfates has been detected and, in sharp contrast to the limpet enzyme, the mammalian enzyme will not hydrolyze cortisone–21–sulfate.

The general enzymatic properties of the steroid sulfatases, both of molluscan and of mammalian tissues, are summarized in Table 2–12. These properties show that the enzymes are quite typical sulfatases, which, unfortunately, means that no very effective specific inhibitors are available, apart from phosphate ions in the case of the molluscan enzyme. This lack of a specific inhibitor of the sulfatases, comparable in efficiency to glucarolactone as an inhibitor of β–glucuronidase, is unfortunate, because it makes difficult adequate verification that the hydrolysis of a steroid conjugate by any given enzyme preparation is in fact due to the sulfatases therein.

C. The Hydrolysis of Cortisone–21–Sulfate

As already pointed out, steroid sulfatase preparations from mollusks, but not from mammalian liver, can hydrolyze cortisone–21–sulfate. Roy [103] suggested that, in preparations from *P. vulgata*, this hydrolysis was not due to androstenolone sulfatase but to a separate enzyme which might be

TABLE 2–12. *The properties of some androstenolone sulfatases*

	Enzymes from					
	Helix pomatia	*Patella vulgata*	Ox liver	Human placenta	Rat liver	Rat testes
Optimum pH	4.5–5.0	5.0	7.7–7.9	7.5	6.5–7.5	6.8–7.7
Optimum substrate concentration	0.6	0.2	0.3			
K_m (mM)	0.028	0.04	0.04	0.04	0.01	0.012
Inhibition by						
Cl^-	±	±	−			
SO_3^{2-}	+++	—	+++			
SO_4^{2-}	++	++	−			
HPO_4^{2-}	+++	+++	+			

identical to glucosulfatase. Dodgson [195] later showed that the participation of glucosulfatase was most unlikely, but also concluded that the results of "competing substrate" experiments with an enzyme preparation from the periwinkle, *Littorina littorea*, although inconclusive, were not inconsistent with cortisone–21–sulfate and dehydroisoandrosterone sulfate being hydrolyzed by separate enzymes. Jarrige [191], on the other hand, was unable to separate the cortisone sulfatase and androstenolone sulfatase activities in a preparation from *H. pomatia*. Nevertheless, examination of the elution pattern of the two activities from DEAE–cellulose shown by Jarrige strongly suggests that there are indeed two separate enzymes, as is also indicated, although rather less convincingly, by inhibition studies [190]. The conclusion therefore seems inescapable that cortisone–21–sulfate must be hydrolyzed by yet another steroid sulfatase which may conveniently be called cortisone sulfatase. Again, only further work on the purification of the enzyme can give a definite answer.

Obviously, little can be said of the properties of the presumed cortisone sulfatase, but Jarrige *et al.* [179] have shown that it is presumably not absolutely specific, because the enzyme preparation from *H. pomatia* could hydrolyze not only cortisone–21–sulfate but also deoxycorticosterone–21–sulfate [21–sulfooxypregn–4–ene–3,20–dione] and 17α–hydroxydeoxycorticosterone–21–sulfate [17α–hydroxy–21–sulfooxypregn–4–ene–3,20–dione] at quite comparable rates. The pH optimum for the hydrolysis of cortisone–21–sulfate has been given as 4.7 [179] and 5.3 [103], and the K_m is approximately 0.2 mM cortisone–21–sulfate.

As stated above, attempts to detect the hydrolysis of cortisone–21–sulfate by preparations of mammalian steroid sulfatase have been unsuccessful. Nevertheless, Dodgson *et al.* [196] have shown that, when cortisone–21–[^{35}S]–sulfate is administered to rats, a very high proportion of the radioactivity appears as inorganic sulfate ions in the urine within 24 hours. Although it is likely that this is not a consequence of a direct hydrolysis of cortisone–21–sulfate, the liberation of sulfate ions is interesting and perhaps suggests that mammalian tissues do contain an enzyme related to the cortisone sulfatase of molluscs.

D. The Function of the Steroid Sulfatases

In light of present ideas on the importance of sulfate esters in the metabolism of steroids following the demonstration of "direct" pathways for the conversion of cholesteryl sulfate to hormonal steroids, there is little difficulty in visualizing appropriate functions for the steroid sulfatases of mammals. At least in theory, and probably in practice, the steroid sulfatases could function as a means of liberating hormonal steroids, or hormone precursors, from the conjugated form in which they occur in blood. This provides an obvious explanation for the high levels of steroid sulfatase found in human placenta, although perhaps not such an obvious one for the enzyme in liver.

In mollusks, however, the situation is quite different. First, these animals appear to contain a more impressive battery of steroid sulfatases than do mammalian tissues; and, second, the steroid sulfatases of the mollusk appear to be digestive enzymes insofar as they are secreted into the digestive juices from the hepatopancreas (digestive gland). What the substrates for those extracellular steroid sulfatases can be is quite unknown, and one is left with the suspicion that some quite unsuspected function must exist for these enzymes. Certainly mollusks contain rather exotic sterols [197] whose sulfate esters would be potential substrates for androstenolone sulfatase, but it is hard to believe that such can be the explanation for the existence of these enzymes even if the sterols were formed from cholesterol through a "sulfate" pathway. Therefore, it can be seen that, in the molluscan steroid sulfatases, we have a group of enzymes looking for a substrate, and a function—unless one accepts that it has evolved solely for the convenience of steroid chemists. Perhaps this search for substrates is nearing an end now that a number of C_{18}, C_{19} and C_{21} steroids have been isolated from the tissues of the slug *Arion ater* [198, 199]: although this species has not been shown to contain steroid sulfatases its near relative *A. empiricorum* certainly has [185].

VI. The Synthesis and Hydrolysis of *N*-Acetylglucosaminides

A. The Biosynthesis of Acetylglucosaminides

Little information is available on the formation of acetylglucosaminides *in vivo* or *in vitro*, but this can be presumed to take place by the transfer of the *N*–acetyl–D–glucosaminyl residue from uridine diphosphate *N*–acetylglucosamine in a reaction analogous to that which occurs in the synthesis of glucuronides. Certainly Jirku and Layne [4] were able to show the production of 17α–estradiol–3–glucuronide–17–β–*N*–acetylglucosaminide when 17β–estradiol was incubated in a rabbit liver homogenate fortified with uridine diphosphate glucuronic acid and uridine diphosphate *N*–acetylglucosamine. Nothing is known of the transferase which must be involved in this reaction, but it is interesting that it has been suggested [4] that its substrate may not be 17β–estradiol itself but rather 17β–estradiol–3–glucuronide because no evidence was found for the formation of 17α–estradiol–17–*N*–acetylglucosaminide. If it be accepted that the structure of the double conjugate is indeed that proposed by Layne [200] and that the glucosamine is present in the β–configuration, then the stereochemistry of its synthesis is quite analogous to that of β–glucuronides: in uridine diphosphate *N*–acetylglucosamine the carbohydrate is, on analogy with other similar nucleotides, presumably present in the α configuration, and during the transfer of the glycosyl residue, inversion must occur (see Section II–C). On the other hand, it must be admitted that the evidence for a β glycosidic linkage in the estradiol acetylglucosaminide is not entirely satisfactory because it is based solely on the ability of emulsion to hydrolyze the conjugate (see Section VI–B)*.

The situation is complicated by the apparently unambiguous characterization of an α–glycosidic linkage in the pregnenediol–3–sulfate–20–α*N*–acetylglucosaminide which is present in human urine [5]. If the synthesis of this glycoside proceeds from the same uridine diphosphate *N*–acetylglucosamine, then it must either occur directly with no inversion of the glycosidic linkage or indirectly by some more complex mechanism. This particular aspect of the biosynthesis of acetylglucosaminides is certainly in need of investigation and some most interesting, and fundamental, problems appear to be involved. If the assigned configurations of both of the steroid acetylglucosaminides which have so far been isolated are correct, then there must be two different synthetic pathways involved. If, on the other hand, only the structure of the pregnenediol derivative has been correctly assigned,

* Collins, Williams and Layne [Arch. Biochem. Biophys. **121**, 609 (1967)] have confirmed the β–glycosidic linkage by ORD measurements.

then its synthesis must take a rather unexpected stereochemical course. (See footnote p. 75.)

The general mode of formation of uridine diphosphate *N*-acetylglucosamine is clearly understood although details are still lacking. Strominger and Smith [201] partially purified from calf liver a UDP-glucosamine pyrophosphorylase (UTP: 2-acetamido-2-deoxy-α-D-glucose-1-phosphate uridylyltransferase, EC 2.7.7.23) which catalyzed the reversible reaction

UTP + *N*-acetylglucosamine 1-phosphate
⇅
UDP-*N*-acetylglucosamine + PP_i.

The enzyme was rather unstable, even at −20°, and was purified only about 20-fold in a yield of some 10% from an extract of acetone-dried liver. Like the corresponding reaction with glucose (see Section II-A), this reaction has been studied mainly in the direction of pyrophosphorolysis, for which the equilibrium constant was about 0.5, the pH optimum was 7.2, and the values of the K_m for uridine diphosphate *N*-acetylglucosamine and for pyrophosphate were 0.26 and 2.0 mM, respectively. As expected, Mg^{2+} ions were required as cofactors. Smith and Mills [202] claimed that this pyrophosphorolysis occurred in only the nuclear fraction of guinea pig liver, as had been previously reported for rat liver [203]. On the other hand, Maley *et al.* [204] were unable to show the synthesis of uridine diphosphate *N*-acetylglucosamine by rat liver nuclei although they could detect it in the soluble fraction of rat liver [205].

Maley *et al.* have also shown that the nuclear fraction [204] and the soluble fraction [206] of rat liver can catalyze the related reaction

UTP + glucosamine 1-phosphate
⇅
UDP-glucosamine + PP_i

but whether or not a separate enzyme is involved here is still not clear. Indeed, the simple aminosugar could conceivably be a substrate for UDP-glucose pyrophosphorylase [207] rather than for the so-called UDP-glucosamine pyrophosphorylase for which a better trivial name would seem to be UDP-acetylglucosamine pyrophosphorylase.

B. α-Acetylglucosaminidase

This enzyme has been little studied and it does not appear in the Enzyme Commission list. Roseman and Dorfman [208] were the first to show the presence of an α-acetylglucosaminidase in higher animals by detecting it in preparations of testicular hyaluronidase. The two activities were, however, obviously due to different enzymes because they could be separated. α–Acetylglucosaminidase was also shown to be present in the testes, spleen, liver, kidney, and lung of rats and rabbits.

More recently, α–acetylglucosaminidase has been partially purified (some 80–fold) from pig liver to give, in rather a poor yield, a preparation devoid of β–acetylglucosaminidase and β–glucuronidase activities [209]. The enzyme was shown to be present in all rat tissues examined and, at least in liver, to be a typical lysosomal acid hydrolase.

Little can be said of the general properties or the specificity of the enzyme. It is rather unstable below pH 5.0, but in citrate buffers the pH optimum for the hydrolysis of phenyl *N*–acetyl–α–glucosaminide was 4.6 and for the *p*–nitro derivative, 4.7. The enzymatic activity was considerably depressed in acetate buffers. The above two substrates, of which the latter is preferred for routine assay, had K_m values of 1.4 and 0.30 mM, respectively. *N*–Acetylglucosamine, *N*–acetylmannosamine, and *N*–acetylglucosaminonolactone were inhibitors of α–acetylglucosaminidase, with K_i values of 0.72, 0.1 and 0.25 mM, respectively. Although *N*–acetylmannosamine was quite a powerful inhibitor of α–acetylglucosaminidase, the latter cannot hydrolyze either α–or β–acetylmannosaminides. α–Acetylglucosaminidase is quite distinct from α–galactosaminidase [210], in sharp contrast to β–acetylglucosaminidase which appears to be identical with β–acetylgalactosaminidase (see Section VI–C).

α–Acetylglucosaminidase occurs in the digestive juice of *Helix pomatia* [211] and the visceral hump of *Patella vulgata* [6]. It was long ago reported to be "practically" absent from emulsin [212] but no figures were given and this statement ("Auf α–Phenyl–*N*–acetyl–D–glucosaminid sind unsere chromatographierten Emulsin-Fraktionen sowie das gesamte Merksche Emulsin praktisch unwirksam") does not appear to have been confirmed using the more sensitive techniques now available. Whether or not α–acetylglucosaminidase is absent from emulsin is of considerable importance, because the assignment of the β configuration to the estradiol acetylglucosaminide of rabbit urine is based only on the hydrolysis of the conjugate by large amounts of emulsin [3, 200]. If this preparation did in fact contain significant amounts of α–acetylglucosaminidase, then the interpretation of the results would be in some doubt. (See footnote p. 75.)

C. β–Acetylglucosaminidase

β–Acetylaminodeoxyglucosidase [β–2–acetylamino–2–deoxy–D–glucoside acetylaminodeoxyglucohydrolase, EC 3.2.1.30] is rather better known than the enzyme considered in the previous section although there is still relatively little information on purified preparations. Its occurrence in emulsin has been clear for many years, and Helfrich and Iloff [213] showed that β–acetylglucosaminidase was quite distinct from β–glucosidase. Watanabe [214] subsequently showed the presence of β–acetylglucosaminidase in cattle tissues, and Neuberger and Pitt Rivers [215] showed that it occurred in the intestinal juices of *H. pomatia*. Much more recently Conchie and Levvy [6] found β–acetylglucosaminidase in the visceral hump of the limpet *P. vulgata*, although only in amounts about 100–fold less than β–glucuronidase.

The work of Neuberger and Pitt Rivers [215] is important because it still remains the most recent investigation of the general specificity of the enzyme. According to these authors, the enzyme is not a general β–acylglucosaminidase as it will hydrolyze only *N*–acetyl– or *N*–formyl–β–D–glucosaminides. It will not attack simple β–glucosaminides, in which the amino group is free, nor β–acetylglucosaminides in which the hydroxyl groups at positions 3, 4, and 6 of the sugar are acetylated or methylated.

Recent studies on this enzyme stem from those of Pugh, Leaback, and Walker [216] on the β–acetylglucosaminidase of rat kidney, and were greatly stimulated by the work of Conchie, Findlay, and Levvy [217] which showed that the epididymis of rat or of mouse was a very rich source of the glycosidase. The even richer source of pig epididymis was utilized by Findlay and Levvy [218] in their preparation of the enzyme. This tissue is ten times richer in β–acetylglucosaminidase than is rat epididymis, and only a 70–fold purification was required to give a preparation with a specific activity of 2,900,000 (μg of *p*–nitrophenol liberated from the glucoside in 1 hr per mg of protein). By selecting only the portion in the caput of the epididymis richest in the enzyme, an even more active preparation could be obtained. No data were given on the presence or absence of possible contaminating enzymes except that *p*–nitrophenyl *N*–acetyl–β–D–galactosaminide was also hydrolyzed by the preparation. The purified enzyme was quite stable when stored at 0° in neutral solution but its activity was rapidly destroyed at pH values much below 4. Weissmann, Hadjiioannou, and Tornheim [219] used the much less rich source of beef liver to prepare β–acetylglucosaminidase, but after a 4,000–fold purification the final material had only about one third of the specific activity of that from pig epididymis. The liver enzyme also showed β–acetylgalactosaminidase activity but was devoid of α–acetylglucosaminidase activity. As the final yield from liver was only about 3,000 units/g tissue compared with about 300,000 units/g from pig epididymis, it is obvious that the latter tissue is the one of choice for the preparation of the enzyme and should find an increasing use in steroid enzymology. Unfortunately, it is not clear whether or not either of these preparations shows β-glucuronidase activity. However, pig tissues are known [57] to be remarkably poor sources of this enzyme, so that the β-acetylglucosaminidase from pig epididymis is unlikely to show much β-glucuronidase activity. Recently, preparations of β-acetylglucosaminidase from ram testis [220] and from bovine uterus [221] have been described, but the use of these sources seems to offer no advantages.

One interesting problem in connection with the enzyme is its specificity toward the sugar moiety of the substrate; β-acetylglucosaminidase activity always seems to be associated with β-acetylgalactosaminidase activity [222], although not in any constant ratio. Even highly purified preparations of β-acetylglucosaminidase show β-acetylgalactosaminidase activity, and it is of obvious interest whether or not only a single enzyme is involved. This problem has been extensively studied by Walker who, from detailed investigations of the kinetics of the reaction with both types of substrate, concluded that β-acetylglucosaminidase of ram testis [223] and of rat kidney [224] could indeed hydrolyze either *N*-acetyl-β-glucosaminides or *N*-acetyl-β-galactosaminides. As already pointed out, this situation is quite different from that existing for the α series of glycosides where two separate glycosidases are involved.

Again little can be said of the kinetic properties of the enzymes and certainly nothing about their behavior toward steroid glycosides. Like β-glucuronidase, β-acetylglucosaminidase is unstable in dilute solution, so that assays are normally carried out in the presence of 0.01% bovine serum albumin, and, further, the reaction mixture is frequently made 0.1 M in NaCl. The latter does not protect the enzyme but seems to increase its activity by about 10%. All studies have shown that the pH optimum for β-acetylglucosaminidases from mammalian sources and from emulsin is in the region of 4.2 to 4.6 in citrate buffers. A considerable decrease in activity occurs in acetate buffers, un-ionized acetic acid probably being an inhibitor of the enzyme [216]. The pH optimum of the enzyme from *H. pomatia* has been reported [215] to be at pH 3.6 where the mammalian enzymes are quite unstable [218]. The specificity has not been exhaustively investigated but both alkyl and aryl glycosides are hydrolyzed [225]; phenyl and *p*-nitrophenyl *N*-acetyl-β-D-glucosaminides are commonly used as substrates, the latter having obvious practical advantages as well as being more rapidly hydrolyzed. A fluorimetric assay has been described [226] with methylumbelliferone *N*-acetyl-β-glucosaminide as substrate but apart from its greater sensitivity this method seems to have no advantages. Values for the K_m cluster around 3 mM phenyl *N*-acetylglucosaminide and 1.5 mM *p*-nitrophenyl *N*-acetylglucosaminide [216, 218, 219, 223, 225];

for the two corresponding *N*–acetylgalactosaminides, the values are 1.2 and 0.28 mM, respectively [225]. As already mentioned, acetic acid is an inhibitor of the enzyme, as is also acetamide, but by far the most important inhibitors are *N*–acetylglucosaminonolactone and *N*–acetylgalactosaminonolactone (2–acetamido–2–deoxygluconolactone and 2–acetamido–2–deoxygalactonolactone, respectively) [227]. The results with these lactones are complicated by the fact that several molecular species are present in their solutions, the composition of which may alter considerably with pH, so that it is difficult to give real values for the K_i of these compounds. Values of 0.5 μM *N*–acetylglucosaminonolactone and 2 μM *N*–acetylgalactosaminonolactone have been reported [218] for the purified enzyme from pig epididymis, and similar figures have been given for the enzyme of rat kidney [224] and beef liver [219]. The corresponding values for the enzyme from *P. vulgata* are much higher, being 4 μM *N*–acetylglucosaminonolactone and 30 μM *N*–acetylgalactosaminonolactone, respectively. Levvy and Conchie [228] have recently given details of the preparation of *N*–acetylglucosaminonolactone; the product is a mixture of 1→4 and 1→5 lactones which gives a 50% inhibition of the *β*–acetylglucosaminidase from rat epididymis at a concentration of 2 μM. It is of interest that α–acetylglucosaminidase is, as already noted above in Section VI–B, also inhibited by *N*–acetylglucosaminonolactone, but only much less powerfully than the *β*–acetylglucosaminidases from either mammalian [209, 227] or molluscan [227] sources.

Of the chemical properties of the enzyme little indeed can be said except that, to judge by its behavior on Sephadex, the molecular weight of *β*–acetylglucosaminidase must be about 150,000 [229].

β–Acetylglucosaminidase is a lysosomal enzyme which is found in the mitochondrial and microsomal fractions of the usual types of homogenate [230]. Few studies have been made of the behavior of *β*–acetylglucosaminidase *in vivo* apart from the very comprehensive studies made by Levvy's group which have shown that this enzyme, like several other glycosidases, closely parallels *β*–glucuronidase in its changes in tumours [231] and in normal tissues following hormonal and other treatment [72].

VII. The Enzymatic Hydrolysis of Urinary Steroid Conjugates

One of the main stimuli to the investigations of the hydrolytic enzymes which have been considered in the previous sections has been their potential value in steroid methodology, particularly as reagents for the hydrolysis of the conjugated steroids in urine. This technique is obviously much gentler and more specific than the alternative of acid hydrolysis, but there are some potential sources of error which must be kept in mind when hydrolases are being used for this purpose. Among the more important of these are:

(1) The purity of the enzyme preparation;
(2) The specificity of the hydrolase;
(3) The interference caused by competing substrates;
(4) The presence of inhibitors in biological fluids.

Each of these points may conveniently be considered in turn.

Except when the desired aim is the quantitative liberation of a steroid from all its conjugated forms, whatever their chemical nature may be, the advantages of using homogeneous enzyme preparations should not, it would be thought, require to be stressed; the dangers of using crude enzyme preparations should be obvious. As an example of these dangers, it need only be pointed out that, if the preparations of β-glucuronidase which were used in the identification of the 17β-estradiol conjugate occurring in rabbit urine had been of molluscan rather than mammalian origin, then the presence of the glucosamide link in the conjugate would have been difficult, if not impossible, to detect. When pure enzymes are not available (as, for instance, is unfortunately the case with the steroid sulfatases) then a knowledge of the contaminating enzymes in any given preparation is essential. When using such impure preparations the technique is sometimes used of adding an inhibitor to suppress the action of a known enzyme and so to "prove" that the hydrolyzed conjugate does indeed have the suspected structure. This approach has been used most frequently to "prove" or to "disprove" the identity of a given conjugate with a glucuronide by making use of the powerful and specific inhibitor of β-glucuronidase, glucarolactone (saccharo-1,4-lactone). Even this procedure is not without its dangers; a definite inhibition of the hydrolysis can be taken as satisfactory evidence for the presence of a glucuronide, but an apparent lack of inhibition is an ambiguous result in the absence of more detailed studies of the hydrolysis than are usually available. This ambiguity arises from the fact that hydrolyses of urinary steroid conjugates are normally carried out with much higher concentrations of enzyme and for much longer times than are used in purely enzymological investigations. For instance, for the hydrolysis of steroid glucuronides in urine, β-glucuronidase is often used in concentrations of 1,000 units/ml and for times up to 72 hours, whereas in the usual kinetic investigations, the enzyme concentration is perhaps 20 units/ml and the time of hydrolysis only 1 hour. An apparent lack of inhibition by glucaro-1,4-lactone in the former could be due simply to the very different conditions, and primarily to the fact that the *initial* rates of hydrolysis were not measured, only the extent of the hydrolysis in a given time being determined; if the time were long, then essentially complete hydrolysis could occur even if the enzyme were appreciably inhibited. A reduction of the enzyme concentration and the use of shorter hydrolysis times might clearly show the expected inhibition.

Now that steroid *N*-acetylglucosaminides have been shown to be of natural occurrence, even greater care should be taken when using unpurified enzymes, even in the presence of conventional inhibitors, because both α- and β-acetylglucosaminidases will be present in extracts of molluscan viscera, and at least the latter is likely to be found in simple extracts of mammalian liver. If such a molluscan preparation were found to hydrolyze a steroid conjugate when amounts of glucaro-1,4-lactone sufficient to inhibit β-glucuronidase were present, then this would now be taken as a strong indication that the steroid conjugate was in fact a sulfate. Such an interpretation could be grossly in error because glucaro-1,4-lactone does not inhibit *N*-acetylglucosaminidases so that the conjugate could equally be an acetylglucosaminide. In principle, these two possibilities could be distinguished by testing for hydrolysis in the presence of *N*-acetylglucosaminonolactone which inhibits *N*-acetylglucosaminidases, although only rather inefficiently in the case of α-acetylglucosaminidase. Perhaps it is not too much to suggest that an "acetylglucosaminonolactone control" should routinely be added to the "glucarolactone control" at present used in steroid methodology.

The specificity of the hydrolases is probably only a problem with the sulfatases; the β-glucuronidases seem to be rather nonspecific although the reports of steroid glucuronides resistant to hydrolysis by β-glucuronidase must be remembered. With the steroid sulfatases, however, it is commonly stated that some steroid sulfates—for instance, androsterone sulfate or testosterone sulfate [17β-sulfooxyandrost-4-en-3-one]—are resistant to enzymatic hydrolysis. Again, some caution is necessary because of the extreme conditions under which the enzymes are often used. Consider the hydrolysis of conjugated steroids in urine by a preparation of the intestinal juice of *Helix pomatia*. Suppose that the concentration of β-glucuronidase were 1,000 units/ml, then the steroid sulfatase concentration might be of the order of 10 units/ml; in theory, therefore, the incubation of 1 liter of urine for 24 hours under such conditions could result in the liberation of 240 mg of dehydroisoandrosterone from its sulfate. If there were present another steroid sulfate which was hydrolyzed at only 0.1% of the rate of dehydroisoandrosterone sulfate, then in 24 hours 240 μg of this second steroid would be liberated. This is far from an insignificant amount when modern methods of steroid analysis are applied, yet none of the kinetic studies so far carried out with androstenolone sulfatase could have detected the hydrolysis of a steroid sulfate reacting at only 0.1% of the rate of dehydroisoandrosterone sulfate. The extrapolation of specificity data from the experiments of the enzymologist to the more drastic conditions used in steroid analysis is therefore a procedure fraught with danger.

The problem of competing substrates is, on the other hand, probably less important than is often thought. Again, this arises from the peculiar conditions of the reaction as carried out in, say, urine. If Michaelis–type kinetics be assumed for the hydrolysis, then the initial velocity (a parameter of little practical interest to the steroid analyst) is given by the expression:

$$v = \frac{Vs}{K_m + s}.$$

and in the presence of a competing substrate at a concentration of s' and with a K_m of K'_m, the corresponding equation for the initial velocity becomes:

$$v_i = \frac{Vs}{(1 + s'/K'_m)\, K_m + s}.$$

In the presence of n such competing substrates, the rate of hydrolysis of the first substrate is given by:

$$v_i = \frac{Vs}{[\Pi(1 + s'/K'_m)]\, K_m + s}.$$

Now if the conditions are such that the concentrations of *all* the substrates are less than the corresponding values of K_m, then the last equation simplifies to:

$$v_i = \frac{V}{K_m} s$$

which is identical to that for the uninhibited reaction, showing normal Michaelis–type kinetics, when $s < K_m$. In other words, no inhibition by competing substrates can occur under such conditions. Consider now androstenolone sulfatase. The data summarized in Table 2–12 show that the values of K_m for most substrates are about 0.02 mM, which corresponds to steroid concentrations of about 6 mg/liter, a figure which is probably much greater than the normal concentrations of most of those substrates in urine (*e.g.*, Fotherby [232] reported that the mean daily excretion of dehydroisoandrosterone was 1.2 and 0.5 mg in males and females respectively, corresponding to a concentration of considerably less than 1 mg/liter). In other words, under the conditions existing during the hydrolysis of steroid sulfates in urine, inhibition by competing substrates may be of little importance. If, however, the conjugate fraction be concentrated before being subjected to enzymatic hydrolysis, then the condition that $s < K_m$ may no longer hold, and the inhibition by competing substrates will become important, as is obvious from a consideration of the appropriate equations given above. That the magnitude of "competing substrate" effects is dependent upon the values of the ratios s/K_m has not been stressed and the fact that very different results may be obtained depending on whether or not the conjugate fraction has been concentrated does not appear to have been noted.

The final problem, that of the presence of inhibitors in many biological fluids, is probably of little importance because the effects can always be minimized by increasing the concentration of the enzyme. Inhibitors certainly do occur in urine, sometimes in quite large amounts. For example, glucaric acid, which readily lactonizes to the much more powerful inhibitor, glucaro-1,4-lactone, occurs in urine in concentrations such that β-glucuronidase is inhibited by some 85% in a reaction mixture containing 25% of human pregnancy urine. Urine also normally contains not only quite high concentrations of phosphate and sulfate ions, which are inhibitors of the type II arylsulfatases and of the molluscan steroid sulfatases, but also compounds such as androsterone sulfate which are specific, although weak, inhibitors of androstenolone sulfatase. Nevertheless, the effects of these inhibitors can be eliminated either by increasing the enzyme concentration or by partially purifying the conjugate fraction prior to its hydrolysis; the latter procedure of course only brings more complex difficulties in its train.

Despite these potential dangers, there is no doubt that the enzymatic hydrolysis of steroid conjugates is a most valuable technique, and the above comments are only a plea that the technique be used more critically than sometimes is the case. In particular, a greater use should be made of homogeneous preparations of β-glucuronidase, and further attempts must be made to obtain more highly purified samples of the steroid sulfatases. The availability of homogeneous steroid sulfatases of well-defined specificity would pay dividends to the steroid chemist, quite apart from the intrinsic interest of such enzymes to the enzymologist.

Finally, it may be worth noting that the conjugating systems have been little used in steroid chemistry, presumably because the steroid sulfates and steroid glucuronides are quite readily accessible by chemical methods. Nevertheless, enzymatic methods might be useful for the preparation of conjugates, and particularly of double conjugates, of labile steroids; again, for this purpose it would be useful to have available much purer preparations of the sulfotransferases and transglucuronylases than are available at present.

VIII. References

1. Oertel, G.W., Eik-Nes, K. B.: Acta Endocrinol. **28**, 293 (1958).
2. — Groot, K.: Z. Physiol. Chem. **341**, 1 (1965).
3. Layne, D. S., Sheth, N. A., Kirdani, R. Y.: J. Biol. Chem. **239**, 3221 (1964).
4. Jirku, H., Layne, D. S.: Biochemistry **4**, 2126 (1965).
5. Arcos, M., Lieberman, S.: Biochemistry **6**, 2032 (1967).
6. Conchie, J., Levvy, G. A.: Biochem. J. **65**, 389 (1957).
7. Noguchi, S., Fukushima, D. K.: J. Biol. Chem. **241**, 761 (1966).
8. Fukushima, D. K., Noguchi, S., Bradlow, H. L., Zumoff, B., Kozuma, K., Hellman, L., Gallagher, T. F.: J. Biol. Chem. **241**, 5336 (1966).
9. Kornel, L., Fujiki, T.: Abstr. 39, 48th Meeting of the Endocrine Soc., Chicago, Ill., June 1966, p. 46.
10. Hecker, E., Walter, G., Marks, F.: Biochim. Biophys. Acta **111**, 546 (1965).
11. Crépy, O.: Arch. Sci. Physiol. **1**, 427 (1947).
12. Dutton, G. J.: Biochem. J. **64**, 693 (1956).
13. — in: Glucuronic acid, free and combined (G. J. Dutton, ed.), p. 185. New York, N. Y.: Academic Press 1966.
14. Munch-Petersen, A., Kalckar, H. M., Cutolo, E., Smith, E. E. B.: Nature **172**, 1036 (1953).
15. Mills, G. T., Ondarza, R., Smith, E. E. B.: Biochim. Biophys. Acta **14**, 159 (1954).
16. Smith, E. E. B., Mills, G. T.: Biochim. Biophys. Acta **18**, 152 (1955).
17. Basu, D. K., Bachhawat, B. K.: J. Neurochem. **7**, 174 (1961).
18. Ginsburg, V.: J. Biol. Chem. **232**, 55 (1958).
19. Munch-Petersen, A.: Acta Chem. Scand. **9**, 1523 (1955).
20. Turner, D. H., Turner, J. F.: Biochem. J. **69**, 448 (1958).
21. Munch-Petersen, A.: Acta Chem. Scand. **11**, 1079 (1957).
22. Strominger, J. L., Maxwell, E. S., Axelrod, J., Kalckar, H. M.: J. Biol. Chem. **224**, 79 (1957).
23. Ballard, F. J., Oliver, I. T.: Biochim. Biophys. Acta **71**, 578 (1963).
24. Wilson, D.: Anal. Biochem. **10**, 472 (1965).
25. Maxwell, E. S., Kalckar, H. M., Strominger, J. L.: Arch. Biochem. Biophys. **65**, 2 (1956).
26. Neufeld, E. F., Hall, C. W.: Biochem. Biophys. Res. Commun. **19**, 456 (1965).
27. Salitis, G., Oliver, I. T.: Biochim. Biophys. Acta **81**, 55 (1964).
28. Silbert, J. E., Hughes, E. F. X.: Biochim. Biophys. Acta **83**, 355 (1964).
29. Goldberg, N. D., Dahl, J. L., Parks, R. E., Jr.: J. Biol. Chem. **238**, 3109 (1963).
30. Krakow, G., Ludowieg, J., Mather, J. H., Normore, W. M., Tosi, L., Udaka, S., Vennesland, B.: Biochemistry **2**, 1009 (1963).
31. Strominger, J. L., Mapson, L. W.: Biochem. J. **66**, 567 (1957).
32. Dutton, G. J., Storey, I. D. E.: Biochem. J. **57**, 275 (1954).
33. Hartiala, K. J. V.: Ann. Med. Exptl. Biol. Fenniae (Helsinki) **33**, 239 (1955).
34. Lehtinen, A., Hartiala, K., Nurmikko, V.: Acta Chem. Scand. **12**, 1589 (1958).
35. Stevenson, I. H., Dutton, G. J.: Biochem. J. **82**, 330 (1962).
36. Dutton, G. J.: Biochem. J. **71**, 141 (1959).

37. Dutton, G. J.: Biochem. Pharmacol. **6**, 65 (1961).
38. — Storey, I. D. E, in: Methods in enzymology, vol. 5 (S. P. Colowick and N. O. Kaplan, eds.), p. 159. New York, N. Y.: Academic Press 1962.
39. Arias, I. M.: Biochem. Biophys. Res. Commun. **6**, 81 (1961).
40. Lathe, G. H., Walker, M.: Biochem. J. **70**, 705 (1958).
41. Dutton, G. J., Greig, C. G.: Biochem. J. **66**, 52 P (1957).
42. Isselbacher, K. J., Chrabas, M. F., Quinn, R. C.: J. Biol. Chem. **237**, 3033 (1962).
43. Storey, I. D. E.: Biochem. J. **95**, 201, 209 (1965).
44. Dutton, G. J., Lawes, J.: Biochem. J. **98**, 30 P (1966).
45. Storey, I. D. E.: Biochem. Soc. Symp. (Cambridge, Engl.) **20**, 80 (1961).
46. Isselbacher, K. J: Recent Prog. Hormone Res. **12**, 134 (1956).
47. Newcomer, W. S., Heninger, R. W.: Proc. Soc. Exptl. Biol. Med. **105**, 32 (1960).
48. Smith, E. R., Breuer, H.: Biochem. J. **88**, 168 (1963).
49. Breuer, H., Wessendorf, D.: Z. Physiol. Chem. **345**, 1 (1966).
50. Dahm, K., Breuer, H., Lindlau, M.: Z. Physiol. Chem. **345**, 139 (1966).
51. — — Biochim. Biophys. Acta **113**, 404 (1966); **128**, 306 (1966).
52. — — Z. Klin. Chem. **4**, 153 (1966).
53. — — Acta Endocrinol. **52**, 43 (1966).
54. Schreifers, H., Ghraf, R., Pohl, F.: Z. Physiol. Chem. **344**, 25 (1966).
55. Slaunwhite, W. R., Jr., Lichtman, M. A., Sandberg, A. A.: J. Clin. Endocrinol. Metab. **24**, 638 (1964).
56. Fishman, W. H.: Advan. Enzymol. **16**, 361 (1955).
57. Levvy, G. A., Marsh, C. A.: Advan. Carbohydrate Chem. **14**, 381 (1959).
58. — Conchie, J., in: Glucuronic acid, free and combined (G. J. Dutton, ed.), p. 301. New York, N. Y.: Academic Press 1966.
59. Graham, A. F.: Biochem. J. **40**, 603 (1946).
60. Levvy, G. A.: Biochem. J. **40**, 396 (1946).
61. Kerr, L. M. H., Graham, A. F., Levvy, G. A.: Biochem. J. **42**, 191 (1948).
62. Fishman, W. H., Springer, B., Brunetti, R.: J. Biol. Chem. **173**, 449 (1948).
63. — Bernfeld, P., in: Methods in enzymology, vol. 1 (S. P. Colowick and N. O. Kaplan, eds.), p. 262. New York, N. Y.: Academic Press 1955. — Fishman, W. H.: Methods Biochem. Analy. **15**, 77 (1967).
64. Bernfeld, P., Nisselbaum, J. S., Fishman, W. H.: J. Biol. Chem. **202**, 763 (1953).
65. Bonnichsen, R.: Acta Chem. Scand. **18**, 1302 (1964).
66. Musa, B. U., Doe, R. P., Seal, U. S.: J. Biol. Chem. **240**, 2811 (1965).
67. Plapp, B. V., Cole, R. D.: Arch. Biochem. Biophys. **116**, 193 (1966).
68. Levvy, G. A., McAllan, A., Marsh, C. A.: Biochem. J. **69**, 22 (1958).
69. — Hay, A. J., Marsh, C. A.: Biochem. J. **65**, 203 (1957).
70. Alfsen, A., Jayle, M. F.: Bull. Soc. Chim. Biol. **40**, 2143 (1958).
71. Wakabayashi, M., Fishman, W. H.: J. Biol. Chem. **236**, 996 (1961).
72. Conchie, J., Findlay, J.: J. Endocrinol. **18**, 132 (1959).
73. Levvy, G. A., Marsh, C. A., in: Enzymes, vol. 4A, 2nd ed. (P. D. Boyer, H. Lardy, and K. Myrbäck, eds.), p. 397. New York, N. Y.: Academic Press 1960.
74. de Duve, C., Pressman, B. C., Gianetto, R., Wattiaux, R., Applemans, F.: Biochem. J. **60**, 604 (1955).

75. Conchie, J., Levvy, G. A.: Biochem. Soc. Symp. (Cambridge, Engl.) **23**, 86 (1963).
76. Bernfeld, P., Bernfeld, H. C., Nisselbaum, J. S., Fishman, W. H.: J. Am. Chem. Soc. **76**, 4872 (1954).
77. — Jacobson, S., Bernfeld, H. C.: Arch. Biochem. Biophys. **69**, 198 (1957).
78. Nichol, L.W., Roy, A. B.: Biochemistry **4**, 386 (1965).
79. Mills, G.T., Paul, J., Smith, E. E. B.: Biochem. J. **53**, 232 (1953).
80. Smith, E. E. B., Mills, G.T.: Biochem. J. **54**, 164 (1953).
81. Marsh, C. A., Levvy, G. A.: Biochem. J. **68**, 610 (1958).
82. Kato, K., Yoshida, K., Tsukamoto, H.: Chem. Pharm. Bull. (Tokyo) **12**, 664 (1964).
83. — — — Chem. Pharm. Bull. (Tokyo) **12**, 670 (1964).
84. Oshima, G.: J. Biochem. (Tokyo) **23**, 305 (1936).
85. Becker, B., Friedenwald, J. S.: Arch. Biochem. Biophys. **22**, 101 (1949).
86. Karunairatnam, M. C., Levvy, G. A.: Biochem. J. **44**, 599 (1949).
87. Wong, E., Rossiter, R. J.: Can. J. Med. Sci. **29**, 195 (1951).
88. Levvy, G. A., Marsh, C. A.: Nature **180**, 197 (1957).
89. — Biochem. J. **52**, 464 (1952).
90. Tappel, A. L., Dillard, C. J.: J. Biol. Chem. **242**, 2463 (1967).
91. Marsh, C. A.: Biochem. J. **86**, 77 (1963).
92. Walker, P. G., Levvy, G. A.: Biochem. J. **54**, 56 (1953).
93. Fishman, W. H., Altman, K. I., Springer, B.: Federation Proc. **7**, 154 (1948).
94. Fialkow, P. J., Fishman, W. H.: J. Biol. Chem. **236**, 2169 (1961).
95. Fishman, W. H., Green, S.: J. Biol. Chem. **225**, 435 (1957).
96. Yoshida, K., Kato, K., Tsukamoto, H.: Chem. Pharm. Bull. (Tokyo) **12**, 656 (1964).
97. Alfsen, A.: Bull. Soc. Chim. Biol. **41**, 1469 (1959).
98. Buehler, H. J., Katzman, P. A., Doisy, E. A.: Proc. Soc. Exptl. Biol. Med. **76**, 672 (1951).
99. Beall, D., Grant, G. A.: Rev. Can. Biol. **11**, 51 (1952).
100. Marsh, C. A.: Biochem. J. **59**, 375 (1955).
101. Gautney, M. C., Barker, S. B., Hill, S. R., Jr.: Science **129**, 1281 (1959).
102. Ryan, M.T., Mavrides, C. A.: Science **131**, 101 (1960).
103. Roy, A. B.: Biochem. J. **62**, 41 (1956).
104. — Biochem. J. **66**, 700 (1957).
105. Wotiz, H. H., Smakula, E., Lichtin, N. N., Leftin, J. H.: J. Am. Chem. Soc. **81**, 1704 (1959).
106. Foggitt, F., Kellie, A. E.: Biochem. J. **91**, 209 (1964).
107. Becker, J. F.: Biochim. Biophys. Acta **100**, 574 (1965).
108. Hobkirk, R., Cohen, J. J.: Can. J. Biochem. Physiol. **38**, 769 (1960).
109. Becker, J. F.: Biochim. Biophys. Acta **100**, 582 (1965).
110. Voigt, K. D., Lemmer, M., Tamm, J.: Biochem. Z. **332**, 550 (1960).
111. Kellie, A. E.: Biochem. J. **100**, 631 (1966).
112. Wakabayashi, M., Wotiz, H. H., Fishman, W. H.: Biochim. Biophys. Acta **48**, 198 (1961).
113. Pasqualini, J. R., Uhrich, F., Jayle, M. F.: Biochim. Biophys. Acta **104**, 515 (1965).
114. De Meio, R. H., Lewycka, C.: Endocrinology **56**, 489 (1955).
115. Robbins, P.W., Lipmann, F.: J. Biol. Chem. **233**, 686 (1958).

116. Akagi, J. M., Campbell, L. L.: J. Bacteriol. **84**, 1194 (1962).
117. Lipmann, F.: Science **128**, 575 (1958).
118. Sundaresan, P. R.: Biochim. Biophys. Acta **113**, 95 (1966).
119. Varandani, P. T., Wolf, G., Johnson, B. C.: Biochem. Biophys. Res. Commun. **3**, 97 (1960).
120. Subba Rao, K., Ganguly, J.: Biochem. J. **90**, 104 (1964).
121. — — Biochem. J. **98**, 693 (1966).
122. Pasternak, C. A., Humphries, S. K., Pirie, A.: Biochem. J. **86**, 382 (1963).
123. Hall, M. O., Straatsma, B. R.: Biochim. Biophys. Acta **124**, 246 (1966).
124. Wilson, L. G., Bandurski, R. S.: J. Biol. Chem. **233**, 975 (1958).
125. Robbins, P. W., Lipmann, F.: J. Biol. Chem. **233**, 681 (1958).
126. Wengle, B.: Acta Soc. Med. Upsalien. **68**, 154 (1963).
127. Carroll, J., Spencer, B.: Biochem. J. **94**, 20 P (1965).
128. — — Biochem. J. **96**, 79 P (1965).
129. Robbins, P. W., Lipmann, F.: J. Biol. Chem. **229**, 837 (1957).
130. Brunngraber, E. G.: J. Biol. Chem. **233**, 472 (1958).
131. Banerjee, R. K., Roy, A. B.: Mol. Pharmacol. **2**, 56 (1966).
132. Baddiley, J., Buchanan, J. G., Letters, R., Sanderson, A. R.: J. Chem. Soc. 1731 (1959).
133. — Sanderson, A. R.: Biochem. Prep. **10**, 3 (1963).
134. Fogarty, L. M., Rees, W. R.: Nature **193**, 1180 (1962).
135. Cherniak, R., Davidson, E. A.: J. Biol. Chem. **239**, 2986 (1964).
136. Schneider, J. J., Lewbart, M. L.: J. Biol. Chem. **222**, 787 (1956).
137. Roy, A. B.: Biochem. J. **63**, 294 (1956).
138. Nose, Y., Lipmann, F.: J. Biol. Chem. **233**, 1348 (1958).
139. Adams, J. B., Poulos, A.: Biochim. Biophys. Acta **146**, 493, 509, 522 (1967).
140. Boström, H., Wengle, B.: Acta Soc. Med. Upsalien. **69**, 41 (1964).
141. — Franksson, C., Wengle, B.: Acta Endocrinol. **47**, 633 (1964).
142. Adams, J. B.: Biochim. Biophys. Acta **82**, 572 (1964).
143. Holcenberg, J. S., Rosen, S. W.: Arch. Biochem. Biophys. **110**, 551 (1965).
144. Boström, H., Wengle, B.: Acta Endocrinol. **56**, 691 (1967).
145. Wallace, E., Silberman, N.: J. Biol. Chem. **239**. 2809 (1964).
146. Dixon, W. R., Vincent, V., Kase, N.: Steroids **6**, 757 (1965).
147. Baulieu, E. E., Fabre-Jung, I., Huis in't Veld, L. G.: Endocrinology **81**, 34 (1967).
148. Wengle, B.: Acta Endocrinol. **52**, 607 (1966).
149. Pulkkinen, M. O.: Acta Physiol. Scand. **66**, 115 (1966).
150. Adams, J. B.: Arch. Biochem. Biophys. **101**, 478 (1963).
151. Banerjee, R. K., Roy, A. B.: Biochim. Biophys. Acta **137**, 211 (1967).
152. Wengle, B., Boström, H.: Acta Chem. Scand. **17**, 1203 (1963).
153. Levitz, M., Katz, J., Twombly, G. H.: Steroids **6**, 553 (1965).
154. Cooke, B. A., Taylor, W.: Biochem. J. **81**, 12 P (1961).
155. Banerjee, R. K., Roy, A. B.: Biochim. Biophys. Acta **151**, 573 (1968).
156. Cleland, W. W.: Biochim. Biophys. Acta **67**, 104 (1963).
157. Roy, A. B.: Biochem. J. **79**, 253 (1961).
158. — Biochem. J. **82**, 66 (1962).

159. Roy, A. B.: J. Mol. Biol. **10**, 176 (1964).
160. Herrmann, I., Repke, K.: Arch. Exptl. Pathol. Pharmakol. **248**, 370 (1964).
161. Higaki, M., Takahashi, M., Suzuki, T., Sahashi, Y.: J. Vitaminol. (Kyoto) **11**, 261, 266 (1965).
162. Vestermark, A., Boström, H.: Exp. Cell. Res. **18**, 174 (1959).
163. Palmer, R. H.: Proc. Natl. Acad. Sci. U.S. **58**, 1047 (1967).
164. Bridgwater, R. J., Ryan, D. A.: Biochem. J. **65**, 24 P (1957).
165. Roy, A. B.: Biochem. J. **68**, 519 (1958).
166. Butenandt, A., Hofstetter, H.: Z. Physiol. Chem. **259**, 222 (1939).
167. Cohen, H., Bates, R. W.: Endocrinology **44**, 317 (1949).
168. Pulkkinen, M. O.: Acta Physiol. Scand. **52**, Supp. 180 (1961).
169. Roy, A. B.: Advan. Enzymol. **22**, 205 (1960).
170. Dodgson, K. S., Spencer, B.: Methods Biochem. Analy. **4**, 211 (1957).
171. Fowler, L. R., Rammler, D. H.: Biochemistry **3**, 230 (1964).
172. Dodgson, K. S., Spencer, B., Wynn, C. H.: Biochem. J. **62**, 500 (1956).
173. Roy, A. B.: Biochem. J. **64**, 651 (1956).
174. French, A. P., Warren, J. C.: Biochem. J. **105**, 233 (1967).
175. Webb, E. C., Morrow, P. F. W.: Biochem. J. **73**, 7 (1959).
176. Dodgson, K. S.: Enzymologia **20**, 301 (1959).
177. Nichol, L. W., Roy, A. B.: J. Biochem. (Tokyo) **55**, 643 (1964).
178. Allen, E., Roy, A. B.: Biochim. Biophys. Acta **168**, 243 (1968).
179. Jarrige, P., Yon, J., Jayle, M. F.: Bull. Soc. Chim. Biol. **45**, 783 (1963).
180. Pulkkinen, M. O., Paunio, I.: Ann. Med. Exptl. Biol. Fenniae (Helsinki) **41**, 283 (1963).
181. Henry, R., Thevenet, M., Jarrige, P.: Bull. Soc. Chim. Biol. **34**, 897 (1952).
182. Savard, H., Bagnoli, E., Dorfman, R. I.: Federation Proc. **13**, 289 (1954).
183. Stitch, S. R., Halkerston, I. D. K., Hillman, J.: Biochem. J. **63**, 705 (1956).
184. Takahashi, N., Egami, F.: J. Biochem. (Tokyo) **49**, 358 (1961).
185. Ney, K. H., Ammon, R.: Z. Physiol. Chem. **315**, 145 (1959).
186. Corner, E. D. S., Leon, Y. A., Bulbrook, R. D.: J. Marine Biol. Assoc. U.K. **39**, 51 (1960).
187. Leon, Y. A., Bulbrook, R. D., Corner, E. D. S.: Biochem. J. **75**, 612 (1960).
188. Gibian, H., Bratfisch, G.: Z. Physiol. Chem. **305**, 265 (1956).
189. Roy, A. B.: Australian J. Exptl. Biol. Med. Sci. **41**, 331 (1963).
190. Jarrige, P.: Purification et Propriétés des Sulfatases du Suc Digestif d'Hélix Pomatia. Paris: R. Foulon & Cie. 1962 (Doctoral Thesis).
191. — Bull. Soc. Chem. Biol. **45**, 761 (1963).
192. Burstein, S., Dorfman, R. I.: J. Biol. Chem. **238**, 1656 (1963).
193. Warren, J. C., French, A. P.: J. Clin. Endocrinol. Metab. **25**, 278 (1965).
194. French, A. P., Warren, J. C.: Steroids **6**, 865 (1965).
195. Dodgson, K. S.: Biochem. J. **78**, 324 (1961).
196. — Gatehouse, P. W., Lloyd, A. G., Powell, G. M.: Biochem. J. **95**, 18 P (1965).
197. Bergmann, W.: Comp. Biochem. **3**, 103 (1962).
198. Gottfried, H., Lusis, O.: Nature **212**, 1488 (1966).
199. — Dorfman, R. I., Wall, P. E.: Nature **215**, 409 (1967).

200. Layne, D. S.: Endocrinology **76**, 600 (1965).
201. Strominger, J. L., Smith, M. S.: J. Biol. Chem. **234**, 1822 (1959).
202. Smith, E. E. B., Mills, G. T.: Biochim. Biophys. Acta **13**, 386 (1954).
203. — Munch-Petersen, A., Mills, G. T.: Nature **172**, 1038 (1953).
204. Maley, F., Maley, G. F., Lardy, H. A.: J. Am. Chem. Soc. **78**, 5303 (1956).
205. — Lardy, H. A.: Science **124**, 1207 (1956).
206. — Maley, G. F.: Biochim. Biophys. Acta **31**, 577 (1959).
207. Leloir, L. F., Cardini, C. E., in: Enzymes, vol. 2, 2nd ed. (P. D. Boyer, H. Lardy, and K. Myrbäck, eds.), p. 39. New York, N. Y.: Academic Press 1960.
208. Roseman, S., Dorfman, A.: J. Biol. Chem. **191**, 607 (1951).
209. Weissmann, B., Rowin, G., Marshall, J., Friederici, D.: Biochemistry **6**, 207 (1967).
210. — Friederici, D.: Biochim. Biophys. Acta **117**, 498 (1966).
211. Zechmeister, L., Tóth, G., Vajda, É.: Enzymologia **7**, 170 (1939).
212. — — Enzymologia **7**, 165 (1939).
213. Helferich, B., Iloff, A.: Z. Physiol. Chem. **221**, 252 (1933).
214. Watanabe, K.: J. Biochem. (Tokyo) **24**, 297, 305, 315 (1936).
215. Neuberger, A., Pitt Rivers, R. V.: Biochem. J. **33**, 1580 (1939).
216. Pugh, D., Leaback, D. H., Walker, P. G.: Biochem. J. **65**, 464 (1957).
217. Conchie, J., Findlay, J., Levvy, G. A.: Biochem. J. **71**, 318 (1959).
218. Findlay, J., Levvy, G. A.: Biochem. J. **77**, 170 (1960).
219. Weissman, B., Hadjiioannou, S., Tornheim, J.: J. Biol. Chem. **239**, 59 (1964).
220. Caygill, J. C., Roston, C. P. J., Jevons, F. R.: Biochem. J. **98**, 405 (1966).
221. Coleman, R. L., Scroggs, R. A., Whittington, A.: Biochim. Biophys. Acta **146**, 290 (1967).
222. Woollen, J. W., Walker, P. G., Heyworth, R.: Biochem. J. **79**, 294 (1961).
223. — Heyworth, R., Walker, P. G.: Biochem. J. **78**, 111 (1961).
224. Walker, P. G., Woollen, J. W., Heyworth, R.: Biochem. J. **79**, 288 (1961).
225. Borooah, J., Leaback, D. H., Walker, P. G.: Biochem. J. **78**, 106 (1961).
226. Leaback, D. H., Walker, P. G.: Biochem. J. **78**, 151 (1961).
227. Findlay, J., Levvy, G. A., Marsh, C. A.: Biochem. J. **69**, 467 (1958).
228. Levvy, G. A., Conchie, J.: Methods Enzymol. **8**, 573 (1966).
229. Robinson, D., Price, R. G., Dance, N.: Biochem. J. **102**, 525 (1967).
230. Conchie, J., Hay, A. J.: Biochem. J. **87**, 354 (1963).
231. — Levvy, G. A.: Brit. J. Cancer **11**, 487 (1957).
232. Fotherby, K.: Biochem. J. **73**, 339 (1959).

Chapter 3

The Hydrolysis of Steroid Conjugates

H. LEON BRADLOW

Contents

I. Introduction

From the very earliest studies in the twenties, when steroid hormone research was in its infancy, the hydrolysis of conjugates has been a major interest of steroid investigators. Except for a tiny fraction excreted as free compounds, steroid hormones and their metabolites are excreted as conjugates.

Although the fecal route of excretion is important in animals [1], it is of minor importance in man [2], in whom urinary excretion is the major pathway. Consequently major attention has been focused on the hydrolysis of urinary conjugates. Because the analysis of intact steroid conjugates presents formidable difficulties, hydrolysis has been a necessary preliminary to almost all work with only a few exceptions, such as the Venning procedure [3] for pregnanediol-3-glucuronide and the Cohen assay for estrogens [4]. Over the years emphasis has shifted from vigorous nonselective procedures to gentle, relatively selective, enzymatic methods. Despite the substantial progress that has been achieved in this forty-year period, considerable controversy still exists about the advantages and faults of the various methods. For any given study, the advantages and disadvantages must be carefully considered in selecting the method.

Funk *et al.* [5, 6], in the earliest reported study of hydrolysis, found that the yield of male hormone was substantially increased when the urine was acidified prior to extraction. Doisy *et al.* [7], in their pioneering studies on estrogens, also found that acidification increased the yield. Hot acid hydrolysis, a technique which survives to the present as a standard procedure, was first reported by Lipschütz and Poch [8] in 1932. This procedure was rapidly adapted by a great many investigators. The use of a solvent overlay to extract the liberated steroids during acid hydrolysis was described by Borchard *et al.* [9] in 1934. In the same year, the first identified artifact of acid hydrolysis, 3β-chloroandrost-5-en-17-one, was reported by Butenandt and Dannenbaum [10]. Acid hydrolysis of pregnanediol conjugates was first reported by Odell and Marrian [11] in 1936.

In the first deliberate attempt to utilize enzyme preparations, Veitch *et al.* [12] described the hydrolysis of pregnanediol-3-glucuronide with emulsin. However, the basis for this approach had already been laid in preliminary observations on the action of bacteria on pregnancy urine. Cohen *et al.* [13] in 1935 noted that estriol glucuronides were completely hydrolyzed by prolonged exposure of pregnancy urine to bacterial action; and Patterson [14] in 1937 used *Escherichia coli* preparations to effect the hydrolysis of conjugated estrogens in early pregnancy urines. In 1940, Bucher and Geschickter [15] demonstrated that bacterial hydrolysis could be a source of error in the Venning assay [3] for pregnanediol–3–glucuronide. They found that incubation of urines with bacteria caused significant hydrolysis of pregnanediol–3–glucuronide, and suggested its possible usefulness. In 1939, Fishman [16] initiated the modern approach to enzyme hydrolysis when he described a calf–spleen glucuronidase preparation capable of hydrolyzing steroid glucuronides. In 1943, Talbot *et al.* [17] reported the hydrolysis of pregnanediol–3–glucuronide in good yield with a rat–liver acetone powder. In 1949, Buehler *et al.* [18] induced increased glucuronidase activity in *E. coli* by culturing the organism in the presence of menthol glucuronide. The bacterial filtrate proved to be a rich source of steroid glucuronidase activity. The first report of enzymatic hydrolysis of a steroid sulfate was made by Butenandt and Hofstetter [19], who used a fungal sulfatase preparation. These developments mark the end of the pioneering phase and the start of the modern era.

From the very beginning, research on the hydrolysis of steroid conjugates has orginated out of two often–conflicting interests. On the one side are the clinically oriented workers interested in rapid methods for the assay of hormones, to whom hydrolysis is a necessary preliminary to the desired assay. Alterations in structure and artifact formation that does not interfere with the desired measurement are of no consequence. Methods like hot acid hydrolysis and oxidative cleavage have been preferred despite the complications of artifact formation. When enzymatic methods have been used, they have favored short incubations using high enzyme levels.

The other group consists of those investigators interested in the structure of metabolites as well as where the conjugating groups are attached to the steroid. One of their principal objectives is the recovery of the intact steroid metabolites free of alterations. Major attention has been focused on enzymatic and solvolytic methods capable of yielding intact metabolites. In recent years interest has turned again to the isolation of intact conjugates, a return to the classical isolation approach of Venning [3].

II. Neutral Hydrolytic Procedure

A. Thermal Solvolysis

Nonacid–catalyzed hydrolysis is a useful specific method for the cleavage of 3β–hydroxy–Δ^5–sulfates, dating back to early observations of Dingemanse *et al.* [20] who found that the recovery of androgens was increased when the urine was heated at neutral pH. Subsequently, Dingemanse and Huis in't Veld [21] reported the isolation of the *cyclo*steroid IV from the urine of tumor patients which had been heated at 100° for 6 hours. This compound aroused considerable interest as the first naturally occurring *cyclo*steroid; indeed, Dingemanse [22] argued that dehydroisoandrosterone [3β–hydroxyandrost–5–en–17–one], isolated from urinary extracts, was a product of the chemical alteration of the sulfate of IV during hydrolysis, with the latter being the natural secretory product. That this was not the case was demonstrated conclusively by Teich *et al.* [23]. They found that, when an aqueous solution of dehydroisoandrosterone sulfate [3β–sulfo-oxyandrost–5–en–17–one] was refluxed, the proportion of dehydroisoandrosterone (II) and the *cyclo*steroid isolated (IV) varied with the pH. At pH 3.2 and lower, only II was formed; at pH 4.7, equal amounts of II and IV were formed; and at pH 6.8, twice as much IV as II was formed. These results clearly show that IV is derived from I. Subsequently, Huis in't Veld [24] isolated dehydroisoandrosterone sulfate from the same urines from which Dingemanse and Huis in't Veld [21] had previously isolated the *cyclo*steroid. The ease of hydrolysis of dehydroisoandrosterone sulfate (I) is due to the presence of the Δ^5–double bond which provides anchimeric assistance [25]. The intermediate *cyclo*steroid ion III reacts to give either the *cyclo*steroid IV or free dehydroisoandrosterone (II) in proportions depending on the exact pH. As the pH increases, the proportion of IV increases.

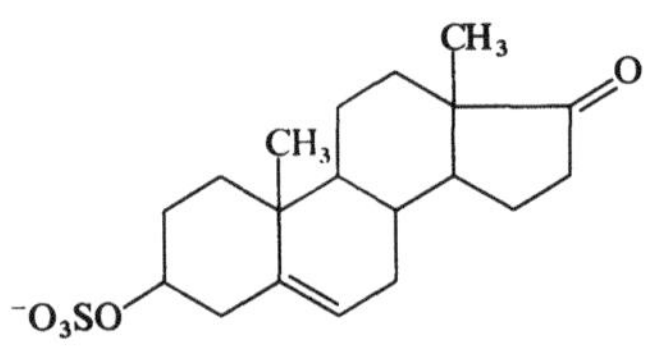

I

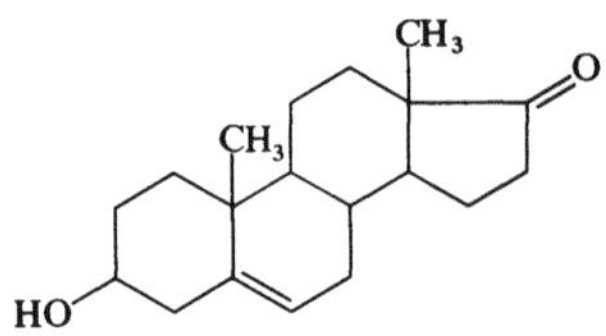

II

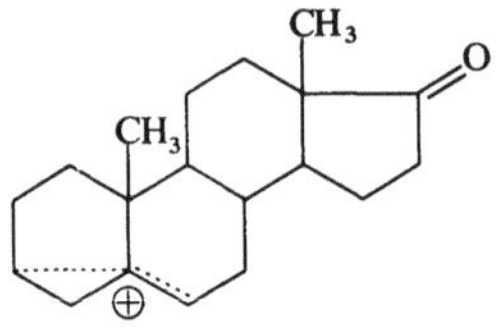

III

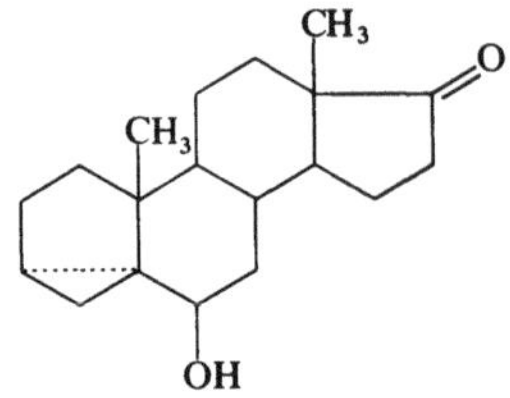

IV

In 1943, Talbot *et al.* [26] employed this thermal lability as the basis for a specific method for the determination of dehydroisoandrosterone. The conjugates were extracted with butanol and then heated at 100° for 4 hours in a pH 5.8 acetate buffer containing barium chloride. The liberated dehydroisoandrosterone was then extracted and analyzed. Barium chloride was included in the hope that precipitation of liberated sulfate would speed up the solvolysis.

The procedure was critically investigated by Bitman and Cohen [27] who confirmed that the original method only hydrolyzed dehydroisoandrosterone sulfate. They also found that barium chloride played no role in the hydrolysis, which could be carried out successfully in a simple buffer solution. The neutral extract contained 25% of the total urinary 17-ketosteroids obtained after acid hydrolysis. Chromatographic analysis showed that dehydroisoandrosterone and a small amount of 3β-chloroandrost-5-en-17-one accounted for virtually all of the 17-ketosteroids present. The latter compound is formed by reaction of chloride ion with the *cyclo*steroid ion III. Buehler *et al.* [28] found that dehydroisoandrosterone sulfate was split to dehydroisoandrosterone by autoclaving at 15 psi for 1 hour at pH 7.3. Roy [29] found that thermal solvolysis of dehydroisoandrosterone sulfate, pregnenolone sulfate [3β-sulfooxypregn-5-en-20-one], and cholesterol sulfate [3β-sulfooxycholest-5-ene] was comparable (over 50% in 2 hours). The stability of steroid glucuronides and saturated sulfates to thermal solvolysis was observed by all three groups.

Fotherby [30] also took advantage of this selective hydrolysis in his recent studies on dehydroisoandrosterone in man. He found that the mixture of II and IV arising from the solvolysis could be assayed by the Zimmermann reaction without prior isomerization of IV to II. Because of the specificity of thermal solvolysis, the assay can be carried out with a minimum of prior purification. Fotherby subsequently applied this procedure in isolating from urine a number of rare 3β-hydroxy-Δ^5-steroids, 16α-hydroxydehydroisoandrosterone [3β,16α-dihydroxyandrost-5-en-17-one], androstenediol [androst-5-ene-3β,17β-diol], androst-5-ene-3β,16β,17β-triol, and pregnenetriol [pregn-5-ene-3β,17α,20α-triol] [31]. Mild acid treatment served to isomerize any of the *cyclo*steroids formed back to the normal isomers.

Stern and Barwell [32] used thermal solvolysis as the initial step in their procedure for the determination of urinary pregnenetriol. Since a substantial fraction of this compound is excreted as the glucuronide [33], the method is actually one for the determination of that fraction excreted as the sulfate.

B. Comment

Thermal solvolysis has been found to be a very specific procedure for the hydrolysis of 3β-hydroxy-Δ^5-steroid sulfates. Because of its unusual selectivity it has a useful role in the study of these compounds. It facilitates their isolation in the presence of large amounts of saturated congeners which would interfere if less selective procedures, like acid-catalyzed solvolysis or sulfatase hydrolysis, were employed.

III. Oxidative Procedures

A. Oxidative Elimination of Glucuronide Conjugates of Corticoids

This approach to the liberation of steroid aglycones was developed by Norymberski *et al.* [34] as a solution to the problem of the analysis of sensitive corticoid metabolites in urine. The extensive literature on this method has been reviewed by Péron [35] and Cope [36]. The oxidation converts the glucuronic acid residue into a formate ester which can be readily removed, if desired, by mild acid or base treatment. At the same time, any glycolic links in the steroid moiety are simultaneously split. This is both a virtue, giving a stable, easily quantified 17–ketone in place of the relatively unstable corticoid side chains, and a limitation in that the individual identity of the various metabolites is lost as a consequence. In the original "indirect" procedure [34] the untreated urine or steroid–conjugate extracts were directly oxidized with sodium bismuthate followed by hot acid hydrolysis. After extraction, the steroids were quantified by the Zimmermann reaction. This 17–ketosteroid value is the sum of the endogenous 17–ketosteroid and those newly formed by degradation of the corticoid side chain. To determine the latter it is necessary to separately measure the endogenous 17–ketosteroids following acid hydrolysis and subtract this value from the total found after bismuthate oxidation. The difference between these two values is the amount of 17–ketogenic steroids present in the extract. The 17–ketosteroids arising from the oxidation include etiocholanolone [3α–hydroxy–5β–androstan–17–one] derived from pregnanetriol [5β–pregnane–3α,17α,20α–triol]; 11β–hydroxyetiocholanolone [3α,11β–dihydroxy–5β–androstan–17–one] and 11–ketoetiocholanolone [3α–hydroxy–5β–androstane–11,17–dione] derived from the various tetrahydro– and hexahydro–metabolites of cortisol [11β,17α,21–trihydroxypregn–4–ene–3,20–dione] and 11–ketopregnane–3α,17,20α–triol. 17–Hydroxy–20–ketones are not cleaved by this procedure. Edwards and Kellie [37] found that oxidation of the glucuronic acid was the rate–limiting step and that the rate could be markedly increased by using dilute trichloroacetic acid as the solvent.

Many groups reported that this original procedure was poorly reproducible [38] and that it presented a number of technical problems. The quality of the sodium bismuthate proved to be especially critical. Diczfalusy *et al.* [39] found that 18 hours was required for complete oxidation using Baker analyzed sodium bismuthate, whereas, with material obtained from British Drug Houses, the reaction was substantially complete within one–half hour. The oxidation period must be restandardized whenever a new batch of bismuthate is used.

Alteration of endogenous 17–ketosteroids by sodium bismuthate represents another souce of error [40] since the artifacts formed in this way have a different Zimmermann equivalent from that of the endogenous compounds. Because of all of these problems, the original "indirect" method has been largely abandoned in favor of the modified direct method of Appleby *et al.* [41]. The urine is initially treated with sodium borohydride to reduce 17–ketosteroids to 17–hydroxy compounds, and 17–hydroxy–20–ketosteroids to 17,20–glycols. Oxidation with sodium bismuthate, followed by acid hydrolysis, is then carried out as in the original method. This pretreatment eliminates the need to subtract endogenous 17–ketosteroids from the final measurement and also permits 17,20–ketols to be included in the assay. Comparative studies by Jørgensen [38], Moxham and Nabarro [42], Butt *et al.* [43], and Raftopoulo *et al.* [44] showed good correlation between 17–ketogenic steroid values obtained in this manner and 17–hydroxycorticoids measured by the Porter–Silber reaction. Since the method measures 20–hydroxy– and 21–deoxy–compounds not measured by the Porter–Silber or Mader–Buck reactions the 17–ketogenic steroid values are always larger. Several sources of error are still possible. Smith and Tompsett [45] pointed out that, during the acid hydrolysis step, sodium bismuthate can generate chlorine, which can reoxidize 17–hydroxy compounds back to 17–ketosteroids, inflating the 17–ketogenic steroid values. Norymberski and Stubbs [46] and Sobel *et al.* [47] pointed out that sufficient urea is ordinarily present in urine to react with all of the chlorine formed. Freislederer [48] found that these oxidative side reactions could also be overcome by the addition of sodium sulfite during the acid hydrolysis step.

Another source of error is the heterogeneity of the reaction product. Sodium bismuthate oxidizes some of the initially formed 11β–hydroxyetiocholanolone to 11–ketoetiocholanolone [49]. Part of the surviving 11β–hydroxy compound is dehydrated to 9–dehydroetiocholanolone [3α–hydroxy–5β–androst–9(11)–en–17–one] during the subsequent acid hydrolysis. Since the relative Zimmermann chromogenicity of these compounds varied from 85 for 11–hydroxyetiocholanolone to 120 for 11–ketoetiocholanolone [50], errors in estimation may arise. A possible solution has been offered by Vestergaard and Sayegh [51], who found that all of these compounds have the same chromogenicity when the Zimmermann reaction is carried out in pyridine solution.

Because of its ability to cleave both glucuronides [52] and glycol side chains [53], periodic acid was an attractive alternative to bismuthate. This was first applied to corticosteroid analysis by Tompsett and Smith [54], but artifact formation was still a problem with their experimental conditions. Few [55] improved the process considerably by carrying out the oxidation step in neutral solution following pretreatment with sodium borohydride. Under these conditions the glucuronic acid residue is removed with minimum artifact formation. All of the 11–oxygenated corticoids are converted into 11β–hydroxyetiocholanolone which can be easily separated and quantitated. Thus, the urinary metabolites of cortisol can be estimated without interference from 11–deoxycorticosteroids. A quicker procedure was described by Metcalf [56] who reported that, by raising the temperature, the entire reduction and oxidation sequence could be carried out within 1 hour. Further improvements were made by James and Caie [57], Menini and Norymberski [58], Murphy and West [59], Caisey and Child [60], and Vestergaard and Sayegh [51]. The first three groups concentrated mainly on the analysis of the derived 17–ketosteroids rather than on the oxidation step itself. Vestergaard and Sayegh reported extensive studies on refinements in the entire procedure designed to increase reproducibility and reliability. Caisey and Child described an automated procedure having increased reliability. The periodate method, as carried out by all of these groups, will not measure 17–hydroxycorticoid–3–sulfates. However, this is not a serious objection since both Edwards and Kellie [37] and Few [55] concluded that such compounds make only a minor contribution.

Decisions concerning which of the two procedures is preferable, or whether either is truly satisfactory, are difficult to make in view of the considerable disagreements between different groups.

As Norymberski [61] himself pointed out, there are four major difficulties with the bismuthate procedure:

1. Serious underestimation occurs in the presence of excessive amounts of glucose, and this may require preliminary removal if present in large quantity.

2. Meprobamate and possibly other drugs are reported to interfere with the assay.

3. Complete accuracy is not possible with a mixture of 17–ketosteroids of unknown composition.

4. At subnormal excretion levels, the method suffers from lack of precision.

The first is a serious problem, but it can be overcome by increasing the amount of sodium bismuthate, by removing the glucose by yeast fermentation, or by extracting the conjugates. The extraction procedure is generally considered the most desirable. In the periodate method, simple dilution or the addition of extra periodate solves the problem [56].

The second problem is common to many assay procedures, and is possibly best handled by withholding drugs during tests. An instance of drug interference in 17–ketogenic steroid determinations was recently reported by

Mattox *et al.* [62] who found spuriously high 17–ketogenic steroid values when analyzing the urine of patients receiving Phenaglycodol [2–(4–chlorophenyl)–3–methyl–2,3–butanediol]. Ketones were formed which contributed to the Zimmermann chromogen.

While the third problem is real for the bismuthate method, in most cases the errors arising from this source are small compared with those derived from other technical factors. As discussed above, when the Zimmermann reaction is carried out in pyridine solution, the problem is eliminated. Since the periodate method gives a single product, 11β–hydroxyetiocholanolone, from the 11–oxygenated corticoids, this problem is minimized.

The fourth problem cited is the source of considerable disagreement. In one of the few direct comparisons, Rutherford and Nelson [63] had great difficulty in achieving reproducible results when using sodium bismuthate, whereas excellent results were obtained with sodium periodate. On the other hand, Vestergaard and Sayegh [51] found that, under their standardized conditions, bismuthate gave excellent reproducibility down to 0.5 mg/day, while the coefficient of variation was greater for periodate at levels below 10 mg/day. They concluded that the bismuthate procedure was somewhat superior for longitudinal studies in which differences in steroid excretion are being studied, whereas periodate was the method of choice for general clinical studies because of its greater simplicity. The results of Thomas [64] are in agreement with this position.

Although the bismuthate method yields slightly more 17–ketogenic steroids than the periodate method, it has not been established that the difference is due to cortisol metabolites. A curious problem, which has never been discussed, is why hot acid hydrolysis continues to be included in the bismuthate procedure. Although required in the original method, once the use of sodium borohydride was introduced, hot acid hydrolysis following the oxidation step became pointless. As Edwards and Kellie [37] and Few [55] have shown, the contribution of corticoid sulfates that might require acid hydrolysis can be ignored. Since the oxidation step converts glucuronide residues to formate esters, which are easily removed by mild acid or base treatment, no useful role remains for hot acid hydrolysis. On the other hand, this step does produce a variety of artifacts, dehydration products, and interfering chromogens which only serve to make the analysis more difficult. Although studies have not been carried out on this point, it seems probable that some of the difficulties in the bismuthate procedure could be eliminated by omitting this step.

As Cope [36] pointed out, comparisons between the 17–ketogenic steroid and 17–hydroxycorticoid measurements showing good agreement may not be meaningful, since both procedures may hide the same fallacies. The real issue is the extent to which either method truly measures corticoid secretion. Studies by Cope and Pearson [65] showed that the correlation between 17–ketogenic steroid values and cortisol production rates was good only in the high normal region (> 35 mg/day) where the clinical need is least. In the normal range and just above, where the need is greatest for a meaningful result, the correlation is so poor that the results might be misleading. Suggested sources for these discrepancies include errors in obtaining complete urine collections, contributions from noncorticoid sources of 17–ketogenic steroids (*e.g.*, pregnanetriol), variations in reaction conditions, changes in metabolites or conjugates, and the like. Among other anomalies, they observed patients with Cushing's syndrome in whom the 17–ketogenic steroid values were in the normal range. On the other hand, James and Caie [57], using periodate oxidation, obtained a regression line indicating reasonable correlation at low and normal values, and increasingly poor correlation at high levels. No satisfactory resolution of these differences has been achieved. The lack of correlation between 17–ketogenic steroid values and cortisol production rates during acute bacterial infections was emphasized by Cornil *et al.* [66], who found that these values did not reflect the magnitude of the increase in cortisol production occurring in febrile states.

Although some workers [47] have argued that inclusion of etiocholanolone derived from pregnanetriol and Compound S [17α,21–dihydroxypregn–4–ene–3,20–dione] is a virtue because it leads to elevated values in patients with the adrenogenital syndrome, where 17–hydroxycorticoid values could be low or normal, its inclusion becomes a source of error when one is interested in using 17–ketogenic steroid values as a measure of cortisol production. Only the original method of Few [55] and the vapor–phase chromatography method of Murphy and West [59] specifically measure such a compound. It is possible that this is the reason for the good correlation observed by Few.

Several groups [67] have utilized oxidative cleavage with periodate [67b–d] or bismuthate [67a] to drastically decrease the time and labor required for cortisol production–rate determinations. All of the C21–reduced metabolites of cortisol were converted to 11β–hydroxyetiocholanolone which was purified chromatographically. From the specific activity of this single derivative, the production rate was calculated in the usual manner.

B. Other Uses of Oxidative Cleavage

Ulick *et al.* [68] found that oxidative hydrolysis was useful in the study of aldosterone [11β,21–dihydroxy–3,20–dioxopregn–4–en–18–al] metabolism. With periodate oxidation, they found that tetrahydroaldosterone glucuronide [11β,21–dihydroxy–18,20–dioxo–5β–pregnan–3α–yl–β–D–

glucopyranosiduronic acid] (I) could be converted into a stable derivative, the etianic acid lactone III. With proper choice of pH, the intermediate formate II was hydrolyzed and only III was isolated. This lactone is specifically derived from a unique metabolite of aldosterone, and readily separated from the oxidation product V of 18–hydroxytetrahydro A [3α,18,21–trihydroxy–5β–pregnane–11,20–dione] (IV). The compound is a convenient intermediate for double isotope derivative assays.

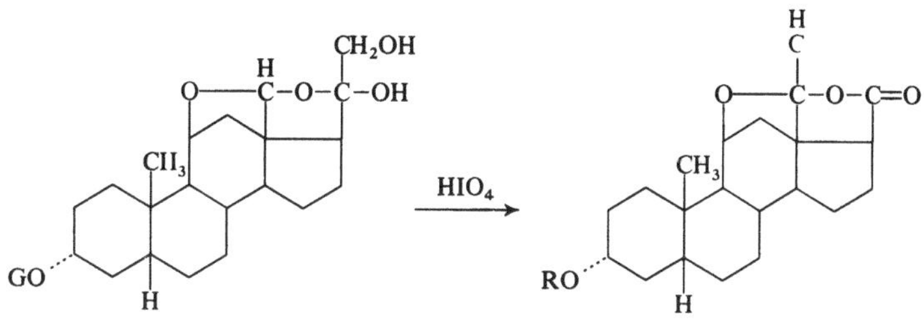

I

II, R=HCO
III, R=H

G≅glucuronic acid

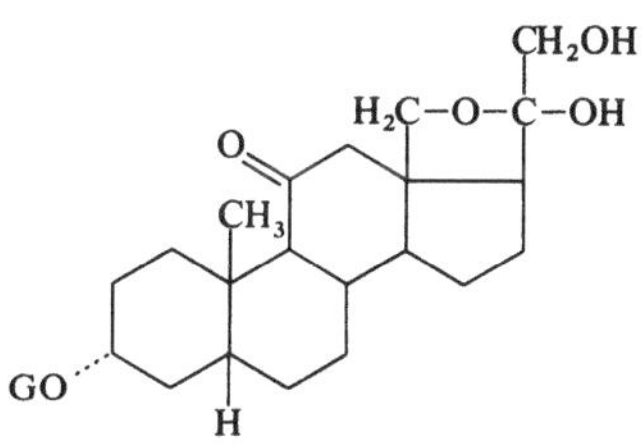

IV

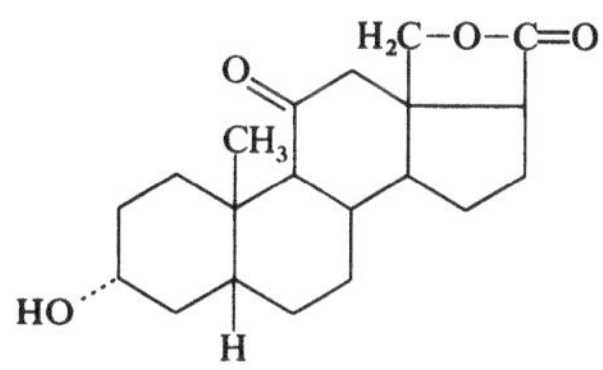

V

G≅glucuronic acid

Morris [69] demonstrated that oxidative cleavage can serve as the starting point of a method for the assay of pregnanetriol. The conjugates were extracted from the urine and carried through the Appleby method [41]. Etiocholanolone was separated by chromatography and assayed. It should be kept in mind that this assay procedure cannot distinguish between pregnanetriol and tetrahydro Compound S [3α,17α,21–trihydroxy–5β–pregnan–20–one] which also yields etiocholanolone. The results agreed favorably with those obtained with other methods. Incidental to their 17–ketogenic steroid measurements, Murphy and West [59] found that a peak with the retention time of etiocholanolone was present in the vapor phase chromatography scans, although no serious attempt was made to quantitate this peak as a measure of pregnanetriol.

Pregnanediol determinations can also be carried out on the extracts obtained after periodate oxidation. Both Valente [70] and Murphy and West [59] reported that the pregnanediol values obtained after oxidative cleavage of the glucuronide moiety were in good agreement with those obtained after enzyme or acid hydrolysis. Because the measurements can be carried out on the same extracts used for 17–ketogenic steroid measurements, this precedure offers advantages in terms of economy and speed.

C. Comment

Despite its many advantages for clinical assay, discussion of the 17–ketogenic steroid method has tended to ignore some of the clinical limitations of this approach. As a result of elimination of the side chain and reduction at C11, the method prevents one from distinguishing between normal patients and those with extra–adrenal disorders in which the proportions of the different cortisol metabolites are altered. Thus the same 17–ketogenic steroid values will be obtained in some cirrhotic patients as in normal subjects, despite the fact that cortolone [3α,17α,20α,21–tetrahydroxy–5β–pregnan–11–one] formation is markedly increased at the expense of tetrahydrocortisone [3α,17α,21–trihydroxy–5β–pregnane–11,20–dione] in this illness [71]. A similar loss of structural information would occur in patients with thyroid disturbances, where the ratio of 11–hydroxy and 11–keto metabolites is shifted [72].

Another serious practical limitation is that oxidative hydrolysis fails when glucuronide formation is depressed. The data in Table 3–1 show that

TABLE 3–1. *Cleavage of glucuronidase–resistant urinary cortisol conjugates by the Few procedure*

Subject	Recovered in the organic extract (%)	Subject	Recovered in the organic extract (%)
I	47.3	VII	59.3
II	59.1	VIII	15.2
III	29.5	IX	42.5
IV	61.1	X	44.1
V	41.0	XI	41.6
VI	25.5		

In each case the conjugates remaining in the urine after hydrolysis with 300 units of Ketodase for 5 days were treated by the method of Few [Few, J. D.: J. Endocrinol. **22**, 31 (1961)]. Paper chromatography of the neutral extracts established that only a small fraction (20% or less) was actually 11β–hydroxyetiocholanolone or 11–ketoetiocholanolone.

oxidative cleavage does not cleave conjugates that are resistant to glucuronidase hydrolysis. Consequently serious underestimation of corticoid production may occur in patients in whom cortisol secretion actually is normal.

Few has also found that from 25 to 50% of the urinary radioactivity derived from cortisol is not converted into ether–extractable compounds by treatment with sodium borohydride followed by periodic acid [73]. This loss is considered to play a major role in accounting for discrepancy between 17–ketogenic steroid values and cortisol production rates.

Therefore, despite the advantages of 17–ketogenic steroid measurements as a screening procedure, one must be alert for those patients in whom precise measurements of the production rate or the metabolite pattern is essential for proper clinical evaluation.

IV. Solvolytic Procedures

Mild acid–catalyzed hydrolysis of steroid conjugates at, or near, room temperature has been studied extensively because it cleaves conjugates with less artifact formation than hot acid hydrolysis. This method is more or less restricted to sulfates, although Jacobsohn and Lieberman [74] described a method applicable to steroid glucuronides.

In early studies, several groups used urinary acidification to increase the release of androgens [5, 6, 75], estrogens [76], and corticoids [77]. The role of the solvent was not appreciated at that time.

A. Nonsolvent–Assisted Hydrolysis

Only three examples of steroid conjugates that are readily hydrolyzed merely on standing in acidified aqueous solution have been described. One is the 18–glucuronide of aldosterone I, which is rapidly cleaved by aqueous acid and substantially unaffected by glucuronidase [78]. Steric factors related to crowding about the 18 position or its unusual structure, which contains a hemiacetal structure linked to an acetal function, have been suggested as the cause of this unusual lability.

Recently Noguchi and Fukushima [79] found that the conjugate of 3,11β–dihydroxyandrost–4–en–17–one (III), tentatively identified as a glucuronide, is exceptionally labile, undergoing quantitative solvolysis on standing at 37° at pH 5. The reaction is even more rapid at lower pH values. However, considerable dehydration to give the diene VI occurs under these conditions. This conjugate is unique, in that the products of solvolysis in urine include the 3α– and 3β–hydroxyl isomers Va and b, as well as the unusual 3α–ureido compound Vc. The resonance–stabilized allylic carbonium ion IV is undoubtedly the driving force behind these unusual reactions.

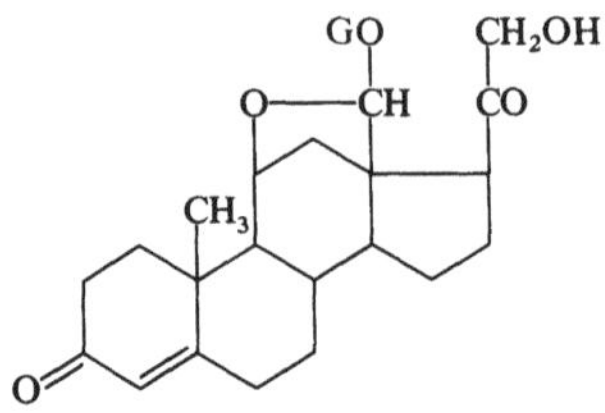

I

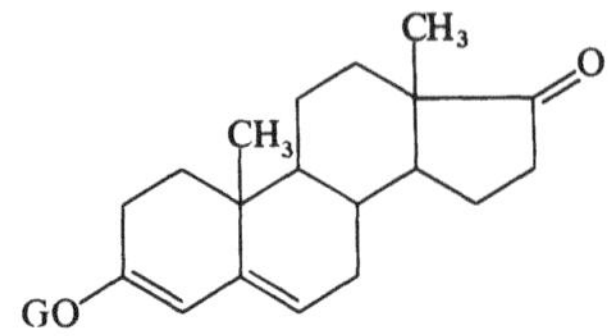

II

G≅glucuronic acid

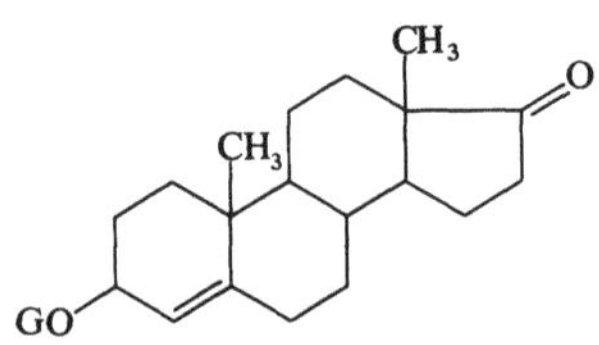

III

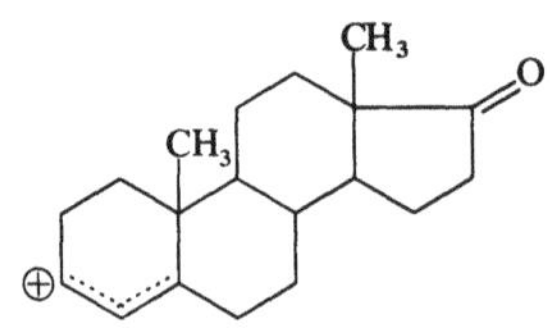

IV

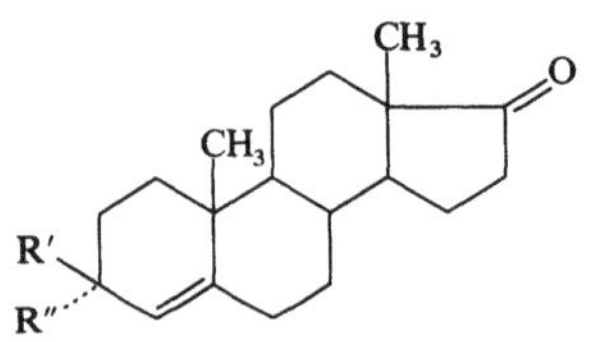

Va, R′=H, R″=HO
Vb, R′=OH, R″=H
Vc, R′=H, R″=NH_2CONH

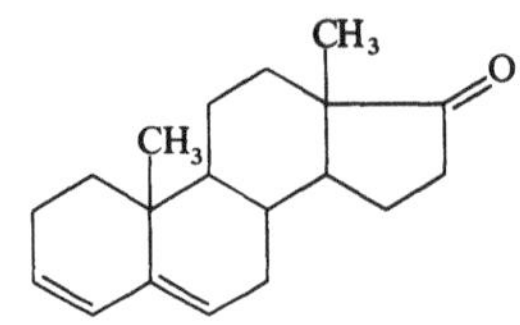

VI

Finally, the enol glucuronide of androst-4-ene-3,17-dione (II) [80] is spontaneously hydrolyzed at pH values below 6. This is of more than academic interest since the urinary excretion of androstenedione enol glucuronide [17-oxoandrosta-3,5-dien-3-yl-β-D-glucopyranosiduronic acid] has been described [81].

B. Solvent-Catalyzed Hydrolysis

1. Continuous Ether Method

Continuous ether extraction of urine or a conjugate extract acidified to pH 1 has proved a useful and popular method for the hydrolysis of steroid sulfates. Studies by Lieberman and Dobriner [82] in 1948 established that, under these conditions, sulfates are rapidly hydrolyzed, while glucuronides are almost totally unaffected. Buehler *et al.* [83] confirmed these findings at a slightly lower pH, 0.7. On the other hand, when the urine was 7.2 *N* in hydrochloric acid, they found that glucuronides were cleaved in good yield.

Cohen and Oneson [84] reported that this latter procedure destroyed a substantial fraction of the dehydroisoandrosterone present. This strong acid method does not appear to have been explored further. Lieberman *et al.* [85] demonstrated that at pH 1 the reaction proceeds with rupture of the S–OR linkage and complete retention of configuration. No signs of epimerization could be detected.

Continuous ether solvolysis has been included in the hydrolytic sequence of a number of methods for the analysis of 17–ketosteroids [86]. Since these studies have contributed relatively little to our knowledge of solvolysis, no attempt will be made to discuss the various minor modifications in these methods. It was assumed in these studies that the hydrolysis occurred in the aqueous phase, and that the ether merely extracted the liberated steroids; indeed, in at least one study, marble chips were included in the ether to neutralize any acids carried over into the ether [86f]. This assumption was proved wrong by Burstein and Lieberman [87]. Dehydroisoandrosterone sulfate was found to be stable in aqueous solution at pH 1 for up to 9 days. Even cortisone sulfate [17α–hydroxy–21–sulfooxypregn–4–ene–3,11,20–trione], which is rapidly solvolyzed on continuous extraction, was stable for 24 hours in acid solution. The hydrolysis proceeds in two stages: first, extraction of the conjugates into ether; second, solvolysis of the sulfates in the ether solution. It is clear, therefore, that any polar sulfates that are not extracted into the ether will not be hydrolyzed. An instance of this arose in a study of the further metabolism of androsterone sulfate [3α–sulfooxy–5α–androstan–17–one] [88]. After continuous ether extraction, androstane–3α,17β–diol disulate [3α,17β–disulfooxy–5α–androstane] was recovered unhydrolyzed in the aqueous phase. Since this conjugate was readily hydrolyzed by the tetrahydrofuran–perchloric acid method of Burstein and Lieberman [87], it is clear that the unfavorable partition coefficient was the cause of the difficulty. This limitation must be kept in mind in any case in which very polar conjugates may be present.

2. Static Procedure

Hydrolysis of steroid conjugates in acidified dioxane, tetrahydrofuran, or ethyl acetate is a very useful procedure with considerable potential. The method dates from the observation of Grant and Beall [89] that purified estrogen sulfates were rapidly hydrolyzed on standing in dioxane solution at room temperature. The solvolysis was blocked by the presence of 10% or more of water or alcohol, or by impurities in the conjugate extracts. Cohen and Oneson [84] found that 17–ketosteroid sulfates were also cleaved under these conditions. Thus, dehydroisoandrosterone sulfate was quantitatively hydrolyzed in 4 hours at 25°, or 10 minutes at 100°. The solvolysis of these sulfates was also blocked by the presence of water (> 1 %), alcohol (> 30 %), or by impurities in the conjugate extracts, but the interference could be overcome by the addition of acid. On balance they felt that trichloroacetic acid was the best choice, since acetic acid was too weak and mineral acids caused some glucuronide hydrolysis. In the presence of 10 % w/v of trichloroacetic acid, the solvolysis of pure dehydroisoandrosterone sulfate was complete in 10 minutes at room temperature although crude urinary extracts still required more than 2 hours.

Some years later, the mechanism was studied by Burstein and Lieberman [90] who found that the reaction proceeded by first-order kinetics in organic solvents of low polarity, especially ethers, and that the rate was unchanged over a 10-fold range of acid concentration from 0.001 to 0.01 M.

$$ROSO_3^{\ominus} + H^{\oplus} \rightleftharpoons \underset{\text{I}}{ROSO_3H} \rightleftharpoons \underset{\text{II}}{R\overset{\oplus}{\underset{H}{O}}\overset{\ominus}{S}O_3}$$

Solvated ion — Undissociated acid — Reactive species

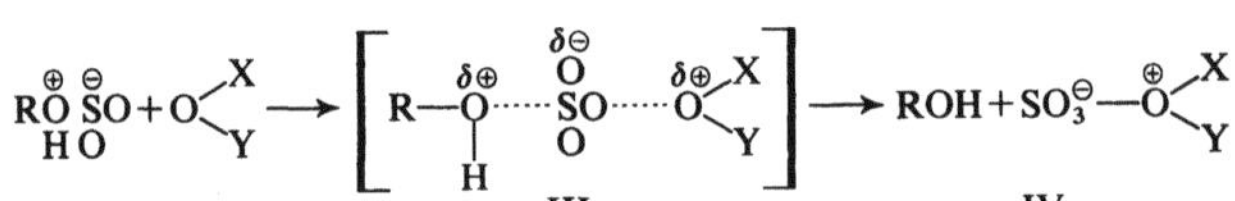

III IV

They suggested the following mechanism: the undissociated free acid I is in rapid equilibrium with the dipole ion II. The latter reacts with a nucleophilic reagent XOY to give the transition state III. This rearranges to give the free alcohol, and the sulfated nucleophilic reagent IV; XOY, depending on circumstances, could be water, alcohol, or ether. This mechanism is consistent with the studies of Lewis and Boozer [91] who found that, when the solvolysis was carried out in $H_2{}^{18}O$, no ^{18}O was incorporated into the isolated steroids. The close agreement of the rates of solvolysis of dehydroisoandrosterone sulfate and isoandrosterone sulfate [3β-sulfooxy-5α-androstan-17-one] is consistent with this mechanism and against one involving a carbonium ion intermediate. If the latter were the correct mechanism, then the solvolysis of dehydroisoandrosterone sulfate should be accelerated by anchimeric assistance from the 5,6 double bond. The solvolysis of steroid sulfates was found to proceed rapidly in tetrahydrofuran containing up to 6 % water and 0.01 *N* in perchloric acid. Ethyl acetate could also be used although there was increased artifact formation and some acetylation. As a consequence, an additional base hydrolysis step is required if the free compounds are to be obtained. Similar results were obtained by

DePaoli *et al.* [92], who used vapor phase chromatography to demonstrate the absence of artifact formation. Extending earlier observations of Segal *et al.* [93], they found that ethyl acetate containing 10^{-4}M sulfuric acid could also be used for the hydrolysis of sulfates.

The yield was seriously affected by the presence of impurities in the conjugate extracts. When the conjugates were extracted with ether-alcohol [94], the recovery of ketosteroids following ethyl acetate–perchloric acid solvolysis was only 55% of that obtained when the conjugates were extracted with ethyl acetate at pH 4 or tetrahydrofuran at pH 7, from urine containing 50 g/100 ml of ammonium sulfate.

Although the rate of solvolysis in ethyl acetate 0.01 M in $HClO_4$ is faster than that in tetrahydrofuran 0.01 M in $HClO_4$, the ultimate yield is the same in both solvents. Tetrahydrofuran was shown to have several important advantages over ethyl acetate: 1. The conjugates are more efficiently extracted from urine. 2. The increased artifact formation and acetylation observed with ethyl acetate are avoided. As a result, the need for a saponification step is eliminated. 3. With minor modifications, glucuronides can also be cleaved.

Jacobsohn and Lieberman [74] found that 17–ketosteroid glucuronides were also solvolyzed in tetrahydrofuran provided the water concentration was reduced below 0.5%, and the perchloric acid concentration was raised from 0.01 to 0.1 M. By appropriate manipulation of water and acid concentration, sulfates and glucuronides could be solvolyzed either sequentially or together. Artifact formation was minimal, and the recovery was as good as that obtained after β–glucuronidase hydrolysis, followed by continuous ether extraction at pH 1.

Preliminary experiments indicate that the method is also applicable to the hydrolysis of pregnanediol–3–glucuronide [20α–hydroxy–5β–pregnan–3α–yl–β–D–glucopyranosiduronic acid]. Recovery studies have shown that even those 17–ketosteroids, dehydroisoandrosterone, 11β–hydroxyandrosterone [3α,11β–dihydroxy–5α–androstan–17–one], and 11β–hydroxyetiocholanolone, known to be sensitive to mineral acid, are recovered quantitatively from tetrahydrofuran 0.1 *N* in perchloric acid. On the other hand, pregnane–3α,17,20α–triol was destroyed, and most corticoids appeared to be rearranged under these conditions.

An unusual variant of this procedure has been described by Eberlein for steroid sulfates [95]. The urine or plasma is treated with Methyl Green, which forms a chloroform–soluble complex with steroid sulfates. The complex, when dissolved in acetic acid containing a trace of perchloric acid and heated briefly, is solvolyzed to give steroid acetates. The method is probably satisfactory for 17–ketosteroid sulfates, but it seems likely that serious artifact formation will occur with 17,20–glycols. The results [96] obtained for cord–blood metabolites using this solvolysis procedure must be treated with caution until this extent of rearrangement has been evaluated.

C. Comment

For those steroid aglycones that are resistant to rearrangement in tetrahydrofuran containing perchloric acid, this procedure has a great many advantages for hydrolysis of the conjugates, and in many cases may well be the method of choice. The extracts obtained are relatively clean and quite suitable for vapor phase chromatographic analysis. In choosing between enzyme hydrolysis and solvolysis, several practical considerations must be evaluated. Although enzyme hydrolysis can be carried out directly on the urine, the conjugates must be extracted before solvolysis can be carried out. The enzyme extract can also be analyzed for its corticoid content as well as for 17–ketosteroids; the solvolytic extract cannot be used for corticoid analysis. On the other hand, solvolysis combines, in a single step, enzyme hydrolysis plus continuous ether extraction at pH 1.

Continuous ether extraction at pH 1 remains a popular method for the hydrolysis of steroid sulfates following enzyme hydrolysis. Although limited to those sulfates extractable by ether, this has not been a serious problem in practice. While continuous ether extraction takes longer, it requires less labor than tetrahydrofuran–perchloric acid solvolysis. The choice between these two methods of solvolysis depends upon the aims of the study and the physical facilities available.

V. Hot Acid Hydrolysis

In addition to its historic significance, hot acid hydrolysis still retains its popularity largely for reasons of speed and cost. Since the procedures have been customized for particular hormone–metabolite groups, each will be discussed separately.

A. Progesterone Metabolites

Because of problems with the gravimetric analysis of pregnanediol–3–glucuronide [15], methods for the hydrolysis of the conjugate have been studied intensively. Hot acid hydrolysis was found to hydrolyze pregnanediol–3–glucuronide in good yield by Odell and Marrian [97] in 1936. Major improvements in this procedure were introduced by several groups [98–100] to improve the yield and decrease artifact formation. This procedure was most systematically studied by Klopper *et al.* [101], who found that the best yield of pregnanediol was obtained using 10% v/v of concentrated hydrochloric acid and a toluene overlay, although the extract required extensive purification before final quantitation. Alternatives to this procedure have been offered by Jensen [102] and Waldi [103], but no direct comparisons have been carried out. Using Klopper's method, van der Molen [104] reported that the recovery of pregnanolone [3α–hydroxy–5β–pregnan–20–one] was greater with acid than with enzyme hydrolysis, and that acid-catalyzed rearrangement of urinary pregnanetriol to pregnanolone was not a serious source of error.

Although Klopper's procedure has become a standard, the limitations have been pointed out by several groups. Talbot *et al.* [17] reported a 30% loss of pregnanediol on acid hydrolysis of pregnancy urine, whereas the recovery was 92% following glucuronidase hydrolysis. Patti *et al.* [105] found that acid hydrolysis gave a product that required extensive purification prior to gas chromatographic analysis, whereas the enzyme–hydrolysis extract was easily analyzed by this procedure. Barry *et al.* [106a], after a comparison of several methods, also concluded that acid hydrolysis was less satisfactory than enzyme hydrolysis from the point of view of specificity, precision, sensitivity, and convenience. Metcalf [106b] has recently described an analytical method for pregnanolone, in which pregnanediol was selectively destroyed by heating with 3.8 *N* hydrochloric acid at 100° for 15 minutes. These results emphasize the possibilities for serious error when hot acid hydrolysis is used in the determination of pregnanediol.

Recent work has suggested several sources of artifact formation. Hirschman and Williams [107] showed that acid hydrolysis of 5α–pregnane–3β,20β–diol–20–sulfate [20β–sulfooxy–5α–pregnan–3β–ol] results in rearrangement to uranediol [17α–methyl–D–homo–5α–androstane–3β,17aβ–diol]. They concluded that all of the uranediol isolated from natural sources was an artifact of hydrolysis. Since Okamura and Tateyama [108] reported that a substantial portion of urinary pregnanediol is excreted as the sulfate, the possibility of significant artifact formation cannot be excluded as the location of the sulfate group has not yet been determined.

1. Comment

Despite these limitations, acid hydrolysis of pregnanediol conjugates is still commonly employed in clinical laboratories because of its low cost, convenience, and speed. However, since vapor–phase chromatography offers considerable advantages with regard to sensitivity and specificity for the analysis of steroid metabolites, it seems likely that acid hydrolysis will be increasingly superseded by milder procedures that introduce fewer artifacts.

B. Androgen Metabolites

Despite the abundant evidence of artifact formation, hot acid hydrolysis remains the overwhelming choice of most clinical chemists for the determination of 17-ketosteroids.

Engel [109] in his review of 17-ketosteroid determinations has summarized the variety of conditions that have been employed by different groups. The major variables that have been explored include:

1. Whether hydrochloric or sulfuric acid should be used and how much?
2. Should the urine be acidified cold or heated before acidification?
3. How long should the heating be continued?
4. Finally, how rapidly should the urine bc cooled and how well-controlled should the cooling be?

1. Acid concentrations between 1.0 and 1.6 N have been used most commonly. Essentially comparable results have been obtained with hydrochloric and sulfuric acids, although the latter has been favored on the grounds that its use will result in less formation of chlorinated artifacts. This argument is not compelling in view of the high concentration of chloride ions in urine.

2. The second point has never been adequately resolved. Some workers prefer preheating because it permits more precise control of the time the urine is exposed to hot acid. Others have argued that the use of a constant volume provides for more uniform heating of the urines.

3. In general, a short hydrolysis (10–30 minutes) has been preferred, and the duration has varied inversely with the temperature employed.

4. No satisfactory method has been devised for controlling the cooling time. The rate of cooling is determined primarily by the volume of the sample and the tap-water temperature. Here, too, the use of a constant sample volume is the best way to control cooling. The importance of this variable has never really been established.

The effect of all of these variables has been studied largely by comparing the gross yield of 17-ketosteroids. Almost nothing has been done on the nature of the 17-ketosteroids released, and the extent of artifact formation. In the absence of this information we cannot tell whether a specific change improves the yield by increasing the extent of hydrolysis or decreasing the extent of destruction.

Problems of artifact formation were recognized early. The sensitivity of dehydroisoandrosterone to acid hydrolysis was shown by Butenandt and Grosse [110] in 1936, who isolated 3β-chloroandrost-5-en-17-one, following acid hydrolysis. Dobriner *et al.* [111] reported dehydration to androst-3,5-dien-17-one. The common intermediate in both reactions is the cyclo-steroid ion (III), which can react with chloride ion to give the 3β–chloro–compound (II) or by losing a proton to give the diene (IV). Consequently, hot acid hydrolysis is precluded when one wishes to accurately measure dehydroisoandrosterone (I). Since the Zimmermann equivalent of III and IV are higher than I, this increased color can compensate for some of the destruction and may at times result in an apparent 17–ketosteroid value greater than the sum of the 17–ketosteroids actually present. Hot acid hydrolysis also increases the concentration of nonsteroidal chromogens leading to serious errors (*vide infra*).

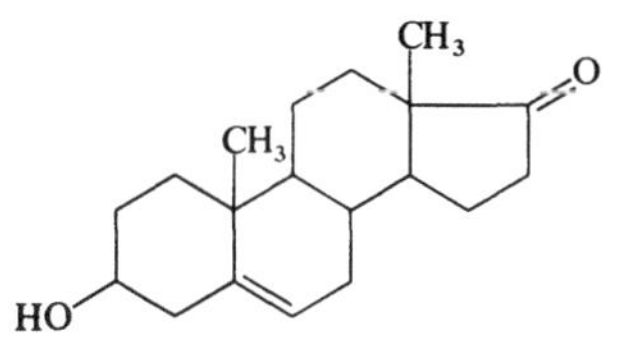

I

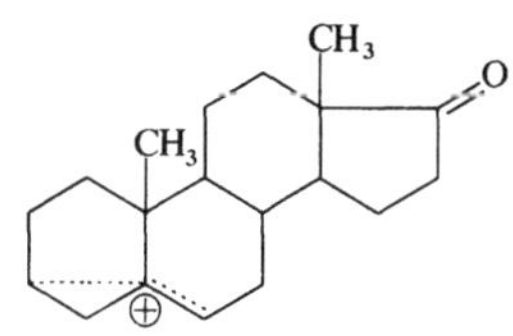

III

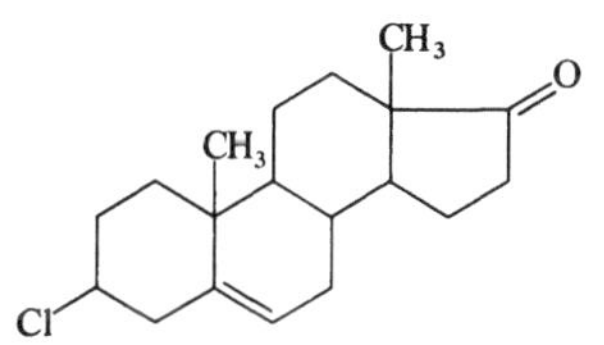

II

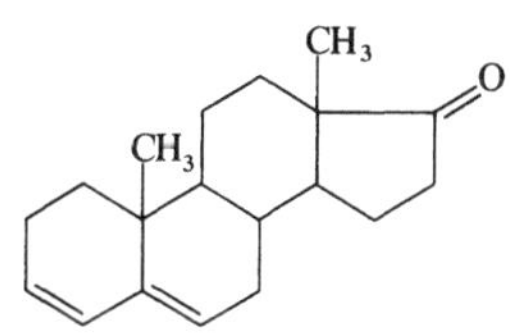

IV

Venning *et al.* [112] in 1942 pointed out that hot acid converts some of the urinary androsterone sulfate to androst–2–en–17–one. More recent work by Ramseyer and Hirschman [113] showed that hot acid hydrolysis of androsterone sulfate leads to the formation of 2α–, 2β–, and 3β–hydroxy epimers in addition to the unsaturated compound. Dehydration of 11β–hydroxy compounds to the $\Delta^{9(11)}$–compound was first reported by Dobriner [114].

Many approaches have been explored in an attempt to minimize these losses. Dingemanse and Laqueur [115] showed that destruction of dehydroisoandrosterone could be minimized by simultaneous extraction of the liberated steroids by a benzene overlayer. Since dehydroisoandrosterone is the principal compound lost in hot acid hydrolysis, Jensen and Tötterman [116] developed a two–stage sequence in which dehydroisoandrosterone sulfate is hydrolyzed by thermal solvolysis followed by hot acid hydrolysis to cleave the other 17–ketosteroid conjugates. The average recovery after this sequence was 120% of that of a single–stage procedure. Birke and Plantin [117] found that the recovery of 17–ketosteroids depended on the pH, with the maximum at values between 0.2 and 0.6. For the isolation of dehydroisoandrosterone alone, the optimum pH was 0.8. In their final procedure, the authors added enough 40% sulfuric acid to bring the pH to 0.4–0.5, and heating the material for 25 minutes. Under these conditions the use of hydrochloric acid resulted in complete destruction of dehydroisoandrosterone.

Lieberman *et al.* [118], in a review of various methods, concluded that hot acid hydrolysis was suitable only for the hydrolysis of the conjugates of those steroids not sensitive to acid (androsterone [3α–hydroxy–5α–androstan–17–one], etiocholanolone, and 11–ketoetiocholanolone), but not for acid–labile compounds like 11β-hydroxyandrosterone, 11β–hydroxy-etiocholanolone, or dehydroisoandrosterone.

The two recommended methods for 17–ketosteroids in volumes 2 and 5 of Standard Methods of Clinical Chemistry [119] used concentrations of hydrochloric acid, which would certainly destroy substantially all of the dehydroisoandrosterone present in the urines. While the total 17–ketosteroid values obtained by either procedure may fortuitously be reasonable (*vide infra*), any attempt to fractionate the extracts obtained using either procedure in even the simplest manner will yield erroneous results. Vestergaard and Claussen [120] concluded that acid hydrolysis with simultaneous benzene extraction, although giving a slightly lower total recovery than enzyme hydrolysis followed by solvolysis, was still adequate for routine clinical purposes. In a comparative study of hydrolytic procedures, Goldzieher and Axelrod [121] found that nonspecific chromogens account for 73% of the total Zimmermann chromogens following acid hydrolysis. As a result, the correlation between the total Zimmermann chromogens and the actual 17–ketosteroids present is exceedingly rough.

Acid hydrolysis has been retained by many research groups [122] as the final step after enzyme hydrolysis and solvolysis because an additional increment of Zimmermann–reactive material is obtained following this step. Jaoudé *et al.* [122a] found that androsterone and 11β–hydroxyandrosterone were the principal compounds recovered from the final hot acid extract. Androsterone was also the only compound recovered in the final hot acid extract, following the administration of 4–^{14}C–testosterone [17β–hydroxyandrost–4–en–3–one] to patients [123]. It is not clear whether this additional increment of androsterone is derived from a different conjugate or from acid hydrolysis of residual androsterone glucuronide remaining after the previous hydrolytic steps. Studies by several groups [124] have shown that androsterone glucuronide [17–oxo–5α–androstan–3α–yl–β–D–glucopyranosiduronic acid] is more slowly hydrolyzed than etiocholanolone glucuronide [17–oxo–5β–androstan–3α–yl–β–D–glucopyranosiduronic acid] by all of the common enzyme preparations. The marked difference in the specific activity of the androsterone derived from the glucuronide and hot acid fractions [123] (Table 3–2) suggests that this latter explanation is not adequate. A substantial fraction of the androsterone in the hot acid fraction must be derived from some precursor other than the material obtained from the glucuronide and sulfate fractions.

TABLE 3–2. *Relative specific activity of androsterone conjugated as the glucuronide, the sulfate, and as hot acid–cleaved conjugates*

Patient	Glucuronide [a, b]	Sulfate [c]	Hot acid [d]
Gr	100	25	29
WI	100	16	22
SZ	100	69	38
GN	100	89	43
FN	100	88	17
LE	100	29	49
KN	100	42	39
AN	100	65	38
DO	100	43	22
BI	100	71	83
HN	100	54	67
GN	100	57	79

[a] Relative specific activity with the glucuronide value set equal to 100.

[b] Glucuronide = that fraction of androsterone liberated following hydrolysis with 300 units/ml of Ketodase for 5 days at 37° followed by continuous ether extraction.

[c] Sulfate = that fraction of androsterone liberated following continuous ether extraction of the acidified residual urine (pH 1) from the enzyme hydrolysis.

[d] Hot acid = that fraction of androsterone liberated by boiling the residual urine from the preceding step for 30 minutes after making it 1 *N* in sulfuric acid, followed by continuous ether extraction.

Attempts to use vapor phase chromatography for the analysis of acid–hydrolyzed extracts have not been successful. Several laboratories [125] have found that artifacts generated during acid hydrolysis seriously interfere with vapor phase chromatography, generating numerous spurious peaks.

1. Comment

The dilemma mentioned in the introduction comes to the fore in the use of acid hydrolysis for 17–ketosteroids. Despite the numerous difficulties which render hot acid hydrolysis unattractive for research studies, it has retained its attractiveness for clinicians seeking a quick total value as a diagnostic tool. In this situation it has been argued that errors up to 20% are not important, since clinicians need to know only whether the value is high or low. However, the results of Goldzieher and Axelrod [121] make it clear that, in many cases, results may be incorrect by as much as 300%. In addition, the advantages of speed are largely illusory since there are almost no situations in which such speed is essential, and the actual labor involved in enzyme hydrolysis is not much greater. The difficulties in carrying out vapor phase chromatographic analysis on the extracts obtained after acid hydrolysis represent a new limitation on the procedure. All of these problems make acid hydrolysis increasingly unattractive even for clinical studies.

Whenever detailed information is needed in a clinical situation or for research studies, acid hydrolysis is completely unacceptable. It causes serious distortions in the metabolic pattern because of the loss of acid–sensitive compounds, and, in addition, precludes the possibility of studying changes in conjugation pattern.

C. Estrogen Metabolites

Acid-catalyzed hydrolysis of estrogen conjugates has been studied intensively and numerous methods have been described. The effects of variations in temperature, time, acid concentration, and protective agents have been studied in great detail. Since the earlier work has been reviewed by Bauld and Greenway [126], no attempt will be made here to recapitulate all of these variations.

The large number of estrogens and their conjugates creates serious difficulties in evaluating these methods. Absolute recovery experiments, based on the recovery of known amounts of an added conjugate, have not been carried out because of the unavailability of these conjugates. Since the estrogen class reactions cannot be applied to unfractionated urine, most investigators have been forced to judge from comparative recoveries that a given hydrolytic procedure gave a higher recovery than another one. One method may appear to give more metabolites, not because it hydrolyzes conjugates better, but because it destroys less. Recovery figures based on the addition of free metabolites to the urine are not necessarily valid since the stability of conjugates to acid may be greater or less than that of the free metabolites, as pointed out by Katzman *et al.* [127]. A serious limitation, which has only recently been appreciated, is that acid hydrolysis permits recovery of only the classical estrogens, estrone [3–hydroxyestra–1,3,5(10)–trien–17–one], 17β–estradiol [estra–1,3,5(10)–triene–3,17β–diol], and estriol [estra–1,3,5(10)–triene–3,16α,17β–triol]. Bugge *et al.* [128] showed that acid hydrolysis destroys 2–methoxyestrone

[3-hydroxy-2-methoxyestra-1,3,5(10)-trien-17-one], 2-hydroxyestrone [2,3-dihydroxyestra-1,3,5(10)-trien-17-one], and the ring-D ketols, all important metabolites, which vary significantly in different diseases. Because of all of these imponderables, it is difficult to assess these methods correctly. These uncertainties can be eliminated if known amounts of radioactive estrogen conjugates are added to the specimen to be hydrolyzed as internal recovery standards. From the radioactivity present in the purified extract one can readily correct for the various losses. Nachtigal *et al.* [129a] applied this technique to the analysis of pregnancy plasma for estriol, using a mixture of estriol conjugates in the same proportions as they are present in plasma. The method has been simplified by Bassett and Levitz [129b] who showed that ^{3}H-estriol-16-glucuronide could be substituted for the conjugate mixture previously used. Jaffe and Levitz [129c] and Schindler *et al.* [129d] described methods for urinary estriol in pregnancy which employ the same technique to correct for hydrolytic destruction. Schindler *et al.* [129d] added estriol or estriol-3-glucuronide [16α,17β-dihydroxyestra-1,3,5(10)-trien-3-yl-β-D-glucopyranosiduronic acid] to pregnancy urine prior to hydrolysis.

Marrian and Bauld [130a, b], and Brown [130c] carried out extensive studies on acid hydrolysis of estrogen conjugates. In the standardized method which these workers devised, hydrolysis was effected by refluxing for 60 minutes with 15% v/v of concentrated hydrochloric acid. This time was found to be a balance point, since, with longer heating, the extent of the decomposition was greater than the increase in hydrolysis, whereas for shorter times, the reverse was true. The phenolic extract obtained after this hydrolysis required careful purification prior to final analysis. Brown and Blair [131] showed that the recovery of the classical estrogens, when corrected for losses, agrees well with the recovery after enzymatic hydrolysis. This procedure has been used extensively in pregnancy studies.

Frandsen [132] stated that the increased speed with which estriol measurements can be made in cases of fetal stress is an important advantage of acid hydrolysis over enzymatic methods. Since estriol is present in large amounts in late pregnancy urines, it has been argued that many of the difficulties that arise when the concentration is lower will not occur. However, a number of precautions still must be kept in mind. If glucose is present, either the urine must be diluted or the conjugates extracted prior to hydrolysis, as suggested by Frandsen *et al.* [133]. Removal of the sugar by yeast fermentation was found to destroy estriol [134]. Serious interference with the recovery of estriol will occur when any of a wide variety of drugs is present. Touchstone *et al.* [135] reported that the presence of methenamine mandelate, a drug used in the treatment of urinary tract infections, resulted in the destruction of estriol by formaldehyde released during acid hydrolysis. Brown *et al.*[136] reported that phenolphthalein, cascara, senna, and diethylstilbestrol [3,4–bis–(4′–hydroxyphenyl)–3–hex–ene], all interfere with acid hydrolysis. Adlercreutz and Schauman [137] found that the acid hydrolysis of urine from patients with liver disease led to negative values for estrogen. This problem also arose during the studies of biliary metabolites. In most of these situations, the interference can be completely eliminated by using enzyme hydrolysis.

The dangers inherent in interference by drugs or their metabolites which react with hormones during analysis were stressed in a recent note by Sele and Frandsen [138]. They emphasized that drugs should be discontinued during the test period and that the analyst should be informed of any that cannot be discontinued.

The magnitude of the destruction of estriol in pregnancy urine has been measured recently by Schindler *et al.* [129d]. Using 15α–^{3}H–estriol or 16–^{14}C–estriol glucuronide as recovery standard, they measured the recovery of estriol with changes in the concentration of some common sugars, the formaldehyde–generating drugs noted by Touchstone *et al.* [135], and total urinary solutes. In the presence of glucose, the losses were greater than 50%, and over 95% in the presence of sucrose, fructose, and inulin, whereas almost complete destruction was found in the presence of the formaldehyde–generating drugs. No loss of estriol was observed in the presence of dimethylamine, albumin, or Urofix® (Biological Products, N.Y.). In the absence of such compounds, the effect of solute concentration, while serious, was not as severe. A linear decrease from 90 to 60% in the recovery of estriol occurred as the specific gravity rose from 1.000 to 1.020. Since changes in estriol excretion are often the basis for decisions on premature delivery of babies, Schindler *et al.* [129d] argued that one must be sure that any changes in estriol values truly reflect changes in excretion and not changes in recovery. This problem is especially serious in cases of toxemia, where diuresis, either spontaneous or drug–induced, is a common occurrence. Previous studies [139] on the association of increased excretion of estriol and diuresis will have to be re–evaluated in the light of these findings to determine the extent to which the apparent increased recovery of estriol was actually a consequence of the decreased specific gravity of the urine. In such situations, the use of a radioactive internal standard provides the most reliable measure of estriol losses during hydrolysis. In the absence of glucose and interfering drugs, simple dilution as suggested by Hobkirk *et al.* [134] seems to minimize the effects of solute concentration. Brown *et al.* [136] concluded that the destruction of estriol observed in concentrated urines was due to a specific substance whose effects could be minimized by dilution. Although the results of Rosenmund [140], Van Bruggen [141], and others have indicated that the addition of reducing agents increased the recovery when estrogen conjugates were hydrolyzed in aqueous solution, the results of Smith and Smith [142] and Venning *et al.* [143] suggested that there are sufficient natural reducing agents present in urine to obviate the need to add additional reducing agents. These results suggested that urine contains both protective reducing agents and other substances that destroy estrogens during acid hydrolysis. Dilution minimizes the effect of the harmful substances without interfering with the protective effect of the reducing agents. Further evidence concerning these interfering substances was reported by Cohen [4b]. He found that, if the estrogen conjugates were isolated by ammonium sulfate precipitation and then subjected to acid hydrolysis. the estriol level was substantially greater than that obtained when the urine was hydrolyzed directly. The solubility of these interfering substances in saturated ammonium sulfate suggests that they must be low molecular-weight substances. He also found that his procedure resulted in less destruction of estrogens than the dilution technique reported by Hobkirk *et al.* [134]. A further advantage of Cohen's procedure is that a direct Kober assay can be carried out on the conjugates if desired. The nature of both groups of substances is completely unknown.

Smith and Smith [142] and Stimmel [144] reported that the presence of zinc dust increased the recovery of estrogen following acid hydrolysis. However, the studies of Marrian and Bauld [130a], Falk and Heard [145], and Reckers and Katzman [146] showed that this method can lead to serious changes in the metabolites, including loss of the oxygen function at C17. Since these conditions are similar in many respects to those employed in the Clemmenson reduction of ketones to deoxy compounds, the results are not surprising. Because of the enormous difficulties in interpreting the results, methods capable of effecting such alterations should not be used, whether one wishes only a crude total value or intends to further fractionate the extracts.

D. Comment

The only situation in which acid hydrolysis of estrogen conjugates can still be justified is fetal distress, where the saving in time may be crucial. Because of the unpredictable losses that arise during acid hydrolysis, the use of an internal standard [129] seems mandatory. The development of a nonhydrolytic method for estimating estrogens in pregnancy urine by Cohen [4] and a rapid enzymatic method by Scholler *et al.* [147] suggests that satisfactory alternatives to acid hydrolysis exist even in those sitations where speed is required.

In other situations, because of the errors and distortions of the estrogen pattern that are caused by acid hydrolysis, this approach can no longer be considered acceptable. Since only the classical estrogens can be recovered after hot acid treatment, declines in these compounds may actually reflect diversion to other estrogens destroyed by acid hydrolysis, rather than an actual decrease in the production of estrogenic substances.

VI. Enzymatic Hydrolysis

A. Sulfatase Hydrolysis

Since the purely enzymological aspects are dealt with in Chapter 2, this section will be restricted to practical applications of sulfatase preparations to the hydrolysis of steroid sulfatases. Despite considerable efforts by many research groups, the results have been very limited in scope.

Efforts by Buehler *et al.* [28] to find a bacterial sulfatase comparable in activity to their *E. coli* β–glucuronidase were completely unsuccessful. Although mammalian sulfatase preparations have been obtained, Roy [148] concluded, after careful study, that none of these preparations was sufficiently active to be useful.

Fungi and mollusks have been the source of all of the useful steroid sulfatase preparations. A high degree of substrate specificity is a characteristic of all of these preparations, greatly limiting their usefulness. The failure to find a broadly active preparation is not surprising, since Fromageot [149] noted in his review that a high degree of substrate specificity is a basic characteristic of all sulfatases.

1. Fungal Preparations

Enzyme preparations from *Aspergillus oryzae* have been employed for more than 25 years. Butenandt and Hofstetter [19] used Taka®–diastase to hydrolyze estrone sulfate [3–sulfooxyestra–1,3,5(10)–trien–17–one], and Abbott [150] reported that phenol sulfatase activity was present in Polidase S®, Clarase®, Diastase vera®, Mylase P®, and related preparations. Although initial reports by Cohen and Bates [151] claimed that a substantial fraction of estrone, 17β–estradiol, and estriol conjugates of human pregnancy urine (5 to 100%) were cleaved by Mylase P, subsequent studies by Stimmel [152], Oneson and Cohen [153], and Katzman *et al.* [127] could not confirm these results. The last group found that Mylase P hydrolyzed estrone sulfate quantitatively, and that this sulfatase and β–glucuronidase could be used in either order [154] for the hydrolysis of estriol double conjugates. They also found [127] that Mylase P and bacterial

β–glucuronidase can be employed simultaneously, since their pH optima were close and there was no interference between the two enzymes. Although the sulfatase activity of the various *A. oryzae* preparations are roughly comparable, the observation of Emerman and Levitz [155] indicates a need for caution. They found that, while Taka–diastase and Mylase P were equally effective hydrolytically, the former contained sufficient dehydrogenase activity to cause substantial interconversion of estrone and 17β–estradiol, while the latter was free of such activity.

2. Molluscan Sulfatases

In the early fifties, several British and French research groups [156] isolated active steroid and phenol sulfatases from a number of molluscan species. Although active preparations have been isolated from many species [157], greatest attention has been given to two marine species, *Patella vulgata*, the common limpet, and *Littorina littorea*, the common periwinkle, and to the common snail, *Helix pomatia*. Since all three have roughly similar activities, they will be discussed together.

a) Phenol Sulfatase Activity. Dodgson and Spencer [158] found that the aryl sulfatase activity of acetone powders obtained from the marine mollusks ranged from 7,000 to 140,000 units/g, from 5 to 100 times greater than the best mammalian or fungal preparations, and that the activity of snail preparations was even greater. Stitch and Halkerston [159] reported that inhibition of this enzymic activity by sulfate and phosphate ions could be readily overcome by precipitation with barium chloride. Since the barium salts do not interfere, they need not be removed. The use of these active preparations gave cleaner extracts which were easier to purify than those obtained when mammalian or fungal preparations were used.

Because the crude preparation also contains substantial glucuronidase activity [160], the material can be used to effect the total hydrolysis of estrogen conjugates [161], cleaving the glucuronides and sulfates simultaneously. Although the pH optima of the two enzyme activities are not identical [162], they are sufficiently close that satisfactory hydrolysis can be achieved at an intermediate pH. Bauld and Greenway [126] found that contaminants derived from enzyme hydrolysis created problems in measuring estriol by their method.

Jarrige [163] described the purification and properties of snail sulfatase. Use of this purified sulfatase for the hydrolysis of estrogen sulfates has been described by Jarrige *et al.* [164].

b) Steroid Sulfatase. Roy [29] suggested that this enzyme in the marine species should be called 3β–steroid sulfatase since the enzyme hydrolyzes only 3β,Δ^5– and 3β,5α–steroid sulfates. The sulfates of testosterone [17β–

sulfooxyandrost–4–en–3–one], androsterone, etiocholanolone [3α–sulfooxy–5β–androstan–17–one], and pregnanolone [3α–sulfooxy–5β–pregnan–20–one] are not hydrolyzed. Several investigators [157, 164, 165] have shown that the *Helix* preparation also hydrolyzes etiocholanolone sulfate in addition to $3\beta,\Delta^5$– and 3β,5α–sulfates. Roy has suggested (Chapter 2) that the hydrolysis of etiocholanolone sulfate is probably due to a specific enzyme differing from the more abundant 3β–steroid sulfatase.

In the light of these substrate limitations [156b, 166a, b] the claim that molluscan preparations could be used for the total hydrolysis of the urinary 17–ketosteroid conjugates was premature. The reduced destruction and artifact formation are more than counterbalanced by the failure to hydrolyze a major urinary component, androsterone sulfate.

c) C21 Steroid Sulfatase. This enzyme, which has been found in several molluscan species [29], is most abundant in the *Helix* preparation. Work by Dodgson [167] and Roy [29] indicates that this enzyme is probably distinct from the other alkyl sulfatases (*cf.* Chapter 2). Pasqualini and Jayle [168] have described its use for the hydrolysis of 17–deoxy–C_{21}–steroid sulfates.

All of the available evidence suggests that sulfation does not play an important role in the conjugation of cortisol metabolites. While studying oxidative cleavage of cortisol metabolites (*cf.* p. 139) both Edwards and Kellie [37] and Few [55] concluded that only a minor fraction of urinary cortisol metabolites were excreted as sulfate conjugates. A comparison of the hydrolysis of the urinary metabolites following an intravenous $4\text{–}^{14}C$–cortisol tracer [169] (Table 3–3) with a mammalian glucuronidase prepara-

TABLE 3–3. *Hydrolysis of the urinary conjugates derived from cortisol by mammalian and snail β–glucuronidase*

Patient	Mammalian[a]	Snail
Rs	76	74
Fn	33	39
Nm	56	60
Sk	67	65
Kn	71	71
Vo	72	72
Hr	69	75

[a] % of the radioactivity present in the original urine recovered in the neutral extract.

In each study, equal portions of a 48–hr urine collection following intravenous administration of $4\text{–}^{14}C$–cortisol was incubated with 300 units/ml of glucuronidase for 5 days at 37° and pH 5.

tion free of sulfatase activity and a snail preparation containing both activities leads to the same conclusion. The recovery of neutral radioactivity was not increased when the mixed enzyme preparation was used.

3. Comment

Both fungal and molluscan preparations have been successfully used for the hydrolysis of estrogen sulfates. Although the snail preparation is as efficient as solvolysis for the hydrolysis of estrogen sulfates, the extract is bulkier when the snail preparation is used. Since the stability of estrogen metabolites toward molluscan preparations has not been established, the use of Mylase P may be preferable as this preparation has been shown not to alter estrogens [155].

The results of Pasqualini and Jayle [168] suggest that the molluscan C21–sulfatase may be of practical value for those 17–deoxy–corticoid metabolites which are often excreted as the sulfates.

In view of the high degree of specificity of molluscan steroid sulfatases, these preparations have only a limited range of application. It should be kept in mind that, under the conditions actually employed in the hydrolysis of steroid conjugates, substantial hydrolysis of steroid conjugates which are nominally resistant to steroid sulfatases can occur (*cf.* Chapter 2 for a detailed discussion of this point). Although many workers have drawn sweeping conclusions from experiments carried out under enzymological conditions, this may lead to serious errors when hydrolyses are carried out under more drastic conditions.

B. β–Glucuronidase Hydrolysis

Although the widespread use of β–glucuronidase hydrolysis is of recent origin, it dates back to the early observation of Cohen *et al.* [13] that bacterial contamination of pregnancy urines resulted in the hydrolysis of estrogen conjugates. Successful hydrolysis of estrogen conjugates following deliberate inoculation with *E. coli* was described by Patterson [14]. Bucher and Geschickter [15] found that bacterial contamination also resulted in the hydrolysis of pregnanediol–3–glucuronide. Veitch *et al.* [12] described the use of a plant enzyme, emulsin, for the hydrolysis of pregnanediol–3–glucuronide. The first active mammalian preparation was described by Fishman [16] in 1939, and Talbot *et al.* [17] showed that crude rat–liver preparations could also be used to hydrolyze pregnanediol–3–glucuronide.

Despite these few early studies, serious application of β–glucuronidase hydrolysis dates from 1950. In rapid succession highly active bacterial [18, 170], mammalian [171], and molluscan [156b, 160b, 172] preparations were described. Buehler *et al.* [170a] described the induction of an active glucuronidase by growing *E. coli* in the presence of menthol glucuronide. This preparation cleaved the glucuronides of estriol, pregnanediol, phenolphthalein, corticoids, and ketosteroids, but was devoid of sulfatase activity. The optimum pH was between 6 and 7. Some inhibition by NH_4^+ was observed.

In 1951, Cohen [171a] described the isolation of a glucuronidase from calf spleen which cleaved a broad spectrum of steroid glucuronides. A short time later, Bernfeld *et al.* [171c] described a beef liver preparation which was much more active than the spleen material. When it became commercially available, this material rapidly supplanted the dry spleen preparation. The commercial product, Ketodase® (Warner–Chilcott Laboratories) is the best known form.

Another rich source of β–glucuronidase was found in the digestive juice of the snail, *Helix pomatia* [156b]. All known glucuronides were readily hydrolyzed by this preparation, which also contained substantial sulfatase activity. If desired, the sulfatase activity can be blocked by the presence of 1% of fluoride ion. Alternatively, the purification procedure of Wakabayashi and Fishman [173] can be used to obtain a sulfatase–free glucuronidase preparation. Preparations with a comparable spectrum of activity were also obtained from the limpet and the periwinkle by Stitch and Halkerston [172] and Dodgson *et al.* [160b]. β–Glucuronidase activity was shown to be present to some extent in a wide variety of marine mollusks [160b]. Other molluscan β–glucuronidase preparations have been described by Billett [174], Cox [175], and Fripp [176]. Since none of the these sources have come into general use, and since they offer no special advantages, further discussion will be confined to the snail and limpet preparations, which are commercially available and have come into general use.

The respective merits of the different enzyme preparations have been the subject of numerous studies by many groups. Despite all this work, the problem has not been adequately resolved.

A good deal of the confusion has originated in the choice of substrate used in the assay procedures. The confusion concerning the role of organic solvents as adjuvants of bacterial glucuronidase and inhibitors of the mammalian enzyme provides a good case history. In 1959, Gautney *et al.* [177], expanding on a Sigma Corporation bulletin, described the activation of phenolphthalein glucuronide hydrolysis by bacterial β–glucuronidase by a variety of organic solvents. Hydrolysis of the same substrate by a spleen preparation was inhibited by these solvents. Ryan and Mavrides [178] showed that the activation was related to the solvent-water interface. As the area of the interface was increased by emulsification or by spreading the solution out, the activity of the bacterial enzyme increased. They suggested that there was some physiochemical property of the bacterial enzyme at the interface not shared by the mammalian enzyme.

Since pre–extraction of free steroids with chlorinated solvents is a step in many sequential procedures for the extraction of steroid conjugates, these results created considerable concern. Smith and Bongiovanni [179] demonstrated that these results were true, but also not relevant to the practical situation involved in steroid conjugate hydrolysis. Hydrolysis of urinary 17–ketosteroid and corticoid glucuronides by the mammalian enzyme was only slightly affected by solvents at all but the lowest enzyme concentration (Table 3–4), unlike the hydrolysis of phenolphthalein

TABLE 3–4. *Influence of halogenated solvents on the hydrolysis of steroid conjugates*

Units/ml of Ketodase	Control	Methylene chloride	Carbon tetrachloride
		17–Ketosteroids	
10	16	9.4	10.7
100	45	43.3	41.0
500	100	96.0	94.0
		Corticoids	
10	16	14	13
100	59	68	62
250	86	81	81
500	100	100	100

pH 4.5, 37°, 48 hr. All hydrolyses were carried out on aliquots of the same urine. The results are the % of the maximum achieved in the control aliquot.

glucuronide which was significantly affected at the typical concentrations employed in the standard activity assay [180], but less so at higher enzyme concentrations. Studies by Kirk *et al.* [181], Slaunwhite and Sandberg [182], and Vestergaard [183] also confirmed that chloroform was without effect on the hydrolysis of steroid conjugates by β–glucuronidase of bacterial, mammalian, or molluscan origin. Thus the entire effect turned out to be an artifact of the assay system of no practical consequence.

Since different systems will hydrolyze test substrates like phenolphthalein glucuronide or 8–hydroxyquinoline glucuronide at very different rates, having no relationship to the rate at which they hydrolyze steroid conjugates [184], it is not meaningful to compare the extent of hydrolysis following the addition of the same number of units of the various enzyme preparations to samples.

Most authors have concluded that one unit of bacterial enzyme, as determined by the Sigma procedure [185], is equivalent to 10 to 20 Fishman units of Ketodase or molluscan material. The amounts of enzymes used have been adjusted accordingly in many studies. Only a few investigators have systematically compared all four sources. In many cases, the results are contradictory, with each group concluding that its favorite enzyme was better. Cohen *et al.* [186] compared the activity of several different, dry, spleen preparations, Ketodase, and a bacterial preparation of unstated origin. All of the preparations yielded about the same amount of 17-ketosteroids and corticoids from samples of pooled urines. Any differences were within experimental error. The yield of pregnanediol, on the other hand, varied greatly. Quantitative hydrolysis was obtained with Ketodase or with their own preparation of spleen enzyme. With commercial spleen preparations, hydrolysis was poor, ranging from 24 to 79% of the total, while the bacterial enzyme gave very erratic results ranging from 18 to 90% of the maximum. The cause of these variations was not explained.

Comparative studies [187] on the hydrolysis of the urinary metabolites of ^{3}H-cortisol in man with Ketodase or bacterial β-glucuronidase demonstrated that the yield was the same for both preparations although the rate for the bacterial enzyme was somewhat faster. Changes in the pH during the extraction of the metabolites and the addition of deoxynucleic acid were both without effect. The extract obtained following use of the mammalian enzyme contained substantially less pigments and other interfering substances than the bacterial extracts. Since no attempt was made to suppress bacterial growth, it is possible that the addition of antibiotics might have improved the results obtained with the bacterial enzyme. Jayle *et al.* [161] compared the ability of snail and bacterial preparations to hydrolyze urinary estrogen conjugates. They concluded that the snail preparation, used at a 2.5- to 6-fold greater concentration than the bacterial preparation (in Fishman units), gave somewhat better results. The increase is as likely to be due to the phenol sulfatase content of the snail preparation as to any increased glucuronidase activity.

Slaunwhite and Sandberg [182], in a comparison of the ability of bacterial, mammalian, and molluscan preparations for the hydrolysis of urinary and biliary conjugates of ^{14}C-labeled estrogens, found that the last mentioned was slightly better for the hydrolysis of urinary conjugates and much better for the hydrolysis of biliary conjugates. Three times as much molluscan enzyme was needed for a comparable degree of hydrolysis when the incubation time was decreased from 96 to 48 hours. The superiority of the molluscan preparation was more marked with estrone and 17β-estradiol conjugates than with estriol conjugates, which were equally hydrolyzed by the different preparations. Artifact formation was found to be a more serious problem with the bacterial preparation (up to 84%) than with the others. Metal ion inhibition of hydrolysis was only serious for Cu^{2+} and Hg^{2+}, and this could be reversed by the addition of ethylenediaminetetraacetic acid. Both urine and bile contained other inhibitors which could not be overcome by ethylenediaminetetraacetic acid, serum albumin, or protamine. An immediate drop in enzyme activity occurred when the various preparations were assayed in the presence of urine, but no further drop occurred on continued incubation. Other glucuronides were considered the most probable source of the inhibition. The evidence presented does not exclude the possibility that the marginal advantage of the molluscan preparation was primarily due to the fact that it was the only one of the three with sulfatase activity. For the hydrolysis of bile conjugates, the molluscan preparation was unequivocally better, hydrolyzing three times as much of the biliary conjugates as the other two enzymes. In the light of present knowledge on the presence of double conjugates in bile [188], these latter results are not surprising.

Hobkirk and Cohen [189] compared all four preparations (bacterial, liver, snail, and limpet) for the hydrolysis of endogenous, urinary 17-ketosteroid conjugates. The rate of hydrolysis of 5β-metabolites was the same with all four preparations; however, the hydrolysis of 5α-metabolites varied markedly with the various preparations. Hydrolysis of 5α-metabolites was most rapid with the bacterial enzyme but the recovery decreased with time after 48 hours. Although the rate was slower, snail and limpet materials produced the same ultimate amounts of 5α-metabolites as the bacterial preparation. Hydrolysis of the 5α-compounds was slowest with the beef liver preparation, and the ultimate yield was substantially less than with the other three preparations. The effect was most pronounced with 11β-hydroxyandrosterone. Poor hydrolysis of 5α-compounds by the liver enzyme was also reported by Wotiz *et al.* [190]. Although it has been suggested that a portion of the 5α-metabolites is conjugated to some unknown, nonlabile conjugate, hydrolyzable by a hypothetical constituent of the other enzyme preparations, the over-all evidence is against this possibility. Sufficiently long hydrolysis with Ketodase will hydrolyze 11β-hydroxyandrosterone glucuronide [191], a conjugate which is only partially hydrolyzed on 5-day incubation with 300 units of Ketodase/ml. Unpublished work from this laboratory has shown that the same is true for urinary metabolites of 4-^{14}C-allotetrahydrocortisone [3α,17α,21-trihydroxy-5α-pregnane-11,20-dione]. As shown in Table 3-5, after hydrolysis for 120 hours, rehydrolysis with Ketodase for another 120 hours released a substantial second increment of radioactive metabolites.

TABLE 3–5. *Hydrolysis of the urinary conjugates derived from allotetrahydrocortisone*

Patient	a	b
GE	37	22
GT	44	23
ST	33	25

[a] % of urinary radioactivity recovered in the neutral extract after incubation for 5 days at 37° with 300 units/ml of Ketodase followed by continuous ether extraction for 48 hours.

[b] % recovered in the neutral extract when the residual urine after the initial enzyme hydrolysis was retreated with an additional 300 units/ml of Ketodase for a second 5–day period.

The first critical study of enzyme hydrolysis on purified steroid conjugates was carried out by Bongiovanni [165]. He compared the hydrolysis of androsterone glucuronide, etiocholanolone glucuronide, and tetrahydrocortisone glucuronide [17α,21–dihydroxy–11,20–dioxo–5β–pregnan–3α–yl–β–D–glucopyranosiduronic acid] by bacterial, mammalian, and snail enzymes. Although the bacterial enzyme was somewhat faster than the other two, the ultimate yield was about the same with all three preparations. Etiocholanolone glucuronide was the most rapidly hydrolyzed of the three. Wakabayashi and Fishman [192] carried out a much more extensive study of the comparative ability of liver, bacterial, snail, and limpet preparations to hydrolyze glucuronides of C_{18}, C_{19}, and C_{21} compounds. These workers adjusted the potency of the different preparations to the same activity level prior to comparing the hydrolytic ability of the enzymes. The pH optimum for the hydrolysis of all of the steroid conjugates was significantly less for all four enzymes than for phenolphthalein glucuronide. Conjugates of the pregnane compounds were hydrolyzed more readily than either the estrogen or androgen conjugates. They concluded that no one enzyme preparation was uniquely superior in efficiency of hydrolysis of the substrates studied, although choices existed for several of the conjugates.

Kirk *et al.* [181], in the following year, reached quite opposite conclusions in a comparison of bacterial and mammalian preparations for the hydrolysis of estrogen and 17–ketosteroid conjugates in pregnancy urines. They concluded that the bacterial enzyme was preferable because less was required and the rate of hydrolysis was greater. They believed that these factors would help minimize the extent of alterations of the liberated steroids. They also concluded that the molluscan preparations were disadvantageous because large amounts were required and knowledge of the type of conjugation is lost. They were unable to reconcile their findings, that at least 300 units/ml of Ketodase were required for maximum yield of 17-ketosteroids, with the earlier report of Wotiz *et al.* [190] that maximum recovery was obtained with only 50 to 100 units of Ketodase/ml. They did concur that hydrolysis of phenolphthalein glucuronide is a poor index of the activity of a glucuronidase preparation toward other glucuronides.

Vestergaard [183], in the same year, reported the comparative hydrolysis of urinary 17-ketosteroid conjugates by liver, bacterial, and limpet preparations. With optimum amounts of each preparation, Ketodase 500 units/ml/5 days, bacterial enzyme 250 units/ml/20 hr, limpet 2,000 units/ml/48 hr, they found that the maximum yield of ketosteroids released from pooled male urine was the same for the three preparations. On the score of time alone, only the bacterial preparation was capable of splitting the 17-ketosteroid conjugates within a 24-hour period; however, frequent adjustments of the pH were required because it tended to drift excessively. The limpet preparation required a 2-day incubation period but showed no tendency toward pH drift. The liver preparation required the longest time, at least 5 days, and also showed no tendency to pH drift. When extracts of urinary conjugates were studied, the bacterial preparation was unsatisfactory, and the other two were equally good, although the limpet preparation had the added advantage of also hydrolyzing dehydroisoandrosterone sulfate. He too found that phenolphthalein glucuronide is a poor test substrate for evaluating the ability of enzymes to hydrolyze steroid glucuronides.

One additional useful feature of this study was a cost analysis for each of the three enzyme preparations for 30-ml aliquots of urine. Ketodase was the cheapest at 36¢, whereas the bacterial preparation was the most expensive ($ 1.01 to 1.73). Since the cost of the enzyme has often been advanced as a reason for not employing enzyme hydrolysis, these figures suggest that cost is not really a limiting factor.

The most recent study in this controversial field was by Becker [124b], who compared the hydrolytic activity of bacterial, molluscan, and mammalian β-glucuronidases on the synthetic glucuronides of 3-hydroxy-17-ketosteroids and phenolphthalein. The pH activity curves, K_m values, and the effects of buffer ions on enzyme activity were determined. Although the rate constants for the three preparations differed, all four of the synthetic conjugates were hydrolyzed by all three. The results suggest that variations in buffer composition might offer advantages in selected cases. The use of a dimethylglutarate buffer markedly enhanced the hydrolysis of androsterone glucuronide by the bacterial enzyme, while in a phosphate buffer the hydrolysis was markedly inhibited. Contrary results were observed with the molluscan preparation where the hydrolysis was best in an acetate buffer and slowest in a dimethylglutarate buffer.

"Transferase" activity has been reported to be a property of many β–glucuronidases [193]. The effect of propylene glycol on the hydrolysis of steroid conjugates has been explored by Cohen. Although the addition of propylene glycol resulted in some activation of the hydrolysis of pregnanediol–3–glucuronide by old spleen enzyme preparations [194], the hydrolysis of estrogen conjugates by mammalian, bacterial, or molluscan glucuronidase was inhibited by the addition of propylene glycol [195]. Becker [124b] in his studies on synthetic conjugates (*vide supra*) found propylene glycol to have no effect on the rate of hydrolysis or the ultimate yield of aglycone. Until these conflicting results are resolved, prudence indicates that the addition of potential acceptors like propylene glycol should be avoided.

From this mass of often conflicting data there are certain reasonably firm conclusions that can be reached.

1. There is no one best source of β–glucuronidase for the hydrolysis of steroid glucuronides. With proper choice of enzyme concentration and time of incubation, all of the common enzyme preparations are capable of satisfactorily hydrolyzing steroid glucuronides.

2. There are marked differences in the rates of hydrolysis by different preparations. The bacterial preparation is fastest, and possibly advantageous in situations where very rapid hydrolysis is required.

3. The rates of hydrolysis of steroid glucuronides vary with the structure. Pregnane derivatives are most rapidly hydrolyzed, followed by corticoid, 17–ketosteroid, and estrogen conjugates, in that order. Within each group the rate depends on the configuration of the hydroxyl group and the adjacent ring junctions. 3–Hydroxy–5β–compounds are hydrolyzed more rapidly than 3–hydroxy–5α–compounds, and 3β–glucuronides better than 3α–glucuronides.

In a few situations, unusual circumstances may determine the choice of enzyme. Only one situation has been observed in which only one preparation can be used. Pasqualini *et al.* [196] reported that the 18–glucuronide of aldosterone is cleaved slowly by the snail preparation, but not by the bacterial or mammalian enzymes. The extent of artifact formation is often a major determinant. Since the most serious alterations have been reported with the bacterial preparation [186], this material should be used with caution whenever fractionation of the hydrolysate is planned. Solomon [197] has also observed serious alterations with the snail preparation. The least amount of alteration has been observed with the mammalian preparation.

Very considerable differences have been observed in the amount of extraneous material present in the extract following the use of different enzyme preparations. As noted above [187], although the recovery of radioactivity was the same when samples containing radioactive cortisol metabolites were incubated with Ketodase or the bacterial preparation, the bacterial extracts were much bulkier and more difficult to analyze. A similar comparison of snail and mammalian preparations revealed that the extracts obtained following the use of the snail material were considerably heavier than those obtained following the use of mammalian material. Fishman [198] observed a similar increase in bulk in the hydrolysis of radioactive estrogen metabolites with the snail preparation.

Interference with further analytical procedures is sometimes a problem. Bauld and Greenway [126] had difficulty in quantitating estriol when they used a limpet preparation instead of acid hydrolysis, and additional purification steps were required as described by Brown and Blair [131]. Metcalf [199] reported that variable amounts of interfering material were present in different batches of Ketodase which seriously altered the blank values in her method for measuring acetaldehydrogenic steroids. Kream [200] found that the use of bacterial enzyme results in a lower blank in 17–hydroxycorticoid assays than when Ketodase was used.

A major obstacle to the widespread use of enzyme hydrolysis has been the "inhibitor problem". Although, as discussed in Chapter 2, there are glucuronidase inhibitors present in urine and bile, the real question is whether serious interference occurs in the hydrolysis of steroid conjugates under the usual conditions of hydrolysis. When assays are made of enzyme activity in the presence of urine, it can be demonstrated that a prompt drop in enzyme activity occurs. However, when adequate amounts of enzyme are used and hydrolysis is carried out for a suitable time, complete hydrolysis of glucuronides can be achieved. Zumoff and Bradlow [201] found that hydrolysis with 300 units of Ketodase/ml. for 5 days released almost all of the glucuronides actually present in the urine samples containing radioactive steroid metabolites. Although it has been shown that glucurono–1,4–lactone will inhibit hydrolysis [202], no evidence has been presented for the common occurrence of this or related substances in urine. Kushinsky and Chen [203] found that the nature and extent of inhibitors was uncertain, although they felt that saccharo–1,4–lactone was most likely the principal inhibitor. However, even in this study, the effect was only significant at low concentrations of enzyme and short periods of hydrolysis. With higher enzyme concentrations and longer incubation periods, inhibition was minor, although they suggested that prolonged enzyme hydrolysis was risky because of the potential lability of some steroids under these conditions.

Cohen [195] reported evidence suggesting that two types of inhibitors are present. One group consists of compounds which remain water–soluble when the conjugates are precipitated by saturation with ammonium sulfate. Saccharo–1,4–lactone belongs to this class. The other consists of substances which can be separated from the steroid conjugates on a protein polypeptide precipitate. The action of both groups of inhibitors is to raise the optimum pH hydrolysis. As might be anticipated, the effect is most marked when low levels of enzyme are employed. On the basis of these results, he concluded that Ketodase was a very poor hydrolytic agent for urinary estrogen conjugates. This is in conflict with the findings of Fishman [198], that substantially all of radioactivity present as glucuronide after the administration of a radioactive estradiol tracer is cleaved by incubation with 300 units/ml of Ketodase for 5 days at 37° (an average of 75% of the urinary radioactivity). The difference is undoubtedly due to the differences in enzyme concentration and incubation time. Stempfel *et al.* [204] reported data suggesting that salicylate glucuronide could act as a competitive inhibitor interfering with the hydrolysis of corticoid metabolite conjugates. When the hydrolysis was carried out at a concentration of 400 units/ml of Ketodase for 48 hours, the recovery of corticoid conjugates was substantially less in the presence of salicylate glucuronides. The inhibition could be completely overcome by using 3,000 units of enzyme/ml for 5 days. Although successful, the authors felt that these latter conditions might result in destruction of some of the metabolites. Extraction of the conjugates prior to hydrolysis was recommended as the choice when large amounts of urinary glucuronides derived from drugs might be present. Table 3–6 compares the hydrolysis of the urinary metabolites of 4–^{14}C–cortisol, 1,2–^{3}H–tetrahydrocortisol

TABLE 3–6. *Noninhibition of β–glucuronidase hydrolysis of urinary conjugates derived from corticoids by aspirin treatment*

	Cortisol [a]	Tetrahydro-cortisol [a]	Tetrahydro-cortisone [a]
Control	63 [b]	70	60
Aspirin [c]	62	75	68

[a] In each case an aliquot of a 48–hr urine collection following intravenous administration of the labeled precursor was hydrolyzed with 300 units/ml of Ketodase for 5 days at 37° and pH 5.

[b] % of the urinary radioactivity recovered in the neutral extract.

[c] 4 g of aspirin was given daily for at least 14 days prior to and during the administration of the radioactive tracer.

[3α,11β,17α,21–tetrahydroxy-5β–pregnan–20–one], and 1,2–^{3}H–tetrahydrocortisone in the same patient before and after treatment with 4 g/day of aspirin. No significant difference in the hydrolysis of any of the conjugates was detected. The ability of Fukushima *et al.* [205] to account for almost all of the urinary neutral extract derived from 4–^{14}C–cortisol as identified compounds following a 5–day incubation suggests that prolonged incubation does not cause serious decomposition.

Much of the confusion concerning the "inhibitor" problem rests upon a failure to properly appreciate the fact that competition is dependent upon the ratio of the substrate concentration "s" to the K_m value (*cf.* Chapter 2). Under conditions in which the substrate concentrations are low relative to K_m, the velocity in the presence of the competitor is identical to that in the absence of the competitor; *i.e.*, inhibition by competing substrates is impossible. If "s" is increased by concentration prior to enzyme hydrolysis, then the ratio "s"/K_m may not be low and competitive inhibition may become important.

In the light of all of these considerations, the practical significance of enzyme inhibition, under the conditions employed by steroid analysts, is small since the effects can always be overcome by increasing the enzyme concentration. This is not to say that, from time to time, such severe inhibition may occur that prior extraction of the conjugates will be required.

An implicit assumption in most studies is that compounds isolated following glucuronidase hydrolysis were originally present as glucuronides. Unless the structure of the conjugates has been verified by isolation or by specific inhibition studies [202], cleavage of a conjugate by β–glucuronidase must be regarded as supplying probable but not conclusive evidence that the conjugate is a glucuronide. The basic problem is the nonavailability of pure glucuronidase preparations free of other hydrolytic enzymes.

Recently, Layne [206] and Arcos and Lieberman [207] reported slow hydrolysis of steroid N–acetylaminoglycosides by Ketodase. In both cases the compounds were originally present as double conjugates and no single N–acetylaminoglycoside conjugate has been isolated from natural sources. The evidence to date suggests that misinterpretation from this source is not likely to be important.

C. Simultaneous Enzyme Hydrolysis of Double Conjugates

Hydrolysis of steroids, doubly conjugated as the sulfate and the glucuronide can be carried out in different sequences. The sulfate can be removed chemically or enzymatically as desired, followed by glucuronidase hydrolysis of the remaining glucuronide. Alternatively the glucuronide can be hydrolyzed first with glucuronidase, followed by sulfatase hydrolysis or solvolysis. Either sequence is satisfactory.

Finally, using either one of the molluscan preparations [160] or a mixture of bacterial glucuronidase and Mylase P [127], simultaneous hydrolysis of both conjugates can be carried out. Practically speaking, this approach is confined to estrogen double conjugates.

D. Comment

Although substantial progress has been made in obtaining glucuronidase preparations of high potency and specificity for glucuronides, only meager progress has been achieved in obtaining comparable sulfatase preparations. All of the sulfatase preparations possess high specificity cleaving only aromatic, $3\beta,\Delta^5$– and $3\beta,5\alpha$–sulfates. The results to date suggest that it is not likely that broadly active preparations will be found.

Thus far, results suggest that all of the common preparations (mammalian, bacterial, or molluscan) can be used for the hydrolysis of all of the known steroid glucuronides except aldosterone glucuronide, which can only be hydrolyzed by the snail enzyme. The choice depends on a variety of factors like cost, local availability, the extent of alterations of the metabolites under investigation by the enzyme preparation and interference with analytical procedures caused by enzyme impurities. In the light of the cost analysis cited by Vestergaard [183] and the lesser tendency to alter steroid metabolites, the mammalian enzyme is probably the choice in most situations, even though its rate of hydrolysis is somewhat slower than the others. Fortunately, few endocrine problems require answers so quickly that speed is a major factor. In such cases the bacterial enzyme may be preferred, although one should keep in mind that this preparation has the greatest tendency to cause alterations in steroid metabolites. Recently, Scholler *et al.* [147] reported that rapid hydrolysis of estrogen conjugates can also be achieved with high concentrations of the snail enzyme.

Many of the problems of inadequate hydrolysis are probably due to the use of inadequate amounts of enzyme. It cannot be emphasized too strongly that enzyme concentrations from 300 to 500 units per ml are required for complete hydrolysis [183]. In some cases, especially with $3\alpha,5\alpha$–steroid glucuronides, prolonged hydrolysis may be necessary to achieve complete recovery of the aglycones. Recent reports of the hydrolysis of N–acetylaminoglycosides by glucuronidase serve as a reminder that hydrolysis by glucuronidase represents probable but not conclusive evidence that the conjugate is a glucuronide.

VII. Summary

Although substantial progress has been made in developing methods for the hydrolysis of steroid metabolite conjugates, many problems remain. Two major areas where further research is needed are the development of methods for the hydrolysis of sulfates of acid–sensitive compounds, and enzymatic procedures for the hydrolysis of conjugates not hydrolyzed by presently available enzyme preparations.

The prospects for finding a nonspecific sulfatase preparation do not seem bright since a high degree of specificity appears to be an inherent characteristic of sulfatases [149]. Practically speaking, all of the currently available preparations are limited to the hydrolysis of estrogen sulfates, $3\beta,\Delta^5$– and $3\beta,5\alpha$–steroid sulfates, and corticosteroid C21–sulfates. For all other sulfates, some form of solvolysis remains the method of choice despite the problems of artifact formation when dealing with the sulfate of metabolites with ketolic groups and tertiary hydroxyls. The possibility of carrying out solvolysis in the absence of acid is suggested by the recent studies of Bayyuk and Juraydini [208] describing the solvolysis of some sterol sulfates on heating in dimethylsulfoxide–water solution.

The problem of conjugates not cleaved by glucuronidases or solvolysis, although not serious when studying normal patients, is often major when studying patients receiving drugs like o,p'DDD[1–(2'–chlorophenyl),1–(4'–chlorophenyl)–2,2–dichloroethane] or Elipten [3–ethyl–(4'–aminophenyl)–2,6–dioxopiperidine] and in patients with liver disease [72]. In such paticnts more than 50% of the urinary radioactive metabolites may be excreted as conjugates resistant to Ketodase or *Helix pomatia* hydrolysis. For obvious reasons, hot acid hydrolysis as a final step after milder procedures is restricted to those metabolites not altered by hot acid. In patients not receiving radioactive tracers, the decrease in hydrolysis may erroneously lead one to think that the formation of the precursor has diminished when only a change in the conjugation pattern has actually occurred. The extent of this effect is illustrated in Table 3–7. In both patients a marked decrease in the glucuronide fraction was evident for all of the tracers. The production rate was not significantly changed by the treatment. Also, column chromatography of the urinary conjugates derived from 4–^{14}C–testosterone in the control and treatment periods confirmed the decrease in the fraction of the urinary radioactivity present as glucuronides.

The existence of this problem has been largely ignored. This may be due in part to the fact that the problem can be recognized only when tracer studies are carried out. The true extent of the alteration of steroid hormone metabolism in disease may be masked if the metabolites conjugated to

TABLE 3–7. *Effects of o,p'DDD on the conjugation of steroid metabolites*

Patient	Hormone given	% of urinary radioactivity					
		"Glucuronide"		"Sulfate"		"Hot Acid"	
		Control	DDD	Control	DDD	Control	DDD
VK	4–^{14}C–Testosterone	57	7	13	28	4	30
	4–^{14}C–Dehydroiso-androsterone	48	13	12	32	10	38
	7α–^{3}H–Dehydroiso-androsterone sulfate	36	3	15	9	9	15
	4–^{14}C–Cortisol	60	12	1	4		
WD	4–^{14}C–Dehydroiso-androsterone	23	17	30	24	13	18
	4–^{14}C–Cortisol	28	14	1	1		

In each study, the radioactive hormone was administered intravenously. "Glucuronide" is that fraction released by incubation with 300 units/ml of Ketodase at pH 5.0 and 37° for 5 days. "Sulfate" is that fraction released by continuous ether extraction of the urine for 48 hours after the enzymatic incubation. "Hot Acid" is that fraction released by adding 5% v/v of concentrated sulfuric acid, refluxing for 30 minutes, and continuously extracting with ether for 48 hours.

these unknown conjugates differ in structure or relative concentration from those in the hydrolyzable fraction. Research into the structure and hydrolysis of these conjugates is needed. Aside from the knowledge of the steroid metabolites in this fraction, the structure of the unknown conjugating moieties may provide additional insight into the effects of illness on metabolic processes.

At present, many enzyme preparations must be used with a certain degree of caution because of the presence of other enzymatic activities capable of altering the released metabolites, or the presence of impurities which can interfere with subsequent analytical procedures [126, 199, 200]. The development of preparations containing only the desired enzyme activity would enormously simplify many research problems. Unlike the sulfatase–specificity problem discussed above, this appears to be inherently solvable.

While nonselective procedures like hot acid hydrolysis and oxidative cleavage will always retain some popularity for clinical assays because of their speed and low cost, it appears likely that the former at least will be increasingly displaced by procedures better adapted to modern analytical methods [105, 106, 126]. As discussed in earlier sections, the limitations to both procedures are built into them, making the prospects for substantial improvement doubtful.

For the present time a combination of enzyme hydrolysis and solvolysis would appear to be the method of choice for most situations other than the clinical assays mentioned above. The exact combination which would be optimal in a given laboratory will not always be the same, since it depends to a great degree on factors like the availability and cost of different enzyme preparations, the nature of the conjugates under investigation, and the facilities available.

VIII. References

1. a) Ashmore, J., Elliott, W. H., Doisy, E. A., Jr., Doisy, E. A.: J. Biol. Chem. **200**, 661 (1953). — b) Valcourt, A. J., Thayer, S. A., Doisy, E. A., Jr., Elliott, W. H., Doisy, E. A.: Endocrinology **57**, 692 (1955). — c) Grady, H. J., Elliott, W. H., Doisy, E. A., Jr., Bocklage, B. C., Doisy, E. A.: J. Biol. Chem. **195**, 755 (1952). — d) Bradlow, H. L., Dobriner, K., Gallagher, T. F.: Endocrinology **54**, 343 (1954).
2. Hellman, L., Bradlow, H. L., Frazell, E. I., Gallagher, T. F.: J. Clin. Invest. **35**, 1033 (1956).
3. Venning, E. H.: J. Biol. Chem. **119**, 473 (1937).
4. a) Cohen, S. L.: J. Clin. Endocrinol. Metab. **26**, 994 (1966). — b) J. Clin. Endocrinol. Metab. **29**, 47 (1969).
5. Funk, C., Harrow, B., Lejwa, A.: Proc. Soc. Exptl. Biol. Med. **26**, 569 (1929).
6. — — Biochem. J. **24**, 1678 (1930).
7. Doisy, E. A., Veler, C. D., Thayer, S.: J. Biol. Chem. **86**, 499 (1930).
8. Lipschütz, A., Poch, E.: Compt. Rend. Soc. Biol. **111**, 856 (1932).
9. Borchardt, H., Dingemanse, E., Laqueur, E.: Naturwissenschaften **22**, 190 (1934).
10. Butenandt, A., Dannenbaum, H.: Z. Physiol. Chem. **229**, 192 (1934).
11. Odell, A. D., Marrian, G. F.: Can. Med. Conf. 1936.
12. Veitch, F. P., Jr., Milone, H. S., Everitt, E. L.: Science **88**, 551 (1938).
13. Cohen, S. L., Marrian, G. F., Watson, M.: Lancet **I**, 674 (1935).
14. Patterson, J.: Brit. Med. J. **II**, 522 (1937).
15. Bucher, N. L. R., Geschickter, C. F.: Endocrinology **27**, 727 (1940).
16. Fishman, W. H.: J. Biol. Chem. **131**, 225 (1939).
17. Talbot, N. B., Ryan, J., Wolfe, J. K.: J. Biol. Chem. **151**, 607 (1943).
18. Buehler, H. J., Katzman, P. A., Doisy, P. P., Doisy, E. A.: Proc. Soc. Exptl. Biol. Med. **72**, 297 (1949).
19. Butenandt, A., Hofstetter, H.: Z. Physiol. Chem. **259**, 222 (1939).
20. Dingemanse, E., Borchardt, H., Laqueur, E.: Biochem. J. **31**, 500 (1937).
21. — Huis in't Veld, L. G.: J. Biol. Chem. **195**, 827 (1952).

22. Dingemanse, E.: Ciba Found. Colloq. Endocrinol. **2**, 251 (1952).
23. Teich, S., Rogers, J., Lieberman, S., Engel, L. L., Davis, J. W.: J. Am. Chem. Soc. **75**, 2523 (1953).
24. Huis in't Veld, L. G.: Rec. Trav. Chim. **76**, 622 (1957).
25. Winstein, S., Adams, R.: J. Am. Chem. Soc. **70**, 838 (1948).
26. Talbot, N. B., Ryan, J., Wolfe, J. K.: J. Biol. Chem. **148**, 593 (1943).
27. a) Bitman, J., Cohen, S. L.: J. Biol. Chem. **179**, 455 (1949). — b) J. Biol. Chem. **191**, 351 (1951).
28. Buehler, H. J., Katzman, P. A., Doisy, E. A.: Federation Proc. **9**, 157 (1950).
29. Roy, A. B.: Biochem. J. **62**, 41 (1956).
30. Fotherby, K.: Biochem. J. **73**, 339 (1959).
31. — Biochem. J. **69**, 596 (1958).
32. Stern, M. I., Barwell, J. O. H.: J. Endocrinol. **27**, 87 (1963).
33. Gallagher, T. F., Fukushima, D. K., Hellman, L.: Metabolism **11**, 1155 (1962).
34. Norymberski, J. K., Stubbs, R. D., West, H. F.: Lancet **I**, 1276 (1935).
35. Péron, F. G., in: Methods in hormone research (R. I. Dorfman, ed.), vol. 1, p. 199. New York: Academic Press 1962.
36. Cope, C. L.: Adrenal steroids and disease, chap. 8. Philadelphia, Pa.: J. B. Lippincott 1964.
37. Edwards, R. W. H., Kellie, A. E.: Acta Endocrinol. **27**, 262 (1958).
38. Jørgensen, M.: Acta Endocrinol. **26**, 424 (1957).
39. Diczfalusy, E., Plantin, L.-O., Birke, G., Ingall, S. C., Norymberski, J. D.: Acta Endocrinol. **27**, 275 (1958).
40. Breuer, H., Nocke, L.: Z. Vitamin-, Hormon- Fermentforsch. **9**, 28 (1958).
41. Appleby, J. I., Gibson, G., Norymberski, J. K., Stubbs, R. D.: Biochem. J. **60**, 453 (1955).
42. Moxham, A., Nabarro, J. D. N.: J. Clin. Pathol. **9**, 351 (1956).
43. Butt, W. R., Kornel, L., Morris, R.: Acta Endocrinol. **26**, 65 (1957).
44. Raftopoulo, R., Staudinger, H., Weissbecker, L.: Clin. Chim. Acta **4**, 463 (1959).
45. Smith, D. C., Tompsett, S. L.: Analyst **80**, 397 (1955).
46. Norymberski, J. K., Stubbs, R. D.: Biochem. J. **64**, 176 (1956).
47. Sobel, C., Golub, O. J., Henry, R. J., Jacobs, S. L., Basu, G. K.: J. Clin. Endocrinol. Metab. **18**, 208 (1958).
48. Freislederer, W.: Acta Endocrinol. **35**, 95 (1960).
49. Diczfalusy, E., Plantin, L.-O., Birke, G., Westman, A.: Acta Endocrinol. **18**, 356 (1955).
50. Wilson, H.: Arch. Biochem. Biophys. **52**, 217 (1954).
51. Vestergaard, P., Sayegh, J. F.: Clin. Chim. Acta **14**, 247 (1966).
52. Huebner, C. F., Lohmar, R., Dimler, R. J., Moore, S., Link, K. P.: J. Biol. Chem. **159**, 503 (1945).
53. Reichstein, T., Shoppee, C. W.: Vitamins Hormones **1**, 345 (1943).
54. Tompsett, S. L., Smith, D. C.: Acta Endocrinol. **20**, 303 (1955).
55. Few, J. D.: J. Endocrinol. **22**, 31 (1961).
56. Metcalf, M. G.: J. Endocrinol. **26**, 415 (1963).
57. James, V. H. T., Caie, E.: J. Clin. Endocrinol. Metab. **24**, 180 (1964).

58. Menini, E., Norymberski, J. K.: Biochem. J. **84**, 195 (1962).
59. Murphy, D., West, H. F.: J. Endocrinol. **36**, 331 (1966).
60. Caisey, J. D., Child, K. J.: J. Endocrinol. **38**, 363 (1967).
61. Norymberski, J. K.: Methods of group corticosteroid estimation in urine. II. The adrenal cortex (G. McGowan and M. Sandler, eds.), p. 88. London: Pitman Medical 1960.
62. Mattox, V. R., Randall, R. V., Goodrich, J.: Metabolism **16**, 752 (1967).
63. Rutherford, E. R., Nelson, D. H.: J. Clin. Endocrinol. Metab. **23**, 533 (1963).
64. Thomas, B. S.: J. Clin. Endocrinol. Metab. **25**, 710 (1965).
65. Cope, C. L., Pearson, J.: J. Clin. Pathol. **18**, 82 (1965).
66. Cornil, A., Copinschi, G., Leclercq, R., Franckson, J. R. M.: Acta Endocrinol. **58**, 1 (1968).
67. a) Copinschi, G., Cornil, A., Leclercq, R., Franckson, J. R. M.: Acta Endocrinol. **51**, 186 (1966). – b) Mathieu-Nast, C., Mathieu, H. P., Blanquet, P.: Clin. Chim. Acta **21**, 91 (1968). – c) Kelly, W. G., Ranucci, S. R., Shaver, J. C.: Steroids **11**, 429 (1968). – d) Makin, H. L. J., Edwards, R. W. H., Clayton, B. E.: Clin. Sci. **34**, 549 (1968).
68. Ulick, S., Nicolis, G. L., Vetter, K. K., in: Aldosterone (E.–E.Baulieu and P. Robel, eds.), p. 3. Philadelphia, Pa.: F. A. Davis Co., 1964.
69. Morris, R.: Acta Endocrinol. **32**, 596 (1959).
70. Valente, D.: Boll. Soc. Ital. Biol. Sper. **41**, 1542 (1965).
71. Zumoff, B., Bradlow, H. L., Gallagher, T. F., Hellman, L.: J. Clin. Invest. **46**, 1735 (1967).
72. Hellman, L., Bradlow, H. L., Zumoff, B., Fukushima, D. K., Gallagher, T. F.: J. Clin. Endocrinol. Metab. **19**, 936 (1959).
73. Few, J. D.: Private Communication.
74. Jacobsohn, G. M., Lieberman, S.: J. Biol. Chem. **237**, 1469 (1962).
75. Siebke, C. H.: Zentr. Gynaekol. 1601 (1960).
76. a) Edson, M., Heard, R. D. H.: J. Biol. Chem. **130**, 579 (1939). – b) Rakoff, A. E., Paschkis, K. E., Cantarow, A.: Am. J. Obstet. Gynecol. **46**, 856 (1943).
77. a) Venning, E. H., Browne, J. S. L.: Federation Proc. **4**, 108 (1945). – b) Paterson, J. Y. F., Marrian, G. F.: Biochem. J. **48**, 33 P (1951).
78. a) Dyrenfurth, I., Venning, E. H.: Endocrinology **60**, 136 (1957). – b) Underwood, R. H., Flood, C. A., Tait, S. A. S., Tait, J. F.: J. Clin. Endocrinol. Metab. **21**, 1092 (1961).
79. Noguchi, S., Fukushima, D. K.: J. Biol. Chem. **241**, 761 (1966).
80. Wakabayashi, M., Wotiz, H. H., Fishman, W. H.: Biochim. Biophys. Acta **48**, 198 (1961).
81. Schubert, K.: Acta Endocrinol. **27**, 36 (1958).
82. Lieberman, S., Dobriner, K.: Recent Progr. Hormone Res. **3**, 71 (1948).
83. Buehler, H. J., Katzman, P. A., Doisy, E. A.: Proc. Soc. Exptl. Biol. Med. **78**, 3 (1951).
84. Cohen, S. L., Oneson, I. B.: J. Biol. Chem. **204**, 245 (1953).
85. Lieberman, S., Hariton, L. B., Fukushima, D. K.: J. Am. Chem. Soc. **70**, 1427 (1948).

86. a) James, V. H. T.: J. Endocrinol. **22**, 195 (1961). – b) Schubert, K., Hobe, G.: Z. Physiol. Chem. **329**, 202 (1962). – c) Zicha, L., Scheiffarth, F., Bergner, D.: Aerztl. Wochenschr. **14**, 845 (1959). – d) Rivera, R., Dorfman, R. I., Forchielli, E.: Acta Endocrinol. **54**, 37 (1967). – e) Staib, W., Schild, W.: Klin. Wochenschr. **36**, 600 (1958). – f) Brooks, R. V.: Biochem. J. **68**, 50 (1958).
87. Burstein, S., Lieberman, S.: J. Biol. Chem. **233**, 331 (1958).
88. Yagi, A.: Private communication.
89. Grant, G. A., Beall, D.: Recent Progr. Hormone Res. **5**, 307 (1950).
90. Burstein, S., Lieberman, S.: J. Am. Chem. Soc. **80**, 5235 (1958).
91. Lewis, E. S., Boozer, C. E.: J. Am. Chem. Soc. **74**, 308 (1952).
92. DePaoli, J. C., Nishizawa, E., Eik-Nes, K. B.: J. Clin. Endocrinol. Metab. **23**, 81 (1963).
93. Segal, L., Segal, B., Nes, W. R.: J. Biol. Chem. **235**, 3108 (1960).
94. Edwards, R. W. H., Kellie, A. E., Wade, A. P.: Mem. Soc. Endocrinol. No 2, Conference Society for Endocrinology, London, 1953, p. 53.
95. Eberlein, W. R.: J. Clin. Endocrinol. Metab. **22**, 963 (1962).
96. — J. Clin. Endocrinol. Metab. **25**, 1101 (1965).
97. Odell, A. D., Marrian, G. F.: Biochem. J. **30**, 1533 (1936).
98. Astwood, E. B., Seegar Jones, G. E.: J. Biol. Chem. **137**, 397 (1941).
99. Stimmel, B. F.: Endocrinology **40**, 446 (1947).
100. Sommerville, I. F., Gough, N., Marrian, G. F.: J. Endocrinol. **5**, 247 (1948).
101. Klopper, A., Michie, E. A., Brown, J. B.: J. Endocrinol. **12**, 209 (1955).
102. Jensen, C.: Acta Endocrinol. **28**, 37 (1958).
103. Waldi, D.: Aerztl. Lab. **11**, 1 (1965).
104. van der Molen, H. J.: Acta Endocrinol. **41**, 234 (1962).
105. Patti, A., Bonanno, P., Frauley, T. F., Stein, A. A.: Obstet. Gynecol. **21**, 302 (1963).
106. a) Barry, R. D., Guarnieri, M., Besch, P. K., Ring, W., Besch, N. F.: Anal. Chem. **38**, 983 (1966). – b) Metcalf, M. G.: Clin. Chem. **15**, 24 (1969).
107. Hirschmann, H., Williams, J. S.: J. Biol. Chem. **238**, 2305 (1963).
108. Okamura, Y., Tateyama, H.: Folia Endocrinol. Jap. **41**, 137 (1965).
109. Engel, L. L., in: Methods of biochemical analysis (D. Glick, ed.), vol. 1, p. 479. New York: Interscience Pub. 1954.
110. Butenandt, A., Grosse, W.: Chem. Ber. **69**, 2776 (1936).
111. Dobriner, K., Rhoads, C. P., Lieberman, S., Hill, B. R., Fieser, L. F.: Science **99**, 494 (1944).
112. Venning, E. H., Hoffman, M. M., Browne, J. S. L.: J. Biol. Chem. **146**, 369 (1942).
113. Ramseyer, J., Hirschman, H.: Federation Proc. **24**, 534 (1965).
114. Dobriner, K.: Univ. Inter Centra. Cancrum Acta **6**, 315 (1948).
115. Dingemanse, E., Laqueur, E.: Biochem. J. **32**, 651 (1938).
116. Jensen, C. C., Tötterman, L. E.: Acta Endocrinol. **10**, 221 (1952).
117. Birke, G., Plantin, L.-O.: Acta Med. Scand. **148**, Suppl. **291**, 1954.
118. Lieberman, S., Mond, B., Smyles, E.: Recent Progr. Hormone Res. **9**, 113 (1954).
119. a) Chaney, A. L., in: Standard methods of clinical chemistry (M. Reiner, ed.), vol. 2, p. 79. New York: Academic Press 1958. – b) Peterson, R.E., in: Standard

methods of clinical chemistry (D. Seligson, ed.), vol. 4, p. 151. New York: Academic Press 1963.
120. Vestergaard, P., Claussen, B.: Acta Endocrinol. **39**, Suppl. **64** (1962).
121. Goldzieher, J. W., Axelrod, L. R.: J. Clin. Endocrinol. Metab. **22**, 1234 (1962).
122. a) Jaoudé, F. A., Baulieu, E.-E., Jayle, M. F.: Acta Endocrinol. **26**, 30 (1957). – b) Brooks, R. V.: Biochem. J. **68**, 50 (1958). – c) Schubert, K.: Acta Endocrinol. **26**, 173 (1957).
123. Bradlow, H. L.: Unpublished studies.
124. a) Kellie, A. E.: Biochem. J. **100**, 631 (1966). – b) Becker, J. F.: Biochim. Biophys. Acta **100**, 582 (1965).
125. a) Cawley, L. P., Musser, B. O., Faucette, W., Beckloff, S., Learned, H.: Clin. Chem. **11**, 1009 (1965). – b) Takuyasu, H., Kinoshita, K., Kumamoto, Y., Ikemoto, N.: Horumon to Rinsho **12**, 895 (1964).
126. Bauld, W. S., Greenway, R. M., in: Methods of biochemical analysis (D. Glick, ed.), vol. 5, p. 479. New York: Interscience Pub. 1957.
127. Katzman, P. A., Straw, R. F., Buehler, H. J., Doisy, E. A.: Recent Progr. Hormone Res. **9**, 45 (1954).
128. Bugge, S., Nilsen, M., Metcalfe-Gibson, A., Hobkirk, R.: Can. J. Biochem. Physiol. **39**, 1501 (1961).
129. a) Nachtigall, L., Bassett, M., Hogsander, U., Slagle, S., Levitz, M.: J. Clin. Endocrinol. Metab. **26**, 941 (1966). – b) Bassett, M., Levitz, M.: J. Clin. Endocrinol. Metab. **28**, 1067 (1968). – c) Jaffe, S. H., Levitz, M.: Am. J. Obstet. Gynecol. **98**, 992 (1967). – d) Schindler, A. E., Ratanasopa, V., Herrmann, W. L.: Clin. Chem. **13**, 186 (1967).
130. a) Marrian, G. F., Bauld, W. S.: Acta Endocrinol. **7**, 240 (1951). – b) — Biochem. J. **59**, 136 (1955). – c) Brown, J. B.: Biochem. J. **60**, 185 (1955).
131. Brown, J. B., Blair, H. A. F.: J. Endocrinol. **17**, 411 (1958).
132. Frandsen, V. A.: Private communication.
133. — Pedersen, J., Stakemann, G.: Acta Endocrinol. **40**, 400 (1962).
134. Hobkirk, R., Alfheim, H., Bugge, S.: J. Clin. Endocrinol. **19**, 1352 (1959).
135. Touchstone, J. C., Stojkewycz, M., Smith, K.: Clin. Chem. **11**, 1019 (1965).
136. Brown, J. B., Bulbrook, R. D., Greenwood, F. C.: J. Endocrinol. **16**, 41 (1957).
137. Adlercreutz, H., Schauman, K.-O.: Acta Endocrinol. **46**, 230 (1964).
138. Sele, V., Frandsen, V. A.: Acta Obstet. Gynecol. Scand. **46**, Suppl. **9**, 43 (1967).
139. a) Timonen, S., Hirvonen, E., Sokkanen, R.: Acta Endocrinol. **49**, 393 (1965). – b) Ittrich, G., Neumann, H., Igel, H.: Zentr. Gynaekol. **85**, 361 (1963).
140. Rosenmund, H.: Helv. Physiol. Pharmacol. Acta **6**, 349 (1948).
141. Van Bruggen, J. T.: J. Lab. Clin. Med. **33**, 207 (1948).
142. Smith, O. W., Smith, G. van S.: Endocrinology **28**, 740 (1941).
143. Venning, E. H., Evelyn, K. A., Harkness, E. V., Browne, J. S. L.: J. Biol. Chem. **120**, 225 (1937).
144. Stimmel, B. F.: J. Biol. Chem. **178**, 217 (1949).
145. Falk, H. L., Heard, R. D. H.: Quoted in Heard, R. D. H., Saffran, J. C., Recent Progr. Hormone Res. **4**, 43 (1949).
146. Reckers, L., Katzman, P. A.: Federation Proc. **15**, 335 (1956).

147. Scholler, R., Métay, S., Herbin, S., Jayle, M. F.: European J. Ster. **1**, 373 (1966).
148. Roy, A. B.: Biochem. J. **66**, 700 (1957).
149. Fromageot, C., in: The enzymes (J. B. Sumner and K. Myrbäck, eds.), vol. 1, part 1, p. 517. New York: Academic Press 1950.
150. Abbott, L. D., Jr.: Arch. Biochem. **15**, 205 (1947).
151. Cohen, H., Bates, R. W.: Endocrinology **45**, 86 (1949).
152. Stimmel, B. F.: J. Biol. Chem. **162**, 99 (1946).
153. Oneson, I. B., Cohen, S. L.: Endocrinology **51**, 173 (1952).
154. Straw, R. F., Katzman, P. A., Doisy, E. A.: Endocrinology **57**, 87 (1955).
155. Emerman, S. L., Levitz, M.: Steroids **3**, 351 (1964).
156. a) Dodgson, K. S., Spencer, B.: Biochem. J. **51**, XXIX (1952). – b) Henry, R., Thevenet, M.: Bull. Soc. Chim. Biol. **34**, 886 (1952). – c) Roy, A. B.: Biochim. Biophys. Acta **15**, 300 (1954).
157. a) Corner, E. D. S., Leon, Y. A., Bulbrook, R. D.: J. Marine Biol. Assoc. U. K. **39**, 51 (1960). – b) Leon, Y. A., Bulbrook, R. D., Corner, E. D. S.: Biochem. J. **75**, 612 (1960).
158. Dodgson, K. S., Spencer, B.: Biochem. J. **57**, 310 (1954).
159. Stitch, S. R., Halkerston, I. D. K.: Biochem. J. **63**, 710 (1956).
160. a) Jarrige, P., Henry, R.: Bull. Soc. Chim. Biol. **34**, 872 (1952). – b) Dodgson, K. S., Lewis, J. I. M., Spencer, B.: Biochem. J. **51**, xlii (1952).
161. Jayle, M. F., Scholler, R., Jarrige, P., Métay, S.: Bull. Soc. Chim. Biol. **41**, 1593 (1959).
162. Stitch, S. R., Halkerston, I. D. K., Hillman, J.: Biochem. J. **63**, 705 (1956).
163. Jarrige, P.: Bull. Soc. Chim. Biol. **45**, 761 (1963).
164. — Yon, J., Jayle, M. F.: Bull. Soc. Chim. Biol. **45**, 783 (1963).
165. Bongiovanni, A. M.: J. Lab. Clin. Med. **55**, 812 (1960).
166. a) Jayle, M. F., Baulieu, E.-E.: Bull. Soc. Chim. Biol. **36**, 1391 (1954). – b) Stitch, S. R., Halkerston, I. D. K.: Nature **172**, 398 (1953).
167. Dodgson, K. S.: Biochem. J. **78**, 324 (1961).
168. Pasqualini, J. R., Jayle, M. F.: J. Clin. Invest. **41**, 981 (1962).
169. Unpublished data from this laboratory.
170. a) Buehler, H. J., Katzman, P. A., Doisy, E. A.: Proc. Soc. Exptl. Biol. Med. **76**, 672 (1951). – b) Doyle, M. L., Katzman, P. A., Doisy, E. A.: J. Biol. Chem. **217**, 921 (1955).
171. a) Cohen, S. L.: J. Biol. Chem. **192**, 147 (1951). – b) Fishman, W. H., Talalay, P.: Science **105**, 131 (1947). – c) Bernfeld, P., Nisselbaum, J. S., Fishman, W. H.: J. Biol. Chem. **202**, 763 (1953). – d) Smith, E. E. B., Mills, G. T.: Biochem. J. **54**, 164 (1953).
172. Stitch, S. R., Halkerston, I. D. K.: J. Endocrinol. **9**, xxxvi (1953).
173. Wakabayashi, M., Fishman, W. H.: Biochim. Biophys. Acta **48**, 195 (1961).
174. Billett, F.: Biochem. J. **57**, 159 (1954).
175. Cox, R. I.: Biochem. J. **71**, 763 (1959).
176. Fripp, P. J.: Comp. Biochem. Physiol. **15**, 361 (1965).
177. Gautney, M. C., Barker, S. B., Hill, S. R., Jr.: Science **129**, 1281 (1959).
178. Ryan, M. T., Mavrides, C. A.: Science **131**, 101 (1960).

179. Smith, J. D., Bongiovanni, A. M.: Science **131**, 101 (1960).
180. Talalay, P., Fishman, W. H., Huggins, C.: J. Biol. Chem. **166**, 757 (1946).
181. Kirk, B. K., Felger, C. B., Katzman, P. A.: Arch. Biochem. Biophys. **98**, 206 (1962).
182. Slaunwhite, W. R., Jr., Sandberg, A. A.: Endocrinology **67**, 815 (1960).
183. Vestergaard, P.: Acta Endocrinol. **39**, Suppl. **64**, 50 (1962).
184. a) Voigt, K. D., Klempien, E.: Acta Endocrinol. **35**, Suppl. **51**, 1043 (1960). – b) Alfsen, A.: Bull. Soc. Chim. Biol. **41**, 1469 (1959).
185. a) Sigma Chemicals Co., St. Louis, Mo., Bulletin No. 4. – b) Silber, R. H., Porter, C. C., in: Methods of biochemical analysis (D. Glick, ed.), vol. 4, p. 139. New York: Interscience Pub. 1957.
186. Cohen, S. L., Goldfine, M. M., Toussaint, F., Friedman, K., Noma, I.: Endocrinology, **54**, 353 (1954).
187. Bradlow, H. L., Kritchevsky, T. H., Katzenellenbogen, E. R., Dobriner, K., Gallagher, T. F.: Federation Proc. **12**, 181 (1953).
188. a) Inoue, N., Slaunwhite, W. R., Jr., Graham, J. B., Sandberg, A. A.: Abstr. 146, 49th Meeting of the Endocrine Soc., Miami, Florida, June 1967, p. 101. – b) Levitz, M., Katz, J.: Abstr. 227, 49th Meeting of the Endocrine Soc., Miami, Florida, June 1967, p. 142.
189. Hobkirk, R., Cohen, J. J.: Can. J. Biochem. Physiol. **38**, 769 (1960).
190. Wotiz, H. H., Lemon, H. M., Marcus, P., Savard, K.: J. Clin. Endocrinol. Metab. **17**, 534 (1957).
191. Bradlow, H. L., Gallagher, T. F.: J. Clin. Endocrinol. Metab. **19**, 1575 (1959).
192. Wakabayashi, M., Fishman, W. H.: J. Biol. Chem. **236**, 996 (1961).
193. a) Beyler, A. L., Szego, C.: Endocrinology **54**, 323 (1954). – b) Isselbacher, K. J., Axelrod, J.: J. Am. Chem. Soc. **77**, 1070 (1955). – c) Arias, I. M.: Advances Clin. Chem. **3**, 35 (1960).
194. Cohen, S. L.: Can. J. Biochem. **42**, 127 (1964).
195. — Can. J. Biochem. **46**, 563 (1968).
196. Pasqualini, J. R., Ulrich, F., Jayle, M. F.: Biochim. Biophys. Acta **104**, 515 (1965).
197. Solomon, S.: Private communication.
198. Fishman, J.: Private communication.
199. Metcalf, M. G.: Anal. Biochem. **13**, 483 (1965).
200. Kream, J.: Private communication.
201. Zumoff, B., Bradlow, H. L.: J. Clin. Endocrinol. Metab. **23**, 799 (1963).
202. a) Levvy, G. A.: Biochem. J. **52**, 464 (1952). – b) Marsh, C. A.: Nature **194**, 974 (1962).
203. Kushinsky, S., Chen, V. L.: Enzymol. Biol. Clin. **8**, 266 (1967).
204. Stempfel, R. S., Jr., Sidbury, J. B., Jr., Migeon, C. J.: J. Clin. Endocrinol. Metab. **20**, 814 (1960).
205. Fukushima, D. K., Bradlow, H. L., Hellman, L., Zumoff, B., Gallagher, T. F.: J. Biol. Chem. **235**, 2246 (1960).
206. Layne, D. S.: Endocrinology **76**, 600 (1965).
207. Arcos, M., Lieberman, S.: Abstr. 152, 49th Meeting of the Endocrine Society, Miami, Florida, June 1967, p. 104.
208. Bayyuk, S. I., Juraydini, A. M.: Steroids **10**, 307 (1967).

Chapter 4

The Isolation of Steroid Conjugates

PENTTI K. SIITERI

Contents

I. Introduction

Most steroid hormones exist in biological tissues and fluids not as the free compounds but as water-soluble conjugates. This fact was recognized by the earliest investigators who found that the yield of biologically active steroids, extractable from urine with organic solvents, was greatly increased by prior hydrolytic treatment with acid. It was not until 1936, however, that the pioneer efforts of Cohen and Marrian [1] and Venning and Browne [2] resulted in the successful isolation from human pregnancy urine of reasonably pure estriol and pregnanediol glucuronides, respectively. Two years later, Schachter and Marrian reported the isolation of estrone sulfate from pregnant mare's urine [3]. Venning *et al.* reported the isolation and full characterization of androsterone sulfate in 1942 [4] while Munson *et al.* reported the isolation of the crystalline semicarbazone of dehydroisoandrosterone sulfate from human urine in 1944 [5]. Nevertheless, as recently as 1953 only 4 intact steroid sulfates and 3 glucuronides had been isolated from urine and fully characterized, while none had yet been obtained from other sources such as plasma or bile. On the other hand, approximately 150 free steroids had been isolated and identified at that time. While interest in the chemical and biological properties of conjugated steroids has increased tremendously during the past 15 years, this disparity remains.

The prolonged neglect of steroid conjugates can be ascribed to several factors. First, the combination of a lipophilic steroid molecule with a hydrophilic molecule such as glucuronic or sulphuric acid results in a compound of greatly increased polarity. The separation of steroid conjugates from other urinary solids therefore presents a much more formidable task than does the isolation of their free steroid counterparts. Also discouraging until recently has been the unavailability of authentic reference compounds, particularly the steroid glucuronides, with which isolated materials can be unequivocally identified. Finally, early investigations demonstrated that conjugated forms of steroids such as androsterone glucuronide possessed little biological activity. Together with observations on the role of conjugation as a detoxication mechanism for drugs, these early findings led to the general view that steroid conjugates were metabolic end products destined only for excretion. It is not surprising, therefore, that this aspect of steroid endocrinology was shunned by all but a few brave investigators until the mid 1950's.

However, the elegant work of Schneider and Lewbart [6] demonstrated that the steroid conjugates could easily be handled by modern methods of isolation such as countercurrent distribution and chromatography, and marked the beginning of a new era. Armed with these and other powerful methods including infrared spectrophotometry and radioactive tracers, other investigators began to seriously reconsider the role of conjugates. The studies of Baulieu and co-workers in Paris and of Lieberman's group in New York soon revealed the participation of conjugated forms in the biosynthesis, secretion, and metabolism of steroid hormones, and opened the way for a full-scale assault on these problems which continues unabated.

Nevertheless, we are far from understanding the true biological role of the known and perhaps yet–to–be–discovered forms of conjugated steroids. The complexity of the problems arising from the great number of metabolically important steroids, most of which have several potential sites of conjugation, and the recent demonstrations of novel conjugating moieties, requires that the utmost sophistication and skill be exercised in the isolation and identification of these compounds. It is the purpose of this chapter to critically assess the present state of the art.

No attempt has been made to cite all of the literature pertaining to this subject. The reader's attention is directed to the recent companion volume by Bernstein *et al.* [7] in which references to the purification methods that have been applied to all of the known individual steroid conjugates may be found. In addition, an excellent review concerning the analysis and identification of steroid conjugates has been published by Pasqualini [8].

II. Chemical Types and Occurrence

The detailed chemistry of the steroid conjugates has been reviewed in Chapter 1 in this volume. Only those chemical aspects which relate to their successful isolation and purification will be considered here. While certain unusual conjugating moieties have recently been described, such as N–acetylglucosamine by Layne [9] and also by Arcos and Lieberman [10], the great majority of the steroids normally occurring in urine or plasma are undoubtedly linked to sulfuric or glucuronic acid.

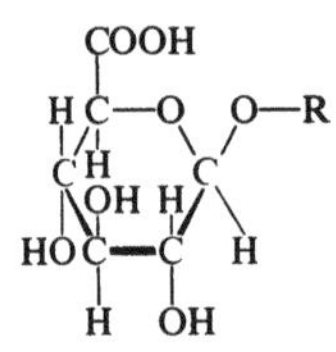
I

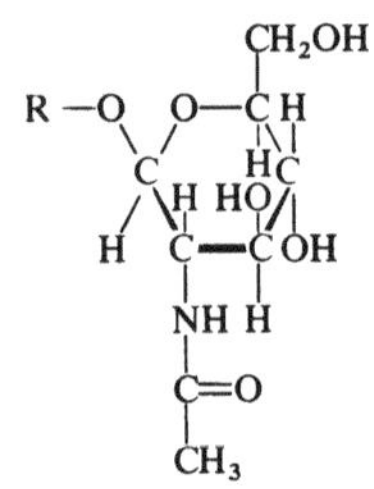
II

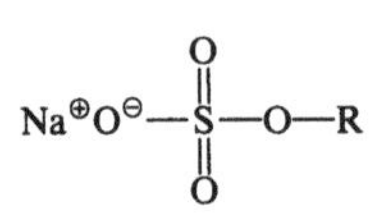
III

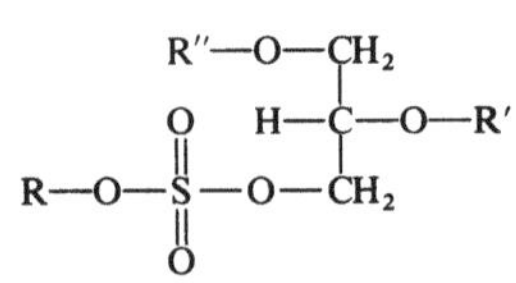
IV

Fig. 4–1. Structures of steroid conjugates

As is evident in Fig. 4–1, conjugation with sulfuric acid results in a simple ester of a secondary alcohol if the hydroxyl group resides on the steroid nucleus (III). Corticosteroid sulfates have been described in which conjugation occurs at C21, which of course bears a primary alcohol. The known glucuronides of steroids, on the other hand, are generally not esters. The steroid aglycones are linked by a glycosidic bond to carbon 1 of glucuronic acid. All of the steroid glucuronides thus far isolated from natural sources and whose structures have been chemically determined possess a 6-membered pyranose structure with the glycosidic bond at C1 in the β–orientation as shown in Fig. 4–1 (I). They are, therefore, correctly designated as β–D–glucopyranosiduronic acids. Conjugation with N–acetylglucosamine, however, may involve either an α– or β–glycosidic linkage. Layne found that the rabbit excretes 17α–estradiol [estra–1,3,5(10)–triene–3,17α–diol] as a 17α–β–D–2–acetamido–2–deoxy–glucopyranoside (II) as shown in Fig. 4–1 [9]. On the other hand, Arcos and Lieberman isolated a conjugate of pregnenediol [pregn–5–ene–3β,20α–diol] from human urine in which N–acetylglucosamine was attached in α–glycosidic linkage at C20 [10]. Although the most common site of conjugation in the steroid is a hydroxyl group, evidence has accumulated which indicates that conjugation with glucuronic acid can occur at C3 of Δ^4–3 ketones *via* the 3,5–dien–3β–ol structure [11]. Precaution must be exercised in attempts to isolate such glucuronides since it has been shown that they are susceptible to non-enzymatic cleavage at acid pH [12]. No evidence for the natural occurrence of the corresponding sulfates has been presented to date, which is not surprising since such compounds would likely be quite unstable.

A number of double conjugates has recently been found which contain more than one conjugating moiety per steroid molecule. Indeed the N–acetylglucosamine conjugates mentioned above also contained either glucuronic [9] or sulfuric acid [10]. Diczfalusy *et al.* were the first to demonstrate the formation of a "mixed" double conjugate, estriol–3–sulfate–16–glucuronide [17β–hydroxy–3–sulfooxyestra–1,3,5(10)–trien–16α–yl–β–D–glucopyranosiduronic acid] [13]. In addition, a number of steroid disulfates have been found in human urine [10, 14, 15]. These compounds are very polar and require extreme measures during extraction and purification to insure their recovery.

Steroid phosphates have been reported to occur naturally, but the evidence available to date has not been convincing. Oertel and his associates, in a long series of papers, have described conjugates of steroids in which diglycerides are linked to steroid sulfates *via* the free hydroxyl group of glycerol to form sulfate diesters, or sulfatides (IV). For example, it has been claimed that the major radioactive product in blood following the administration of isotope-labeled dehydroisoandrosterone sulfate [3β-sulfooxy-androst-5-en-17-one] to humans is the sulfatide [16]. In contrast to conjugation with glucuronic or sulfuric acids, this mode of conjugation results in compounds that are considerably less polar than the simple sulfates. However, the presence of such compounds in nature has yet to be confirmed in other laboratories. It might be mentioned that the author has performed a large number of similar experiments with radioactive dehydroisoandrosterone sulfate in which blood samples were extracted by a variety of procedures including acetone-ethanol and chloroform-methanol. The carefully dried extracts were then subjected to the mild technique of liquid-liquid partition chromatography, which clearly separates the free 17-ketosteroids from their monosulfates and monoglucuronides. In no instance has a significant amount of radioactivity less polar than dehydroisoandrosterone sulfate been observed.

It is evident that the major change in the properties of any steroid upon conjugation with either glucuronic or sulfuric acids is a marked increase in polarity. This term is used here in its simplest context and serves only to indicate the relative affinity of a compound for the more polar phase of a 2-phase system. Thus, while many free steroids can be extracted from aqueous media such as urine with relatively nonpolar solvents, *e.g.*, ether or methylene chloride, their conjugates require the use of hydroxylic solvents such as *n*-butanol. The position in the steroid molecule to which a conjugating group is attached is of some importance in determining the chromatographic properties of the conjugate. The most important factors with respect to glucuronides would appear to arise not from interactions between the two portions of the molecule but rather from the nature of the functional groups of the steroid which remain free and/or the configuration of the steroid molecule. For example, both the 16α- and 17β-monoglucuronides of estriol [estra-1,3,5(10)-triene-3,16α,17β-triol] are readily separated from the 3-monoglucuronide since the former retain the phenolic hydroxyl which is quite different in terms of polarity than are the secondary alcohols. On the other hand it is exceedingly difficult to separate the 16α- and 17β-monoglucuronides. Another example, which will be described in detail below, are the glucuronides of the simple 17-ketosteroids which can be separated by partition chromatography in acid systems with about the same ease as the free compounds. This would indicate that variation in the orientation of the bulky glucuronic acid moiety with respect to the steroid molecule, which results from attachment at the 3α-(axial) or 3β-(equatorial) position in 5α- or 5β-steroids, has little effect on the relative polarity of the compound. The sulfates of such steroids, however, are not as easily separated in either acidic or basic systems. This suggests that the structural features which normally determine polarity of the steroid adjacent to the point of attachment may be masked in the sulfates. In basic systems this may arise from solvated ion effects due to clustering of solvent molecules around the sulfate moiety which is in close proximity to the steroid. However, other factors such as the inability to completely suppress ionization of the free proton in acid systems may also account for the difficulties encountered in the separation of steroid sulfates.

In addition to increasing the polarity of the steroid molecule it is evident that the conjugates will behave as weak acids. Therefore, the extractability and chromatographic behavior of sulfates and glucuronides will be dependent upon the pH of the system. For example, glucuronides can be extracted much more readily from acidic than from basic solutions. While the steroid sulfates are esters of a strong acid, they apparently are less acid in character than the glucuronides since they behave as weak bases in many chromatographic systems. This difference has been used to advantage by several workers. Solvent partition systems containing ammonium hydroxide are available which readily separate sulfates from glucuronides providing that no large overlap in polarity exists due to the steroid molecule itself. A further basis for group separation of the two conjugate types is provided by the fact that the sulfates can form relatively stable "complexes" with certain organic bases which have great solubility in chlorinated hydrocarbons such as chloroform. Whether or not this phenomenon is due to complex formation or simply nonpolar salt formation has not been ascertained.

As indicated above, the successful separation of conjugates within a class of sulfates or glucuronides by partition processes depends largely upon the same factors which determine the separation of the free steroids. Since both classes of conjugates may be converted to ionized forms, other methods that depend upon this property, such as electrophoresis and ion exchange chromatography, which are extremely effective in other areas, have naturally been investigated. As might be expected, electrophoresis is very efficient for the group separation of sulfates, glucuronides, and double conjugates, but are relatively inefficient for the separation of individual conjugates of steroids having subtle differences in structure. Similarly, ion exchange chromatography has generally been disappointing when applied to individual or even group separations. Most investigations of this technique have shown that the adsorption of conjugates to the exchange materials is due to nonspecific hydrophobic forces as well as the anticipated electrostatic interactions.

The most important chemical aspect of steroid conjugates for purposes of the present discussion is the polar nature of these compounds. This property necessitates the expenditure of a considerable amount of effort in order to obtain extracts from biological sources which are amenable to chromatographic or other purification methods. Since the success or failure of any of these procedures is largely dependent upon the load that is placed upon the system, the extent of preliminary purification during the initial extraction from biological sources is crucial to the successful isolation of intact steroid conjugates. The majority of the steroid conjugates thus far isolated in pure form have been obtained from urine. However, a number of papers has recently appeared in which conjugated steroids have been isolated from blood or plasma, human bile, amniotic fluid, and fetal tissues. While each of these sources presents slightly different problems in terms of eliminating contaminating materials, most of the general principles of conjugate isolation have been elaborated in the study of urine.

III. Measurement and Detection

As in the isolation of any compound from complex, naturally occurring mixtures, one must have at hand sensitive and specific methods for the detection of steroid sulfates and glucuronides. Conjugation usually does not interfere with those functional groups which give rise to the commonly used identifying reactions for the free steroids. Therefore, it is useful both for the purposes of isolation and identification to perform such tests as the Zimmermann reaction for ketosteroids, the antimony trichloride reaction, reaction with anisaldehyde, and the various procedures used for corticoids such as blue tetrazolium reduction and reaction with isonicotinic acid hydrochloride. However, it should be pointed out that unusual conjugates may be encountered in which the reactivity toward these reagents may be reduced or completely eliminated.

The specific detection of conjugating groups is essential for following the isolation of intact conjugates and for distinguishing between glucuronides, sulfates, and other possible forms. The sulfate group can be detected easily on chromatograms or quantitated in solution by means of "complex" formation with methylene blue. Procedures have been described by Roy [17] and Crépy and Rulleau-Meslin [18] in which steroid sulfates in acidic solution containing methylene blue are quantitatively extracted into chloroform. Spectrophotometric analysis of the blue chloroform extract yields accurate measurements of 2–50 µg of steroid sulfate. Steroid glucuronides do not form a complex, whereas phosphates give a much weaker blue color. However, other organic sulfates in biological samples form complexes so that this test can be safely used only after preliminary separation of the steroid sulfates. Steroid sulfate spots on paper or thin layer chromatograms are detected easily by this method simply by immersing the chromatogram in the reagent followed by washing with chloroform leaving a white spot on a blue background [19]. Steroid sulfates can also be detected easily on chromatograms by the rhodizonate test as described by Burma [20] and Schneider and Lewbart [21]. Steroid phosphates can be dectected easily by the Hanes–Isherwood Test [22].

Steroid glucuronides are not as easily detected since the classical trichloroacetic acid–naphthoresorcinol reagent reacts only with free glucuronic acid or glucuronolactone. Preliminary hydrolysis of steroid glucuronides on chromatograms must therefore be performed. This can be accomplished on thin layer plates by acid treatment but is more difficult on paper chromatograms. However, *in situ* enzymatic hydrolysis on the chromatogram with β–glucuronidase overcomes this problem. Unpublished observations from this laboratory have shown that steroid glucuronides on TLC plates are readily detected simply by spraying with 10% H_2SO_4 in ethanol followed by brief heating at 100°. Characteristic colors are obtained for various glucuronides which are helpful in identifying unknown compounds. Crépy *et al.* devised a simple test for steroid glucuronides in which solutions of 1–(2–pyridylazo)–2–naphthol (PAN) in 95% ethanol followed by $Co(NO_3)_2$ in ethanol–dichloromethane are sprayed on the plate or paper [23]. One to 2 µg of glucuronides can be detected as red spots which fade to green on a yellow background. However, this test must be used with caution since positive reactions were obtained with some free steroids as well as steroid sulfates [24]. Susceptibility to hydrolysis by specific chemical or enzymatic methods followed by analysis of the free steroid serves as a useful method for the identification of conjugates. However, positive identification of a conjugating group by its enzymatic removal can be accomplished only with pure enzymes. Commercially available β–glucuronidase preparations cannot be used for this purpose since they contain β–glucosidase activity as evidenced by extensive hydrolysis of conjugates containing N–acetylglucosamine [9].

Finally, it should be mentioned that a large number of radioactive steroid conjugates have been prepared by a variety of methods. For example, many studies with dehydroisoandrosterone sulfate labeled either with tritium or sulfur–35, or both, have been reported. The establishment of the radiochemical purity of such conjugates is extremely important whether they are to be used in chemical or metabolic studies. The ease with which this can be accomplished depends largely upon the method of labeling. For example, the author studied the metabolism of tritium–labeled androsterone glucuronide [17–oxo–5α–androstan–3α–yl–β–D–glucopyranosiduronic acid] prepared by the Wilzbach method [25]. After extensive chromatography, less than 2% of the starting material was recovered as pure, tritiated androsterone glucuronide. More recently a number of complex radioactive estrogen glucuronides and sulfates have been synthesized by Levitz [26] and others. Baulieu and his co-workers have also prepared radioactive glucuronides of testosterone [17β-hydroxyandrost-4-en-3-one], dehydroisoandrosterone [3β-hydroxyandrost-5-en-17-one] and other important 17-ketosteroids [27]. A number of these labeled conjugates are now commercially available and lend themselves ideally to the investigation and development of new methods for the isolation of steroid conjugates.

IV. Extraction of Steroid Conjugates

A. Solvent Extraction

The most widely employed approach to the preliminary fractionation of steroid conjugates from crude biological samples has been organic solvent extraction. Of the many solvents employed for this purpose, *n*-butanol has been most often used. While this solvent is quite efficient for the extraction of most steroid conjugates, important exceptions have been noted. For example, Grant and Beall found that *n*-butanol extracted only 75% of the estrogen conjugates from pregnant mare's urine whereas a 95% recovery could be obtained with cyclohexanol [28]. Furthermore, *n*-butanol suffers from several practical disadvantages such as its tendency to produce troublesome emulsions when used to extract urine. In addition, while the very polar nature of *n*-butanol promotes good recovery of most steroid conjugates, it also results in the extraction of large amounts of extraneous urinary solids. The dried extracts are therefore very crude and usually are quite unsuitable for paper or thin layer chromatography. Finally, the high boiling point of *n*-butanol makes its removal not only a nuisance but a real danger when large volumes containing labile conjugates must be dealt with. The latter problem can be alleviated somewhat by recognizing that *n*-butanol and water form a lower boiling azeotropic mixture.

It is possible to obtain less weighty extracts with *n*-butanol if relatively nonpolar conjugates are to be isolated. Adjustment of the pH of the urine to alkaline values prior to extraction or back-washing the extract with $NaHCO_3$ or even Na_2CO_3 solutions yields much cleaner extracts. For example, Munson *et al.* [5] first isolated dehydroisoandrosterone sulfate from urine by extraction with *n*-butanol followed by alkaline and water washes of the extract. Further refinements of the *n*-butanol extraction method were achieved by Jayle and co-workers [29–32] who showed that

17–ketosteroid sulfates and glucuronides could be selectively extracted from urine at pH values of 11 and 5.8, respectively. However, it has been shown repeatedly that a low pH [1–3] must be employed if quantitative recoveries of very polar compounds, such as estriol glucuronide, are to be achieved.

A marked improvement in extraction methods for steroid conjugates was introduced in 1953 by Edwards *et al.*, who took advantage of the "salting out" principle [33]. These workers utilized the general method of Kamil *et al.* [34] in which ammonium sulfate was added to urine to a final concentration of 50 g per 100 ml. Extraction of the urine with a 3:1 mixture of ether and ethanol gave not only good recoveries, but also yielded much less weighty extracts than those obtained with *n*–butanol. Kellie has recently evaluated the use of other solvents at various pH values in this extraction procedure [35]. It was demonstrated that authentic dehydroisoandrosterone and 17β–estradiol [estra–1,3,5(10)–triene–3,17β–diol] glucuronides are quantitatively removed at all pH values between 2 and 10 by the original procedure. Surprisingly, tetrahydrocortisone glucuronide [17α,21–dihydroxy–11,20–dioxo–5β–pregnan–3α–yl–β–D–glucopyranosiduronic acid] was also extractable at pH 6. Ethyl acetate and tetrahydrofuran were less efficient at neutral pH for the extraction of the estrogen glucuronides. Measurements of the amounts of urinary solids extracted from urine by the ether–ethanol mixture, ethyl acetate, or tetrahydrofuran indicated that the former were approximately equivalent whereas tetrahydrofuran extracted considerably more extraneous solids at pH 3.8 or 5.9. These findings were in contrast to those of De Paoli *et al.* [36] who claimed that extraction with tetrahydrofuran produces 10% of the residue obtained with ether–ethanol. Several variations of the "salting out" technique have been reported. For example, Tamm *et al.* extracted conjugates of synthetic corticoids with ethyl acetate: *n*–butanol (3:1) from plasma which had been adjusted to 3M NaCl [37]. A comprehensive evaluation of various extraction methods for urinary conjugated corticosteroids has also been carried out by Kornel [38]. Of the various methods tested, the most efficient appeared to be extraction of urine which had been adjusted previously to pH 1 and saturated with ammonium sulfate with a 1:10 mixture of *n*–butanol and chloroform. Essentially quantitative recovery was obtained with the synthetic conjugates tetrahydrocortisone glucuronide and cortisol–21–sulfate [11β,17α–dihydroxy–21–sulfooxypregn–4–ene–3,20–dione]. The dried extracts contained minimal amounts of extraneous impurities. However, only a small fraction of the corticosteroid disulfates such as tetrahydrocortisone–3,21–disulfate [17α–hydroxy–3α,21–disulfooxy–5β–pregnane–3,20–dione] were extracted by this procedure. These results also illustrated the advantages of extraction with less polar solvent mixtures as compared with *n*–butanol. It was also demonstrated that, if *n*–butanol is used for extraction of corticosteroid conjugates, it is not possible to wash the extract with alkali or sodium bicarbonate since considerable losses were incurred. A modification of Kornel's procedure for extraction of all of the conjugates present in human urine has been introduced by Devis and Noiret in which the urine saturated with ammonium sulfate is adjusted to pH 1 and extracted twice with ethyl acetate [39]. More recently, Nielson has used *n*–propyl alcohol for the extraction of 17α–hydroxycorticosteroid conjugates from human urine [40]. It is claimed by this author that less solvent is required as compared with *n*–butanol and that the tendency to emulsify is decreased by the use of *n*–propyl alcohol. Again, a mixture of solvents such as ethyl acetate and *n*–propyl alcohol gave a cleaner extract.

It is of considerable importance in many studies to be able to examine separately those conjugates present as sulfates and glucuronides. Separation of these two classes of conjugates of nonpolar steroids is a simple task. For example, 17–ketosteroid sulfates can be obtained in relatively pure form by taking advantage of the nonpolar character of their pyridinium salts. McKenna and Norymberski [41] first utilized this principle for the isolation of urinary dehydroisoandrosterone sulfate and related compounds by the addition of 0.3 molar pyridinium sulfate to urine followed by extraction with chloroform. It was shown that 80 to 100% of the sulfates were recovered under these conditions. McKenna and Rippon [42] extended this procedure to the measurement of the principle 17–ketosteroid sulfates in human plasma. That the use of a nonpolar solvent such as chloroform yields extracts which are markedly free of other urinary constituents has no doubt been responsible for the widespread use of this method. This principle has also been applied by Eberlein who utilized methyl green for complex formation rather than pyridinium ion. The extraction procedure forms the basis for relatively simple procedures for the quantitation of 17–ketosteroid sulfates in both blood [43] and urine [44]. The steroids are measured as their acetates following transesterification which destroys the colored complex.

Kellie and Wade earlier had devised a chemical procedure in which sulfates were separated from glucuronides [45]. The extract obtained by the ether–ethanol procedure was dried and then acetylated with pyridine and acetic anhydride. The glucuronide acetates thus formed were separated by benzene extraction of aqueous solutions which retained the sulfates. This procedure is not entirely satisfactory in that polyhydroxy steroids which are excreted as sulfates will also be acetylated and may well be extracted along with the glucuronide fraction.

B. Gel Filtration

In 1961 Beling reported the observation that passage of urine through a column of Sephadex, a polymerized dextran, resulted in elution of the estrogen conjugates in two distinct, concentrated peaks, I and II [46]. Subsequent studies revealed that the separation was dependent upon the nature of the conjugating moiety and the exact position at which it was attached to the steroid molecule [47]. For example, estriol-3-glucuronide [16α,17β-dihydroxyestra-1,3,5(10)-trien-3-yl-β-D-glucopyranosiduronic acid] was found in peak I, whereas the 16-glucuronide [3,17β-dihydroxyestra-1,3,5(10)-trien-16α-yl-β-D-glucopyranosiduronic acid] was found in peak II. While the physical basis for this separation was not apparent, it was dependent upon other substances present in urine, since separation was not observed when experiments with pure conjugates in distilled water were performed. Nevertheless, the technique proved to be useful in that much extraneous material was eliminated and, more importantly, inhibitors of β-glucuronidase were removed during the process. This permitted a more nearly quantitative analysis of estriol in pregnancy urine when enzymatic hydrolysis of the conjugates was employed. Kushinsky and Otterness also investigated gel filtration as a means of eliminating β-glucuronidase inhibitors and concluded that these compounds, principally saccharo-1,4-lactone, are not completely removed by gel filtration of urine [48]. However, their study was complicated by the fact that subdivision of Beling's peak I into 5 peaks was achieved by lengthening the column and other procedural manipulations. The possibility that these zones may not be representative of conjugates orginally present must also be considered since the urine was concentrated 5-fold by evaporation at reduced pressure and temperature prior to filtration. It has been the author's experience that extensive hydrolysis of estrogen conjugates accompanies this "mild" procedure (see below).

More recently, Smith and Kellie have made extensive use of gel filtration in the isolation of intact estrogen conjugates of late-pregnancy urine [49]. With the aid of synthetic conjugates they made the surprising observation that the separation of peaks I and II of Beling was dependent upon the presence of urate. Indeed, spectrophotometric analysis of the eluates from columns loaded with pregnancy urine showed a marked drop in uric acid concentration which coincided with the emergence of peak II. Furthermore, peak I could be further subdivided into two peaks by removing the uric acid and re-running in acetate buffer. The mechanism involved in this separation is not clear although some evidence for the participation of acidic urinary components such as *p*-hydroxyphenylacetic acid was obtained. A further indication of the complexity of the mechanisms involved in separation of estrogen conjugates by gel filtration was the fact that 16α–hydroxyestrone–16–glucuronide [3–hydroxy–17–oxoestra–1,3,5(10)–trien–16α–yl–β–D–glucopyranosiduronic acid] a major conjugate isolated in these studies, was found in both peaks I and II. It is evident that gel filtration of very crude mixtures, such as urine, is a complicated process, and that simple molecular sieving is inadequate to explain the separations which have been observed. Nevertheless, this would appear to be a useful preliminary purification technique in quantitative methods for conjugates such as the GLC method for plasma steroid sulfates described by Sjövall and Vihko [50].

C. Ion Exchange Resins

Numerous attempts have been made to take advantage of the ionic character of steroid conjugates in developing isolation methods based on ion exchange phenomena. However, this approach has been disappointing. It has been demonstrated repeatedly that absorption of conjugates to the commonly employed ion exchange resins does not result simply from ionic interactions [51]. For example, Hähnel found it necessary to use ethanolic buffer solutions at 50° in order to elute urinary estrogen conjugates from columns of Deacidite FF® (Permutit, Paramus, N.J.), a synthetic anion exchange resin [52]. The same author investigated the use of the recently available anion exchanger, diethylaminoethyl (DEAE)—Sephadex® (Pharmacia, Uppsala), which was expected to display reduced nonspecific absorption toward steroid conjugates [53]. While the estrogen conjugates were eluted from this material more readily, the role of ion exchange phenomenon in the process was not clear. For example, it was observed that changes in the pH of solution from which the estrogen conjugates were absorbed in the range of 5.5 to 7.0 had no effect on the amount absorbed. Furthermore, separation of some conjugates occurred similar to those obtained by Beling using ordinary Sephadex columns. Nevertheless, elution of urinary conjugates by means of NaCl gradients yielded sufficiently pure fractions to allow direct estrogen analysis by the Kober reaction. The procedure also separates free from conjugated estrogens but does not appear to be applicable to large volumes of urine.

An unusual ion exchange method has been devised by Kushinsky and Tang whereby a solution of the liquid ion exchanger, Amberlite LA–1® (Rohm & Haas, Phila.), in ethyl acetate is used to extract urine containing estrogen conjugates [54]. The conjugates are then back–extracted with ammonia solutions from which they are recovered by evaporation. While relatively clean extracts were obtained and the recovery of sulfates or glucuronides was adequate, the procedure is tedious in that multiple extractions are required and large volumes of solvent must be concentrated. The same authors utilized Celite® (Johns–Manville, N.Y.) columns coated with LA–1 for the separation of free and conjugated estrogens in urine [55].

The difficulty experienced by many workers in removing steroid conjugates from true anion exchange resins undoubtedly arises from the combination of strong electrostatic and hydrophobic interactions which tightly bind the conjugates to the resin. It might be anticipated that a useful material for the purpose at hand would result from the elimination of one of these factors. Indeed, the recent work of Bradlow with the neutral polystyrene resin, Amberlite XAD–2, fulfills this expectation [56]. It was demonstrated that conjugates of metabolites derived from cortisol [11β,17α,21–trihydroxypregn–4–ene–3,20–dione], testosterone, progesterone [pregn–4–ene–3,20–dione], 17β–estradiol, and aldosterone [11β,21–dihydroxy–3,20–dioxopregn–4–en–18–al] were quantitatively removed from urine simply by passing it through a column of resin. Extreme variations in the pH of the urine had no effect on the efficiency of adsorption nor were the conjugates eluted by 0.1 *N* solution of acid or base. Following adsorption, distilled water was passed through the column which removed most of the urinary solids. The conjugates were then eluted with methanol (2.5 liters per liter of urine). Concentration of the methanol produced an extract which contained nearly all of the conjugates but 10% or less of the total urinary solids. The column can apparently be re–used indefinitely simply by washing with water between each run. As much as 4 liters of urine per day can be processed with a single column containing 1 kg of resin. That much of the operation can be performed unattended is an additional advantage. Indeed, it appears to be the most suitable method available to date for the isolation of conjugates from large volumes of urine. However, the present author has attempted to apply this technique to the extraction of radioactive dehydroisoandrosterone sulfate from plasma without success. It would seem that binding of dehydroisoandrosterone sulfate to serum albumin is sufficiently great to prevent complete adsorption to the resin even when plasma is diluted 100–fold.

D. Adsorption Methods

Relatively little attention has been given to methods by which conjugated steroids are removed from urine or other fluids by batchwise adsorption techniques. Under appropriate conditions, such methods should have considerable advantages over solvent extraction. Generally speaking, this approach should require smaller volumes of organic solvents and thus minimize the difficulties of evaporation and possible destruction of steroid conjugates during the concentration process.

The fact that steroid conjugates are strongly bound to proteins such as albumin forms the basis of extraction methods reported by two groups of investigators. Slaunwhite and Sandberg dialyzed urine against several changes of 5% bovine serum albumin (one–fifth volume) causing the conjugates to pass into the albumin solution. The conjugate albumin complex was then precipitated by the addition of alcohol [57]. The method was satisfactory for estrogen and androgen conjugates but not for corticoid conjugates. On the other hand, Zumoff and Bradlow added egg albumin directly to the urine (4 g/100 ml) and the albumin–conjugate complex was precipitated with ammonium sulfate [58]. The precipitated protein was then separated by centrifugation and redissolved in water. The addition of 5 volumes of 1:1 acetone–methanol disrupted the complex and also reprecipitated the protein. The resulting organic solvent mixture was found to contain more than 90% of the conjugated steroids present in the initial urine and less than 10% of the nonsteroidal solids. Both of these groups have studied the recovery of conjugated metabolites from human urine following the injection of various radioactive steroids. Comparison of the proportion of metabolites obtained following β–glucuronidase treatment of raw urine or purified conjugate fractions suggested that artifacts were not formed by these procedures.

A more classical adsorption method was reported by Arcos and Lieberman, who devised a procedure for concentrating urinary steroid conjugates based on their adsorption to charcoal [59]. The method consists of dilution of the urine with an equal volume of water and adjustment of the pH to 5–6 with sulfuric acid. Four g each of charcoal and Celite per liter of diluted urine was then added to adsorb the steroid conjugates. Following separation of the charcoal–Celite mixture by filtration, the conjugates were eluted with ammoniacal ethanol. Again, studies performed on urine containing radioactive steroid conjugates demonstrated the efficiency of the procedure. Good recovery of the commonly occurring conjugates, including metabolites of cortisol, was demonstrated by subsequent chromatography on Celite. The results were comparable to those obtained by the ether–ethanol procedure of Edwards *et al.* [33]. This simple and efficient method would appear to be quite valuable, particularly when working with very large volumes of urine.

E. Miscellaneous Methods

Many of the above described methods, which have been successfully applied to urine, have been adapted to the extraction of steroid conjugates from plasma or tissue. Other approaches to obtaining conjugates from the latter sources are available. For example, Baulieu subjected lyophylized plasma to solvent extraction in a Soxhlet–type apparatus [60]. Much of the lipid was first removed by extraction with hexane and the free steroids were obtained with methylene chloride. Steroid conjugates were finally removed from the protein with methanol. Paper and column chromatography together with paper electrophoresis were then used to demonstrate that the latter steroids were indeed conjugated. A unique method for the direct isolation of small quantities of conjugated steroids from tissue has been devised by Cortes and Péron [61]. Corticosterone sulfate [11β–hydroxy–21–sulfooxypregn–4–ene–3,20–dione] and corticosterone [11β,21–dihydroxypregn–4–ene–3,20–dione]–protein complexes were extracted by subjecting the tissue to conditions of electrophoresis.

Unpublished experiments in the author's laboratory have been carried out in which urine has been concentrated by the addition of *n*–butanol and evaporation under reduced pressure at 30°. The concentrated urine was mixed with Celite and applied to large columns of Celite containing 20 to 30% ammonium sulfate solution as stationary phase. Gradient elution with solvent systems consisting of ethyl acetate and increasing amounts of *t*–butanol was then carried out to elute the conjugates. This procedure was devised in order to avoid losses of conjugates during extraction and to insure detection of unknown types of conjugates in urine following the administration of radioactive precursors. While the procedure has successfully been used for the isolation of simple 17–ketosteroid glucuronides and sulfates, it cannot be used for estrogen conjugates. It has been observed that concentration of urine under the above conditions results in extensive hydrolysis of some estrogen conjugates. Of the total urinary radioactivity, 20–30% appeared in the first few fractions in the above system. This material was shown to be unconjugated by subsequent chromatography on Celite in published systems [62].

It should be pointed out that a number of the earlier workers utilized the insoluble heavy metal salts of glucuronides for their separation from crude mixtures. For example, Westphal precipitated pregnanediol glucuronide [20α–hydroxy–5β–pregnan–3α–yl–β–D–glucopyranosiduronic acid] from urine extracts by forming its barium salt in ethanolic solutions [63]. Others have used lead salts for the same purpose [64]. On the other hand, some compounds such as pregnanediol glucuronide, can be precipitated from aqueous solution by the addition of organic solvents such as acetone [65]. While these methods may seem crude by today's standards, they are still quite useful for gross separations when large amounts of material are available, such as in pregnancy urine extracts or synthetic reaction mixtures.

V. Purification of Individual Conjugates

Having reduced the bulk of an extract containing steroid conjugates to manageable proportions, a number of further purification procedures are available for the isolation of individual components. Since both the nature of the remaining impurities and the chemical properties of the conjugates vary considerably, no single method exists which is clearly superior to another. Among the possible choices are countercurrent distribution (CCD), partition or adsorption column chromatography, paper or thin layer chromatography (TLC), and electrophoresis. Each of these approaches has its advantages or limitations for a particular problem. However, only in rare situations has it been possible to isolate a pure steroid conjugate by the use of a single technique. Multiple procedures are usually required and indeed are mandatory if minute quantities are to be identified with certainty.

A. Countercurrent Distribution

This powerful technique was developed by Craig and Craig for the analysis of naturally occurring peptides [66]. In principle, the method involves the repeated partition of a compound in a two–phase solvent system and can be carried out in a series of separatory funnels. In its most simple form, the compound to be purified is mixed with equal volumes of pre–equilibrated solvents. The particular mixture of solvents is usually selected to give an equal distribution of the substance between the two phases. The system can be altered, of course, depending upon the particular problem, but this condition provides the maximum theoretical separation of an individual component. After the first equilibration, the mobile solvent, usually the upper phase, is transferred to a second vessel containing stationary phase, and fresh mobile phase is added to the first vessel. This process can be repeated many hundreds of times with the aid of automatic equipment and ideally results in Gaussian shaped curves when tube content is plotted against tube number. Separation of individual components is thus achieved if their distribution coefficients,

$$K = \frac{\text{conc. mobile phase}}{\text{conc. stationary phase}}$$

are sufficiently different. Not only can the behavior of a pure solute in CCD be mathematically predicted from its distribution coefficient, but the number of transfers required for effective separation of two solutes can be calculated from their K values [66]. Thus, the selection of a solvent system which will achieve the desired separation can be made on the basis of simple preliminary experiments. This property of CCD affords a distinct advantage over many purification methods, such as adsorption chromatography, in which systems are based on empirical findings.

The capacity of CCD for crude extracts is dependent upon the volume of solvent used in the apparatus and to some extent upon the choice of solvent system. Ordinarily these systems can handle larger quantities of crude extracts than conventional column, paper, or thin layer chromatographic systems, but practical limits are imposed by the solubility of the solutes and other factors. For example, overloading a CCD system can lead to aberrant behavior of a compound if its distribution coefficient changes with concentration. In extreme cases overloading with crude samples may disrupt the equilibrium between the two solvent phases and lead to the formation of stable emulsions. Solvent systems containing *n*–butanol are notorious for their tendency to reduce a beautifully clean apparatus to a frightful mess. This is a disastrous result, indeed, since the cleansing of a 100–tube machine together with the preparation of the solvent system requires several days effort. Slow operation of the apparatus is also essential for success, particularly with steroid conjugates because of their detergent properties. A week or more may therefore be required to accomplish several hundred transfers of crude samples. The CCD apparatus itself is costly and requires a large amount of space in a constant temperature laboratory for optimum results.

An equally important application of CCD is its use as an analytical tool. It has been widely used for the determination of radiochemical homogeneity of purified metabolites and for the positive identification of isolated compounds by comparisons of their behavior with that of authentic material. Agreement between experimental and theoretical CCD curves for both mass and radioactivity in several solvent systems constitutes an extremely powerful criterion of purity with the important proviso that a sufficient number of transfers are made so that closely related substances can be expected to be at least partially resolved. Constancy of specific activity across a peak is the usual criterion of purity. Rigorous statistical methods for the analysis of such data obtained with estrone sulfate [3–sulfooxyestra–1,3,5(10)–trien–17–one] have been presented by Sheps *et al.* [67].

Schneider and Lewbart were the first to report on the extensive application of CCD to the isolation of steroid conjugates [6]. They studied the metabolic fate of androsterone [3α–hydroxy–5α–androstan–17–one], dehydroisoandrosterone, tetrahydrocortisone [3α,17α,21–trihydroxy–5β–pregnane–11,20–dione], and cortolone [3α,17α,20α,21–tetrahydroxy–5β–pregnan–11–one] in humans by isolating intact conjugated metabolites from urine. Following the administration of the 17–ketosteroids, crude butanol extracts of 30-hour urine collections were subjected to CCD using 150 transfers in a solvent system consisting of 0.2% NH_4OH: 100, ethyl acetate: 50, *n*-butanol: 50. The fastest moving simple sulfates were readily separated from the bulk of the glucuronides. The latter partially overlapped with more polar sulfates but were separated from the bulk of the extraneous very polar material. Crystalline androsterone glucuronide was easily obtained following an additional 450 transfers of the glucuronide fraction in the system ethyl acetate: 71, *n*-hexane: 29, acetic acid: 10, water: 90. The glucuronides of tetrahydrocortisone and β-cortolone [3α,17α,20β,21-tetrahydroxy-5β-pregnan-11-one] were also isolated in essentially pure form by various combinations of CCD and column chromatography. As with the simpler compounds, acidic systems afforded the best resolution of the glucuronides of the more polar steroids. These results can be generalized in that the group separation of sulfates and glucuronides are best achieved in basic solvent systems, and individual steroid glucuronides are best separated in acid systems in all methods which utilize solvent partition. More recently, Hadd and Dorfman isolated pure androsterone and etiocholanolone [3α-hydroxy-5β-androstan-17-one] glucuronides from the urine of a woman with an adrenal adenoma by using similar CCD techniques [68].

CCD has been extensively applied to estrogen conjugates in both isolation and characterization studies. Purdy *et al.* rigorously characterized radioactive estrone sulfate isolated from plasma of women who had received ^{14}C-17β-estradiol largely on the basis of critical CCD analysis [69]. Diczfalusy and his associates have made wide use of CCD in their studies of estrogen metabolism during pregnancy. Many CCD solvent systems were devised for both the isolation of estrogen conjugates such as estriol glucuronides and sulfates [70] and also for characterization of new compounds such as 15α-hydroxy-17β-estradiol [estra-1,3,5(10)-triene-3,15α,17β-triol] [71]. Levitz *et al.* have also utilized CCD for studies of complex estrogen conjugates such as estriol-3-sulfate-16-glucuronide [26].

B. Column Chromatography

Several distinctly different forms of column chromatography have been applied to the purification of steroid conjugates. The earliest applications utilized a variety of adsorbents with which only moderate degrees of selectivity were realized. More recently, liquid-liquid partition chromatography on an inert support such as Celite has been used for the purification of steroid conjugates in many laboratories. Gel filtration and ion exchange techniques have also been applied as indicated earlier. Since these forms of chromatography depend upon totally different physical principles, they shall be considered separately in some detail.

1. Adsorption Chromatography

A variety of adsorbents such as alumina, Florisil® (Floridin Co., Tallahassee, Fla.), and silica gel have been used for the chromatographic separation of steroid conjugates. In general, these methods consist of the introduction of a crude extract dissolved in a solvent of low polarity on to a column of the adsorbent whereby all of the steroid conjugates are adsorbed. Solvents, or mixtures of solvents, of increasing polarity are sequentially introduced into the column causing the elution of the desired compounds. While such systems possess the advantages of having a relatively large capacity for crude samples and simplicity, they have a number of serious drawbacks. Generally speaking, the resolution of adsorption columns is poor, so that separation of closely related compounds is seldom achieved. This, of course, can be considered an advantage if it is desired to separate broad classes of compounds such as the sulfates and glucuronides of the 17–ketosteroids. A further disadvantage of such systems is the ever–present danger of alterations in the structure of complex conjugates due to interactions with potentially destructive adsorbents such as alumina. Incomplete recoveries may also arise from such interactions, since rarely, if ever, are these systems operated under conditions of reversible equilibrium. That systems for adsorption chromatography must be selected on an empirical trial–and–error basis is also of considerable disadvantage when compared with CCD. This problem is accentuated by the inherent variability in the properties of adsorbents which can result in lost time, and occasionally tempers, when published procedures fail to perform in one's own laboratory.

Nevertheless, adsorption columns have been very useful for the separation of classes of conjugates. The first application of adsorption chromatography to steroid conjugate isolation appears to be that of Venning *et al.* [4]. These workers utilized columns of alumina for the purification of urinary extracts in their pioneering studies which led to the first isolation of a crystalline conjugate, androsterone sulfate [3α–sulfooxy–5α–androstan–17–one]. Pure estrone sulfate was later obtained by Grant and Beall from pregnant mare's urine with the aid of columns containing activated alumina and Celite [28]. These authors utilized solvent systems consisting of benzene and increasing amounts of methanol for elution of the conjugates.

The usefulness of alumina columns for the separation of 17–ketosteroid sulfates and glucuronides was described in 1957 by Crépy *et al.* [72] and also by Barlow and Kellie in 1959 [73]. The former workers applied the concentrated *n*–butanol extract of a 24–hour urine collection to a column of alumina which was then successively eluted with butanol containing 2, 6, and 10% water, and finally 15% 0.1 *N* NH_4OH. The 17–ketosteroid sulfate fraction was completely eluted with 6% aqueous butanol and cleanly separated from the glucuronides which were not removed until the ammoniacal butanol solution was used. In addition to separation of the conjugates, considerable purification of each class was achieved. Barlow and Kellie, working with much smaller volumes of urine (10–200 ml), applied ether–ethanol extracts directly to alumina columns which were then washed with ethanol. Elution of the 17-ketosteroid sulfates was accomplished with 50% aqueous ethanol following which the 17-ketosteroid glucuronides were eluted with 1.0 *M* acetate buffer, pH 5.0. While both of these procedures are useful, it would appear that the method of Crépy *et al.* gives somewhat cleaner separations. Other applications of alumina columns have been reviewed by Jayle *et al.* [32].

The use of Florisil, a synthetic magnesium silicate, as a chromatographic medium for steroid conjugates was first reported by Brancoccio *et al.* for the separation of free and conjugated corticosteroids [74]. More recently, Herrman utilized Florisil columns for the separation and measurement of 17-ketosteroids and their sulfates in plasma [75]. Alcoholic extracts of plasma were applied to Florisil columns in benzene and the free steroids eluted with benzene–ethyl acetate. Sulfates were eluted with ethyl acetate–ethanol mixtures and glucuronides with aqueous ethanol. Baulieu has described the use of Florisil columns for the separation of urinary conjugates of important C_{19}-steroids [76]. Much of the work of Baulieu's group concerning the metabolism of steroid conjugates and the methodology they have used recently has been reviewed [77, 78]. Sarfaty and Summer [79] have also utilized Florisil columns for 17-ketosteroid conjugates using essentially the system first described by Herrman. Pasqualini has also made extensive use of Florisil and alumina columns for the isolation of more polar conjugates such as aldosterone glucuronide and other corticosteroid sulfates [80, 14].

2. *Liquid–Liquid Partition Chromatography*

The technique of column liquid–liquid partition chromatography (CLLC) was first introduced by Martin and Synge in 1932 [81]. Under ideal conditions, this procedure offers most of the theoretical advantages of CCD, yet is much simpler to perform. However, ideal behavior can only be achieved if the material used for supporting the stationary phase is completely inert toward the solutes which are to be isolated or characterized. Diatomaceous earth, such as Celite, is one of the few materials which approaches this criterion for a wide variety of relatively polar substances, including the steroids and their conjugates. A clear distinction can therefore be drawn between CLLC as performed with Celite and with materials such as cellulose, silica gel, or paper. While the literature contains many examples of the successful use of the latter supports it should be recognized that these are complex systems involving both adsorption and partition processes. Consequently they have been empirically devised. On the other hand, when Celite columns are used, solvent systems can almost always be selected on the basis of distribution coefficients measured in the absence of the support. Moreover, the behavior of a solute in such a system can be predicted as will be shown below.

Among other advantages of this technique are large capacity and the aforementioned powerful criterion of purity afforded by multiple analyses across a peak. In view of these advantages it is perhaps surprising that this method has been so little used until recently. Much of the recently described theory and practical aspects of gas–liquid chromatography has been known in terms of the liquid–liquid process for many years [82]. The neglect of CLLC appears to stem from a general lack of understanding of its virtues and of its simplicity both from a theoretical and practical point of view. It therefore seems appropriate to briefly describe some of the general principles underlying the successful application of CLLC. Several terms which are used below are defined as follows:

Distribution coefficient (K)—concentration of compound in mobile phase/concentration of compound in stationary phase.

Hold–back volume (HBV)—the volume of mobile phase required to fill the column bed containing both support and stationary phase.

Stationary phase volume (V_s)—the volume of stationary phase contained in the column.

Elution volume (V_e)—the volume of mobile phase which emerges from the column before the peak of a particular compound is eluted.

The basic relationship between these parameters can be expressed as follows:

$$V_e = 1/K(V_s) + \text{HBV}.$$

However, as will be shown later, it is more meaningful to express the volume required to elute a compound, V_e, in terms of HBV's. Furthermore, the above expression was derived for columns which are "slurry" packed and therefore contain one HBV of mobile phase when the sample is applied. If no mobile phase is present, as when columns are prepared by the "dry pack" method, the following simple equation can be used: Number of HBV's to peak $= 1/K \cdot V_s/\text{HBV}$.

It is seen that the relative position of emergence of a compound from a column is a function only of its distribution coefficient and the volume ratio of the two phases contained in the column. The analogy to CCD is immediately apparent since these are the same factors which determine the rate of migration of a solute in that process. Also as in CCD, the shape of a peak eluted from a partition column will be that of a normal distribution, providing that K does not vary with concentration and that the operation is carried out without perturbing factors such as adsorption, nonequilibrium flow rates, or serious column imperfections.

It has been shown that the resolving power of a partition column increases exponentially and then approaches a maximum as the number of HBV's required to elute a compound becomes larger [82]. Thus, only 10–20% of the maximum theoretical resolution is achieved for a compound which is eluted in the first HBV, whereas about 90% can be realized when the compound is eluted in the fifth HBV. However, it is a characteristic of any partition process for bands to become broader and peak concentrations smaller as the number of transfers increases. This fact, together with the increasing importance of deviations from ideal behavior due to column-bed imperfections, laminar flow, etc., plus time limitations, place upper limits on the useful number of HBV's used to elute a compound, and hence on the resolution obtainable. The normal useful working range of elution volumes is about 3–15 HBV's.

The optimum ratio of stationary phase to mobile phase, V_s/HBV, is approximately 1/3 for moderate size columns prepared by the "dry pack" method using Johns–Manville Celite 545. With this ratio in the above equation it can be seen that the range of elution volumes of 3–15 HBV's corresponds to k values of about 0.1 to 0.02, respectively. Selection of a solvent system for a particular problem involves no more than the measurement of distribution coefficients in trial systems until one is found which gives a value in this range. This is easily and quickly done with small amounts of material in test tube experiments, particularly if a sensitive method of assay is available. It is evident from the above relationships that the ratio of mobile phase to stationary phase should be kept small in order to achieve good resolution and economy of operation. Obviously, the greater the amount of stationary phase contained in a column, the larger will be its capacity. However, the amount of stationary phase which can be held by the support is limited. The optimum ratio of stationary phase to Celite is 1:2 for most applications. A 1:1 ratio can occasionally be used without excessive bleeding of stationary phase.

The importance of the proper selection of solvent system to give a suitable value of K and of the use of HBV's rather than total elution volume (V_e) can be illustrated by the following example. Suppose two compounds, A and B, with K values of 0.1 and 1.0 in a particular system were chromatographed on two separate columns which contained 10 ml and 100 ml of stationary phase and which also had 30 ml and 300 ml HBV's, respectively. Both A and B would be eluted after a total of 300 ml of mobile phase passed through the columns, but A would be in the tenth HBV and B in the first HBV. Both compounds might exhibit very symmetrical elution curves, but B could be grossly impure. If only the total effluent volumes were reported for these columns it would be impossible to assess their relative performance. Therefore, if chromatographic homogeneity, as evidenced by a symmetrical elution curve, is to be used as a criterion of purity, it is essential to provide complete data so that the chromatogram can be interpreted properly.

The theoretical treatment from which the above, or similar equations, have been derived is based on certain assumptions of ideal behavior. Some of these and their practical consequences are as follows:

(a) As indicated before, the chromatographic process must consist only of a reversible partition of solute between two immiscible liquid phases.

(b) Compositional equilibrium must be maintained between the two liquid phases throughout the length of the column.

Since the composition of all solvent systems is to a greater or lesser degree dependent on the environmental temperature, extreme fluctuations in temperature must be avoided during the course of an experiment. Obviously, K values and column results from different laboratories can be compared only if obtained at the same temperature. Maintainence of compositional equilibrium also requires that the column contain the solute (*s*) in concentrations less than their limiting solubility in either phase. "Overloading" with either crude extracts or pure compound can disrupt the two-phase equilibrium by a "salting-out" effect and result in column failure. This is the most common cause of poor results with column partition chromatography, but is easily recognized by the appearance of two phases in the column effluent. Termination of the experiment by stripping the column with alcohol or other solvent and starting over with either a smaller load or larger column is recommended in such cases.

(c) The flow rate of mobile phase through the column must be slow enough to permit complete, reversible transfer of solute between the two phases and yet fast enough so that diffusion of solute does not lead to band spreading. The use of very low flow rates for optimum results was advocated by the originators of the method. However, recent experience has shown that ordinary columns can be completed in a few hours rather than days. Indeed, with the use of high column pressures satisfactory separations have been obtained in minutes.

(d) The theoretical plate concept, as applied to partition columns, was based on an infinitely thin layer of solute which is successively partitioned between mobile and stationary phase as it progresses down the length of the column. Although this situation cannot exist, it is obvious that sample band widths should be kept as narrow as practicable in order to minimize spreading of bands and loss of resolution.

(e) The compound being chromatographed must have a distribution coefficient which is independent of concentration. A skewed or non-Gaussian shaped peak is observed if the distribution coefficient of a compound varies with its absolute concentration. This phenomenon rarely occurs, but it can be minimized by using small column loads.

Up to this point, this discussion has been concerned with chromatography which is designated as "elution analysis". This term is applicable to all forms of chromatography in which a solute is partitioned between two phases whose compositions remain constant throughout the process. The number of compounds which can be separated by elution analysis is limited by the factors previously described. However, a broader range of compounds can be chromatographed in a single operation by increasing the eluting power of the mobile phase either in a stepwise or a continuous manner. The latter technique, known as gradient elution chromatography, also overcomes the tendency of slow-moving solute bands to broaden. Good resolution can therefore be obtained throughout the chromatogram if the technique is used properly. A variety of free steroids including 17-ketosteroids [83] and estrogens [62] have been subjected to gradient elution chromatography on alumina and Celite, respectively. Gradient elution chromatography of steroid glucuronides on Celite has been described by Baulieu *et al.* [77].

Major drawbacks to the CLLC technique originally described by Martin and Synge were the tedious column-packing method and the very slow flow rates which they advocated. As a result, little advantage over CCD methods except simplicity of equipment was offered by the original method. Furthermore, the scale of operation was limited, since it is virtually impossible to pack large columns by the slurry-tamping rod method. The alternative "dry pack" method of column preparation, whereby portions of Celite equilibrated only with stationary phase are tamped into the column with the aid of a close fitting cork attached to a dowel, was said to yield "inelegant results". This notion, unsupported by data, has persisted throughout the years and has probably been a major deterrant to the widespread acceptance of CLLC. The "dry pack" method permits small or large columns to be prepared in a relatively short time and extends the useful range of analysis many-fold. While some loss of resolving power may be incurred by irregularities in solvent flow due to nonuniform packing, these effects are small and the resolution obtainable is adequate for most problems. The fact that experiments involving 100-mg quantities of compounds can be completed in one or two days far outweights any small loss in resolution.

The application of CLLC to the purification of steroid conjugates was briefly described by Bush in 1957 as an extension of the development of paper chromatographic systems for this purpose [84]. In 1964 Foggit and Kellie described a number of CLLC solvent systems for the isolation of individual C_{19}– and C_{21}–glucuronides [85]. The author has applied this technique to a number of steroid problems beginning in 1959 in the laboratory of Professor Lieberman at Columbia [86]. From these and many subsequent studies it has become apparent that CLLC is a most valuable tool for handling steroid conjugate isolation problems. Fig. 4–2 compares the results obtained when CLLC on Celite was carried out with androsterone, etiocholanolone, and dehydroisoandrosterone and with their glucuronides. The chromatogram in Fig. 4–2 B was obtained as follows. A mixture of 2 mg of each steroid was dissolved in 0.5 ml of stationary phase and mixed with 1 g of Celite. The sample was then packed on top of a 20 g column which contained 10 ml of stationary phase and had a measured HBV of 24.0 ml. The solvent system consisted of isooctane: 500, *t*–butanol: 100, methanol: 350, water: 50 [87]. Alternate fractions were analyzed by the Zimmermann reaction, and it can be seen that the compounds were well resolved into perfectly symmetrical peaks. It is of interest to note that the measured *K*

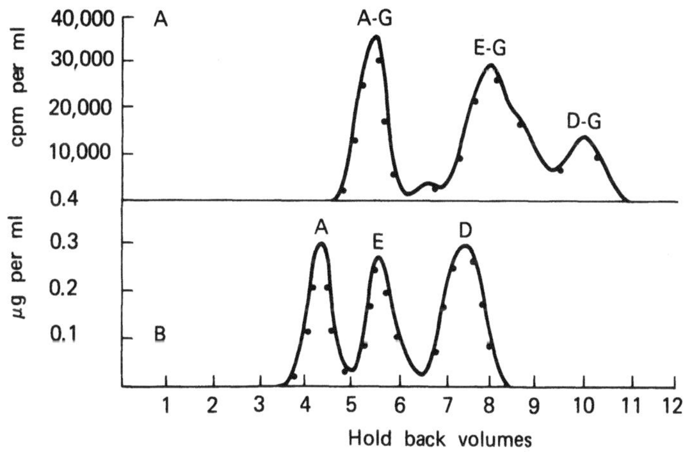

Fig. 4–2. A. Celite column chromatographic separation of radioactive glucuronides of androsterone (A–G), etiocholanolone (E–G), and dehydroisoandrosterone (D–G). B. Celite column chromatography of free steroids shown in A. See text for details

value for dehydroisoandrosterone in this system is 0.056 and therefore the calculated position of emergence was:

$$\frac{1}{0.056} \times \frac{10}{24}, \quad \text{or} \quad 7.45 \text{ HBV}.$$

As seen in the chromatogram, the peak concentration of dehydroisoandrosterone occurred at 7.4 HBV.

Fig. 4–2A illustrates the results obtained when radioactive 17–ketosteroid glucuronides were subjected to partition chromatography on Celite. The conjugates were isolated from a 24–hour urine sample collected from an individual who had received an injection of radioactive dehydroisoandrosterone. Following extraction by the Edwards *et al.* procedure [33], preliminary separation of steroid sulfates and glucuronides was achieved by CLLC in the system, isooctane:200, *t*–butanol:500, water:475, conc. NH_4OH:25 [25]. The glucuronide fraction was then further resolved on a 50–g Celite column using the system, isooctane:300, *t*–butanol:500, water:500, acetic acid:125 [87]. Radioassay of the effluent fractions was followed by hydrolysis with β–glucuronidase and identification of the liberated steroids by further chromatography. It can be seen that the degree of separation of the glucuronides obtained is equal to, if not better than, that of the corresponding free steroids. The efficiency of such column systems is illustrated by the fact that crystalline glucuronides of androsterone and etiocholanolone have been obtained in similar experiments simply by slow evaporation of the appropriate fractions.

As indicated earlier, the separation of individual steroid sulfates is more difficult than the corresponding glucuronides. However, Calvin *et al.* have devised a number of ingenious solvent systems which perform quite well in the separation of many important steroid sulfates [88]. For example, solvent systems containing pyridine sulfate were used for the separation of the monosulfates of dehydroisoandrosterone, pregnenediol, 17α–hydroxypregnenolone [3β,17α–dihydroxypregn–5–en–20–one], and androstenediol [androst–5–ene–3β,17β–diol]. Nonetheless, complete resolution of the sulfates of the three 17–ketosteroids shown in Fig. 4–2B required chromatography in two systems. The behavior of a wide variety of steroidal sulfates in the solvent systems developed by these authors is schematically illustrated in Fig. 4–3. The compositions of the solvent systems are given in Table 4–1. Of particular interest are the AET systems which have high capacity and good resolving power for nonpolar compounds such as cholesterol sulfate [3β–sulfooxycholest–5–ene].

Pasqualini has also tabulated the partition coefficients of many steroid conjugates in a variety of published solvent systems [8]. Such data are

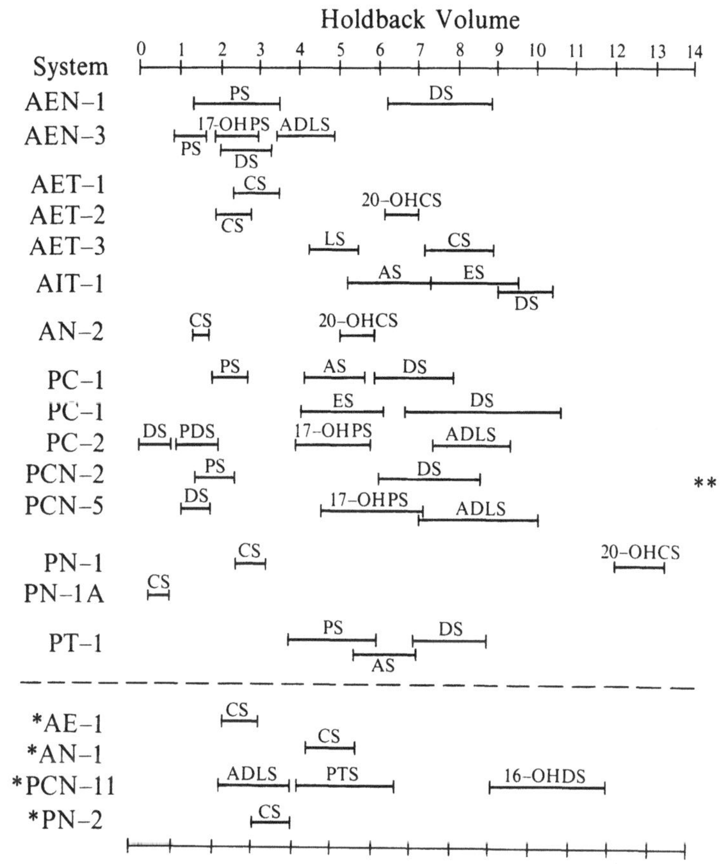

Fig. 4-3. Elution behavior of steroidal sulfates on Celite columns: (*) elution patterns of steroids extracted from biological sources—all others are those obtained with crystalline standards. The bands indicate the approximate boundaries of the chromatographic peaks. The volume of the eluate is expressed in units of 1 hold-back volume and is indicated by the scale at the top of the figure. (**) PDS was obtained in the methanol wash. The following trivial names and abbreviations are used: dehydroisoandrosterone sulfate (DS); 16α-hydroxydehydroisoandrosterone sulfate (16-OHDS) = 16α-hydroxy-3β-sulfooxyandrost-5-en-17-one; androstenediol sulfate (ADLS) = 3β-sulfooxyandrost-5-en-17β-ol; pregnenolone sulfate (PS) = 3β-sulfooxy-pregn-5-en-20-one; 17α-hydroxypregnenolone sulfate (17-OHPS) = 17α-hydroxy-3β-sulfooxypregn-5-en-20-one; pregnenediol sulfate (PDS) = 3β-sulfooxypregn-5-en-20α-ol; pregnenetriol sulfate (PTS) = 3β-sulfo-oxypregn-5-ene-17α,20α-diol; androsterone sulfate (AS); etiocholanolone sulfate (ES); cholesterol sulfate (CS); lanosterol sulfate (LS); 20α-hydroxy-cholesterol sulfate (20-OHCS). (From Calvin, H. I., Roberts, K. D., Weiss, C., Bandi, L., Cos, J. J., Lieberman, S.: Anal. Biochem. **15**, 426 (1966). Reproduced with the permission of the publisher of Analytical Biochemistry.)

obviously helpful for isolation work and, further, can be extremely useful in the characterization and identification of new conjugates.

TABLE 4–1. *Systems for the chromatography of steroid monosulfates on celite columns*

Code name	Composition
AE–1	Methylcyclohexane 30, ethyl acetate 20, methanol 30, 1 *M* NH_4OH 30
AEN–1	Methylcyclohexane 25, ethyl acetate 40, *n*–butanol 8, methanol 20, 1 *M* NH_4OH 30
AEN–3	Isooctane 20, ethyl acetate 40, *n*–butanol 7, methanol 20, 0.1 *M* $(NH_4)_2SO_4$ 30, conc. NH_4OH 0.1
AET–1	Isooctane 40, ethyl acetate 10, *t*–butanol 20, methanol 20, 1 *M* NH_4OH 30
AET–2	Isooctane 30, ethyl acetate 30, *t*–butanol 20, methanol 20, 1 *M* NH_4OH 30
AET–3	Isooctane 50, ethyl acetate 10, *t*–butanol 20, methanol 20, 1 *M* NH_4OH 30
AIT–1	Isooctane 5, isopropyl ether 10, *t*–butanol 8, concentrated NH_4OH 2, water 18
AN–1	*n*–Heptane 40, *n*–butanol 7, methanol 15, 1 *M* NH_4OH 20
AN–2	*n*–Heptane 40, *n*-butanol 7, methanol 10, 1 *M* NH_4OH 20
PC–1	Isooctane 25, chloroform 20, methanol 20, 0.3 *M* pyridinium sulfate 19, pyridine 1
PC–2	Isooctane 20, chloroform 30, methanol 20, 0.3 *M* pyridinium sulfate 20
PCN–2	Isooctane 40, chloroform 20, *n*–butanol 4, methanol 20, 0.3 *M* pyridinium sulfate 19, pyridine 1
PCN–5	Isooctane 40, chloroform 20, *n*–butanol 8, methanol 20, 0.3 *M* pyridinium sulfate 19, pyridine 1
PCN–11	Isooctane 40, chloroform 20, *n*–butanol 10, methanol 20, 0.3 *M* pyridinium sulfate 29, pyridine 1.5
PN–1	*n*–Heptane 40, *n*–butanol 10, methanol 20, 0.3 *M* pyridinium sulfate 19, pyridine 1
PN–1A	*n*–Heptane 40, *n*–butanol 10, methanol 20, 0.3 *M* pyridinium sulfate 20
PN–2	*n*–Heptane 40, *n*–butanol 6, methanol 20, 0.3 *M* pyridinium sulfate 20
PT–1	Isooctane 20, *t*–butanol 4, methanol 20, 0.3 *M* pyridinium chloride 19, pyridine 1

From Calvin, H. I., Roberts, K. D., Weiss, C., Bandi, L., Cos, J. J., Lieberman, S.: Anal. Biochem. **15**, 426 (1966). Reproduced with the permission of the publisher of Analytical Biochemistry.

C. Paper Chromatography

The first published reports concerning the use of paper chromatography in studies of steroid conjugates appeared in 1955. Cavina [89] described three solvent systems (butanol:water, ethyl acetate:butanol:0.2*N* acetic acid, 90:10:100; and isoamyl alcohol:NH_4OH:water, 55:27:18) as applied to separation of various synthetic conjugates. Lewbart and Schneider also described a number of systems containing butyl acetate for use with polar glucuronides and simple sulfates [90]. In general, these early systems had relatively poor resolving power. Improved systems were soon devised by Bush [84], Schneider and Lewbart [6], Baulieu [76], and others.

Again, the separation of sulfates from glucuronides is best accomplished with alkaline solvent systems, whereas acid solvent systems usually yield better separations of individual conjugates. However, alkaline systems are preferred when dealing with steroid sulfates because of the ease with which they are hydrolyzed in organic solvents containing acid [91]. Schneider and Lewbart [6] greatly improved the resolving power of alkaline systems for steroid sulfates by the incorporation of a solvent such as isopropyl ether as a major component of the mobile phase. Indeed, the sulfates of androsterone, etiocholanolone [3α-sulfooxy-5β-androstan-17-one] and dehydroisoandrosterone are readily separated on paper in the system, isopropyl ether:*t*-butanol:NH_4OH:water, 120:80:20:180, with a running time of 15–20 hours. The reasons for the superiority of such systems over otherwise equivalent mixtures containing solvents such as ethyl acetate or ethylene dichloride are not clear although subtle steric interactions between the ether and the steroid portion of the molecule have been implicated by Bush [51]. Caution must be exercised when using these systems because of the strong tendency of isopropyl ether to form peroxides.

Solvent systems containing *t*-butanol for paper chromatography, first described by Eberlein and Bongiovanni [92] for free steroids, are ideally suited for chromatography of steroid conjugates. The fact that *t*-butanol is not esterified in acidic systems as are primary or secondary alcohols permits the use of a wide range of stable systems. In addition, the much greater ease of removal by distillation as compared to *n*-butanol is also important particularly when systems developed for paper chromatography are adapted to the column process. Finally, the unique reversal of polarity of steroidal ketone and hydroxyl groups that occurs with free steroids, such as cortisol and aldosterone, in systems containing *t*-butanol also obtains with conjugated steroids. While the effects may not be as dramatic, they can be important in obtaining good separations. For example, reversed phase chromatography on cellulose acetate foil as described by Carr and Reddy [93] yielded a remarkably good separation of androsterone, testosterone [3–oxoandrost–4–en–17β–yl–β–D–glucopyranosiduronic acid], and 17β–estradiol [3–hydroxyestra–1,3,5(10)–trien–17β–yl–β–D–glucopyranosiduronic acid] glucuronides with the solvent system, water:*t*–butanol:benzene, 2:1:3.

Pasqualini *et al.* [94] have described a number of solvent systems for paper chromatography of several important C_{21}–steroid conjugates. Acid and alkaline systems containing butyl acetate were used for the separation of the 21–sulfates of deoxycorticosterone [21–sulfooxypregn–4–ene–3,20–dione], corticosterone, cortisol, etc. These authors hydrolyzed the sulfates after chromatography by heating the paper in dioxane containing either hydrochloric or trichloroacetic acid. The free steroids could then be detected by blue tetrazolium reduction. Good separation of very polar compounds such as tetrahydrocortisone and cortisol [11β,17α–dihydroxy–3.20–dioxopregn–4–en–21–yl–β–D–glucopyranosiduronic acid] glucuronides was also achieved in the system butyl acetate:toluene:butanol:10% acetic acid:methanol, 50:40:10:50:50.

Some practical problems which arise in the application of paper chromatography to steroid conjugates should be pointed out. Double spot formation due to the presence of two cations can occur if relatively large amounts of a conjugate are being chromatographed. This occurs only with metal salts and is usually not a problem when dealing with pyridinium salts or complexes with dyes such as methylene blue. Indeed, the latter cannot be chromatographed as such in alkaline or acid systems since they disassociate immediately. As indicated earlier, the time of exposure of sulfates to acid systems should be as brief as possible and care must be exercised when the compounds are eluted and the extracts concentrated, in order to avoid hydrolysis. Elution of most conjugates from paper is readily accomplished with methanol but aqueous alcohol may be required for polar glucuronides. Kornel has recommended a mixture of *n*–butanol:70% aqueous methanol, 1:3, which is said to elute even the most polar compounds in a relatively small volume [95].

Specific applications of paper chromatography to steroid conjugates may be found in references cited earlier [6–8, 51].

D. Thin Layer Chromatography

The great attractiveness of TLC arises from the great speed with which separations can be achieved and its high capacity as compared with paper chromatography. Generally speaking, however, the resolution obtainable is not always equal to that of paper chromatography despite the availability of a wide variety of supporting media. For example, the separation of free 17–ketosteroids is quite difficult on either silica gel or alumina. The sulfates and glucuronides of the same steroids are even less well resolved on these as well as other materials which have ionic properties, such as DEAE–cellulose. For example, thin layers of the latter were used by Oertel *et al.* [96] for the separation of androsterone, etiocholanolone and dehydroisoandrosterone sulfates and glucuronides. Seven systems including acetate buffers at various pH values between 4.25 and 5.00 and solvent systems such as isopropanol:water:formic acid, 65:33:2, were used. While these authors stated that satisfactory resolution was obtained with all of these systems, the results were not impressive. For example, the R_f values for the glucuronides of androsterone, etiocholanolone, and dehydroisoandrosterone in the formic acid system were 0.77, 0.75, and 0.72, respectively, while androsterone and dehydroisoandrosterone sulfates had R_f values of 0.22 and 0.33, respectively.

Equally disappointing results have been obtained with silica gel. For example, Wusteman *et al.* examined the behavior of steroid and other organic sulfates on silica gel G (Merck) in both acid and alkaline systems [93]. The migration of such widely different compounds as cholesterol sulfate (R_f 0.58) and cortisone–21–sulfate [17α–hydroxy–21–sulfooxypregn–4–ene–3,11,20–trione] (R_f 0.44) was remarkably similar. Compounds such as dehydroisoandrosterone sulfate and 17β–estradiol–3–sulfate [3–sulfooxy-estra–1,3,5(10)–trien–17β–ol] clearly were not resolved (R_f values of 0.49 and 0.48, respectively). Indeed the poor resolution of conjugates on silica gel was taken advantage of by Sarfaty and Lipsett [98] who utilized silica gel H (Merck) for the group separation of urinary free 17–ketosteroids, their sulfates and their glucuronides. The solvent system used consisted of ethyl acetate:ethanol:15*N* NH_4OH, 5:5:1, and separations required 70–75 min. In addition to using specific spray reagents for detection of sulfates and glucuronides described earlier, the authors eluted the spots with an appropriate solvent which could be utilized in the hydrolysis step, *e.g.*, pH 5 acetate buffer for glucuronides.

Two–dimensional TLC has been applied to the separation of steroid conjugates with more success. Crépy *et al.* [24] separated the 5α– and 5β–isomers of reduced progesterone metabolites on layers composed of an equal mixture of Kieselgur G® and Kieselgel G® (E. Merck, AG, Darmstadt, Ger.), using the system, toluene:*t*–butanol:acetic acid:water, 82:18:30:70. Fishman *et al.* [99] also used two–dimensional chromatography with development in three different solvent systems for the separation of urinary 17β–estradiol and estriol glucuronides.

E. Electrophoresis

As indicated at the outset, electrophoresis is an efficient method only for the separation of conjugates which have different ionizable functions or a different number of the same group, such as mono- and di-sulfates. Therefore, the fact that an isolated conjugate behaves as a single zone during electrophoresis provides little evidence for the homogeneity of the material with regard to its steroid content.

Cavina [89, 100] and Cavina and Tentori [101, 102] have made extensive use of electrophoresis in conjunction with paper chromatography for the analysis of urinary 17-ketosteroids. In a typical procedure, the conjugates obtained from 200 ml of urine by butanol extraction were subjected to electrophoresis on 4 × 35 cm strips of Whatman No. 1 filter paper. Separations were achieved using 0.1 *M* phosphate buffer, pH 7.4 at 200 volts and 0.5 ma for 5–6 hours. The well-separated glucuronide and sulfate zones were detected by the Zimmermann reagent and eluted with aqueous ethanol for quantitation without prior hydrolysis. More recently, a number of workers have used high voltage techniques (2,000–5,000 volts) to achieve the separation of conjugated estrogens [103], androgens [104], and corticosteroids. Kornel has studied extensively the high voltage electrophoresis of mono-, di-, and tri-sulfates of cortisol as well as other C_{21}-steroids [105]. While the use of high voltages yielded rapid separations, wide variations in the choice of buffer or pH had little influence on the degree of separation obtained. It would appear that there is little to recommend this technique in view of the considerable potential hazards involved in the performance of electrophoresis at high voltages.

VI. Concluding Remarks

The history of progress in biochemistry could be accurately displayed on a time scale which denoted the appearance of new methods. Particularly prominent in such a hypothetical plot would be key developments in isolation methods for small organic molecules and macromolecules such as proteins and nucleic acids. While much of our present knowledge has been obtained inferentially, the "chemistry" in biochemistry demands that the ultimate proof of hypotheses be obtained with pure substances by classical methods. There is a tendency among biochemists to use what is "fashionable" at any one time, due in part perhaps to the increasing sophistication of equipment suppliers. However, those techniques and methods which are destined to be milestones are quickly added to the chemist's armamentarium while others fall by the wayside. Although methods are devised for particular problems, they soon find general applicability for many classes of compounds. It should be apparent that steroid conjugates do not represent a uniquely different or difficult group of compounds to deal with. That the techniques ordinarily used for ionized compounds have not been very useful for conjugates is understandable from the above discussion. Future developments are difficult to predict, but as in the past, they will likely arise from some other field. The author is unaware of any serious attempts to apply gas liquid chromatography to intact steroid conjugates or their derivatives. Suitable derivatives of the glucuronides might be expected to be amenable to this powerful technique.

In conclusion, the following is offered as a general guide to the isolation of steroid conjugates. The order of selection of methods is strictly personal and undoubtedly at variance with the preferances of others. However, it is hoped that those readers without experience in this area may benefit from past successes and failures and, possibly, save some of that most precious commodity, time. This, after all, is the primary function of a review article.

A. Extraction of steroid sulfates.
   1. Extraction of pyridinium salts with chloroform.

B. Extraction of steroid sulfates and glucuronides.
   1. Small samples (50–100 ml). Extraction with ether–ethanol following the addition of ammonium sulfate.
   2. Intermediate samples (0.5–2 liter). Absorption with neutral resin, Amberlite, XAD-2.
   3. Large samples (1–10 liter) Absorption with charcoal.

C. Preliminary purification.
   1. Celite column chromatography.
   2. Absorption chromatography.
   3. Gel filtration.

D. Final purification (decreasing order of resolving power).
   1. Paper chromatography.
   2. Celite column chromatography.
   3. Countercurrent distribution.
   4. Thin layer chromatography.
   5. Electrophoresis.

Acknowledgments. The experimental work reported here was supported in part by USPHS Grant No. HD00107, and a grant from the Robert A. Welch Foundation, Houston, Texas.

VII. References

1. Cohen, S.L., Marrian, G.F.: Biochem. J. **30**, 57 (1936).
2. Venning, E.M., Browne, J.S.L.: Proc. Soc. Exptl. Biol. Med. **34**, 792 (1936).
3. Schachter, B., Marrian, G.F.: J. Biol. Chem. **126**, 663 (1938).
4. Venning, E. H., Hoffman, M. M., Browne, J. S. L.: J. Biol. Chem. **146**, 369 (1942).
5. Munson, P. L., Gallagher, T. F., Koch, J. C.: J. Biol. Chem. **152**, 67 (1944).
6. Schneider, J. J., Lewbart, M. L.: Recent Progr. Hormone Res. **15**, 201 (1959).
7. Bernstein, S., Dusza, J. P., Joseph, J. P.: Physical properties of steroid conjugates. Berlin-Heidelberg-New York: Springer 1968.
8. Pasqualini, J. R., in: Steroid hormone analysis, vol. 1 (H. Carstenson, ed.), p. 407. New York, N.Y.: Marcel Dekker, Inc. 1967.
9. Layne, D. S.: Endocrinology **76**, 600 (1965).
10. Arcos, M., Lieberman, S.: Biochemistry **6**, 2032 (1967).
11. Wotiz, H. H., Fishman, W. H.: Steroids **1**, 211 (1963).
12. Wakabayashi, M., Wotiz, H. H., Fishman, W. H.: Biochim. Biophys. Acta **48**, 198 (1961).
13. Diczfalusy, E., Barr, M., Lind, J.: Acta Endocrinol. **46**, 511 (1964).
14. Pasqualini, J. R., Jayle, J. F.: J. Clin. Invest. **41**, 981 (1962).
15. Baulieu, E.-E., Corpéchot, C.: Bull. Soc. Chim. Biol. **47**, 443 (1965).
16. Oertel, G.W., Knapstein, P., Treiber, L.: Z. Physiol. Chem. **345**, 221 (1966).
17. Roy, A. B.: Biochem. J. **62**, 41 (1956).
18. Crépy, O., Rulleau-Meslin, F.: Rev. Franc. Etudes Clin. Biol. **5**, 283 (1960).
19. — Judas, O.: Rev. Franc. Etudes Clin. Biol. **5**, 284 (1960).
20. Burma, D. P.: Anal. Chim. Acta **9**, 513 (1953).
21. Schneider, J. J., Lewbart, M. L.: J. Biol. Chem. **222**, 787 (1956).
22. Hanes, C. S., Isherwood, F. A.: Nature **164**, 1107 (1949).
23. Crépy, O., Lachese, B., Judas, O.: Rev. Franc. Etudes Clin. Biol. **6**, 601 (1961).
24. — Judas, O., Lachese, B.: J. Chromatog. **16**, 340 (1964).
25. Siiteri, P. K., Lieberman, S.: Biochemistry **2**, 1171 (1963).
26. Levitz, M., Katz, J., Twombly, G. H.: Steroids **6**, 553 (1965).
27. Robel, P., Emiliozzi, R., Baulieu, E.-E.: J. Biol. Chem. **241**, 20 (1966).
28. Grant, G. A., Beall, D.: Recent Progr. Hormone Res. **5**, 307 (1950).
29. Jayle, M. F., Baulieu, E.-E.: Bull. Soc. Chim. Biol. **34**, 1200 (1952).
30. — — Gonzalez-Flores, L.: Ann. Endocrinol. (Paris) **14**, 642 (1953).
31. — Crépy, O.: Ann. Endocrinol. (Paris) **14**, 647 (1953).
32. Analyse Des Stéroïdes Hormonaux (M. F. Jayle, ed.), vol. I, p. 30. Paris: Masson & Cie. 1961.
33. Edwards, R.W. H., Kellie, A. E., Wade, A. P.: Mem. Soc. Endocrinol. No. 2, Conference Society for Endocrinology, London, 1953, p. 53.
34. Kamil, I. A., Smith, J. N., Williams, R.T.: Biochem. J. **53**, 129 (1953).
35. Kellie, A. E.: Biochem. J. **100**, 631 (1966).
36. DePaoli, J. C., Nishizawa, E., Eik-Nes, K. B.: J. Clin. Endocrinol. Metab. **23**, 81 (1963).

37. Tamm, J., Voigt, K. D., Volkwein, U.: Steroids **2**, 271 (1963).
38. Kornel, L.: J. Clin. Endocrinol. Metab. **23**, 1192 (1963).
39. Devis, R., Noiret, R.: Ann. Biol. Clin. (Paris) **24**, 1171 (1966).
40. Nielsen, J.: Acta Endocrinol. **45**, Suppl. **90**, 173 (1964).
41. McKenna, J., Norymberski, J. K.: Biochem. J. **76**, 60 P (1960).
42. — Rippon, A. E.: Biochem. J. **95**, 107 (1965).
43. Eberlein, W. R.: J. Clin. Endocrinol. Metab. **23**, 990 (1963).
44. — J. Clin. Endocrinol. Metab. **22**, 963 (1962).
45. Kellie, A. E., Wade, A. P.: Acta Endocrinol. **23**, 357 (1956).
46. Beling, C. G.: Nature **192**, 326 (1961).
47. — Acta Endocrinol. **43**, Suppl., **79**, 98 (1963).
48. Kushinsky, S., Otterness, I.: Steroids **3**, 311 (1964).
49. Smith, E. R., Kellie, A. E.: Biochem. J. **104**, 83 (1967).
50. Sjövall, J., Vihko, R.: Steroids **6**, 597 (1965).
51. Bush, I. E.: The chromatography of steroids, p. 336. New York: Pergamon Press 1961.
52. Hähnel, R.: Clin. Chim. Acta **7**, 768 (1962).
53. — Anal. Biochem. **10**, 184 (1965).
54. Kushinsky, S., Tang, J.: Acta Endocrinol. **43**, 345 (1963).
55. — — Acta Endocrinol. **43**, 361 (1963).
56. Bradlow, H. L.: Steroids **11**, 265 (1968).
57. Slaunwhite, W. R., Jr., Sandberg, A. A.: Endocrinology **62**, 283 (1958).
58. Zumoff, B., Bradlow, H. L.: J. Clin. Endocrinol. Metab. **23**, 799 (1963).
59. Arcos, M., Lieberman, S.: J. Clin. Endocrinol. Metab. **25**, 808 (1965).
60. Baulieu, E.-E.: J. Clin. Endocrinol. Metab. **20**, 900 (1960).
61. Cortes, J. M., Péron, F. G.: Biochim. Biophys. Acta **126**, 43 (1966).
62. Siiteri, P. K.: Steroids **2**, 687 (1963).
63. Westphal, U.: Z. Physiol. Chem. **281**, 14 (1944).
64. Allen, W. M., Viergiver, E.: J. Biol. Chem. **141**, 837 (1941).
65. Venning, E. M.: J. Biol. Chem. **126**, 593 (1937).
66. Craig, L. C., Craig, D., in: Techniques of organic chemistry, vol. 3 (A. Weissberger, ed.), ch. 4. New York: Interscience 1950.
67. Sheps, M. C., Purdy, R. H., Engel, L. L., Oncley, J. L.: J. Biol. Chem. **235**, 3033 (1960).
68. Hadd, H. E., Dorfman, R. I.: J. Biol. Chem. **238**, 907 (1963).
69. Purdy, R. H., Engel, L. L., Oncley, J. L.: J. Biol. Chem. **236**, 1043 (1961).
70. Goebelsman, U., Eriksson, G., Wiqvist, N., Diczfalusy, E.: Acta Endocrinol. **50**, 273 (1965).
71. Schwers, J., Eriksson, G., Wiqvist, N., Diczfalusy, E.: Biochim. Biophys. Acta **100**, 313 (1965).
72. Crépy, O., Jayle, M. F., Meslin, F.: Compt. Rend. Soc. Biol. **151**, 234 (1957).
73. Barlow, J. J., Kellie, A. E.: Biochem. J. **71**, 86 (1959).
74. Brancaccio, B. D., d'Alessandro, B., DeLuca, R., Jacono, G.: Ital. J. Biochem. **8**, 143 (1959).
75. Herrman, W.: Gynaecologia **151**, 156 (1961).

76. Baulieu, E.-E.: J. Clin. Endocrinol. Metab. **22**, 501 (1962).
77. — Emiliozzi, R., Lebeau, M. C., Corpéchot, C., Robel, P.: Advances in tracer methodology, vol. III (Seymour Rothchild, ed.), p. 285. New York: Plenum Press 1966.
78. — Corpéchot, C., Dray, F., Emiliozzi, R., Lebeau, M. C., Mauvais-Jarvis, P., Robel, P.: Recent Progr. Hormone Res. **21**, 411 (1965).
79. Sarfaty, G. A., Summers, M. B.: Clin. Chim. Acta **10**, 505 (1964).
80. Pasqualini, J. R.: Compt. Rend. **259**. 934 (1964).
81. Martin, A. J. P., Synge, R. L. M.: Biochem. J. **35**, 1358 (1941).
82. Johnson, M. J.: Manometric techniques, 3rd. ed. (W. W. Umbreit, R. H. Burris and J. F. Stauffer, eds.), p. 243. Minneapolis, Minn.: Burgess Publishing Co. 1959.
83. Lakshmanan, T. K., Lieberman, S.: Arch. Biochem. Biophys. **55**, 258 (1954).
84. Bush, I. E.: Biochem. J. **67**, 23P (1957).
85. Foggitt, F., Kellie, A. E.: Biochem. J. **91**, 209 (1964).
86. Siiteri, P. K.: Dissertation, Columbia Univ. (1963).
87. — VandeWiele, R. L., Lieberman, S.: J. Clin. Endocrinol. Metab. **23**, 588 (1963).
88. Calvin, H. I., Roberts, K. D., Weiss, C., Bandi, L., Cos, J. J., Lieberman, S.: Anal. Biochem. **15**, 426 (1966).
89. Cavina, G.: Boll. Soc. Ital. Biol. Sper. **31**, 1668 (1955).
90. Lewbart, M. L., Schneider, J. J.: Nature **176**, 1175 (1955).
91. Burstein, S., Lieberman, S.: J. Am. Chem. Soc. **80**, 5235 (1958).
92. Eberlein, W., Bongiovanni, A. M.: Arch. Biochem. Biophys. **59**, 90 (1955).
93. Carr, H. E., Reddy, W. J.: Anal. Biochem. **2**, 152 (1961).
94. Pasqualini, J., Zelnik, R., Jayle, M. F.: Experientia **16**, 317 (1960).
95. Kornel, L.: Anal. Chem. **36**, 443 (1964).
96. Oertel, G. W., Tornero, M. L., Grott, K.: J. Chromatog. **14**, 509 (1964).
97. Wusteman, F. S., Dodgson, K. S., Lloyd, A. G., Rose, F. A., Tudkael, N.: J. Chromatog. **16**, 334 (1964).
98. Sarfaty, G. A., Lipsett, M. B.: Anal. Biochem. **15**, 184 (1966).
99. Fishman, W. H., Harris, F., Green, S.: Steroids **5**, 375 (1965).
100. Cavina, G.: Rend. Ist. Super. Sanita **20**, 923 (1957).
101. — Tentori, L.: Clin. Chim. Acta. **3**, 160 (1958).
102. — — Rend. Ist. Super. Sanita **26**, 56 (1963).
103. Fujii, K., Mizota, S., Takama, T., Mijamoto, S., Ozaki, T.: J. Biochem. (Tokyo) **51**, 167 (1962).
104. Pelzer, H., Staib, W.: Clin. Chim. Acta **2**, 407 (1957).
105. Kornel, L.: J. Clin. Endocrinol. Metab. **27**, 956 (1964).

Chapter 5

The Biochemistry of the 3β–Hydroxy–Δ^5–Steroid Sulfates

KENNETH D. ROBERTS and SEYMOUR LIEBERMAN

Contents

I. Introduction

The very bulk of our knowledge of the biochemistry of the sulfates of the 3β-hydroxy-Δ^5-steroids would appear to be ample justification for including in this volume a separate chapter devoted especially to this kind of steroid conjugate. However, it seems to us that there are other, more trenchant, reasons for this exceptional treatment. Although scores of steroid sulfates have been isolated or detected in body tissues and fluids, only those of the 3β-hydroxy-Δ^5-series seem to possess a uniquely different metabolic role. While the evidence existing at this time does not establish unequivocally the exact nature and importance of their function, there is available now sufficient information about their unusual character to warrant a distinct chapter.

It is generally accepted that steroid sulfates are not inert end-products of metabolism whose sole fate is removal by the excreta. Metabolic reactions that use sulfates as substrates, that is, where both reactants and the products are sulfurylated, are well known but, as of now, only the sulfates of the 3β-hydroxy-Δ^5-steroids participate in reactions generally considered to be biosynthetic. Thus, it has been shown that pregnenolone sulfate can be 17-hydroxylated by homogenates of human adrenal tissue to yield 17α-hydroxypregnenolone-3-sulfate [1]. The enzymatic conversion of 17α-hydroxypregnenolone-3-sulfate to the adrenal secretory product, dehydroisoandrosterone sulfate has also been reported [2]. Of greater consequence is the suggestion that cholesterol sulfate (a recently discovered conjugate which occurs in numerous body tissues) may be converted, *via* sulfated intermediates, into dehydroisoandrosterone sulfate [3] (Fig. 5-1). Efforts to compare the importance of these sulfate pathways for the biosynthesis of the secretory products of the endocrine glands with that of other known routes involving, for example, α,β-unsaturated ketones or the unconjugated 3β-hydroxy-Δ^5-steroids, have been, until now, notably unrewarding.

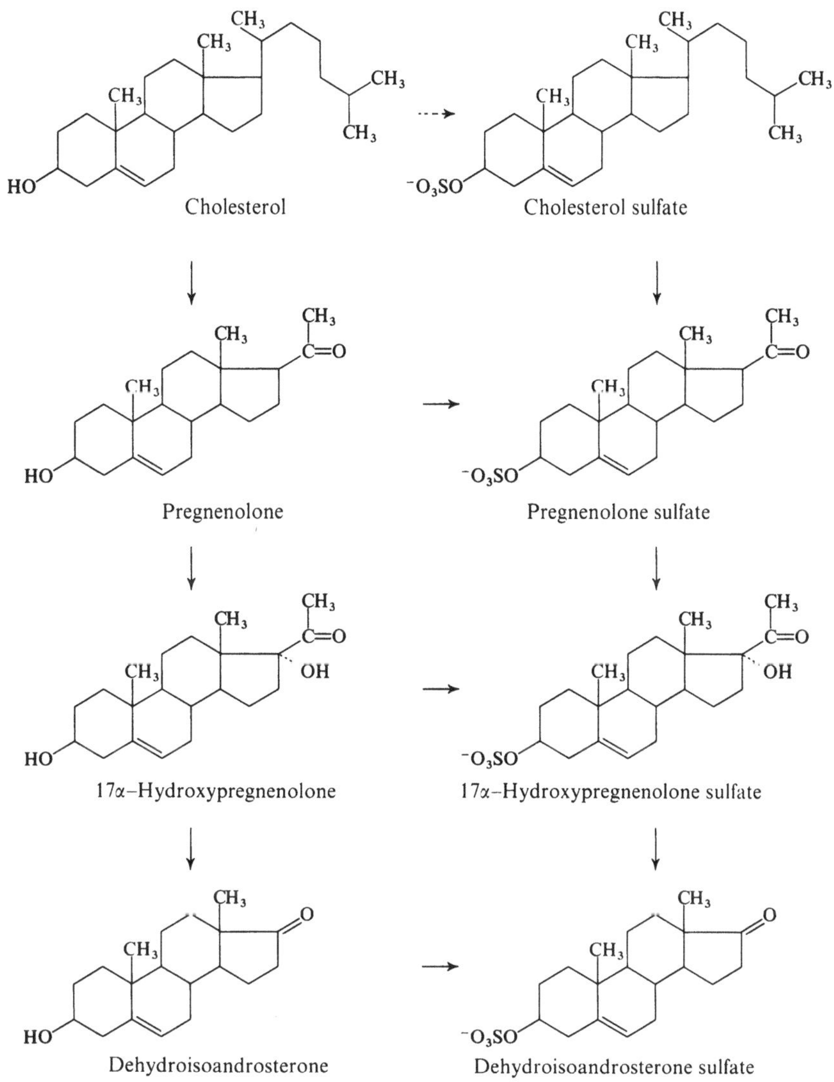

Fig. 5–1. Biogenesis of dehydroisoandrosterone and of dehydroisoandrosterone sulfate *via* pathways involving Δ^5–3β–hydroxy and Δ^5–3β–sulfated intermediates

There is still another reason for this special chapter and that is to call attention to the problem presented by dehydroisoandrosterone sulfate. Notwithstanding the sizable quantities secreted by normal adrenals, its abundance in human plasma and the fact that it is a principal precursor of the urinary 11–deoxy–17–ketosteroids, its physiological significance in the normal state is unknown. Only during pregnancy, where dehydroisoandrosterone sulfate, secreted by the maternal adrenal gland (in early pregnancy) or by the fetal adrenal gland (in late pregnancy), serves as an efficient precursor of the estrogens, has a function been assigned to it. Although much is known about its biochemistry in normal and pathological states, its physiological role in these conditions remains to be uncovered.

This chapter, then, will attempt to summarize our knowledge of the biochemistry of the sulfates of the 3β–hydroxy–Δ^5–steroids as it existed by April, 1968. Greater emphasis will be placed on the recent findings of the last decade; information known for a longer time will be referred to only when necessary for clarity. The individual sulfates are discussed in order of decreasing number of carbon atoms (C_{27}, C_{21}, and C_{19}) following the presumed biosynthetic sequence.

II. Cholesterol Sulfate [3β–Sulfooxycholest–5–ene]

A. Occurrence of Cholesterol Sulfate

Cholesterol sulfate was first isolated from natural sources by Drayer *et al.* [4] who estimated that this sterol conjugate was present in bovine adrenal glands in amounts approximating 1.5 mg/kg of tissue. In subsequent publications, Drayer and Lieberman [5] reported the isolation of cholesterol sulfate from human blood, gallstones, aortic placques, and normal and malignant adrenal tissue. Moser *et al.* [6] have also reported isolating this conjugate from human brain, liver, kidney, plasma, urine, bile, and feces. In this investigation, definitive identification of the conjugate was possible only with the sample recovered from the feces where relatively large quantities occur. The identification of the sterol conjugate isolated from the other tissues was based solely on chromatographic mobility on thin–layer plates. With an internal standard and a double isotope dilution technique, one sample of urine from an adult male was found to contain 120 μg of cholesterol sulfate/liter [10]. The sources from which cholesterol sulfate has been isolated and its concentration in these tissues are summarized in Table 5–1.

In a recent publication, Eneroth and Nystrom [7a] reported levels of 14 to 85 mg of cholesterol sulfate/day in the feces of four subjects. This amount, determined by using labeled cholesterol sulfate as internal standard, represents approximately 2 to 4% of the fecal excretory products of cholesterol [cholest–5–en–3β–ol]. While the quantities of other metabolites of the sterol varied over a wide range, the amount of cholesterol sulfate in the feces was remarkably constant. These authors, using data for the plasma production rates reported by Gurpide *et al.* [8] as the basis for their calcula-

TABLE 5–1. *Natural occurrence of cholesterol sulfate*

Source	Concentration	References
Bovine		
Adrenal	1.5 µg/g	4
Human		
Plasma	195–925 µg/100 ml	8
	< 500 µg/100 ml[a, b]	6
Umbilical cord blood	30 µg/100 ml[b]	11
Urine	120 µg/l	10
Gallstones	500 µg/g	5a
Adrenal tumor	1) 40 µg/g	5b
	2) 200 µg/g	5b
Aorta	15 µg/g	5b
Feces	Not reported	6
Kidney	200 µg/g[a, b]	6
Liver	60 µg/g[a, b]	6
Bile	Not reported[b]	6
Brain	Not reported[b]	6

[a] Approximation.
[b] Tentative identification.

tions, concluded that plasma cholesterol sulfate may be excreted primarily through the feces. Evidence was also obtained for the presence of several other sterol sulfates in human feces. Of this fraction, cholesterol sulfate was estimated to represent at least 85% of the total fecal sulfate excretion [7b].

In studies with normal subjects and patients with neurological disorders, Moser *et al.* [9] observed a rapid incorporation of ^{35}S into biliary cholesterol sulfate following the injection of inorganic ^{35}S–sulfate. In two of these patients, as well as in two control individuals, up to 90% of the radioactivity present in bile was found to be associated with this sterol sulfate, the specific activity of which was highest 40 to 60 hours after injection of the radioisotope.

B. Biosynthetic Origin of Cholesterol Sulfate

Several investigators, using various *in vitro* preparations, have attempted, without success, to demonstrate the sulfurylation of cholesterol either by liver or by adrenal tissue [2, 10, 12–14]. Whereas the sulfurylation of various 3β–hydroxy–Δ^5–steroids of the C_{19}– and C_{21}–series was readily effected, cholesterol could not be converted into its sulfate in these *in vitro* experiments. More recently, however, Banerjee and Roy [15a] have succeeded in sulfurylating cholesterol using a partially purified enzyme preparation obtained from guinea pig liver and 3′-phosphoadenosine-5′-phosphosulfate [PAPS] as the sulfate donor. In this study, the conjugation of the sterol was evidenced by the chloroform-soluble complex produced with methylene blue. By examining the specificity of various partially purified sulfotransferases, these authors proved that cholesterol, androsterone [3α-hydroxy-5α-androstan-17-one] and dehydroisoandrosterone [3β-hydroxyandrost-5-en-17-one] are substrates for the same sulfotransferase. However, the rate of sulfate ester formation with cholesterol as substrate, was about 30% that of the rate of formation of dehydroisoandrosterone sulfate [3β-sulfooxyandrost-5-en-17-one]. The K_m values for cholesterol and dehydroisoandrosterone were identical, 0.02 mM [15b, 16].

This work demonstrated for the first time that cholesterol can be sulfurylated by enzyme preparations from mammalian sources. Why Banerjee and Roy succeeded where other investigators had failed is not clear. It is, of course, possible that the sulfokinase(s) of liver can accept C_{19}- and C_{21}-compounds as well as cholesterol as substrates whereas the sulfokinase(s) of adrenal tissue (used by most other investigators) can sulfurylate the C_{19}- and C_{21}-steroids but not the C_{27}-sterol. Another explanation for the failure of previous workers to effect the conjugation of cholesterol may reside in the fact that crude preparations of the enzyme obtained from liver or adrenal tissue contain sufficiently large amounts of endogenous cholesterol sulfate to have acted as an end-product inhibitor. Another possibility is that endogenous cholesterol could have diluted the tracer cholesterol so that a small conversion to sulfate was obscured. Banerjee and Roy, it will be recalled, employed a partially purified enzyme preparation which probably contained neither cholesterol nor its sulfate. Furthermore, they have pointed out that cholesterol sulfate is rather firmly bound to proteins and that if nonlabeled cholesterol were used as substrate, small yields of cholesterol sulfate may not have been observed due to failure to recover the reaction product absorbed on protein. However, we have attempted to sulfurylate cholesterol with adrenal tissue using radioactive cholesterol. Under these conditions, tracer amounts of the radioactive product should have been easily recovered; nevertheless, no cholesterol sulfate was formed. That the adrenal gland is not a quantitatively important source of circulating cholesterol sulfate, is indicated by the observation of Roberts and Lieberman [10]. They measured the level of the conjugated sterol in the plasma of a patient before and after adrenalectomy and found little or no difference (175 μg *vs.* 195 μg of cholesterol sulfate/100 ml of plasma).

Adams and Wong [17] have reported that cholesterol ^{35}S–sulfate was produced when PAP^{35}S was incubated with supernatant fractions prepared by high speed centrifugation of human breast carcinoma tissue. In agreement with findings obtained with guinea pig liver [15a] the evidence suggested that the sulfurylation of both cholesterol and dehydroisoandrosterone was catalyzed by the same sulfokinase since the addition of dehydroisoandrosterone to the incubation medium inhibited the formation of cholesterol ^{35}S–sulfate. The biosynthesis of cholesterol sulfate from 1–^{14}C–acetate by slices of rabbit testes was reported recently by Young and Hall [18].

C. Cholesterol Sulfate as a Precursor of Adrenal Steroids

Even before the natural occurrence of cholesterol sulfate had been established, an experiment was undertaken to determine whether this sterol sulfate could serve as a precursor of dehydroisoandrosterone sulfate through pathways involving sulfated intermediates. The stimulus for this experiment came from the demonstration that ^{35}S, administered intravenously as pregnenolone–^{35}S–sulfate [3β–sulfooxypregn–5–en–20–one], appeared in the urine as dehydroisoandrosterone–^{35}S–sulfate [19] (*vide infra*). Moreover, pregnenolone sulfate had been shown to be 17–hydroxylated by homogenates of adrenal tissue yielding 17α–hydroxypregnenolone–3–sulfate [17α–hydroxy–3β–sulfooxypregn–5–en–20–one] [1]. The logical extrapolation of these findings led to an experiment in which cholesterol sulfate labeled with ^{3}H at C 7 and with ^{35}S in the sulfate radical was injected *via* catheter into the splenic artery of a patient with a massive adrenal carcinoma [3b]. Since chromatographic systems with high resolving power for steroid sulfates had been developed [20], it was possible to isolate the four metabolites, shown in Fig. 5–2, from the first 24–hour urine collection. The conjugates, pregnenetriol–3–sulfate [3β–sulfooxypregn–5–ene–17α,20α–diol], dehydroisoandrosterone sulfate, androstenediol–3–sulfate [3β–sulfooxyandrost–5–en–17β–ol] and 16α–hydroxydehydroisoandrosterone sulfate [16α–hydroxy–3β–sulfooxyandrost–5–en–17–one] were purified to constant specific activity and the results revealed that the specific activities with respect to both ^{3}H and ^{35}S, were identical for each metabolite. In addition, the ^{3}H/^{35}S ratios for each urinary conjugate were identical and, most significantly, were approximately equal to the ratio present in the injected cholesterol sulfate. The similarity of the specific activities of the metabolites with respect to ^{3}H, indicated that all were derived from a common steroid precursor; *e.g.*, 17α–hydroxypregnenolone [3β,17α–dihydroxypregn–5–en–20–one], pregnenolone [3β–hydroxypregn–5–en–20–one],

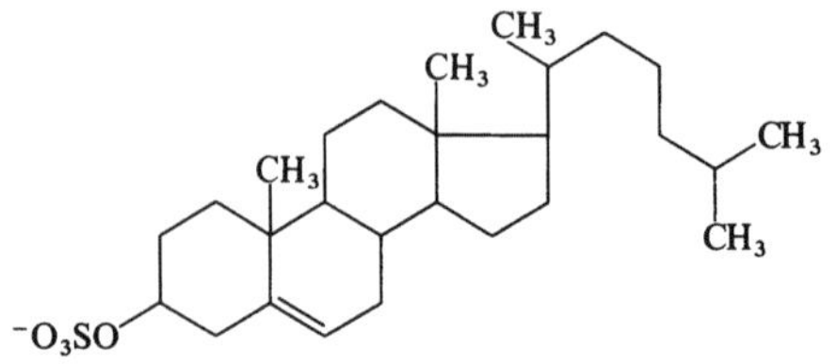

Cholesterol sulfate

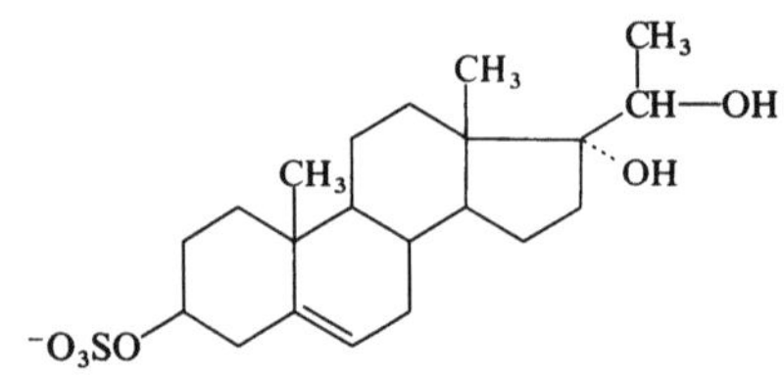

Pregn-5-ene-3β,17α,20α-triol-3-monosulfate

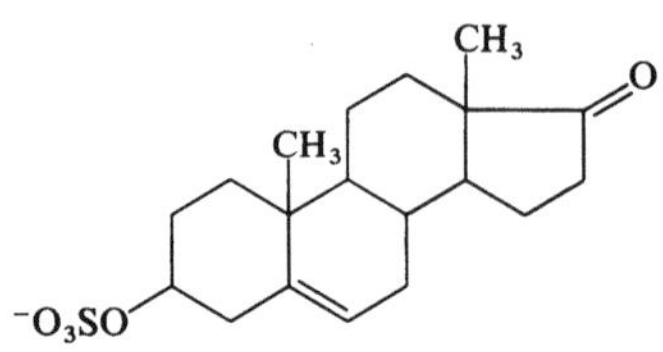

Dehydroisoandrosterone sulfate

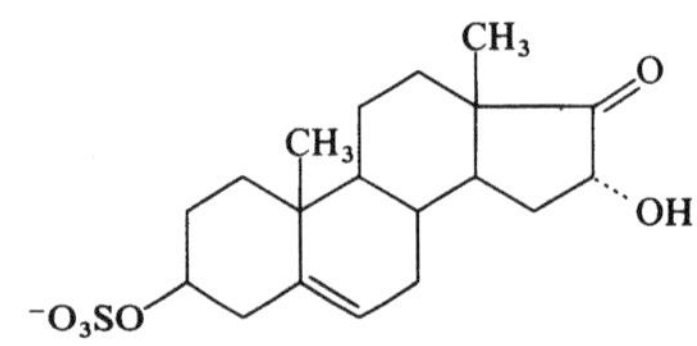

16α-Hydroxydehydroisoandrosterone-3-monosulfate

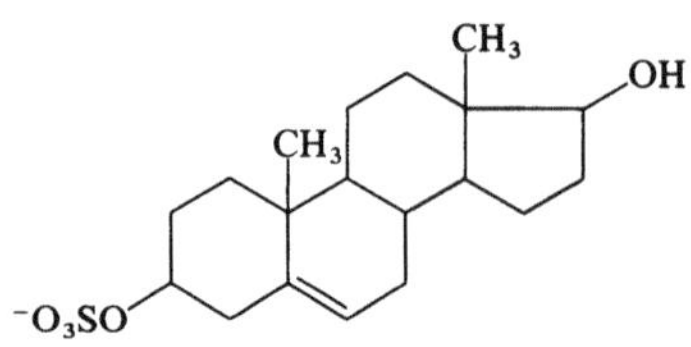

Androst-5-ene-3β,17β-diol-3-monosulfate

Compound	Spec. Act. (cpm/mg)		$^3H/^{35}S$
	^3H	^{35}S	
Cholesterol sulfate	—	—	5.7
Pregn-5-ene-3β,17α,20α-triol-3-sulfate	1,120	230	4.9
Dehydroisoandrosterone sulfate	1,141	204	5.6
Androst-5-ene-3β,17β-diol-3-sulfate	1,200	192	6.2

Fig. 5-2. Urinary 3β-Δ^5-sulfated metabolites isolated after adrenal perfusion of doubly labeled cholesterol sulfate. The specific activities of precursor and products as well as their $^3H/^{35}S$ ratios are indicated in the table

or cholesterol. That the metabolites all had the same specific activity with respect to ^{35}S meant either that they were derived from a common sulfated precursor (such as cholesterol sulfate or pregnenolone sulfate) or that they were all sulfurylated by a common pool of $PAP^{35}S$. The fact that the $^3H/^{35}S$ ratios in pregnenetriol-3-sulfate and in dehydroisoandrosterone sulfate were identical with each other and, more importantly, with that of the injected cholesterol sulfate, provided strong support for the idea that at least some of the injected tracer was converted into the sulfates of the C_{21}–triol and the C_{19}–steroid without cleavage of the sulfate grouping. These findings do not require that the bulk of the C_{21}–sulfate and of the C_{19}–sulfate be derived only from cholesterol sulfate since the results are also consistent with the idea that pregnenolone sulfate was the common sulfated intermediate. Obviously, this C_{21}–sulfate could be formed in the tumor by sulfurylation of pregnenolone. To account for the observed facts, it would be necessary for only some of the tracer to be converted *via* pathways involving sulfated intermediates into pregnenetriol sulfate and dehydroisoandrosterone sulfate. Fig. 5–3 serves to illustrate the route by which the

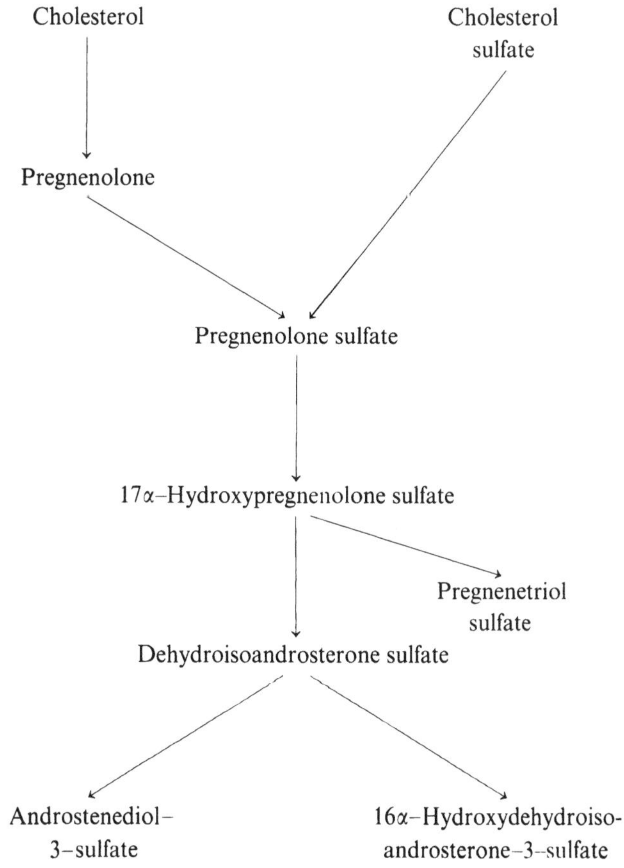

Fig. 5–3. Biogenesis of C_{19}– and C_{21}–steroidal sulfates *via* 3β–Δ^5 hydroxy and 3β–Δ^5–sulfated intermediates

administered radioactive tracer could have labeled pregnenolone sulfate, dehydroisoandrosterone sulfate, and pregnenetriol sulfate without endogenous cholesterol sulfate serving as a precursor. Nevertheless, no matter what the *in vivo* situation may be, the results of this experiment establish with some certainty the existence of the sulfate pathway: cholesterol sulfate → dehydroisoandrosterone sulfate.

In a recent publication, Gurpide *et al.* [8] attempted to evaluate the quantitative importance of blood–borne cholesterol sulfate as a precursor of the two most abundant adrenal secretory products, dehydroisoandrosterone sulfate and cortisol [11β,17α,21–trihydroxypregn–4–ene–3,20–dione]. In *in vivo* experiments, these workers also measured several metabolic parameters which characterize circulating cholesterol sulfate in human subjects. Using isotopic dilution techniques, the following variables were quantified: plasma concentration, metabolic clearance rate, production rate in plasma, and half–life. Their findings are presented in Table 5–2. In addition, these investigators were led to conclude that little, if any, of these adrenal secretory products was derived from blood–borne cholesterol sulfate. The major metabolic fate of the sterol sulfate was its conversion into plasma cholesterol, free as well as esterified. As noted by these workers, their results may not truly reveal the role that cholesterol sulfate, produced within the adrenals, may play in the formation of the sulfates of various 3β–hydroxy–Δ^5–steroids, for it is conceivable that adrenal intracellular

Table 5–2. *Parameters of cholesterol sulfate in human subjects*

Metabolic parameter	Normal males		Normal females			
	Subject 1	Subject 2	Subject 1	Subject 2	Subject 3	Subject 4
Half–life (hours)	14	15	–	–	–	–
Volume of distribution (liters)	9	16	–	–	–	–
Metabolic clearance rate (liters/day)	11	17	–	–	–	–
Plasma concentration (μg/100 ml)	325 460[a]	925 730[a]	270	240	195	165
Production rate in plasma (mg/day)	35	163	–	–	–	–

[a] Analysis of plasma from same subject one month later.

cholesterol sulfate, not in equilibrium with the circulating conjugate, serves as an intermediate in steroidogenesis.

Le Beau and Baulieu [21] have obtained further information about the relative merits of circulating cholesterol and its sulfate as precursors of urinary steroids. They injected intravenously a mixture of ^{3}H–cholesterol sulfate and ^{14}C–cholesterol into a male subject and determined the ^{3}H/^{14}C ratios in various urinary metabolites. They observed that free cholesterol appeared to be converted more efficiently into the urinary metabolites, androsterone, etiocholanolone [3α–hydroxy–5β–androstan–17–one], dehydroisoandrosterone, tetrahydrocortisone [3α,17α,21–trihydroxy–5β–pregnane–11,20–dione], and tetrahydrocortisol [3α,11β,17α,21 tetrahydroxy–5β–pregnan–20–one], than was cholesterol sulfate. In spite of the fact that the injected cholesterol suffered far greater dilution by its endogenous counterpart (the amount of plasma free and fatty acid–esterified cholesterol is about one thousand times greater than the amount of plasma cholesterol sulfate), the conversion of the free sterol into the urinary products mentioned above was larger than that obtained when cholesterol sulfate was the precursor.

Another attempt to determine whether circulating cholesterol sulfate could serve as a precursor in steroid hormone formation, was made by Solomon [22]. This author perfused the labeled conjugated sterol through an intact fetus, *in vitro*, and found radioactive products neither in the perfusate nor in the tissues (adrenals, liver); only the unchanged tracer was recovered. While this finding agrees with that of Gurpide *et al.* [8], neither experiment precludes the possibility that adrenal cholesterol sulfate, not in equilibrium with the circulating conjugate, is involved in the formation of steroid secretory products.

A variety of attempts have been made in the authors' laboratory to effect the conversion of cholesterol sulfate into pregnenolone sulfate using *in vitro* techniques. With ^{3}H–cholesterol–^{35}S–sulfate as a test substance, neither whole homogenates nor intact mitochondria (where the side–chain cleaving enzyme is known to reside) of bovine adrenals were able to bring about this transformation. However, when mitochondrial membranes, ruptured by sonication, were incubated with the precursor, a 10% yield of doubly–labeled pregnenolone sulfate, having both isotopes in the same proportion as did the starting material, was easily recovered [23]. Thus, as suspected from the results of Gurpide *et al.* [8], cholesterol sulfate appears to be transferred through membranes rather poorly and, unless these are disrupted, conversion into other products along the biosynthetic pathway may be difficult to demonstrate. The success Roberts *et al.* [3b] had in demonstrating the conversion of cholesterol sulfate into dehydroisoandrosterone sulfate by the *in situ* perfusion of an adrenal tumor may have been due to the well–recognized "leakiness" exhibited by cancer cells, especially those that are present in poorly encapsulated tumors.

That cholesterol sulfate can serve as a substrate for the side–chain cleaving enzyme system of bovine adrenal mitochondria, was recently demonstrated by Raggatt *et al.* [24]. These workers also reported that the inhibitor specificity for this system was the same whether free cholesterol or cholesterol sulfate was used as substrate. This suggested that the cholesterol–oxidase system is not composed of two separate units, one for free cholesterol and the other for the sulfated sterol, but, rather, is a single system with the capacity to oxidize either substrate. These authors have also measured the K_m values of both substrates; that for cholesterol was found to be smaller than was the K_m for the sulfate. However, these values cannot be accepted as representing the true affinity of the two substrates for the enzyme since they were calculated without taking into account the amount of endogenous cholesterol present in the system. Raggatt and his coworkers observed high levels of cholesterol in mitochondria, and this makes the interpretation of kinetic data, obtained with exogenous radioactive tracers, difficult. Dilution and inhibition caused by the endogenously present substrate would have the effect of altering drastically the observed kinetic constants and velocities. Consequently, the difference between the actual K_m values for the two substrates is probably not as great as that stated. Although the sterol sulfate was reportedly oxidized to ^{14}C–isocaproic acid much less readily than was cholesterol, it is difficult under the *in vitro* conditions employed to obtain kinetic data meaningful for *in vivo* states. *In vitro* experiments that are designed to assess the relative quantitative importance of two pathways are fraught with numerous hazards. Factors such as permeability, disturbance of the natural arrangement of cellular organelles and cofactors, endogenous pool sizes, inhibition of certain reactions by product accumulation, etc., militate against the assessment of valid quantitative data from this type of experiment. The qualitative findings of Raggatt *et al.* [24] remain, however; the mitochondria of their preparation had the capacity to cleave the side–chain of the sterol sulfate under conditions where no sulfatase activity was demonstrable.

III. Pregnenolone Sulfate

A. Occurrence of Pregnenolone Sulfate

Using a double isotope–dilution technique, Bègue [25], in work carried out in this laboratory, was first to provide evidence for the presence of pregnenolone sulfate in normal human plasma. He estimated the concentration to be about 5 μg/100 ml. More recently, Sjövall and Vihko [26] used a combination of gas chromatography and mass spectrometry to estimate a level of 3 μg pregnenolone sulfate/100 ml in pooled serum. Essentially, the same technique as Bègue's was used by Conrad *et al.* [27] for the analysis of human pregnancy plasma at various gestational periods. The range of values estimated (3 to 20 μg/100 ml) did not appear to correlate with the time of gestation. These values, it should be noted, are of the same magnitude as those reported for plasma progesterone [pregn–4–ene–3,20–dione] [28]. Conrad *et al.* compared the pregnenolone sulfate levels in the maternal antecubital vein with those in the umbilical vein and artery. They found that the concentration of pregnenolone sulfate in the umbilical vein plasma was higher than that in the maternal circulation (140 μg *vs.* 20 μg/100 ml and 150 μg *vs.* 60 μg/100 ml), and also observed that the umbilical artery contained more pregnenolone sulfate than did the umbilical vein (210 μg/100 ml *vs.* 140 μg/100 ml). These data thus indicate that pregnenolone sulfate is a product of fetal metabolism. Eberlein [11] has also reported the presence of pregnenolone sulfate in umbilical vein blood, but the concentrations found were much smaller (17 27 μg/100 ml) than those reported by Conrad *et al.* [27]. This discrepancy may be due to the fact that the values obtained by Conrad were corrected for technical losses, whereas those of Eberlein were not.

B. Biosynthetic Origin of Pregnenolone Sulfate

The findings of Arcos *et al.* [29] provided the first indication that pregnenolone sulfate is secreted by the adrenal. After the intravenous administration of ^{3}H–pregnenolone to a female subject, these authors found that the specific activity of urinary pregnenediol [pregn–5–ene–3β,20α–diol] was lower than that of pregnanediol [5β–pregnane–3α,20α–diol]. This suggested that the enediol was not derived solely from the substance administered as tracer, for, if the scheme: urinary pregnenediol ← injected pregnenolone → progesterone → urinary pregnanediol were the only available route to the two diols, both would have the same specific activity. By analogy with the findings obtained when dehydroisoandrosterone was injected [30] *(vide infra)*, the result is consistent with the notion that pregnenolone sulfate, as well as pregnenolone, is a secretory product and that both are progenitors of urinary pregnenediol. Shimizu [31] reported that pregnenolone sulfate and dehydroisoandrosterone sulfate may be formed when 20α–hydroxycholesterol [cholest–5–ene–3β,20α–diol] is incubated with human adrenal slices, adding further support for the adrenal origin of the C_{21}–conjugate.

Several investigators have shown that the human adrenal gland has the capacity to sulfurylate pregnenolone *in vitro* [32–35]. As expected from the data obtained with cord blood, the adrenal of the newborn can readily effect this conjugation [36]. The sulfurylation of pregnenolone has also been reported to occur with testicular tissue [37].

C. Pregnenolone Sulfate as a Substrate in Biosynthetic Reactions

The first suggestion that a steroid sulfate could serve as a substrate in biosynthetic processes, was made by Calvin *et al.* in 1963 [19]. These authors administered pregnenolone ^{35}S–sulfate to a female who bore an adrenal tumor and who was excreting 300 mg of urinary 17–ketosteroids/day. The urinary dehydroisoandrosterone sulfate obtained from the first 24–hour collection of urine was found to contain 0.33% of the injected ^{35}S; and this led the authors to conclude that pregnenolone sulfate had been converted into dehydroisoandrosterone sulfate *via* intermediates which retained the sulfate group. Later work [38] showed that this result was equivocal since the cleavage and reformation of sulfates under *in vivo* conditions are sufficiently important that it becomes difficult to draw conclusions about sulfate pathways from such evidence. Nevertheless, it seems unlikely that the original finding [19] was due to a spurious exchange of ^{35}S–sulfate from pregnenolone to dehydroisoandrosterone. To have found even this small conversion would have required that the transfer occur in a confined area of the body, since any labeled SO_4^{2-} liberated into the general circulation would probably have been greatly diluted by unlabeled ions. In this eventuality, the urinary product would probably have contained undetectable amounts of ^{35}S. The absence of ^{35}S in urinary androsterone sulfate [3α–sulfooxy–5α–androstan–17–one] after the administration of dehydroisoandrosterone ^{35}S–sulfate gives credence to this assumption [39].

Nevertheless, in an effort to overcome this uncertainty, Calvin and Lieberman [1] subsequently prepared doubly–labeled pregnenolone sulfate bearing ^{3}H in the steroid moiety and ^{35}S in the sulfate group. Upon incubation of this substrate with a homogenate of human hyperplastic adrenal tissue, doubly–labeled 17α–hydroxypregnenolone–3–sulfate, bearing the same isotopic ratio as the incubated precursor, was isolated. Thus, under *in vitro* conditions, where it was possible to control competing reactions such as desulfation, unambiguous evidence for the existence of a sulfate pathway was obtained. Further work by Usui and Lieberman [40] demonstrated that pregnenolone sulfate could be hydroxylated at C17 when this conjugate was incubated with bovine adrenal microsomes. Under these conditions only 17α–hydroxypregnenolone–3–sulfate could be isolated. On the other hand, when unconjugated pregnenolone was the substrate, it was hydroxylated at

C21 as well as at C17 and, in addition, was dehydrogenated to yield the Δ^4–3–ketone.

Voigt *et al.* [41] perfused large quantities of pregnenolone sulfate through isolated dog livers and, surprisingly, demonstrated the presence of dehydroisoandrosterone sulfate in the perfusate. Since a labeled precursor was not used in these experiments, it is not certain whether conversions involving sulfated intermediates occurred. Although pregnenolone sulfate was hydroxylated at C7 by rat liver preparations [42], the yield of 7–hydroxylated product was three–fold higher when the free compound was used as substrate. Introduction of the oxygen atom at C7 did not appear to result from auto–oxidation since incubation with boiled tissue failed to yield this metabolite. Killinger and Solomon [43] found that pregnenolone sulfate, when incubated with homogenates of normal human adrenal tissue, remains unchanged. Under these same conditions, unconjugated pregnenolone was converted into pregnenolone sulfate, 17α–hydroxypregnenolone–3–sulfate and pregnenetriol–3–sulfate. In view of the *in vitro* experiments of Usui and Lieberman [40], and those of Calvin and Lieberman [1] described earlier, it is unclear why the sulfated precursor was not utilized in these incubations.

Further studies on the *in vitro* and *in vivo* metabolism of pregnenolone sulfate have been published [38]. The *in vitro* data corroborated the earlier finding [1] that this conjugated steroid, labeled with ^3H and ^{35}S, could serve as a substrate for the 17–hydroxylation enzyme of neoplastic adrenal tissue. In addition to 17α–hydroxypregnenolone–3–sulfate, pregnenediol–3–sulfate [3β–sulfooxypregn–5–en–20α–ol] was also isolated and was found to contain a $^3H/^{35}S$ ratio that was identical with that of the incubated precursor. Thus, enzymatic reduction at C20 had occurred using the sulfated steroid as substrate. *In vivo* data have also demonstrated that pregnenolone sulfate, far from being a metabolic end product, is extensively metabolized. Among the radioactive products found in the urine following the intravenous administration of pregnenolone sulfate were: pregnenediol–3–sulfate, 17α–hydroxypregnenolone–3–sulfate, pregnenetriol–3–sulfate, dehydroisoandrosterone sulfate, and androstenediol–3–sulfate.

In contrast to the findings of Killinger and Solomon [43] who studied normal human adrenal tissue, Griffiths *et al.* [44], recently reported that pregnenolone sulfate was metabolized by an adenomatous adrenal. Griffiths and his coworkers attempted to compare the *in vitro* metabolism by a clear cell adrenal adenoma mince of pregnenolone and its sulfate with that of various other steroids. As was already mentioned, quantitative comparisons from such *in vitro* experiments may not be meaningful but it is clear that in this study both substrates were converted by the tissue into a wide variety of steroids and steroid sulfates. That a homogenate of human fetal adrenal tissue has the capacity to metabolize pregnenolone sulfate to more polar products was demonstrated by Perez–Palacios *et al.* [45]. Since this tissue is virtually devoid of sulfatase activity and, since the incubation was carried out in phosphate buffer, a known inhibitor of that enzyme, it may be assumed that the demonstrated conversions occurred using intact sulfated intermediates. In fact, the sulfates of 17α–hydroxypregnenolone, dehydroisoandrosterone, and 16α–hydroxydehydroisoandrosterone [3β,16α–dihydroxyandrost–5–en–17–one] were obtained in good yield.

The metabolic clearance rate (MCR), volume of distribution, and half–life of pregnenolone and of its sulfate have been determined by Wang *et al.* [46]. Both the MCR and the volume of distribution of the sulfate were smaller than those of the free compound. In both man and rabbit, pregnenolone was cleared three times more rapidly than was the sulfurylated derivative. A metabolic clearance rate of approximately 280 liters of whole blood/day was calculated, and this implies a more rapid clearance than that observed for dehydroisoandrosterone sulfate or for cholesterol sulfate. Assuming a hematocrit value of 50% and a level of pregnenolone sulfate in male plasma to be 10 µg/100 ml [27], these investigators calculated a blood production rate of pregnenolone sulfate of roughly 14 mg/day.

Pion *et al.* [47] advanced the idea that pregnenolone sulfate, produced in the fetus, may be a precursor for the progesterone produced in the placenta. These authors found that an adrenalectomized patient had detectable levels of pregnenolone sulfate in her plasma during the mid-trimester of pregnancy, while none was found post partum. To evaluate the capacity of term placenta to utilize circulating pregnenolone sulfate as a precursor for progesterone production, Pion *et al.* [47] injected labeled pregnenolone sulfate into the umbilical artery following delivery and prior to placental detachment. Since approximately 50% of the radioactivity recovered from the placenta was identified as progesterone, these authors were led to conclude that the conversion of pregnenolone sulfate to progesterone had been significant. This transformation was also observed by Palmer *et al.* [48], who perfused an isolated placenta with labeled pregnenolone sulfate. Of the radioactivity recovered in the unconjugated fraction of the perfusate, approximately half was progesterone and one–third was pregnenolone. This represented a conversion of 1 to 3% of the total radioactivity administered. In view of the high concentrations of sulfatase and 3β–hydroxysteroid dehydrogenase in placenta, this finding is not unexpected for it is quite possible that some fraction of the pregnenolone sulfate arising in the fetus is conveyed to the placenta (its occurrence in cord blood was referred to earlier) where it is converted into progesterone. This substance may be returned to the fetus where it may be employed as a starting material for the synthesis of cortisol [49] and other corticosteroids [50]. It has been shown that the fetal adrenal cortex is relatively deficient in 3β-hydroxysteroid dehydrogenase [51] and so cannot readily convert 3β-hydroxy-Δ^5-steroids into the corresponding α,β-unsaturated ketones [52]. From these considerations, Palmer *et al.* [48] suggested that the progesterone required by fetal adrenals for corticosteroid synthesis is elaborated by the placenta. It remains to be determined what fraction of placental progesterone production arises *de novo* from cholesterol, and what part is derived from pregnenolone sulfate circulating in maternal or fetal blood.

IV. The 3-Sulfates of Pregn-5-ene-3β,20α-diol and Pregn-5-ene-3β,20β-diol

The recovery of the 3-sulfate of pregn-5-ene-3β,20α-diol, in crystalline form, from the urine of a patient who had received labeled pregnenolone sulfate intravenously [38] was the first instance in which the natural occurrence of this conjugate was rigorously demonstrated. In previous investigations, its presence was suggested by the isolation of pregnenediol after the hydrolysis of urine of patients with adrenal tumors [53]. More recently, Wilson *et al.* [54] found that urinary levels of this conjugate in a group of patients with adrenal carcinoma were as much as 4.1 mg/day. These investigators have also reported that from 0 to 0.5 mg of this steroid may be excreted daily in the urine of normal individuals. Likewise, Bongiovanni [55a] estimated that the amount of pregn-5-ene-3β,20α-diol present in normal male urine was 0.38 mg/liter. Earlier [55b, c] he had provided evidence (solvolysis of urinary conjugates) for the excretion of pregnenediol sulfate in the urine of patients with congenital adrenal hyperplasia of the salt-losing variety. In several infants, the urinary levels ranged between 1.8 and 2.5 mg/day. Jacobsohn [56] also isolated large amounts of the same diol (estimated to be 10 mg/day) following the solvolysis of the urine of an adult male with congenital adrenal hyperplasia without salt-losing tendencies.

A novel conjugate of the 20α-isomer was recently isolated from human urine by Arcos and Lieberman [57]. After the intravenous administration of tracer ^{3}H-pregnenolone to a normal female, approximately 8% of the injected radioactivity was excreted in the form of two conjugates more polar than the monoglucuronides. One was characterized as the 3-sulfate of pregnenediol bearing an N-acetylglucosamine residue joined by a β-glycosidic linkage to C 20. The stereochemistry of the glycosidic linkage was established by Matsui and Fukushima [240]. Commercial β-glucuronidase was found to liberate 50% of the N-acetylglucosamine residue after incubation for three days. The other polar conjugate was identified as the disulfate of pregn-5-ene-3β,20α-diol [3β,20α-disulfooxypregn-5-ene] *(vide infra)*.

The presence of pregn-5-ene-3β,20α-diol as well as of its 20β-isomer in human umbilical cord blood was reported by Eberlein [11]. Identification was based on their retention times in gas-liquid chromatography. That these two isomers were originally present in the cord blood as their sulfuric acid esters was indirectly proved by Eberlein's transesterification procedure (conversion of the presumed steroid sulfates into their corresponding acetates by boiling with acetic acid). The concentration reported for the 20α-isomer was 43 μg/100 ml and for the 20β-isomer, 5 μg/100 ml of cord blood. The 3β,20α-diol-3-sulfate was also found in human peripheral plasma by Vihko [58]. The presence of the sulfate moiety was suggested by the fact that the conjugated steroid migrated on methylated Sephadex® (Pharmacia, Uppsala) with a mobility equal to that of a monosulfate. Pregnenediol was liberated only after solvolysis of the plasma extract. Mean values for the 3β,20α-diol conjugate, in male and female plasma, are given in Table 5-3. Evidence for the existence of this diol in plasma as a disulfate was also suggested by its chromatographic behavior on Sephadex [26].

TABLE 5-3. *Concentration of 3β-hydroxy-Δ^5-steroid sulfates in human peripheral plasma*[a]

Sulfate	μg/100 ml plasma			
	Males		Females	
	Mean	Stand. dev.	Mean	Stand. dev.
Androstenediol	28 ±	26	11 ±	8
Pregnene-3β,20α-diol	20 ±	11	13 ±	10
Dehydroisoandrosterone	99 ±	65	60 ±	40

[a] Vihko, R.: Acta Endocrinol. **52**, Suppl. **109** (1966).

The *in vitro* formation of the mono- and disulfates of the pregnenediols by rat liver extracts has been demonstrated by Wengle and Boström [59]. With the 3β,20α-isomer, the predominant product was the 3β-sulfate; only small amounts of the disulfate were formed. The 3β,20β-isomer, on the other hand, yielded principally the disulfate. The 20β-sulfokinase system appeared to have a greater affinity for the 3-sulfate than it had for the free diol since no 3β-hydroxy-20β-sulfate was found.

The most abundant radioactive metabolite isolated from the urine following the intravenous administration of doubly-labeled pregnenolone sulfate to a patient with adrenal carcinoma was found to be pregnenediol-3-monosulfate [38]. Analysis of the $^{3}H/^{35}S$ ratio revealed that the diol monosulfate was formed from the administered conjugate without cleavage of the sulfate radical.

V. Pregn-5-ene-3β,20α-diol-disulfate

The isolation of this conjugate, in crystalline form, from the urine of normal subjects has been reported by Arcos and Lieberman [57]. Following the intravenous administration of ^{3}H-pregnenolone and the ingestion of 1.5 g of the nonradioactive steroid, the presence of this urinary conjugate was indicated by a Crépy-positive substance having chromatographic mobility characteristic of a disulfate. Identification of the crystalline metabolite was established by comparison of its infrared spectrum and its melting point with those of an authentic standard prepared synthetically.

VI. 3β,21-Dihydroxypregn-5-en-20-one-disulfate [3β,21-Disulfooxypregn-5-en-20-one]

The occurrence of this polar conjugate in human urine has been reported by Pasqualini and Jayle [60]. The presence of the two sulfate groups was suggested by its chromatographic and electrophoretic characteristics. The conjugate, prior to solvolysis, did not reduce blue tetrazolium. Its identity was confirmed by establishing that its chromatographic properties were similar to those of the most polar compound obtained after the sulfurylation of 21-hydroxypregnenolone [3β,21-dihydroxypregn-5-en-20-one] with an excess of pyridine-sulfur trioxide. The urinary material was hydrolyzed by the *Helix pomatia* enzyme and yielded 21-hydroxypregnenolone and SO_4^{2-}. In addition, the compound was isolated from blood and urine after the oral administration of large amounts (*ca.* 300 mg) of 21-hydroxypregnenolone to a normal male [61]. When 17α,21-dihydroxypregnenolone [3β,17α,21-trihydroxypregn-5-en-20-one] was injected intravenously, only the 3-sulfate of the corresponding 17,21-dihydroxy metabolite was isolated. These findings seem to be in accord with other studies that have demonstrated that the presence of a 17-hydroxy group substituent interferes with the *in vitro* sulfurylation of the 21-hydroxyl group [62, 63].

VII. 16α–Hydroxypregnenolone–3–sulfate [16α–Hydroxy–3β–sulfooxypregn–5–en–20–one]

The presence of 16α–hydroxypregnenolone [3β,16α–dihydroxypregn–5–en–20–one] in human urine was first reported by Bongiovanni [55c] who isolated this compound after solvolysis of urine from infants with the adrenogenital syndrome characterized by a deficiency of the enzyme, 3β–hydroxysteroid dehydrogenase. One infant with the salt–losing form of the disease excreted 3 mg of this metabolite in the urine daily; another excreted 4.5 mg. Since 16α–hydroxypregnenolone was isolated only after solvolysis, it was assumed that this metabolite was excreted as the 3–sulfate. Identification of the free steroid was confirmed by infrared spectroscopy. That this 16α–hydroxysteroid was also present in the urine of normal infants was established by Reynolds [64], who found, after solvolysis of the urine of infants 1 to 7 days old, a mean level of 0.98 mg/day (corrected for procedural losses). While this steroid was found to be present in all urine samples from infants 1 to 2 months old, its excretion diminished to undetectable levels after the age of 6 months.

The only report of the isolation of 16α–hydroxypregnenolone from blood, is that by Eberlein [11] who provisionally identified this steroid in the sulfate fraction of umbilical cord plasma. In one pool of cord plasma, a value of 1.5 μg/100 ml was reported, while a second pool contained 19 μg/100 ml. Neither value was corrected for losses incurred during the isolation procedure. With *in vitro* techniques, Killinger [65] showed that pregnenolone sulfate may be hydroxylated at C16 by incubation with normal human adrenal tissue.

VIII. 17α–Hydroxypregnenolone–3–sulfate

Although many investigators had previously provided indirect evidence for the presence of this conjugate in urine, its isolation *per se* was first described by Calvin and Lieberman [38]. The first suggestion that 17α–hydroxypregnenolone–3–sulfate may be secreted by the adrenal, was made by Fukushima *et al.* [66]. These investigators administered ^{3}H–17α–hydroxypregnenolone to a female patient with an adrenal tumor and isolated pregnenetriol [pregn–5–ene–3β,17α,20α–triol] from her urine after both glucuronidase and acid hydrolysis. The specific activity of the triol, excreted as the sulfate, was found to be lower than that of the triol excreted as the glucuronide. This is the result that would obtain if a portion of the 17–hydroxypregnenolone made by the tumor was secreted as the sulfate. Evidence for the secretion of this conjugate by the nontumorous adrenal gland was given by Wieland *et al.* [67]. By employing a catheterization method which subjected the patient to a minimum of stress, these authors were able to sample the adrenal venous effluent and so to demonstrate the secretion of 17α–hydroxypregnenolone–3–sulfate by the normal gland. In one subject, 18 µg of the conjugate was found in 100 ml of adrenal venous plasma, whereas the peripheral venous level was less than 1 µg/100 ml. It is noteworthy that the concentration of unconjugated 17α–hydroxypregnenolone in the adrenal venous blood was higher (123 µg/100 ml) than that of its sulfate. The human fetal adrenal gland may also secrete this conjugate; Eberlein [11] measured its concentration in a pool of umbilical cord blood and found it to be 15 µg/100 ml.

Mention has already been made of the fact that 17α–hydroxypregnenolone sulfate may originate *via* hydroxylation of pregnenolone sulfate. In addition, it may also arise *via* sulfurylation of the free compound. Both the adrenal [33] and the testes [37] have been shown to possess sulfokinases that can sulfurylate 17α–hydroxypregnenolone. Le Beau *et al.* [2] studied the *in vitro* metabolism of doubly–labeled 17α–hydroxypregnenolone sulfate by slices of adrenal tumor. The isolation of doubly–labeled dehydroisoandrosterone sulfate indicated the presence of an adrenal side–chain cleaving enzyme with the capacity to utilize a sulfated compound as substrate. Under the same conditions, the percentage conversion of 17α–hydroxypregnenolone to dehydroisoandrosterone was greater than that for 17α–hydroxypregnenolone sulfate to dehydroisoandrosterone sulfate. However, without knowing the concentration of the two substrates endogenously present in the tissue slices, such data do not permit a comparison of the actual rates (weight of substrate transformed per unit time) which are the more meaningful parameters.

An interesting report by Deghenghi and Revesz [68] has described the high oral activity of 17α–acetoxypregnenolone–3–sulfate [17α–acetoxy–3–sulfooxypregn–5–en–20–one] as a progestational agent. When administered orally to rabbits, this sulfate was found to be a more potent gestagen than was its corresponding Δ^4–3–ketone, 17α–acetoxyprogesterone [17α–acetoxypregn–4–ene–3,20–dione].

The utilization of 17α–hydroxypregnenolone–3–sulfate as a substrate for the 3β–hydroxysteroid dehydrogenase system has been studied in a variety of tissues by Baillie *et al.* [69, 70]. The histochemical results obtained with the sulfated steroid were quantitatively different from those obtained with the free steroid and this would suggest that the presence of the sulfate moiety may have some effect on the binding of the enzyme–substrate complex. With 17α–hydroxypregnenolone–3–sulfate as substrate, and a histochemical assay (diformazan deposition) for 3β–hydroxysteroid dehydrogenase activity, the enzyme was found in the zona fasciculata of adrenal tissue and in the placental syntrophoblast. Surprisingly, high activity was present even in a medium containing phosphate ions which are known to inhibit sulfatase. It was shown that 17α–hydroxypregnenolone–3–sulfate was better utilized by the enzyme present in the Leydig cell of rat testes than was the unconjugated steroid.

IX. Pregn–5–ene–3β,16α,20α–triol Sulfate

This conjugate has yet to be isolated as such, but the free triol was first recovered in crystalline form by Hirschmann and Hirschmann [71], following hydrolysis of the urine of a young male with an adrenal tumor. That this triol sulfate was a constituent of normal male urine was demonstrated by Fotherby [72]. Bongiovanni has confirmed this finding and has estimated that 2.5 mg of the triol was present in 100 liters of pooled normal adult male urine [55a]. Identity was established by infrared spectroscopy. In addition, the triol was recently shown to be a urinary metabolite of pregnenolone since it was isolated after solvolysis of the urine of normal males administered this labeled precursor [73].

X. Pregn–5–ene–3β,17α,20α–triol–3–sulfate

The isolation and characterization of the intact triol–3–sulfate from human urine was reported by Roberts *et al.* [3]. Since 1950, when it was isolated after acid hydrolysis by Hirschmann and Hirschmann [74], pregnenetriol has been known to be a major constituent of the complex mixture of steroids that are excreted in the urine of patients with tumors of the adrenal gland. The triol has also been isolated from the urine of normal men by Fotherby [75] who used a method of hydrolysis which provided good, albeit indirect, evidence that the metabolite was excreted in the urine as a sulfate. Wilson *et al.* [54] studied the urinary levels of this triol, isolated from the sulfate fraction, in subjects with non–endocrine disorders and found that the values ranged from 0.1–0.4 mg/day. In patients with adrenal carcinoma, these investigators found titers varying from 4 to 53 mg/day.

Riley [76] reported that the most abundant acetaldehydogenic steroid present in the sulfate fraction of normal human urine was this triol. Surprisingly, he found that 60% of this 3β–hydroxy–Δ^5–steroid is found in the glucuronide fraction. In contrast, in patients with adrenal carcinoma, the major portion of the triol is found in the solvolyzable fraction.

Urinary levels of the triol in normal individuals were also reported by Nowaczynski *et al.* [77] and their results agree with those of others [54]. In the study of twenty-one normal subjects, the mean level of pregnenetriol was found to be 0.3 mg/24 hrs. These investigators have noted that patients with hypertension excrete pregnenetriol in significantly smaller amounts than do normal individuals. Bongiovanni [55c] studied the excretion of 3β-hydroxy-Δ^5-steroids in the urine of a group of infants with the adrenogenital syndrome, and found that those with 3β-hydroxysteroid dehydrogenase deficiency excreted between 3.2 and 8.2 mg of pregnenetriol per day. Infants with 21-hydroxylase deficiency excreted 0.4-2.3 mg of pregnenetriol per day. Presumably, the pregnenetriol levels measured in this study refer to an excretion of pregnenetriol sulfate since the methods of isolation and hydrolysis were specific for sulfates.

After the administration of cholesterol sulfate labeled with ^3H and ^{35}S, Roberts *et al.* [3b] isolated urinary pregnenetriol sulfate bearing a $^3H/^{35}S$ ratio similar to that of the injected tracer. Since the labeled precursor was administered *via* catheter close to the adrenal tumor tissue, the adrenal origin of this conjugate appears likely. Confirmation that the adrenal is a source of pregnenetriol-3-sulfate comes from the researches of Killinger and Solomon [43] who isolated this conjugate after the incubation of normal human adrenal tissue with pregnenolone. Furthermore, the amount of pregnenetriol that can be isolated from urine after solvolysis is increased four times above control levels by the administration of ACTH [78].

It is apparent from the studies of Bradlow *et al.* [79] that most of the pregnenetriol sulfate excreted in urine is not derived from circulating pregnenetriol. Following the intravenous injection of the labeled triol into normal subjects, some 90% of the excreted tracer was found in the urinary glucuronide fraction; the remainder was in the sulfate fraction. Obviously, then, the urinary triol sulfate excreted under endogenous conditions must, for the most part, be derived from precursors other than pregnenetriol. The possibility must, therefore, be considered that 17-hydroxypregnenolone sulfate, produced by the adrenal [66, 67], is directly reduced to the triol either within the adrenal [3b, 43] or in the periphery.

Schneider and Lewbart [12] demonstrated that pregnenetriol, as well as other 3β-hydroxy-Δ^5-steroids, can be sulfurylated in the presence of ATP, SO_4^{2-} and Mg^{2+} by a microsome-free supernatant obtained from rabbit liver.

XI. Pregn-5-ene-3β,17α,20β,21-tetrol-3-sulfate [3β-Sulfooxypregn-5-ene-17α,20β,21-triol]

Bongiovanni [55c] isolated this tetrol from a solvolyzable fraction of the urine of children with the adrenogenital syndrome. The identity of the free tetrol was established by comparison with authentic material prepared by reduction of the corresponding 3β,17α,21-trihydroxypregn-5-en-20-one. Levels up to 5 mg/day were reported in the urine of these children, whereas a pool of urine from eight normal individuals yielded only 0.2 mg/day.

XII. Dehydroisoandrosterone Sulfate

A. Urinary Dehydroisoandrosterone Sulfate

As early as 1934, Butenandt and Tscherning [80] revealed that dehydroisoandrosterone occurs in human urine in conjugated form. It was not until ten years later, however, that Munson *et al.* [81] succeeded in isolating the intact conjugate and proved it to be a sulfate. Since that time, numerous quantitative reports of the urinary excretion of this conjugate have appeared; but many of these are of questionable value and, therefore, only a few recent reports will be mentioned here. At elevated temperatures, dehydroisoandrosterone sulfate is particularly labile toward acid and, under these conditions of hydrolysis, recoveries of the liberated steroid are variable and usually poor. Thus, many of the quantitative determinations made before this fact was recognized are subject to considerable doubt. Not until milder methods of hydrolysis, suitable for the sulfates of 3β-hydroxy-Δ^5-steroids became available (continuous ether extraction of urines acidified to pH 1, solvolysis, boiling urine at neutral pH, or treatment with sulfatase) and not until methods were developed for estimating dehydroisoandrosterone, specifically, were meaningful results obtained [82–88].

Jayle *et al.* [85] reported urinary dehydroisoandrosterone sulfate levels in normal adults of various ages. The hydrolysis method used by these authors involved heating urine in an autoclave at 120° C for one hour at pH 4.7. It had already been demonstrated [89] that the process of boiling urine for 4 hours at this pH is relatively specific for the hydrolysis of the sulfates of 3β-hydroxy-Δ^5-steroids. Recoveries of 80% have been reported. The method of Jayle *et al.* [85] involves the extraction of the liberated dehydroisoandrosterone into ether and the quantification of the sodium hydroxide-washed organic extract using the Zimmermann reaction. The data acquired in this way are presented in Table 5–4. In spite of considerable

TABLE 5-4. *Urinary dehydroisoandrosterone levels in normal subjects*[a]

Subjects	Age range	Number of subjects	mg/24 hours	
			D[b] (mean)	Range
Males	16–29	10	1.6	0.5–2.7
	30–39	18	2.1	1.1–4.3
	40–49	7	1.6	0.6–2.8
	50–59	4	1.4	1.1–1.5
	70+	9	0.9	0.3–1.7
Females	14–29	20	0.9	0.3–1.6
	30–39	25	1.1	0.6–2.2
	40–49	11	0.9	0.3–1.3
	50–59	7	0.5	0.1–0.9
	60+	8	0.4	0.2–0.7

[a] Jayle, M.F., Malassis, D., Pinaud, H.: Acta Endocrinol. **31**, 1 (1959).
[b] Dehydroisoandrosterone.

overlap, it is apparent that the mean value for dehydroisoandrosterone sulfate in the urine of normal females is significantly lower than that for males. They also found that the mean dehydroisoandrosterone sulfate level in both sexes reaches a maximum between 30 and 40 years of age and declines slowly thereafter. In a recent publication, Keutmann and Mason [90] reported the levels of the individual urinary 17-ketosteroids, including dehydroisoandrosterone sulfate, excreted by normal persons ranging in age between 2 and 87 years. Their procedure involved the use of an internal standard to correct for losses, and gas chromatography to separate the individual components. Since the method involved correction for procedural losses, their values are somewhat higher than those reported by previous investigators. Their findings may be summarized as follows: Negligible levels of dehydroisoandrosterone sulfate (0–10 μg/day) were found in children up to 7 years of age. In women, the amounts varied from traces to 3.8 mg/day; while, in men, they ranged from traces to 12.8 mg/day. In accord with the observations of Jayle *et al.* [85] both sexes attained a maximum level between the ages of 25 and 35. The majority of elderly patients excreted very small amounts of the sulfate. Numerous repeat determinations were performed on eight normal individuals and, in every case, wide variations were observed, thus emphasizing the care required to obtain control values.

Urinary dehydroisoandrosterone sulfate levels have been reported to be elevated in various endocrinopathies, particularly in patients with adrenal tumors. In such cases, the levels may range between 200 mg to 1 g/day. In a study of 180 patients with either hirsutism or acquired virilism, Jayle *et al.* [85] found an increased urinary excretion of dehydroisoandrosterone sulfate in most instances. Later, in 1962, Mauvais-Jarvis and Baulieu [88], using a technique in which the 17-ketosteroids were fractionated prior to their measurement, found that the dehydroisoandrosterone sulfate levels were significantly elevated in patients with hirsutism and in patients with the Stein-Leventhal syndrome. In some hirsute patients, no urinary dehydroisoandrosterone sulfate could be detected following adrenal suppression and ovarian stimulation, thus suggesting that the adrenal was the major site of production of this conjugate. In support of this idea are the numerous reports of low urinary dehydroisoandrosterone levels in Addison's disease and after corticosteroid therapy, and increased urinary levels after ACTH [adrenocorticotrophic hormone] administration. In a recent study, Heinrichs and Colás [91] also observed significantly higher than normal levels of urinary dehydroisoandrosterone sulfate in hirsute females. In addition, they noted a decrease in the urinary levels during the late stages of pregnancy.

B. Plasma Dehydroisoandrosterone Sulfate

The assay of plasma 17-ketosteroids, of which dehydroisoandrosterone is the major component, has received a great deal of attention since the first attempts by McCullagh *et al.* [92] to measure the concentration of androgenic hormones in human blood. The first suggestion that steroids occurred in plasma as water-soluble conjugates came from Mühlbock [93] who found that a hydrolytic step was necessary to release completely the estrogenic factor circulating in human pregnancy plasma. Later, several workers found that the most abundant steroid (other than cholesterol) in human plasma is dehydroisoandrosterone sulfate [84, 94]. Originally, only the free steroid was positively identified; the presence of the sulfate grouping as the conjugating moiety was inferred from its acid lability. Later, in attempts to isolate dehydroisoandrosterone sulfate, as such, from normal human plasma, Staib *et al.* [95] obtained a Zimmermann- and methylene blue-positive material (160 μg/100 ml) that had the same chromatographic and electrophoretic mobility as authentic dehydroisoandrosterone sulfate.

An excellent review of the literature on the measurement of the steroid sulfates of human plasma has recently been written by Vihko [58]. Normal values for plasma dehydroisoandrosterone sulfate have been reported by several groups and some of these are recorded in Table 5-5. Although there is considerable overlap, the average value for plasma dehydroisoandrosterone sulfate is significantly higher for men than for women. The effect of age on plasma 17-ketosteroid levels in normal individuals was studied

TABLE 5-5. *Levels of dehydroisoandrosterone sulfate in the plasma of normal subjects*

Males		Females		References
Range	D[a] (mean)	Range	D[a] (mean)	
121–232	169	68–129	94	96
67–160	101	44–106	72	97
38–167	110	73–209	118	98
103–221[b]	167	41–147	101	99

[a] Values are expressed as μg free dehydroisoandrosterone per 100 ml plasma.
[b] Calculated from the values for dehydroisoandrosterone sulfate sodium salt reported by these investigators.

by Migeon *et al.* [100]. Immediately following birth, the concentrations of dehydroisoandrosterone sulfate and androsterone sulfate fell rapidly to undetectable levels and remained there until between the ages of 5 and 7, when they began to rise progressively to normal adult levels. The levels of these plasma conjugates were at their maximum between the ages of 20 and 40. Thereafter, with increasing age, they decreased to very low levels. These observations have recently been confirmed by Deshpande and Bulbrooke [101] and by De Moor and Heyns [102] (Table 5-6). Whether the plasma levels of dehydroisoandrosterone vary diurnally appears to be somewhat in doubt. Migeon *et al.* [100] and Eik-Nes *et al.* [103] found a diurnal variation with the highest values occurring in thc morning. Contrariwise, Lamb *et al.* [104] were unable to demonstrate any cyclic change in the plasma levels of this conjugate.

In 1955, Migeon *et al.* [105] offered evidence (based on a hydrolytic procedure considered to be specific for sulfates) for the occurrence of dehydroisoandrosterone sulfate in pooled umbilical cord plasma. More recently, Eberlein [11] reported the level of this conjugate in umbilical cord blood to be approximately 60 μg/100 ml. Two groups of investigators have attempted to determine whether an arteriovenous difference, for dehydroisoandrosterone sulfate, exists in cord blood. To establish such a difference would be of considerable significance since it would contribute to the current concept that estrogens are synthesized in the placenta from sulfated C_{19}-precursors provided by the fetus. Colás *et al.* [106] measured a constituent of fetal plasma which was positive in the Pettenkofer test and which, after solvolysis, migrated with chromatographic properties similar to those of authentic dehydroisoandrosterone. They found that, measured in this way, the levels of dehydroisoandrosterone sulfate in the blood drawn from the cord artery were not significantly different from those present in

TABLE 5–6. *Levels of dehydroisoandrosterone sulfate in the plasma of normal subjects of various ages*[a]

Subjects	Age range	Number of subjects	D[b] (mean)
Males	16–25	9	88.8±43
	26–35	18	97.4±43
	36–45	12	81.1±60
	46–55	8	59.7±22
	56–65	10	44.8±29
	>65	1	35.2
		58	77.4±46
Females	16–25	33	95.5±39
	26–35	10	74.1±34
	36–45	13	53.3±19
	46–55	10	35.3±10
	56–65	11	30.5±14
	>65	2	20.6
		79	68.2±40

[a] DeMoor, P., Heyns, W., in: Androgens in normal and pathological conditions (A. Vermeulen, ed.), p. 54. New York: Excerpta Medica Foundation 1966.

[b] Values are expressed as μg free dehydroisoandrosterone per 100 ml plasma ± standard deviation.

cord venous blood. In agreement with this finding is the recent report of De Moor and Heyns [102], who, using a gas chromatographic technique, also found that the conjugate concentration in arterial cord plasma did not differ significantly from that found in venous umbilical cord plasma. Simmer *et al.* [107], on the other hand, compared values for dehydroisoandrosterone sulfate in cord arterial and venous blood with the levels found in maternal plasma, and found an apparent average arteriovenous difference between the cord arterial and venous blood. No significant differences were found between fetal and maternal blood levels. After appropriate corrections, the average arteriovenous difference in cord blood was reported to be 15 μg of dehydroisoandrosterone sulfate/100 ml. Probably due to the fact that the latter workers used a tritiated internal standard to correct for procedural losses, the values they reported are several times higher than those observed by previous investigators. In the cord blood of anencephalic fetuses, where the fetal zone of the adrenals is absent, markedly low concentrations of dehydroisoandrosterone sulfate are present [108, 109].

Peripheral levels of dehydroisoandrosterone sulfate in plasma, throughout the gestational period, have been determined by Siiteri and MacDonald [110]. These investigators found that, from the 30th week before delivery to the last week of gestation, the plasma levels of dehydroisoandrosterone sulfate declined generally from a mean of approximately 80 μg/100 ml to about 30 μg/100 ml at term. This was interpreted as probably being due to an increased conversion of circulating dehydroisoandrosterone sulfate to 17β-estradiol, by the placenta, as pregnancy proceeds.

According to Dirscherl and Breuer [111], a conjugated form of dehydroisoandrosterone is present in human seminal fluid. After removal of free steroids, the conjugate was extracted from the semen with butanol. Hydrolysis by *Helix pomatia* enzymes yielded dehydroisoandrosterone which was obtained in crystalline form after microsublimation. Although only 7 μg of dehydroisoandrosterone was found in 100 ml of semen, its identity was established by melting point, R_f, and sulfuric acid chromogen. Confirmatory data were supplied by Steeno *et al.* [112] who, using a gas chromatographic method, found still larger quantities of dehydroisoandrosterone sulfate in seminal plasma. After correction for procedural losses, values of 32 to 38 μg of dehydroisoandrosterone sulfate/100 ml of semen were observed. The concentration of dehydroisoandrosterone sulfate in the seminal plasma of normal males did not differ from that found in the semen of males with various states of infertility. In this investigation, unconjugated dehydroisoandrosterone was isolated from semen only after a solvolytic procedure. In no case was the presence of the free steroid indicated.

Over the past six years, Oertel *et al.* [113–116] have published a score or more of papers relating to the occurrence, composition, and metabolic properties of an unusual form of plasma-borne lipophilic conjugate of various 3β-hydroxy-Δ^5-steroids, particularly dehydroisoandrosterone. These conjugates were extracted from various biological media either with mixtures of ethanol and acetone or with chloroform and methanol. While the reported lipophilic form of dehydroisoandrosterone sulfate was eluted from alumina columns with a 4:1 mixture of acetone:ethanol, the reverse proportion (1:4) was necessary to elute dehydroisoandrosterone sulfate [113]. These conjugates were isolated from various biological sources and appeared to be exceptionally labile. They have been termed "sulfatidyl" compounds since the claim has been made that they are molecules composed of steroid sulfates, glycerol, and various fatty acids joined in a fashion analogous to the known phosphatides. Although alkaline hydrolysis yielded these components in the correct molar ratio [115], none of the intact conjugates has been characterized in an acceptable manner. For example, it is not clear whether the steroid sulfate is covalently linked with the other components or whether the "sulfatide" is merely a loose association between the various constituents of the mixture. The sulfatide of dehydroisoandrosterone is said to be present in human plasma, adrenal tissue, and spinal fluid. It is unstable in the presence of acid or alkali and in the conventional chromatographic systems used for conjugate separation. This extraordinary lability may be the reason why other investigators have failed to encounter these substances. Undoubtedly, the existence of these novel conjugates is of great importance and it is unfortunate that the findings of Oertel have not been confirmed by other investigators. Until these conjugates are characterized in more satisfactory ways, it seems prudent to defer further discussion of the subject.

C. Plasma Dehydroisoandrosterone Sulfate in Endocrine Disorders

The highest concentrations of plasma dehydroisoandrosterone sulfate have been found in patients with adrenal tumors. Baulieu [117a] reported that the peripheral plasma of a patient with an adrenal tumor contained 340 µg of dehydroisoandrosterone sulfate/100 ml of plasma. Adrenal vein plasma obtained at operation from adrenal vein blood contained 535 µg of the conjugate/100 ml. In a subsequent paper, Baulieu [117b] reported the analyses of the plasma of two additional patients with adrenal tumors. Values of 400 and 3,000 µg of the sulfate/100 ml of adrenal venous plasma were observed, while the corresponding values for the peripheral venous plasma were, respectively, 255 and 1,700 µg/100 ml. Baulieu was unable to detect any unconjugated dehydroisoandrosterone either in the plasma or in the tumorous tissue from these subjects.

On the other hand, Saez *et al.* [118] recently reported that unconjugated dehydroisoandrosterone was present in adrenal venous plasma in much higher concentrations than that found in a peripheral vein of patients with adrenocortical tumors (17 µg *vs.* 2 µg/100 ml; 14 µg *vs.* 2 µg/100 ml). In confirmation of previous findings [117], Saez *et al.* [118] also reported much higher amounts of dehydroisoandrosterone sulfate in the adrenal vein than those found in a peripheral vein of patients with adrenal tumors. The dehydroisoandrosterone concentrations were two- to three-fold greater in the adrenal vein than were the concentrations in the periphery (2,600 *vs.* 900; 2,400 *vs.* 1,400; 200 *vs.* 100 µg/100 ml of plasma).

Conrad *et al.* [99] discussed the significance of plasma dehydroisoandrosterone sulfate and androsterone sulfate in the diagnosis of virilizing disorders. They noted that the plasma levels of both sulfates were frequently elevated in patients with various endocrine disorders accompanied by hirsutism. In all the hirsute patients they studied, these investigators found dehydroisoandrosterone sulfate concentrations in plasma significantly higher than those existing in normal males or females. This finding is consistent with the urinary data of Mauvais-Jarvis and Baulieu [88]. However, not all the available evidence is in agreement. In an analysis of the adrenal venous plasma of twenty patients, with and without hirsutism, Weinheimer *et al.* [119] found only slightly elevated dehydroisoandrosterone sulfate levels in some of the hirsute women; the differences between these titers and those for normal women were too small to be conclusive. McKenna and Rippon [98] studied five patients with idiopathic hirsutism and found that none had plasma dehydroisoandrosterone sulfate concentrations beyond the range for normal females. Similarly, De Moor and Heyns [102] found that hirsute women do not have excessive amounts of the conjugate in their plasma. Obviously, the relationship between dehydroisoandrosterone and hirsutism has yet to be resolved. Plasma dehydroisoandrosterone sulfate levels in patients with congenital adrenal hyperplasia were recently measured by Rivarola *et al.* [120]. In four patients with the salt-losing form of the disease, plasma dehydroisoandrosterone sulfate levels were found to be below normal, while in patients with the virilizing form, the concentrations of the sulfate were elevated in four out of six patients.

D. Biosynthetic Origin of Dehydroisoandrosterone Sulfate

The very magnitude of the adrenal secretion of dehydroisoandrosterone sulfate gives special significance to the biosynthetic origin of this conjugate. As was outlined earlier, one of the pathways leading to the formation of this sulfate appears to involve unconjugated steroids as precursors. Beginning with cholesterol and employing 3β-hydroxy-Δ^5-intermediates throughout, the C_{19}-17-ketosteroid is sulfurylated in the terminal step prior to secretion. Thus, the route: cholesterol → pregnenolone → 17α-hydroxypregnenolone → dehydroisoandrosterone → dehydroisoandrosterone sulfate, has been well established. Sulfokinases having the capacity to sulfurylate dehydroisoandrosterone and other 3β-hydroxy-Δ^5-steroids have been demonstrated in tissue from adult adrenal glands [33–35, 121, 122], fetal adrenal glands [62, 123], testes [37, 124] and ovaries [125].

Sulfurylation of dehydroisoandrosterone by human skin preparations has been demonstrated by Gallegos and Berliner [126]. Considering the huge area represented by the skin, this organ may be responsible for the production of significant amounts of dehydroisoandrosterone sulfate. Cohn *et al.* [121], as well as LeBeau *et al.* [2], have shown that dehydroisoandrosterone sulfate may also be produced in good yield from 17α-hydroxypregnenolone by incubation with adrenal tumor tissue.

An alternative route for the formation of dehydroisoandrosterone sulfate is one that involves sulfurylated intermediates with cholesterol sulfate as the starting material. In the section of this chapter dealing with cholesterol sulfate, the conversion of this sterol conjugate into dehydroisoandrosterone sulfate, *via* sulfated intermediates, was discussed. In addition, mention has already been made of the *in vitro* conversions: pregnenolone sulfate → 17α–hydroxypregnenolone sulfate [1] → dehydroisoandrosterone sulfate [2] by slices of tumorous adrenal tissue. There is little doubt that both pathways, one using unconjugated and the other sulfated intermediates, can lead to the formation of dehydroisoandrosterone sulfate. On the other hand, there is considerable uncertainty about the significance of these two routes. The reason for the existence of two routes, the quantitative importance of one route relative to that of the other, the relation of these two routes to others concerned with the biogenesis of the steroid hormones, these are but a few of the important problems that are, at present, unsolved.

1. Adrenal

For at least a quarter of a century, the adrenal origin of dehydroisoandrosterone has been considered to be established. However, not until the arteriovenous difference of dehydroisoandrosterone sulfate across an adrenal tumor was observed [117a] was the possibility even suspected that the sulfate, itself, could be secreted by adrenal tissue. Further evidence that dehydroisoandrosterone sulfate was secreted by the adrenals was provided by Vande Wiele *et al.* [30]. These investigators attempted to measure the secretory rate of dehydroisoandrosterone in human subjects by measuring the specific activity of the compound excreted in urine following the administration of tritiated dehydroisoandrosterone. Their analysis involved extraction of the conjugates from a three–day urine specimen, separation of the sulfates from the glucuronides by partition chromatography on Celite® (Johns-Manville, N.Y.), cleavage of the sulfates by solvolysis, and liberation of the glucuronides by glucuronidase. The individual 17–ketosteroids in each group were isolated by chromatography and their specific activities were measured. In each instance, the specific activity of dehydroisoandrosterone, isolated from the glucuronide fraction, was greater than that of the same steroid isolated from the sulfate fraction. These results suggested that urinary dehydroisoandrosterone sulfate was derived not only from secreted dehydroisoandrosterone, but also from some other precursor which was excreted as dehydroisoandrosterone sulfate without having first been converted into the unconjugated ketosteroid. The observation was rationalized by assuming that dehydroisoandrosterone sulfate, itself, was secreted by the adrenal gland. Confirmatory data were subsequently presented by Vande Wiele *et al.* [30] and by MacDonald *et al.* [127] when the secretory rates of both dehydroisoandrosterone and its sulfate were estimated. These experiments not only confirmed that dehydroisoandrosterone sulfate was secreted, but they also proved that this conjugate was a major steroid secretory product of the human adrenal. In two normal males, the secretory rates of dehydroisoandrosterone sulfate were calculated to be 10.1 and 5.9 mg/day, while the secretory rates of the unconjugated ketosteroid were 7.2 and 3.0 mg/day. In a normal female, these values were found to be 7.7 mg of the sulfate/day and 0.7 mg free dehydroisoandrosterone/day [30, 127] (see Table 5–7). In each case, the secretion of the sulfated C_{19}–compound exceeded the secretion of the unconjugated steroid.

The finding that dehydroisoandrosterone sulfate was quantitatively an important secretory product of the adrenal gland furnished an understanding of a variety of earlier experimental observations. Thus, it explained how, after the administration of labeled dehydroisoandrosterone, the specific activity of the urinary sulfate could be lower than those of androsterone and etiocholanolone [87]. It accounted for the low levels of unconjugated C_{19}–steroids present in adrenal vein blood, when at the same time large amounts of 17–ketosteroids were being excreted in urine. Finally, the previously observed, and unexpectedly high, secretion rates of dehydroisoandrosterone, calculated from the specific activity of urinary dehydroisoandrosterone sulfate following the injection of labeled dehydroisoandrosterone [30, 87], were explained. These were obviously overestimated since the secretion of the 17–ketosteroid sulfate had not been taken into account.

Recognizing that dehydroisoandrosterone sulfate is an adrenal secretory product and that it may be interconverted in peripheral tissue with dehydroisoandrosterone which, in turn, may be transformed into androstenedione [androst–4–ene–3,17–dione] and testosterone [17β–hydroxyandrost–4–en–3–one], Chapdelaine *et al.* [128], following the procedures of Vande Wiele *et al.* [30] and of MacDonald *et al.* [127], attempted to estimate the secretion rates and production rates of dehydroisoandrosterone and its sulfate at different levels of adrenal and testicular steroidogenic activities. They assumed that the situation depicted in Fig. 5–4 prevailed and that, by administering three tracers: ^{3}H–dehydroisoandrosterone sulfate, ^{14}C–dehydroisoandrosterone, and ^{3}H–testosterone, they could estimate the various parameters characterizing the system. Three tracers were administered during four intervals: a period of control, a period during which the adrenal was stimulated by ACTH, a period during which the adrenal was suppressed by dexamethasone [9α–fluoro–11β,17α21–trihydroxy–16α–methylpregna–1,4–diene–3,20–dione] and, finally, a period during which the adrenals were suppressed and the testes stimulated by human chorionic gonadotrophin. The secretory rates of dehydroisoandrosterone and its sulfate were found

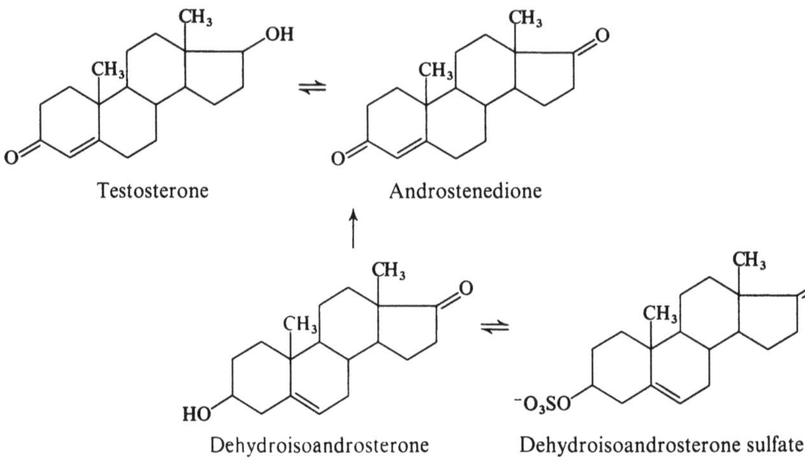

Fig. 5–4. Interconversion of the C_{19}–androgens

to increase approximately two– to three–fold with ACTH administration, and were decreased two– to twelve–fold during adrenal suppression with dexamethasone. During the period of adrenal suppression, the administration of chorionic gonadotrophin failed to increase either the secretory rate or the production rate of dehydroisoandrosterone sulfate. On the other hand, both the secretion rate and the production rate of free dehydroisoandrosterone were doubled under this regimen (see Table 5–8). It is apparent that the major portion of dehydroisoandrosterone and dehydroisoandrosterone sulfate, produced in the normal male, is of adrenal origin. The earlier findings of MacDonald *et al.* [127] and Sandberg *et al.* [129] that dehydroisoandrosterone sulfate is secreted in larger quantities than is free dehydroisoandrosterone were confirmed [128]. These results also suggested that dehydroisoandrosterone is secreted by the testes but that the sulfated compound is not. The findings are also consistent with the recent observations that ACTH administration causes an elevation in the concentration of dehydroisoandrosterone sulfate in peripheral plasma [130] and in the adrenal venous effluent [67]. On the contrary, DeMoor and Heyns [102] reported that the infusion of maximal doses of ACTH into nine subjects did not increase the plasma level of dehydroisoandrosterone sulfate; and this was also the experience of Tamm *et al.* [131].

Wieland *et al.* [67, 132] compared the concentration of dehydroisoandrosterone sulfate in adrenal venous blood with that circulating in the periphery. In a 50–year–old man, with no evidence of endocrine disease, the amount of the conjugate was found to be 356 μg in 100 ml of adrenal venous plasma, whereas 77 μg was found in 100 ml of peripheral plasma. In a second subject, the adrenal venous plasma concentration was found to be 78 μg/100 ml, whereas that in the peripheral plasma was 44 μg/100 ml. Baulieu *et al.* [39], studying patients without endocrine disease, also observed that the levels of plasma dehydroisoandrosterone sulfate in the adrenal venous effluent were significantly higher than the levels found in the peripheral vein. In a recent publication, Saez *et al.* [118], using labeled internal standards to correct for losses, also reported that the levels of dehydroisoandrosterone sulfate found in adrenal venous plasma of patients bearing adrenal tumors were higher than those found in a peripheral vein. Production rates (determined from urinary data) of dehydroisoandrosterone sulfate during pregnancy have been calculated by Siiteri and MacDonald [110]. From the results obtained, it was suggested that there exists a peak of dehydroisoandrosterone sulfate production between the 27th and 32nd weeks of gestation, followed by a decrease to, or below, nonpregnancy levels near term.

2. Gonadal

Although the secretion of dehydroisoandrosterone sulfate by the gonads of normal individuals has not been established, it is reasonably certain that this conjugate is produced by pathological gonadal tissues. For example, Kase and Conrad [133] examined the sulfate fraction of the venous effluent of a virilizing arrhenoblastoma of the ovary and reported a value of 378 μg of dehydroisoandrosterone/100 ml of plasma, whereas plasma obtained from the brachial artery contained only 98 μg/100 ml. Furthermore, O'Malley *et al.* [134] have isolated the sulfate from a virilizing luteoma of pregnancy. On the other hand, Baulieu *et al.* [78] were unable to find this conjugate in ovaries obtained from patients with Stein–Leventhal syndrome when, from the same tissue, unconjugated dehydroisoandrosterone was isolated. As far as testicular production is concerned, Mahesh and Greenblatt [135] reported that spermatic vein blood obtained from patients with testicular feminization contained a significantly higher level of dehydroisoandrosterone sulfate than did arterial blood. Pion *et al.* [136], using a radioactive internal standard to correct for losses, have confirmed these findings. These authors measured concentrations of 272 and 294 μg/100 ml in samples of spermatic vein plasma from two patients with testicular feminization, whereas only 180 and 167 μg/100 ml were found in the plasma from the spermatic artery. Contrariwise, Rivarola *et al.* [137] found that, while the levels of testosterone, androstenedione, and dehydroisoandrosterone in the spermatic vein blood of a male pseudohermaphrodite with testicular feminization were greater than those present in peripheral plasma, no comparable difference could be demonstrated for dehydroisoandrosterone sulfate.

It is evident, then, that while the secretion of dehydroisoandrosterone sulfate by the human adrenal has been established beyond doubt, evidence for the gonadal origin of this conjugate is based only on limited data obtained from subjects with gonadal pathology. However, it seems certain that, if dehydroisoandrosterone sulfate does originate in the gonad, it is of minor quantitative importance, since castration hardly alters the urinary levels of dehydroisoandrosterone sulfate [138] nor are these levels significantly altered by the administration of chorionic gonadotrophin [139–141].

In the boar, in contrast to the human, the testes appear to be the main, if not the sole source, of dehydroisoandrosterone sulfate. Experiments performed by Huis in't Veld *et al.* [142] showed that treatment with human chorionic gonadotrophin increased the level of urinary dehydroisoandrosterone in a boar, while Clark *et al.* [143] reported the detection of dehydroisoandrosterone sulfate in boar spermatic vein blood. In a recent publication, Baulieu *et al.* [144] reported a high concentration of dehydroisoandrosterone sulfate in spermatic vein plasma from the boar (200 μg/100 ml plasma) following treatment with human chorionic gonadotrophin. Dehydroisoandrosterone sulfate was also isolated from the testicular tissue (1.5 μg/g tissue).

E. The Role of Dehydroisoandrosterone Sulfate in the Production of Gonadal Hormones

1. Estrogens

The surprising discovery that blood-borne dehydroisoandrosterone sulfate serves as an efficient precursor of the estrogen produced by the placenta, represents one of the most important advances made in endocrinology in recent years. Several investigators have provided evidence to support the idea that placental estrogen production requires externally supplied precursors. For example, Frandsen and Stakemann [145] suggested that the fetus can make such precursors available to the placenta. They observed that mothers bearing anencephalic fetuses excrete in their urine one-tenth as much estrogen as do mothers bearing normal fetuses. Since these abnormal fetuses lack the "fetal zone" of the adrenal cortex, Frandsen and Stakemann reasoned that the fetal adrenals, in normal pregnancy, furnish steroidal precursors which the placenta converts into estrogens. That the fetal adrenal was not the only source of estrogen precursor is suggested by the fact that the urinary estrogen levels in mothers with anencephalic babies, although low compared with normal pregnancy levels, are still far in excess of those found in nonpregnant females.

Migeon *et al.* [105] and Simmer *et al.* [107] demonstrated that the concentration of dehydroisoandrosterone sulfate in the plasma of fetuses is equal to, or higher than that present in the maternal cirulation. Pulkkinen [146] found that a high concentration of steroid sulfatase is present in human placenta and, more recently, Warren and Timberlake [147, 148] showed that this sulfatase can cleave dehydroisoandrosterone sulfate whether the conjugate approaches the placenta from the maternal or from the fetal side. Finally, Ryan [149] demonstrated that the placenta contained aromatizing enzymes capable of transforming C_{19}–steroids efficiently into estrogens. These discoveries prompted Siiteri and MacDonald [150] to investigate the role of plasma dehydroisoandrosterone sulfate as an estrogen progenitor. They used the novel experimental design of administering, simultaneously, to pregnant women, both the labeled precursor, ^{14}C–dehydroisoandrosterone sulfate and, the labeled product, ^{3}H–17β–estradiol. The latter tracer served as an *in vivo* internal standard which corrects for the losses incurred from the time the 17β–estradiol is formed in the placenta until the isolated and purified urinary product is analyzed. Thus, if the fate of the labeled 17β–estradiol is the same as that of the endogenous 17β–estradiol secreted by the placenta into the blood, the ratio of ^{3}H to ^{14}C in the urinary 17β–estradiol is a measure of the extent of conversion of dehydroisoandrosterone sulfate, present in the maternal circulation, to estrogen; *i.e.*, % conversion of dehydroisoandrosterone sulfate to 17β estradiol $= 100 \times (^{3}H/^{14}C)_{\text{injected tracers}}/(^{3}H/^{14}C)_{\text{urinary 17}\beta\text{-estradiol}}$. Using this experimental approach, it is also possible to estimate the fraction of the 17β–estradiol production rate that is derived from circulating dehydroisoandrosterone sulfate ($PR_{E2_{DS}}$). $PR_{E2_{DS}}$ = production rate of 17β–estradiol × f, where f = the specific activity of urinary 17β–estradiol with respect to ^{14}C, divided by the specific activity of urinary dehydroisoandrosterone sulfate with respect to the same isotope. The production rate of 17β–estradiol was estimated from the specific activity of urinary 17β–estradiol with respect to tritium.

The data obtained during normal pregnancy showed that the percentage conversion of injected dehydroisoandrosterone sulfate to 17β–estradiol rose from 1 to 2% at seven weeks of gestation to 28 to 45% near term. The results from fifteen subjects are presented in Fig. 5–5. In addition, that fraction of the 17β–estradiol production rate which is derived from maternal dehydroisoandrosterone sulfate ($PR_{E2_{DS}}$) was found to vary widely from one individual to another. The remainder presumably comes from precursors, furnished by the fetus, of which dehydroisoandrosterone sulfate is probably

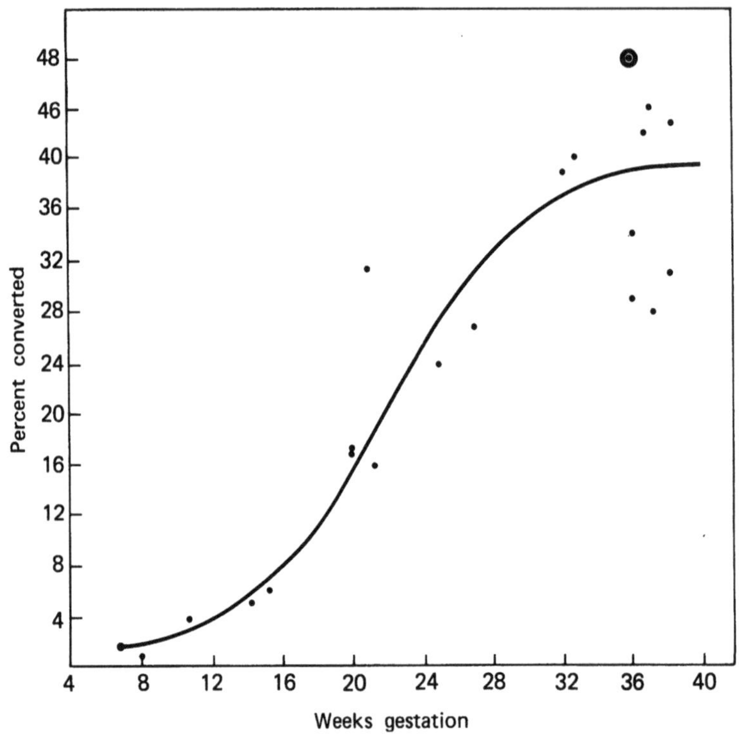

Fig. 5–5. Placental conversion of intravenously administered 4–^{14}C–dehydroisoandrosterone sulfate to 17β–estradiol during normal pregnancy

the principal one. This receives support from the results, obtained from an adrenalectomized pregnant woman, in which the production rate of 17β–estradiol was a fraction (one–third to one–half, depending upon the time of gestation) of that of normal pregnant women [110]. In abnormal pregnancies such as those involving anencephaly and hydatidiform moles, where functioning fetal adrenal tissue is absent, it was found that the major portion of endogenous estrogen production arises from dehydroisoandrosterone sulfate present in the maternal circulation [150b, 151]. That the adrenal glands were the source of the circulating dehydroisoandrosterone sulfate was demonstrated by administration of dexamethasone to one subject, which caused a three–fold decrease in 17β–estradiol production. Administration of ACTH to another subject produced a four–fold increase in the production of the estrogen [150b].

In experiments using several labeled C_{19}–steroids, Baulieu and Dray [152] estimated the fraction of administered tracers that appeared in urinary 17β–estradiol. Note that this fraction is not the same as that determined by

Siiteri and MacDonald who, by administering radioactive 17β-estradiol simultaneously with labeled dehydroisoandrosterone sulfate, obtained a measure of the conversion occurring in the placenta. Baulieu and Dray [152] found that dehydroisoandrosterone sulfate was converted into urinary estrogens more efficiently than were either testosterone or androstenedione when these steroids were administered intravenously. Warren and Timberlake [148] and Barlow *et al.* [153] have also demonstrated that dehydroisoandrosterone sulfate was a precursor of estrogens in pregnant women. All workers agree that this conversion is either absent, or of minor quantitative importance, in nonpregnant women.

Perfusion experiments carried out by Warren and Timberlake [148] showed that the isolated perfused placenta can cleave dehydroisoandrosterone sulfate and convert the liberated dehydroisoandrosterone into 17β-estradiol and estrone. Their results also seemed to indicate that the placenta is less permeable to dehydroisoandrosterone sulfate than it is to unconjugated steroids. Similar results were obtained by Bolté *et al.* [154] who, with a one-pass *in situ* perfusion technique, compared the conversion of dehydroisoandrosterone, androstenedione, testosterone, and dehydroisoandrosterone sulfate into 17β-estradiol. All three unconjugated compounds were aromatized to a much greater extent than was the sulfate. Similarly, when labeled dehydroisoandrosterone and dehydroisoandrosterone sulfate were injected into the uterine artery of a patient whose pregnancy was interrupted 12 to 15 minutes later, the results obtained led Bolté *et al.* [155] to conclude that dehydroisoandrosterone was a more efficient precursor of estrogens that was its sulfate. In these experiments, most of the administered sulfate was recovered unchanged in the perfusate, suggesting that the passage of the conjugate into the placental cell was a limiting step in its utilization for the biosynthesis of the estrogens. It is apparent, then, that when dehydroisoandrosterone sulfate is presented to the placenta in a short-term experiment, its transformation to estrogens is poorer than when dehydroisoandrosterone is the precursor. Nevertheless, the concentration of dehydroisoandrosterone sulfate in the blood supplying the placenta is much greater than that of the unconjugated steroid and, as has been pointed out by Bolté and his colleagues, it is entirely conceivable that the conjugate furnished by the process of slow continuous perfusion of that organ, as it occurs under physiological conditions, may well be the principal precursor of the huge quantities of estrogen produced during pregnancy. In a further experiment [156] designed to approximate physiological events more closely—that is, a long-term, *in situ* placental perfusion—^{14}C-dehydroisoandrosterone sulfate was injected into the umbilical vein while 7-^{3}H-dehydroisoandrosterone sulfate was administered *via* the antecubital vein of the mother. In this instance, a striking difference was observed in the utilization of the two isotopes for estrogen biosynthesis. Dehydroisoandrosterone sulfate, circulating on the fetal side of the placenta, was aromatized to a far greater extent than was the sulfate borne by the maternal circulation.

2. Androgens

Our knowledge of the biochemistry of the androgens has been advanced significantly in the last few years by the abandonment of the simplistic idea that it is sufficient to explain all aspects of androgenicity in terms of one secreted hormone, testosterone. In a presentation given at the 1962 Laurentian Hormone Conference by Vande Wiele and his coworkers [30], it was pointed out that the androgen problem is exceedingly complex since the production, and the peripheral conversion, of at least four substances, testosterone, androstenedione, dehydroisoandrosterone, and dehydroisoandrosterone sulfate and, possibly, also several others, such as androstenediol [androst-5-ene-3β,17β-diol] and its 3-sulfate, must be considered before the biochemical aspects of androgen production can be properly understood. Each of these substances may be secreted by either the adrenals or the gonads or both, and this complication adds to the difficulty of simple generalizations. For example, the secretion by adrenal carcinomas of large amounts of dehydroisoandrosterone sulfate, a compound which, *per se*, may not be an androgen, may result in virilizing effects through the formation by peripheral conversion of sufficient amounts of the more potent androgen, testosterone [157]. Thus, not only may dehydroisoandrosterone sulfate serve as a progenitor of estrogens as it does in pregnancy but it may also be converted into more potent androgenic substances in various circumstances.

By 1962, the metabolic conversions: androstenedione → testosterone, testosterone → androstenedione, dehydroisoandrosterone → androstenedione, and dehydroisoandrosterone → testosterone had been known for some time (for early references see: [158, 159]). Moreover, the suggestion had already been made that dehydroisoandrosterone sulfate was a secretory product of the adrenal [117a]. However, not until the interconvertibility of dehydroisoandrosterone and its sulfate was demonstrated [160, 161], was the possibility recognized that this conjugate, as well as the unconjugated ketosteroid, could be a precursor of the androgens, androstenedione and testosterone. The first attempt to study this question quantitatively was made by Vande Wiele and his coworkers [30] who attempted to estimate the rates of secretion of the four compounds, dehydroisoandrosterone sulfate, dehydroisoandrosterone, androstenedione, and testosterone, and their rates of peripheral interconversion. They administered two or, in some cases, three isotopically labeled steroids to normal subjects, and measured the specific activities of various urinary metabolites of these steroids. These experiments were based on a theoretical analysis of the problem presented by the interconvertibility of dehydroisoandrosterone and its sulfate, made by Gurpide *et al.* [162, 163], which proved that the secretory rates of both compounds and the rates of their interconversion could be evaluated if tracers of both were administered intravenously. When this was done [127], the rate of interconversion between dehydroisoandrosterone and its sulfate was found to be sufficiently slow that the two compounds could be distinguished kinetically from each other and, as a result, the kinetic parameters characterizing the secretion and metabolism of both compounds could be estimated by this technique. The general similarity between $^3H/^{14}C$ (the two isotopes used) ratios of urinary dehydroisoandrosterone glucuronide [17–oxoandrost–5–en–3β–yl–β–D–glucopyranosiduronic acid], androsterone glucuronide [17–oxo–5α–androstan–3α–yl–β–D–glucopyranosiduronic acid], etiocholanolone glucuronide [17–oxo–5β–androstan–3α–yl–β–D–glucopyranosiduronic acid], androsterone sulfate, and etiocholanolone sulfate [3α–sulfooxy–5β–androstan–17–one] indicated that both androsterone and etiocholanolone were derived from dehydroisoandrosterone sulfate by pathways involving only free dehydroisoandrosterone. The results obtained from four subjects indicated that each secreted more dehydroisoandrosterone, as its sulfate, than as the free alcohol. With this design, these workers were also able to estimate the rates of conversion and the transfer factors (ρ) [162] for both dehydroisoandrosterone and its sulfate. Some of their results are given in Table 5–7.

These investigators also administered intravenously a mixture of ^{14}C–testosterone and ^3H–androstenedione and, by this means, found in one normal subject that testosterone was metabolized to androsterone entirely *via* androstenedione. However, the results obtained from two other subjects indicated that this was not so; testosterone was, to a significant extent (approximately 25%) metabolized through pathways not involving androstenedione. Using this estimate and a model (Fig. 5–4) based on the results mentioned above, MacDonald *et al.* [127] attempted to evaluate in the same individual, at the same time, the secretory rates, the production rates, the rates of interconversion, the transfer factors, and the rates of irreversible removal of all four components of the system. The results they obtained specifically for dehydroisoandrosterone and its sulfate are also given in Table 5–7.

As was mentioned previously, Chapdelaine *et al.* [128] used the same experimental design to determine what effect the experimental alteration

TABLE 5–7. *Some metabolic parameters characterizing dehydroisoandrosterone (D) and dehydroisoandrosterone sulfate (DS)*

Metabolic parameter	Male 31 years	Female 28 years	Male[a] 27 years
Secretory rate (mg/day)			
D	7.2	0.7	3.0
DS	10.1	7.7	5.9
Production rate (mg/day)			
D	10.1	6.9	6.6
DS	14.8	8.1	6.9
Rate of interconversion (mg/day)			
D → DS	7.6	6.7	2.5
DS → D	5.7	11.6	5.0
Transfer factors (ρ)			
D → DS	0.59	0.55	0.11
DS → D	0.31	0.81	0.60
DS → Testosterone	—	0.06[a]	0.03
Rate of irreversible metabolism (mg/day)			
D	5.3	5.6	5.5
DS	12.0	2.8	3.4

[a] Results obtained by administering three tracers [MacDonald, P.C., Chapdelaine, A., Gonzales, O., Gurpide, E., Vande Wiele, R. L., Lieberman, S.: J. Clin. Endocrinol. Metab. **25**, 1557 (1965)]. For the definitions of these parameters and the means by which they were estimated, see: Gurpide, E., MacDonald, P.C., Vande Wiele, R. L., Lieberman, S.: J. Clin. Endocrinol. Metab. **23**, 346 (1963); Gurpide, E., MacDonald, P.C., Chapdelaine, A., Vande Wiele, R. L., Lieberman, S.: J. Clin. Endocrinol. Metab. **25**, 1537 (1965); Gurpide, E., Mann, J., Lieberman, S.: J. Clin. Endocrinol. Metab. **23**, 1155 (1963).

of the gonadal and adrenal function of a normal man would have on the estimates of these kinetic parameters. As stated above, adrenal stimulation was achieved by the administration of ACTH; adrenal suppression by the administration of dexamethasone; and adrenal suppression with testicular stimulation by the simultaneous administration of dexamethasone and human chorionic gonadotrophin. Some of the results pertinent for dehydroisoandrosterone and its sulfate are given in Table 5–8. In spite of the fact that the numerical values obtained were quite in accord with expectations based on physiological considerations, the results obtained with this design cannot be considered to be precise, due, principally, to the oversimplification of the model. Nevertheless, these experiments yielded values which established the order of magnitude not only for secretory rates but also for the

TABLE 5-8. *Changes in various metabolic parameters characterizing dehydroisoandrosterone (D) and dehydroisoandrosterone sulfate (DS) produced by altering endocrine status*

Endocrine status	S_D	S_{DS}	PR_D	PR_{DS}	$\rho_{DS \to T}$	Fraction of T derived from S_{DS}
Control	5	24	17	19	0.12	0.29
Adrenal stimulation	17	43	34	51	0.04	0.16
Adrenal suppression	3	2	4	3	0.06	0.02
Adrenal suppression and testicular stimulation	7	0.5	8	3	0.07	0.00

For the definitions of these parameters and the means by which they were estimated, see: Chapdelaine, A., MacDonald, P.C., Gonzalez, O., Gurpide, E., Vande Wiele, R.L., Lieberman, S.: J. Clin. Endocrinol. Metab. **25**, 1569 (1965); Gurpide, E., MacDonald, P.C., Vande Wiele, R.L., Lieberman, S.: J. Clin. Endocrinol. Metab. **23**, 346 (1963); Gurpide, E., MacDonald, P.C., Chapdelaine, A., Vande Wiele, R.L., Lieberman, S.: J. Clin. Endocrinol. Metab. **25**, 1537 (1965); Gurpide, E., Mann, J., Lieberman, S.: J. Clin. Endocrinol. Metab. **23**, 1155 (1963).

rates of peripheral interconversion of the compounds and for the contribution of secreted precursor to the various urinary metabolites.

However, these experimental designs could not identify the sites where the conversions occurred, nor could they establish the actual contribution to the circulation of a compound formed metabolically. For instance, the value for ρ_{DST} (the fraction of dehydroisoandrosterone sulfate converted into testosterone) was evaluated from the specific activity of samples of urinary testosterone glucuronide [3-oxoandrost-4-en-17β-yl-β-D-glucopyranosiduronic acid] rather than from plasma testosterone and led, in the experiments of MacDonald *et al.* [127], to values of 3 and 6%. The values estimated by Chapdelaine *et al.* [128], for this conversion of dehydroisoandrosterone sulfate into testosterone, were between 4 and 12%. However, by analyzing the isotope contents of urinary testosterone glucuronide, as was done in these studies, there is no assurance that testosterone formed from dehydroisoandrosterone sulfate actually enters into the circulation. The conversions: dehydroisoandrosterone sulfate → dehydroisoandrosterone → androstenedione → testosterone → testosterone glucuronide, could all occur in one organ, *e.g.*, the liver, and the testosterone so produced would never appear in the circulation to exert its biological effect at some distant target site. In fact, in 1964, Korenman and Lipsett [164] administered tracers of androstenedione and testosterone and, from the ratio of the two isotopes in urinary testosterone glucuronide, concluded that androstenedione had been converted into testosterone glucuronide without free testosterone ever having entered into the blood. Still later, Korenman and Lipsett [165] administered tracers of dehydroisoandrosterone and testosterone and, from estimations made on plasma levels, concluded that plasma testosterone glucuronide arising from dehydroisoandrosterone, likewise, did not come solely from the plasma testosterone pool. Thus, the experimental design used by MacDonald *et al.* [127] and Chapdelaine *et al.* [128] probably resulted in an overestimation of the percentage conversion of dehydroisoandrosterone sulfate into circulating testosterone. However, even if the testosterone derived from dehydroisoandrosterone sulfate is metabolized further, *in situ*, in the liver, to less androgenic products before its release into the hepatic vein, important physiological effects might result. It is conceivable that this testosterone, produced in the liver, could alter the hepatic processes usually considered to be influenced by this hormone, without, at the same time, contributing to the level of circulating androgens.

Mauvais-Jarvis and Baulieu [166] also attempted to measure the percentage conversion of dehydroisoandrosterone sulfate into testosterone by injecting tracers of the sulfate and testosterone into normal men, normal women, and hirsute women. They measured the isotope ratios in various urinary metabolites and estimated a 5% conversion. Like those estimated by MacDonald *et al.* [127] and Chapdelaine *et al.* [128], this value is probably an overestimation of the contribution of dehydroisoandrosterone sulfate to plasma testosterone since, here too, urinary products were analyzed.

There is another assumption implicit in the four-pool model as depicted in Fig. 5-4 which, if false, would result in inaccurate estimates. This has already been mentioned and concerns the fact that the scheme does not allow for the involvement of other secretory products, *e.g.*, androstenediol, its sulfate or other, still unknown, intermediates. The experiments of Sandberg *et al.* [129] suggested the possible existence of another sulfate (conceivably androstenediol sulfate) which exchanges with dehydroisoandrosterone sulfate in plasma. In addition, results obtained by Mauvais-Jarvis and Baulieu [166] and by Loras and Migeon [167a] have shown that in some, if not in most, instances the secretion and metabolism of intermediates other than the four indicated in Fig. 5-4 cannot be ignored. In addition, the results recently obtained by Jeffcoate and his co-workers [167b] are also inconsistent with the two-dimensional scheme depicted in Fig. 5-4. These investigators acquired evidence that suggests that androstenedione, injected into the peripheral circulation, is metabolized to testosterone glucuronide to a greater extent than is the androstenedione produced from other steroids at other sites.

With all its shortcomings, the approach first suggested by Vande Wiele *et al.* [30] for the quantitative study of the interrelationships between dehydroisoandrosterone sulfate and its C_{19}-relatives had important heuristic value. It exposed the complexity of the whole problem and, by so doing, effected a radical change in our thinking about androgen biochemistry. The scheme outlined could not be relied upon to be applicable to every physiological and pathological circumstance for its very complexity, and the great diversity of natural situations that exist militate against any universal application. Its virtue, however, was that it exposed the problem in a constructive way, making evident the assumptions and limitations and the need for experimental modifications which could obviate its shortcomings. It made clearly evident the means by which the assumptions inherent in the model could be tested experimentally. It discussed the possibility of precursors other than the four shown in Fig. 5–4, and indicated clearly that estimations made on urinary products could not give information about the sites where interconversions occur. Finally, it explained for the first time the distinction between secretory rates and urinary production rates.

At any rate, the evidence presently available indicates that the amount of circulating testosterone that can be derived by peripheral conversion, from dehydroisoandrosterone or its sulfate, is, in the *normal subject*, of little significance. The most pertinent determinations are those of Horton and Tait [168] who avoided the difficulties inherent in the four-pool model and the estimation of isotope contents in urinary metabolites, by measuring the transfer factor for the conversion of blood dehydroisoandrosterone to blood testosterone ($\rho_{D \to DT}$). This factor [162, 169] is estimated from the metabolic clearance rate of dehydroisoandrosterone (MCR^D), the metabolic clearance rate of testosterone (MCR^T), the plasma concentrations of radioactive dehydroisoandrosterone (X^D), and radioactive testosterone (X^T) attained at equilibrium after the constant infusion of labeled dehydroisoandrosterone ($\rho_{D \to T} = MCR^T/MCR^D \times X^T/X^D$). In eight normal subjects, the values for $\rho_{D \to T}$ ranged from 0.5–0.9%. Obviously, the $\rho_{DS \to T}$, had it been measured by this technique, would have been much smaller, probably half. These investigators also estimated that, at most, 15% of plasma testosterone in women arose from peripheral conversion of dehydroisoandrosterone. In men, the value was less than 1%. What these values are in patients with various endocrinopathies remains to be determined.

One noteworthy example has already been reported. In 1966, Lipsett and his coworkers [157] described some results obtained on a patient with a massive interstitial cell carcinoma of the testes. The blood testosterone level in this patient was four times the normal male value. These workers infused tritiated testosterone into this subject and determined the blood testosterone production rate (from PR = MCR × c) and the urinary testosterone production rate from the specific activity of urinary testosterone glucuronide. The urinary testosterone production rate was found to be 53 mg/day whereas the blood testosterone production rate was 48 mg/day. It also appeared that all of the urinary testosterone glucuronide had arisen from testosterone circulating in the plasma. These investigators then infused ^{3}H–dehydroisoandrosterone sulfate and ^{14}C–dehydroisoandrosterone simultaneously into this patient and observed that 90% of the testosterone glucuronide excreted in the urine was derived from blood dehydroisoandrosterone sulfate. Consequently, the plasma sulfate must also have been the unique precursor of the plasma testosterone. Since it had already been shown that all of the testosterone glucuronide excreted in urine had come from dehydroisoandrosterone sulfate, Lipsett and his colleagues reached the amazing conclusion that, in this patient, circulating dehydroisoandrosterone sulfate was the sole source of circulating testosterone.

Lindberg *et al.* [170] investigated the relationship between testosterone, dehydroisoandrosterone, and its sulfate following the intravenous injection, into normal adults, of the labeled C_{19}–compounds. By analyzing the isotope ratios found in plasma at various time intervals, it was clearly shown that, throughout the period under study (4 hours), plasma testosterone was continuously formed from circulating dehydroisoandrosterone sulfate. However, it is difficult to estimate the actual percentage conversion of dehydroisoandrosterone sulfate into testosterone, in plasma, from the data presented by these authors. Slaunwhite *et al.* [171] studied the metabolism of ^{14}C–dehydroisoandrosterone and of ^{3}H–dehydroisoandrosterone sulfate by administering both tracers intravenously into six normal subjects and into five subjects with biliary fistulas. In agreement with previous investigators [127] they found that these two compounds were slowly interconvertible and that free dehydroisoandrosterone was a better precursor of testosterone than was its conjugate. During the first four hours, less than 10% of urinary testosterone glucuronide had come from dehydroisoandrosterone sulfate. This finding is in accord with the data of Chapdelaine *et al.* [128] who also found that free dehydroisoandrosterone, over a period of five days, was a better precursor of testosterone than was the sulfated steroid.

O'Malley *et al.* [134] studied a virilizing luteoma which appeared to be producing sizable quantities of dehydroisoandrosterone sulfate. *In vitro* incubations with labeled precursors led them to conclude that the sulfate was converted into testosterone more efficiently than was dehydroisoandrosterone. They also considered that their results suggested that the conjugate was converted into testosterone *via* androstenediol sulfate and not *via* dehydroisoandrosterone and androstenedione. Unfortunately, *in vitro* studies of this kind cannot provide unequivocal proof of such conclusions, and other evidence will be required before these conclusions can be considered as established. Nevertheless, they are in accord with the findings of Aakvaag *et al.* [172] who showed that canine testes can convert dehydroisoandrosterone sulfate into testosterone, probably *via* routes involving androstenediol sulfate.

To determine whether dehydroisoandrosterone sulfate may serve as a physiological substrate for active androgen production by virilizing ovarian tissue, Sandberg *et al.* [173] incubated homogenates of this tissue with the labeled sulfate. With arrhenoblastomatous tissue, small yields of androstenedione and testosterone were obtained, while polycystic ovarian tissue formed only androstenedione from the labeled conjugate. Gonadotrophin and ACTH had little effect on these conversions nor did they influence the hydrolysis of dehydroisoandrosterone sulfate by homogenates of tissue from normal human ovaries, placenta, or endometrium. Further evidence that dehydroisoandrosterone sulfate may be an important precursor of androgens in ovarian tumor tissue has been provided by Fahmy *et al.* [174]. Following the incubation of labeled dehydroisoandrosterone sulfate with a Hilus-cell tumor mince, radioactive dehydroisoandrosterone, androstenedione, and testosterone were isolated in good yield. Particularly noteworthy was the 41% conversion of dehydroisoandrosterone sulfate into androstenedione.

Human skin, which constitutes some 15% of the total body weight, may also play an important role in the formation of active androgens. Recently, Kim and Herrmann [175] reported that foreskin and abdominal skin obtained from infants had the capacity to hydrolyze dehydroisoandrosterone sulfate to the free alcohol. They also demonstrated that the liberated steroid was further metabolized to androstenedione and testosterone. Abdominal skin obtained from adults, on the other hand, was found to be devoid of sulfatase activity [176]. The ability of vaginal mucosa to utilize dehydroisoandrosterone sulfate for the synthesis of androstenedione and testosterone has also recently been reported [175].

F. Metabolic Activity

In spite of the fact that dehydroisoandrosterone sulfate is one of the principal secretory products of normal adrenals, the biological significance of the secretion of this conjugate is presently unknown. One of the first efforts to associate this conjugate with some metabolic activity was that made by Howard [177] who, using a biological assay that measured the effect of various steroids upon bone maturation, found that the sulfate was much less active than was the free steroid. In this assay, dehydroisoandrosterone exhibited 42% of the potency of testosterone even though it is much less active as an androgen. In the few experiments that have been performed to evaluate the influence of dehydroisoandrosterone sulfate on isolated enzyme systems, the results also have been unrewarding. Thus, while low concentrations of unconjugated dehydroisoandrosterone have a highly specific inhibitory effect on adrenal glucose–6–phosphate dehydrogenase [D-glucose-6-phosphate: NaDPoxidoreductase, EC 1.1.1.49] activity [178, 179], its sulfate is inactive. Likewise, Nielson and Warren [180], working with purified glucose–6–phosphate dehydrogenase from bovine corpus luteum, found that dehydroisoandrosterone sulfate, at a concentration of 1 micromolar, had no inhibitory effect on this enzyme. On the other hand, the activity of the dehydrogenase obtained from rat mammary tissue [181] was strongly inhibited by unconjugated dehydroisoandrosterone. Kowal *et al.* [182] reported that the free ketosteroid produced a marked inhibition of the 3β–hydroxysteroid dehydrogenase obtained from either adrenal or corpus luteum tissue while the sulfate was ineffective. In their study of the inhibition of steroid 11–hydroxylase by androgens, Sharma *et al.* [183] found a slight inhibition of this enzyme at relatively high concentrations of both dehydroisoandrosterone and dehydroisoandrosterone sulfate. This inhibition was less than 50% of that produced by an equimolar concentration of testosterone. In experiments with rat liver slices, Oelkers and Dulce [184] found that, although dehydroisoandrosterone sulfate in concentrations of 10^{-4} M inhibited the activity of tryptophan oxidase, it was much less active than was estrone sulfate [3–sulfooxyestra–1,3,5(10)–trien–17–one]. Since cortisol–21–sulfate [11β,17α–dihydroxy–21–sulfooxypregn–4–ene–3,20–dione] had no effect, the inhibitory action of these conjugates did not appear to be due merely to the presence of the anionic group. It should be noted, however, that to produce its inhibitory effect, concentrations of dehydroisoandrosterone sulfate far above the plasma level are required. Its normal plasma concentrations are of the order of 10^{-6} M.

The most elaborate effort to determine whether dehydroisoandrosterone sulfate is of physiological significance was made by Lipsett *et al.* [185]. Dehydroisoandrosterone, in one experiment, and dehydroisoandrosterone sulfate, in another, were infused intravenously into normal males at a rate of 50 mg/day for 5 days. No alterations in the following parameters were found: urinary nitrogen, calcium, phosphorus, sodium, potassium, chloride, creatinine, uric acid, and hydroxyproline; stool nitrogen, calcium, phosphorus, sodium, and potassium; plasma sodium, potassium, chloride, creatinine, uric acid, calcium, phosphorus, hemoglobin, or hematocrit. This lack of metabolic activity, at least in adults, is consistent with the well-recognized fact that patients with Addison's disease may be adequately maintained with only corticoids and a sodium–retaining hormone. In spite of the fact that the daily production of some 10 to 20 mg of dehydroiso-androsterone sulfate is absent in these patients, the males are normally masculinized and both sexes are fertile. In a recent publication, Drucker *et al.* [186] reported that the prolonged administration of large doses of dehydroisoandrosterone sulfate to two sexually immature patients produced definite androgenic and estrogenic effects. After the administration of the conjugate to a male, the mean sebum production was increased more than two–fold; and when given to a female infant, the vaginal epithelium showed definite maturation of the type associated with estrogen therapy.

Peillon and Racadot [187] studied the influence of testosterone, dehydro-isoandrosterone, and dehydroisoandrosterone sulfate on ovarian and on pituitary gonadotrophic function in prepubertal rats. After administration for extended periods, both dehydroisoandrosterone and testosterone inhibited ovulation and produced involutional changes in the cells secreting luteinizing hormone. Under the same conditions, dehydroisoandrosterone sulfate had no effect. Administration of the free C_{19}–steroid caused a decrease in ovarian weight, degeneration of oocytes, and production of cystic follicles, but again the sulfate exhibited no comparable activity. In their studies on 3β–hydroxysteroid dehydrogenase activity, Baillie and Griffiths [69 a] observed that dehydroisoandrosterone sulfate gives a very weak histochemical reaction in the mouse Leydig cell, whereas heavy formazan deposition occurs in the seminiferous epithelium of the mature gonad. These authors have suggested that the seminiferous tubules may be a target organ for circulating dehydroisoandrosterone sulfate. When pregnenolone sulfate, 17α–hydroxypregnenolone sulfate, and dehydroiso-androsterone sulfate were used as substrates, intense 3β–hydroxysteroid dehydrogenase activity was observed in the syntrophoblast of human placental tissue [69 b]. The corresponding unconjugated steroids were less well utilized, an observation that is surprising, since one would expect that the sulfates could serve as substrates for the enzyme only if they were first cleaved to the free compounds. In the zona fasciculata of adrenal tissue, dehydroisoandrosterone sulfate exhibited an extremely weak histo-chemical reaction, but both pregnenolone sulfate and 17α–hydroxy-pregnenolone–3–sulfate reacted strongly. Not unexpectedly, Goldman *et al.* [188] found 3β–hydroxysteroid dehydrogenase activity in the placental trophoblast when dehydroisoandrosterone sulfate was used as substrate.

Attempts to establish an interrelationship between obesity and derange-ments in dehydroisoandrosterone sulfate metabolism have resulted in the accumulation of conflicting data. Thus, while Šonka and Gregorová [189] and Lopez and Krehl [190] have reported below normal values for urinary dehydroisoandrosterone sulfate in obese patients, Hendrikx *et al.* [191] have observed wide variations; some subjects excreted low levels of dehydroisoandrosterone sulfate while others excreted elevated amounts of the conjugate. Similarly, during fasting, one group [190] observed an increase in the dehydroisoandrosterone sulfate levels, while another [191] found a marked decrease in its excretion. Further investigation is evidently necessary to determine whether or not a specific relationship between dehydroisoandrosterone sulfate metabolism and obesity exists.

G. Protein Binding

Both Bongiovanni and Eberlein [192], and Kellie and Smith [84] observed that the renal clearance of dehydroisoandrosterone sulfate is small, and this was interpreted as being due to extensive binding of the conjugate to plasma proteins. The binding of the sulfate to serum albumin was also studied by Puche and Nes [193] who found that 84% of the circulating conjugate is bound to serum albumin under physiological conditions where the sulfate exists in anionic form. Sulfate ion could reverse the binding, and this observation led to the recommendation that sodium sulfate be added to plasma before extraction of the steroid to avoid error due to binding to protein. Further binding studies were reported by Plager [194], who, using equilibrium dialysis and ultrafiltration techniques, found that, at low concentrations of dehydroisoandrosterone sulfate, about 95% was bound to serum proteins, principally albumin. The data suggested the existence of two kinds of binding sites per albumin molecule. A Scatchard plot of the data obtained from equilibrium dialysis revealed 2 binding sites per albumin molecule for the initial binding component and about 18 sites for the second type. At physiologic concentrations of the conjugate, approximately four to five times as much steroid sulfate is associated with the initial binding component as with the second set. The first group of albumin-binding sites for dehydroisoandrosterone sulfate was found to be different from those for androsterone sulfate and etiocholanolone sulfate. The second kind (of binding sites) may be nonspecific for it appears to be similar for all three steroid sulfates.

Using equilibrium dialysis techniques, Wang and Bulbrook [195] have compared the binding by plasma from man, rabbit, and rat of the sulfate esters of dehydroisoandrosterone, testosterone, 17α-acetoxypregnenolone, and pregnenolone. Plasma of all species was found to have a high binding capacity for these compounds and the results suggested that the sulfates of testosterone, dehydroisoandrosterone and pregnenolone are bound to common sites. This finding prompted these investigators to study the displacement of steroidal sulfates by nonsteroidal organic sulfates. They found that displacement did, indeed, occur, thus indicating that the binding sites are not specific for steroids.

H. Catabolic Fate

The experiments of Roberts *et al.* [160] and of Kellie [161] revealed that dehydroisoandrosterone sulfate is not an end product of catabolism and may be further metabolized. Thus, following the intravenous administration of the tritiated sulfate, the former authors found a significant fraction of the label in the urinary glucuronides of androsterone and etiocholanolone. The latter investigators found a significant increase in the weight of these urinary conjugates following the oral administration of a large dose of unlabeled dehydroisoandrosterone sulfate. From subsequent studies of its fate, it has become apparent that this conjugate can be metabolized by two routes: one, in which the conjugate is transformed into another sulfated product without cleavage of the sulfate ester; and the other, in which the conjugating moiety is cleaved and the resulting liberated steroid is further metabolized. Several examples of the latter route have already been mentioned. Evidence for the existence of the first route involving the catabolism of intact dehydroisoandrosterone sulfate was obtained by Roberts *et al.* [3b] while studying the metabolism of cholesterol sulfate labeled with ^{3}H and ^{35}S; two metabolites of dehydroisoandrosterone sulfate, *i.e.*, androstenediol sulfate and 16α–hydroxydehydroisoandrosterone–3–sulfate, were isolated from the urine bearing a $^{3}H/^{35}S$ ratio similar to that of the isolated urinary dehydroisoandrosterone sulfate. In all probability, the reduction of the C17–ketone and hydroxylation at C16 had occurred using the intact steroid conjugate as substrate. In addition, after the intravenous administration of dehydroisoandrosterone sulfate labeled with both ^{3}H and ^{35}S, Baulieu and Corpéchot [196] were able to isolate from the urine the 3–sulfates of androstenediol, 16α–hydroxydehydroisoandrosterone, and androstenetriol [androst–5–ene–3β,16α,17β–triol]. Each conjugate bore a $^{3}H/^{35}S$ ratio similar to that of the administered tracer. From the isotope contents, it was calculated that the pathway involving intact sulfate esters was of considerable quantitative importance. In another experiment in which ^{3}H–dehydroisoandrosterone sulfate and ^{14}C-dehydroisoandrosterone were injected simultaneously, it was found that the sulfated precursor was more readily converted into androstenediol–3–sulfate and 16α–hydroxydehydroisoandrosterone–3–sulfate than was the free compound. It is evident, then, that steroid sulfates may serve as substrates for enzyme systems catalyzing reduction of the 17–ketone and hydroxylation at C16. Reduction of the

17–ketone group of dehydroisoandrosterone sulfate was also observed by Crépy *et al.* [197] who incubated this conjugate with rabbit liver slices. The 3–sulfates of both the 17α– and the 17β–hydroxy isomers were obtained in equal amounts. A comparison of the metabolism of dehydroisoandrosterone by the perfused previable human fetus with that of its sulfate, was undertaken by Bolté *et al.* [199]. In addition to dehydroisoandrosterone sulfate, the 3–sulfates of androstene–3β,17α–diol, androstenetriol, and 16α–hydroxydehydroisoandrosterone were isolated from the fetal liver. From the radioactivity data it was concluded that very little, if any, dehydroisoandrosterone sulfate is cleaved by the fetus. This supports the observation [196] that reduction of the 17–ketone as well as hydroxylation at C16 can occur using the intact sulfate as substrate.

Many efforts have been made to compare, in quantitative terms, the metabolism of a steroid sulfate with that of the unconjugated counterpart. These were, for the most part, *in vitro* experiments and were carried out in the hope that the quantitative results so obtained would be meaningful for *in vivo* processes. For example, Payne and Mason [199] isolated labeled androstenediol–3–sulfate after incubating ^{3}H-dehydroisoandrosterone sulfate with a soluble extract of rat testes; the reduction at C17 appeared to have occurred without cleavage of the sulfate moiety. When the unconjugated compound was incubated, one–fourth as much product (diol) was isolated. As was previously pointed out [200], the value of such *in vitro* experiments is unclear since these observations can be rationalized by various explanations and these need not have any relevance to *in vivo* circumstances. A similar reservation may be made for the experiment of Morato *et al.* [201] who incubated ^{3}H–dehydroisoandrosterone sulfate and ^{14}C–dehydroisoandrosterone with a placental microsome preparation and examined the isotope ratios in the dehydroisoandrosterone, androstenedione, and estrone isolated after one hour. A higher ^{3}H/^{14}C ratio was found in estrone than was found in androstenedione or in the reisolated dehydroisoandrosterone. This finding is consistent with the existence of a pathway of biosynthesis of estrone from the conjugate which is independent of dehydroisoandrosterone and androstenedione; and, therefore, the authors concluded that hydroxylation at C19 had occurred using dehydroisoandrosterone sulfate as a reactant. Unfortunately, the experimental design employed does not permit a unique explanation for these findings. Therefore, this conclusion must be considered tentative.

An extensive quantitative study of various rates which characterize the synthesis and catabolism of dehydroisoandrosterone and its sulfate in a normal man and in a normal woman was carried out by Sandberg *et al.* [129]. By the intravenous administration of three tracers (^{3}H–dehydroiso-

TABLE 5–9. *Metabolic paramters characterizing dehydroisoandrosterone sulfate in a male and in a female subject*

	Male	Female
Volume of distribution (liters)	5	4.7[a]
Metabolic clearance rate (liters/day)		7.7
Production rate in plasma (mg/day)		10
"Urinary" production rate (mg/day)		6.9[b]
Plasma concentration (μg/100 ml)	268	132
Half–life (hours)		6.7
Pool size (mg)		6.2
Transfer factors (ρ)		
Dehydroisoandrosterone sulfate to dehydroisoandrosterone		{0.63[c] {0.62[d]
Dehydroisoandrosterone to dehydroisoandrosterone sulfate		0.48[d]

[a] These are approximations and may be overestimated.
[b] Calculated from specific activity of dehydroisoandrosterone sulfate.
[c] Estimated from plasma data.
[d] Estimated from urinary data.

androsterone sulfate, dehydroisoandrosterone–^{35}S–sulfate, ^{14}C–dehydroisoandrosterone) and the analysis of blood and urine specimens, these investigators were able to estimate the various parameters listed in Table 5–9. Some figures were evaluated from analysis of blood and others were estimated from urine specimens. Where comparisons were possible, it was found that the values obtained from plasma were comparable to those obtained from urine. In the female subject, the data were consistent with a two–pool model: (Pool I) dehydroisoandrosterone ⇄ dehydroisoandrosterone sulfate (Pool II). However, this conclusion may not be correct since determinations of radioactivity were made only on plasma samples collected 2 hours and longer after injection of the tracers. The data secured from these samples were found to lie on the "tail" of the disappearance curve and, therefore, the existence of other metabolic pools could not be determined. On the other hand, in the experiment with the male subject, several plasma samples were drawn between 11 and 120 minutes after the injection, and the analysis of data derived from these pointed to the existence of a third pool: *i.e.*, one exchanging with dehydroisoandrosterone in plasma. The third pool could be either another anatomical space of distribution of the sulfate or another interconverting steroid sulfate, such as androstenediol–3–sulfate. Other important findings of this study are: the small volume of distribution of dehydroisoandrosterone sulfate (*ca.* 5 liters compared with 15 liters for cortisol and 40 liters for aldosterone [11β,21-dihydroxy-3,20-dioxopregn-4-en-18-al]); the low metabolic clearance rate (7 liters/day, compared with 1,600 liters/day for aldosterone and 3,000 liters/day for progesterone); and the confirmation of previous results [30] which revealed that dehydroisoandrosterone sulfate was secreted by the adrenal. Approximately 50% of the secreted dehydroisoandrosterone was found to be converted into dehydroisoandrosterone sulfate; the reverse reaction was of the same order of magnitude.

From the disappearance curve of the ^{35}S-labeled dehydroisoandrosterone sulfate from plasma, a half-life value of 6.7 hours was calculated. Other investigators [39] have also measured the half-life and observed that the value obtained depends upon which tracer is used. Thus, when ^{3}H-dehydroisoandrosterone sulfate was used, a value of 9 hours was calculated, whereas when ^{35}S-dehydroisoandrosterone sulfate was used, a value of 6 hours was estimated. This difference is due to cleavage of the conjugate and resulfation. Once the labeled sulfate group is split off, the resulfation process involves SO_4^{2-} from the large body pool in which the labeled anion has been diluted.

As mentioned earlier, Slaunwhite *et al.* [171] studied the disposition and metabolic fate of ^{14}C-dehydroisoandrosterone and its tritium-labeled sulfate in human subjects. After simultaneous injection of both compounds, the rate of disappearance of radioactivity from the plasma and the rate of its appearance in the bile and urine were followed. It was found that, of the radioactive label originally associated with free dehydroisoandrosterone, about 2% appeared in the bile in the first 12 hours, and that, after 48 hours, there was no detectable ^{14}C in the bile. In contrast, some 9–16% of the tritium label was found in the bile 12 hours after injection with some radioactivity persisting up to 72 hours postinjection. Even so, this biliary excretion is relatively small when compared with that of estrone and 17β-estradiol [202] where approximately 50% of the intravenously administered estrogen appeared in the bile during the first 12 hours.

Following the intravenous administration of ^{3}H-dehydroisoandrosterone sulfate to normal males, only 15% of the administered tracer was found in the urinary sulfate. The same percentage of the radioactivity appeared in the glucuronide fraction [160]. A more detailed investigation of the percentage conversion of exogenously administered ^{14}C-dehydroisoandrosterone and ^{3}H-dehydroisoandrosterone sulfate to the individual urinary conjugates was made by Baulieu *et al.* [39]. In this study, less than 10% of the tritium was reisolated from the urine as dehydroisoandrosterone sulfate. Other tritiated metabolites of quantitative importance were the expected androsterone glucuronide, etiocholanolone glucuronide, and androstenediol-

3–sulfate. Less than 25 % of the administered tritium label could be accounted for in the 4–day urine collection. In another series of metabolic studies, this same group found that 0.9 to 15 % (mean 10 %) of radioactive dehydroisoandrosterone sulfate, when administered to men, appeared in urine as labeled dehydroisoandrosterone sulfate, whereas when administered to women, the corresponding value was 0.3 to 9 % (mean 4 %) [39].

In contrast to the low recovery of isotope in the urine following the intravenous administration of labeled dehydroisoandrosterone sulfate, MacDonald *et al.* [203] found that 93 % of the isotope was recovered in the various urinary conjugates after the administration of the unconjugated steroid.

In a study designed to determine the influence of the level of secretion of dehydroisoandrosterone upon its metabolism, MacDonald *et al.* [204] found that the fraction of the free steroid converted into urinary dehydroisoandrosterone sulfate was dependent upon the secretory rate of dehydroisoandrosterone; the greater the secretory rate, the greater the fraction of dehydroisoandrosterone excreted as dehydroisoandrosterone sulfate. Thus, the results showed that the secretory rate of a hormone may be a determinant of the excretion pattern of its metabolite.

The first report of a microbiological transformation of a steroid sulfate was presented by YoungLai and Solomon [205]. Following the incubation of dehydroisoandrosterone sulfate with a strain of *Streptomyces roseochromogenus* (ATCC 3347), these investigators were able to isolate 16α–hydroxydehydroisoandrosterone 3 sulfate in 44 % yield. This result is of importance since it makes possible the facile preparation of the labeled counterpart of this 16α–hydroxylated steroid sulfate which is an important precursor of the estriol produced during pregnancy.

XIII. Androst–5–enediol–3–sulfate

Although traces of androstenediol are known to be secreted by the adrenal [206], the sulfated conjugate has not been shown to be a secretory product. The presence of the 3–sulfate in human peripheral blood has been indicated by the findings of Sjövall and Vihko [207]. After extraction of unconjugated materials, a compound, behaving as a monosulfate on columns of methylated Sephadex, was isolated. This substance, after solvolysis and trimethylsilyl ether formation, was analyzed by gas chromatography and mass spectrometry. The retention time and the mass spectrum of the isolated compound were identical with those of the authentic bis–trimethylsilyl ether of androstenediol.

An extensive quantitative examination of 116 samples of human plasma, for the presence of the sulfates of androstenediol, pregn-5-ene-3β,20α-diol, androsterone, dehydroisoandrosterone and isoandrosterone [3β-hydroxy-5α-androstan-17-one] has been carried out by Vihko [58]. Of the sulfates examined, those of androstenediol and pregnenediol were detected, for the first time, in human plasma. The mean values for the 3β-hydroxysteroid sulfates, for males and females, are recorded in Table 5-3. The concentrations of all five steroid sulfates are significantly higher in males than in females. In addition, pregnenediol, androsterone, dehydroisoandrosterone, and isoandrosterone [3β-sulfooxy-5α-androst-17-one] sulfates were found to decrease with age while no significant change in the androstenediol sulfate levels was apparent.

The *in vivo* metabolism of androstenediol sulfate, labeled with both ^3H and ^{35}S, has been studied [208]. The intravenously administered conjugate was converted into urinary dehydroisoandrosterone sulfate without loss of the labeled sulfate radical. It was also demonstrated that the conversion into urinary dehydroisoandrosterone sulfate was greater when the diol-3-sulfate was injected than when the free diol was administered. From the $^3H/^{35}S$ ratios of urinary dehydroisoandrosterone sulfate and androstenediol sulfate, it was possible to estimate that as much as 90% of the oxidation at C17 occurred with the sulfated compound [39]. In this instance, the substrate, androstenediol-3-sulfate, was dehydrogenated by the enzyme, 17β-hydroxysteroid dehydrogenase [17β-hydroxysteroid: NAD(P)oxido-reductase, EC 1.1.1.51], to yield the 17-ketosteroid sulfate. In an earlier section, it has already been pointed out that androstenediol-3-sulfate may arise metabolically from dehydroisoandrosterone sulfate [3b, 39].

Following the injection of doubly-labeled dehydroisoandrosterone sulfate into a human subject, Baulieu *et al.* [39] isolated a urinary metabolite they believed to be, on the basis of its chromatographic mobility, the 3β,17β-di-sulfate. This conjugate, obtained in 0.2% yield, contained a $^3H/^{35}S$ ratio similar to that of the injected precursor, thus indicating that a steroid sulfate (androstenediol-3-sulfate) served *in vivo* as a substrate for the sulfokinase enzyme. Evidence for the presence of the 3β,17β-disulfate in human plasma has been adduced by Sjövall and Vihko [26].

XIV. 7α-Hydroxydehydroisoandrosterone-3-sulfate [7α-Hydroxy-3β-sulfooxyandrost-5-en-17-one]

Although the sulfate of the 7α-hydroxylated derivative of dehydroiso-androsterone has not yet been isolated from human urine, the unconjugated metabolite has been recovered after acid hydrolysis [209-211]. High concentrations of this metabolite (5 to 50 mg/24 hours) have also been isolated from acid-hydrolyzed urine from stallions and mares [212]. In normal human male urine, values of 0.015 to 0.030 mg/24 hours have been detected. These levels are increased after ACTH administration. In patients with adrenal disorders, levels up to 1.3 mg/24 hours have been observed. It is probable that these estimates are minimal since the urine specimens used for these determinations were boiled with acid and this procedure could result in elimination of the allylic 7α-hydroxyl (axial) group. Evidence for the presence in urine of the disulfate of this dihydroxyketone has also been reported [211].

The 3-sulfate of 7α-hydroxydehydroisoandrosterone has been detected in human plasma by Stárka and Hampl [213], who estimated values of 1.5 to 9 μg/100 ml of mixed plasma. Identity of the conjugate was established by its chromatographic mobility, by the characteristic blue color produced by 7-hydroxylated compounds when treated with antimony trichloride, and by the formation of appropriate derivatives; *i.e.*, following oxidation, reduction, and dehydration of the isolated conjugate, the three products had, respectively, identical chromatographic mobilities and chromogenicity as had authentic samples of the 7,17-diketone, the triol, and the $\Delta^{5,7}$ compounds.

The biosynthetic pathways leading to the formation of the ring B unsaturated estrogens, equilin [3-hydroxyestra-1,3,5(10),7-tetraen-17-one] and equilenin [3-hydroxyestra-1,3,5(10),6,8-pentaen-17-one], have been in doubt for some time. Stárka and his coworkers [214] seem to have solved this problem for they were able to convert 7α-hydroxydehydroisoandrosterone-3-sulfate into these two estrogens by incubation with cultures of human placental tissue. They also effected this conversion using the unconjugated compound as substrate; however, the yield of estrogens obtained was higher when the sulfate was the precursor. The aromatization, *in vitro*, of this conjugate to the 7-oxygenated estrogens, by human placenta and ovary, has also been described [215].

XV. 7-Ketodehydroisoandrosterone-3-sulfate [3β-Sulfooxyandrost-5-ene-7,17-dione]

The first suggestion that 7-ketodehydroisoandrosterone-3-sulfate was excreted in human urine came from the work of Lieberman *et al.* [216], and of Fukushima *et al.* [217] who isolated and characterized the free steroid after acid hydrolysis of urine. Its origin is suggested by the fact that the compound is excreted in excessive amounts in the urine of patients with adrenal pathological conditions [218, 219]. Recently, Baulieu *et al.* [117b,

220], and Guillon *et al.* [221] have provided further evidence for its site of formation when they isolated this conjugate from adrenal tumor tissue. In addition, blood samples from two patients bearing adrenal tumors were examined and larger quantities of the conjugate were found in adrenal venous plasma than were present in peripheral plasma (75 *vs.* 45 μg/100 ml [117b] and 100 *vs.* 60 μg/100 ml [220]). Quantification, however, was achieved only by visual estimation and, therefore, was only an approximation. The unconjugated steroid was not detected in any of the samples. The observation that the 7–keto derivative is probably an adrenal secretory product and not a peripheral metabolite of dehydroisoandrosterone is in accord with the findings of Gallagher [219]. Schneider and Lewbart [209] found that the oxygen at C7 was not introduced artifactitiously by auto-oxidation during isolation.

XVI. 16α–Hydroxydehydroisoandrosterone–3–sulfate

Several reports have appeared which indicate the presence of this conjugate in biological fluids. Fotherby *et al.* [222] were the first to isolate a relative of 16α–hydroxydehydroisoandrosterone from urine that was boiled at neutral pH for one hour. The compound isolated was identified as the *cyclo*steroid, 6β,16α–dihydroxy–3,5–*cyclo*androstan–17–one, the formation of which, from the corresponding 3β–hydroxy–Δ^5–steroid conjugate, was anticipated from earlier experience with dehydroisoandrosterone sulfate [223]. The 16–hydroxylated steroid was also found by Bongiovanni [55c] in the urine of infants with the adrenogenital syndrome associated with a deficiency of the 3β–hydroxysteroid dehydrogenase. The urine was subjected to a procedure which removed the free and glucuronic acid–conjugated steroids prior to solvolysis. Thus, the presumption is that 16α–hydroxydehydroisoandrosterone originally existed in the urine as a sulfate. A range of 1.7 to 3.5 mg/day, of this metabolite, was found in urine from these patients while a pool of urine obtained from normal infants contained only 0.2 mg/day. Reynolds [64] confirmed this finding by examining pooled urine from newborn infants. After correction for losses, he found a urinary excretion of 0.29 mg/day which fell to undetectable amounts after the age of six months.

In their search for 16α–hydroxylated steroids which could serve as precursors of estriol [estra–1,3,5(10)–triene–3,16α,17β–triol], Magendantz and Ryan [224a] obtained good evidence for the presence of 16α–hydroxydehydroisoandrosterone in the pooled sera of newborn infants. In addition, these authors [224b] also analyzed sixteen 50–ml samples of umbilical venous sera, which had been subjected to a solvolytic procedure, and found values between 44 and 139 μg (mean 80 ± 26 μg) per 50 ml of serum. Determinations were also carried out on arterial blood samples but the values observed varied so widely that it was not possible to establish the fetus as the site of formation of this compound. Colás *et al.* [106] also estimated the blood levels of 16α-hydroxydehydroisoandrosterone. In some, but not all, of the plasma samples obtained from the umbilical artery, the values measured were significantly higher than those obtained from the umbilical vein, thus suggesting a fetal contribution of this 16-hydroxylated conjugate to the placenta. Following solvolysis, identity was established by mixed melting point and by infrared spectroscopy.

The hydroxylation of dehydroisoandrosterone sulfate at C16 has been demonstrated *in vivo*. As was mentioned earlier, Roberts *et al.* [3b] administered doubly-labeled cholesterol sulfate to a human subject and isolated from her urine 16α-hydroxydehydroisoandrosterone sulfate bearing the same $^3H/^{35}S$ ratio as the injected precursor. The similarity in isotope ratio implies that at least some fraction of the 16α-hydroxylated conjugate arose from sulfated precursors. This was the first report of the isolation and characterization of the intact conjugate from a biological source.

Most of the estriol excreted by normal nonpregnant females is derived from the peripheral metabolism of estrone and 17β-estradiol [225, 226]. In order to explain the synthesis of the grossly disproportionate amounts of estriol produced during pregnancy, other biosynthetic pathways have been considered. For example, estriol may be formed from 16α hydroxydehydroisoandrosterone since, as Magendantz and Ryan [224] first showed, using *in vitro* techniques, the placenta has the capacity to aromatize this 16-hydroxylated-C_{19}-steroid as well as other neutral compounds. Further indirect evidence that estriol found in pregnancy urine is produced through an alternate pathway which involves neither estrone nor 17β-estradiol was presented by Siiteri and MacDonald [150a]. Following the intravenous administration of ^3H-dehydroisoandrosterone sulfate to a patient pregnant with a hydatidiform mole, these authors found that the specific activity of the urinary estriol was higher than either that of estrone or 17β-estradiol. This finding is consistent with the existence of a route for the biosynthesis of estriol which uses neutral 16α-hydroxylated-C_{19}-precursors without the intermediacy of estrone and 17β-estradiol. Additional evidence for this alternate route to estriol was obtained by injecting a mixture of ^{14}C-dehydroisoandrosterone sulfate and ^3H-16α-hydroxydehydroisoandrosterone into pregnant subjects [150b]. Analysis of the isotope contents of the urinary 16α-hydroxylated neutral metabolites and estriol suggested the existence of the pathway: dehydroisoandrosterone sulfate → 16α-hydroxydehydroisoandrosterone (or its sulfate) → estriol. It was found that 15% of the administered dehydroisoandrosterone sulfate and 27% of the 16α-hydroxydehydroisoandrosterone had been converted into urinary estriol by a patient in late pregnancy. Whether androstenetriol was an obligatory intermediate between 16α-hydroxydehydroisoandrosterone and estriol was not ascertained. Siiteri and MacDonald [150b] also showed that dehydroisoandrosterone sulfate, produced by the mother during pregnancy, is a relatively minor precursor of estriol. During the early stages of pregnancy, ovarian 17β-estradiol probably accounts for a large fraction of the estriol formed but, as pregnancy progresses, this decreases and that fraction formed in the placenta from neutral precursors supplied by the fetus increases to 90% or more.

Although not yet confirmed by experiment, considerable evidence has accumulated which suggests that the neutral steroid which originates in the fetus and which serves as the principal precursor of estriol in late pregnancy, is 16α-hydroxydehydroisoandrosterone. *In vitro* experiments with placenta indicate that this tissue effects 16-hydroxylation poorly [227, 228]. The concentration of 16α-hydroxydehydroisoandrosterone sulfate in umbilical cord blood is relatively high [106, 224b] as is the concentration of sulfatase in the placenta [146, 148]. As was mentioned earlier, aromatization of 16α-hydroxylated neutral precursors can be effected by placental preparations *in vitro* [149, 229]. The levels of estriol excreted in the urine of mothers bearing an anencephalic fetus, a hydatidiform mole, or a dead fetus are low. That a pathway for estriol biosynthesis independent of maternally circulating estrone or 17β-estradiol exists has been shown [150b]. This finding, however, did not exclude the possibility that the triol may arise through 16α-hydroxylation occurring in the fetus, using estrone or 17β-estradiol which was formed in the placenta and was conveyed to the fetus without exchanging with the estrogens of the maternal circulation [230]. However, by administering ^{3}H-17β-estradiol directly into a fetus during transfusion, *in utero*, for erythroblastosis fetalis, at the same time that ^{14}C-17β-estradiol was being injected into the maternal peripheral circulation, Gurpide *et al.* [231] were able to estimate that only about 4% of the urinary estriol was derived from 17β-estradiol presented to the fetus and only about 10% arose from maternally circulating 17β-estradiol. It is obvious, then, that the major portion of the urinary estriol must be derived from neutral 16-hydroxylated precursors of fetal and/or maternal origin. While the fetus plays an important role in providing C_{19}-neutral precursors of estriol during normal pregnancy, the presence of a fetus is not an absolute requirement since the conversion of 16α-hydroxydehydroisoandrosterone to the triol has been shown to occur in patients pregnant with a hydatidiform mole [150b].

XVII. 16–Ketoandrost–5–enediol–3–sulfate [3β,17β–Dihydroxyandrost–5–en–16–one sulfate]

This conjugate has not been isolated as such, but the free compound, 16–ketoandrostenediol, has been recovered from the urine of newborn infants [232]. After the free and glucuronidase–liberated steroids were removed from the urine, a solvolytic procedure was used to liberate the metabolite from its presumed sulfate. The dihydroxyketone is probably a metabolite of an adrenal secretory product since it could not be detected in urine after dexamethasone treatment. Normal excretion values were said to be of the order of 0.1 to 0.3 mg/day. The possibility that the 16–keto compound is a tautomeric artifact arising from the 16–hydroxy–17–ketone was eliminated by control experiments.

Evidence for the presence of the sulfate of this α–ketol in umbilical cord blood has been supplied by Colás and Heinrichs [109]. Although a definitive identification was not achieved, a substance having chromatographic properties and a sulfuric acid spectrum similar to those of authentic material was detected.

XVIII. Androst–5–enetriol Sulfate [Androst–5–ene–3β,16α,17β–triol sulfate]

Although isolation of the intact sulfate of this triol has not been described, its presence in human urine may be surmized from the kind of hydrolysis required to liberate it from the conjugated form in which it is excreted. Hirschmann [233] and Marian and Butler [234] isolated this triol only after acid hydrolysis of urine. The latter authors estimated that normal individuals excrete 0.1 mg/liter of urine. Similarly, Reynolds [235] isolated the triol after solvolysis of the urine of premature infants. Since no decrease in the excretion of the triol occurred with increasing age of these infants, he concluded that the fetal zone of the adrenal cortex was not the sole source of this compound. Weinman *et al.* [236] investigated the quantities and modes of conjugation of androstenediol and androstenetriol occurring in human urine and reported that the sulfates of these two steroids represent approximately one–third of the urinary metabolites of dehydroisoandrosterone. The presence of a sulfate group in the conjugates was assumed from the method of hydrolysis employed. This consisted of boiling the urine for four hours in 0.1 molar acetic acid, pH 4.0, a procedure which, according to Bitman and Cohen [89], is specific for the hydrolysis of 3β–sulfated–Δ^5–compounds.

The sulfurylation of the triol by fetal tissues has recently been described by Dell'Acqua *et al.* [237]. Following the intravenous administration of the labeled triol into an intact "fetoplacental unit", the conjugated triol was isolated from various fetal tissues.

In the section on dehydroisoandrosterone sulfate, it was pointed out that androstenetriol sulfate may arise *in vivo* from the 17–ketosteroid conjugate. Corpéchot and Baulieu [238], who carried out these experiments, concluded that most, if not all, of the 16–oxygenated sulfates isolated were formed from dehydroisoandrosterone sulfate *via* a pathway involving intact sulfates. Also mentioned previously was the observation that the biosynthesis of estriol during pregnancy may proceed *via* neutral 16α–hydroxylated–C_{19}–precursors. The intermediacy of androstenetriol, or of its sulfate, was suggested from the findings of an experiment in which ^{14}C–dehydroisoandrosterone and ^{3}H–dehydroisoandrosterone sulfate were injected intravenously into a patient with a hydatidiform mole [150a]. Androstenetriol, isolated from the urinary fraction, had a specific activity, with respect to tritium, similar to that of the urinary dehydroisoandrosterone sulfate and a $^{3}H/^{14}C$ ratio equal to that of the urinary estrogens. This is the result that would be expected if estriol were directly derived from androstenetriol or if both androstenetriol and estriol were derived from a common precursor such as 16α–hydroxydehydroisoandrosterone.

XIX. Concluding Remarks

From the data summarized in this chapter, there can be little doubt that the sulfates of the 3*β*–hydroxy-Δ^5–steroids play a physiological role beyond that concerned with detoxification. The sulfates are not merely waste products to be excreted immediately after formation. From the evidence presented, it is clear they may serve as substrates for at least a dozen enzymes which are involved in the metabolism of the steroid hormones. The enzymatic reactions now considered to use sulfates as reactants are those involved in: hydroxylation at C20 of cholesterol, hydroxylation at C22 of cholesterol, cleavage of the 20,22–glycol to pregnenolone, hydroxylation at C17 of the latter compound, cleavage of the C_{21}–steroid oxygenated at both C17 and C20 to the C_{19}–ketone, reduction of the ketone at C20 to the alcohol, reduction of the ketone at C17 to the alcohol, oxidation of the hydroxyl group at C17 to the ketone, hydroxylation at C16, hydroxylation at C19, hydroxylation at C15 [239], and sulfurylation of the hydroxyl group at C20. About half of these reactions are concerned with what are generally considered to be synthetic reactions; *i.e.*, those reactions necessary for the synthesis of glandular secretory products. In spite of this impressive list, one can only speculate at this time as to the importance of steroid sulfates as biosynthetic intermediates. The most obvious difference between these conjugates and those steroids presently accepted as intermediates for hormone synthesis is the strongly polar sulfate group which, under physiological conditions, exists in its ionized form. In this respect, these conjugates are typical of most reactants involved in biochemical processes in that they, too, are ionic species. Whether reactions of the sulfated intermediates are catalyzed by enzymes that are different from those using the neutral steroids as substrates is undecided. Although it might be expected that an enzyme would recognize the presence (or absence) of the sulfate ion, the meager evidence existing now suggests that at least in one instance the same enzyme can accept either the ionized or the neutral substrate [24]. Much additional work is required before this question can be considered settled.

The processes that are concerned with the transport of molecules across membranes is another area where the characteristic of charge might be expected to be important. Unlike phosphorylated intermediates, steroid sulfates are secreted out of various kinds of cells (hepatic, adrenal, and probably other endocrine glands, kidney etc.) and may, therefore, be found in blood and urine. They also are transported into some cells, *i.e.*, those of the placenta and intestines; but the work of Gurpide *et al.* [8] has clearly shown that cholesterol sulfate, at least, enters into adrenal cells poorly. Further work is necessary before the influence of sulfate conjugation upon the permeability of steroids can be understood adequately.

Even the question of where steroid sulfates fit in the scheme of biosynthesis of hormones is unanswered at the present time. Fig. 5–3 outlines an abbreviated scheme which illustrates how sulfated intermediates might be integrated into previously considered routes of steroid hormone biosynthesis. It is evident from this scheme that, now, three C_{21}–steroids, pregnenolone sulfate, pregnenolone, and progesterone, may be considered to reside in a central position, since from them the remaining hormones can be derived. From these intermediates, the diverging stream of biosynthesis can proceed either in the direction of the C_{21}–adrenocortical hormones, such as cortisol and aldosterone, or toward other products, such as testosterone, 17β–estradiol, dehydroisoandrosterone and its sulfate. The regulatory processes determining which end–products will be formed, and in what quantities, are not yet understood. Obviously, the mechanisms controlling the nature and the amount of the secreted hormones are intimately involved with the enzyme systems responsible for the various conversions along the chain of biosynthesis. Among those systems previously considered as possible control sites are the 3β– and the 17β–hydroxysteroid dehydrogenases, the enzymes responsible for the cleavage of the ketol side chain, –CHOH–CO–CH_3, to a 17-ketosteroid, the several hydroxylases acting at C11, C17, and C21, and the enzymes concerned with the aromatization of the C_{19}-steroids to the C_{18}-estrogens. (In pathological situations, derangements of one or more of these enzyme systems have been invoked to rationalize various clinical syndromes and the formation of abnormal intermediates.) With the demonstration that sulfated intermediates may be involved in steroidogenic processes, the enzymes concerned with the formation of the sulfates, or with their cleavage, must now also be considered as loci where control may be exerted and where, therefore, derangements may occur.

Steroid sulfates may serve as a reservoir for nonconjugated intermediates, in which case the process of sulfation and cleavage could play an important role in the regulation of the rates of synthesis of the steroid hormones. The results of Siiteri and MacDonald [150a], Bolté *et al.* [154], Baulieu and Dray [152], and Warren and Timberlake [148] have revealed a most noteworthy role for dehydroisoandrosterone sulfate; *i.e.*, its conversion into estrogen in pregnant women. It thus becomes possible that the metabolic function of the sulfate ester in the over-all process of estrogen biosynthesis may serve as a biochemical protective group which is ultimately removed at the placenta in order to allow aromatization to estrogens. Thus, sulfation of a steroid can serve to favor its conversion to an active hormone rather than facilitate its excretion into urine.

It is also possible that the pathways involving sulfated intermediates are independent of those employing α,β-unsaturated ketones, perhaps by virtue of localization in the cell or of zonation in the glands, and that each of these routes possesses different physiological roles. It is known, for instance, that the rate of production of the androgens by the adrenals varies independently from that of cortisol. The assumption of a pathway for the production of the androgens separate from that leading to the corticosteroids would offer an explanation for this phenomenon.

We have attempted to summarize in this chapter our present knowledge of the biochemistry of the sulfates of 3β-hydroxy-Δ^5-steroids. It is clear that, in spite of the bulk of this knowledge, the nature of the metabolic function of this class of compounds is still largely unknown. It is noteworthy that information about this outstanding feature is also lacking for many other sulfated conjugates known to exist in nature. Sulfated derivatives of thyroxine, fibrinogen, serotonin, Vitamin D, catecholamines, and ascorbic acid, as well as of various mucopolysaccharides have been described. For none of these products has a role been attributed to the sulfate moiety. Perhaps, by accumulating our knowledge of the biochemistry of the steroid sulfates in this volume, further research on the significance and function of these conjugates will be stimulated.

XX. References

1. Calvin, H. I., Lieberman, S.: Biochemistry **3**, 259 (1964).
2. LeBeau, M.C., Alberga, A., Baulieu, E.-E.: Biochem. Biophys. Res. Commun. **17**, 570 (1964).
3. a) Roberts, K. D., Bandi, L., Calvin, H. I., Drucker, W. D., Lieberman, S.: J. Am. Chem. Soc. **86**, 958 (1964); — b) Biochemistry **3**, 1983 (1964).
4. Drayer, N. M., Roberts, K. D., Bandi, L., Lieberman, S.: J. Biol. Chem. **239**, PC 3112 (1964).
5. a) – Lieberman, S.: Biochem. Biophys. Res. Commun. **18**, 126 (1965); – b) J. Clin. Endocrinol. Metab. **27**, 136 (1967).
6. Moser, H.W., Moser, A. B., Orr, J. C.: Biochim. Biophys. Acta **116**, 146 (1966).
7. a) Eneroth, P., Nystrom, E.: Steroids **11**, 187 (1968); b) Steroids **11**, 417 (1968).
8. Gurpide, E., Roberts, K. D., Welch, M.T., Bandy, L., Lieberman, S.: Biochemistry **5**, 3352 (1966).
9. Moser, H.W., Moser, A. B., McKhann, G. M.: Arch. Neurol. **17**, 494 (1967).
10. Roberts, K. D., Lieberman, S.: Unpublished data.
11. Eberlein, W. R.: J. Clin. Endocrinol. Metab. **25**, 1101 (1965).
12. Schneider, J. J., Lewbart, M. L.: J. Biol. Chem. **222**, 787 (1956).
13. Nose, Y., Lipmann, F.: J. Biol. Chem. **233**, 1348 (1958).
14. Keller, H., Blennemann, H.: Z. Physiol. Chem. **324**, 125 (1961).
15. a) Banerjee, R. K., Roy, A. B.: Mol. Pharmacol. **2**, 56 (1966); — b) Biochim. Biophys. Acta **137**, 211 (1967).
16. Roy, A. B., Banerjee, R. K.: Proc. Intern. Congr. Hormonal Steroids, 2nd, Milan, Italy, May 1966, p. 397 (Pub. 1967).
17. Adams, J. B., Wong, M. S. F.: Steroids **11**, 313 (1968).
18. Young, D. G., Hall, P. F.: Endocrinology **82**, 291 (1968).
19. Calvin, H. I., Vande Wiele, R. L., Lieberman, S.: Biochemistry **2**, 648 (1963).
20. — Roberts, K.D., Weiss, C., Bandi, L., Cos, J.J., Lieberman, S.: Anal. Biochem. **15**, 426 (1966).
21. LeBeau, M.C., Baulieu, E.-E.: Compt. Rend. **263**, 158 (1966).
22. Solomon, S.: J. Clin. Endocrinol. Metab. **26**, 762 (1966).
23. Roberts, K. D., Bandi, L., Lieberman, S.: Biochem. Biophys. Res. Commun. **29**, 741 (1967).
24. a) Raggatt, P. R., Dean, P. D. G., Whitehouse, M.W.: Biochem. J. **96**, 26P. (1965); b) — Whitehouse, M.W.: Biochem. J. **101**, 819 (1966).
25. Bègue, J. A.: Compt. Rend. **260**, 3777 (1965).
26. Sjövall, J., Vihko, R.: Proc. Intern. Congr. Hormonal Steroids, 2nd, Milan, Italy, May 1966, p. 210 (Pub. 1967).
27. Conrad, S. H., Pion, R. J., Kitchin, J. D., III: J. Clin. Endocrinol. Metab. **27**, 114 (1967).
28. Fotherby, K.: Vitamins Hormones **22**, 153 (1964).
29. Arcos, M., Gurpide, E., Vande Wiele, R. L., Lieberman, S.: J. Clin. Endocrinol. Metab. **24**, 237 (1964).

30. Vande Wiele, R. L., MacDonald, P. C., Gurpide, E., Lieberman, S.: Recent Progr. Hormone Res. **19**, 275 (1963).
31. Shimizu, K.: J. Biochem. **59**, 430 (1966).
32. Gual, C., Rojo, B., Lemus, A. E., Rivera, R.: Proc. 3rd Ann. Meeting, Soc. Mex. Nutr. Endocrinol., Mexico City, 1962, p. 126.
33. LeBeau, M. C., Baulieu, E.-E.: Endocrinology **73**, 832 (1963).
34. Adams, J. B.: Biochim. Biophys. Acta **82**, 572 (1964).
35. Boström, H., Franksson, C., Wengle, B.: Acta Endocrinol. **47**, 633 (1964).
36. Klein, G. P., Giroud, C. J. P.: Steroids **9**, 113 (1967).
37. Pierrepoint, C. G., Griffiths, K., Grant, J. K., Stewart, J. S. S.: J. Endocrinol. **35**, 409 (1966).
38. Calvin, H. I., Lieberman, S.: J. Clin. Endocrinol. Metab. **26**, 402 (1966).
39. Baulieu, E.-E., Corpéchot, C., Dray, F., Emiliozzi, R., LeBeau, M. C., Mauvais-Jarvis, P., Robel, P.: Recent Progr. Hormone Res. **21**, 411 (1965).
40. Usui, T., Lieberman, S.: Unpublished data.
41. Voigt, K. D., Tamm, J., Volkswein, U., Schedewie, H.: Acta Endocrinol. **49**, 427 (1965).
42. Stárka, L., Šulcová, J., Dahm, K., Döllefeld, E., Breuer, H.: Biochim. Biophys. Acta **115**, 228 (1966).
43. Killinger, D. W., Solomon, S.: J. Clin. Endocrinol. Metab. **25**, 290 (1965).
44. Griffiths, K., Cunningham, D., Cameron, E. H. D.: J. Endocrinol. **40**, 49 (1968).
45. Perez-Palacios, G., Perez, A. E., Jaffe, R. B.: J. Clin. Endocrinol. Metab. **28**, 19 (1968).
46. Wang, D. Y., Bulbrook, R. D., Ellis, F., Coombs, M. M.: J. Endocrinol. **39**, 395 (1967).
47. Pion, R. J., Conrad, S. H., Wolf, B. J.: J. Clin. Endocrinol. Metab. **26**, 225 (1966).
48. Palmer, R., Eriksson, G., Wiqvist, N., Diczfalusy, E.: Acta Endocrinol. **52**, 598 (1966).
49. a) Solomon, S., Lanman, J. T., Lind, J., Lieberman, S.: J. Biol. Chem. **233**, 1084 (1958); — b) Hillman, D. A., Stachenko, J., Giroud, C. J. P., in: The human adrenal cortex (A. R. Currie, T. Symington, and J. K. Grant, eds.), p. 596. Baltimore, Md.: Williams & Wilkins 1962.
50. Bird, C. E., Wiqvist, N., Diczfalusy, E., Solomon, S.: J. Clin. Endocrinol. Metab. **26**, 1144 (1966).
51. Bloch, E., Benirschke, K., in: The human adrenal cortex (A. R. Currie, T. Symington, and J. K. Grant, eds.), p. 589. Baltimore, Md.: Williams & Wilkins 1962.
52. Mancuso, S., Dell'Acqua, S., Eriksson, G., Wiqvist, N., Diczfalusy, E.: Steroids **5**, 183 (1965).
53. a) Hirschmann, H., Hirschmann, F. B.: J. Biol. Chem. **157**, 601 (1945); — b) Schiller, S., Miller, A. M., Dorfman, R. I., Severinghaus, E. L., McCullagh, E. P.: Endocrinology **37**, 322 (1945).
54. Wilson, H., Lipsett, M. B., Ryan, D. W.: J. Clin. Endocrinol. Metab, **21**, 1304 (1961).
55. a) Bongiovanni, A. M.: J. Clin. Endocrinol. Metab. **26**, 1240 (1966); — b) J. Clin. Endocrinol. Metab. **21**, 860 (1961); — c) J. Clin. Invest. **41**, 2086 (1962).

56. Jacobsohn, G. M.: J. Clin. Endocrinol. Metab. **22**, 859 (1962).
57. Arcos, M., Lieberman, S.: Biochemistry **6**, 2032 (1967).
58. Vihko, R.: Acta Endocrinol. **52**, Suppl. 109 (1966).
59. Wengle, B., Boström, H.: Acta Chem. Scand. **17**, 1203 (1963).
60. Pasqualini, J. R., Jayle, M. F.: J. Clin. Invest. **41**, 981 (1962).
61. — Dutter, F., Jayle, M. F.: Biochim. Biophys. Acta **69**, 331 (1963).
62. Wengle, B.: Acta Soc. Med. Upsalien. **69**, 105 (1964).
63. Boström, H., Wengle, B.: Acta Soc. Med. Upsalien. **69**, 41 (1964).
64. Reynolds, J. W.: J. Clin. Endocrinol. Metab. **25**, 416 (1965).
65. Killinger, D. W.: Dissertation, Department of Biochemistry, McGill University, Montreal, Canada (1967).
66. Fukushima, D. K., Bradlow, H. L., Hellman, L., Gallagher, T. F.: J. Clin. Endocrinol. Metab. **23**, 266 (1963).
67. Wieland, R. G., DeCourcy, C., Levy, R. P., Zala, A. P., Hirschmann, H.: J. Clin. Invest. **44**, 159 (1965).
68. Deghenghi, R., Revesz, C.: J. Endocrinol. **31**, 301 (1965).
69. a) Baillie, A. H., Griffiths, K.: J. Endocrinol. **31**, 207 (1965); — b) — Cameron, E. H. D., Griffiths, K., Hart, D. M.: J. Endocrinol. **31**, 227 (1965).
70. — Ferguson, M. M., Hart, D. M.: Developments in steroid histochemistry. London and New York: Academic Press 1966.
71. Hirschmann, H., Hirschmann, F. B.: J. Biol. Chem. **184**, 259 (1950).
72. Fotherby, K.: Biochem. J. **71**, 209 (1959).
73. Romanoff, L. P., Baxter, M. N., Malhotra, K. K., Thomas, A. W.: J. Clin. Endocrinol. Metab. **27**, 605 (1967).
74. Hirschmann, H., Hirschmann, F. B.: J. Biol. Chem. **187**, 137 (1950).
75. Fotherby, K.: Biochem. J. **69**, 596 (1958).
76. Riley, W. J.: J. Clin. Endocrinol. Metab. **28**, 83 (1968).
77. Nowaczynski, W., Koiw, E., Genest, J.: Can. J. Physiol. Pharmacol. **42**, 547 (1964).
78. Baulieu, E.-E., Mauvais-Jarvis, P., Corpéchot, C.: J. Clin. Endocrinol. Metab. **23**, 374 (1963).
79. Bradlow, H. L., Fukushima, D. K., Hellman, L., Gallagher, T. F.: J. Clin. Endocrinol. Metab. **24**, 359 (1964).
80. Butenandt, A., Tscherning, K.: Z. Physiol. Chem. **229**, 167 (1934).
81. Munson, P. L., Gallagher, T. F., Koch, F. C.: J. Biol. Chem. **152**, 67 (1944).
82. Kellie, A. E., Wade, A. P.: Acta Endocrinol. **23**, 357 (1956).
83. Malassis, D., Baulieu, E.-E., Weinmann, S. H., Crépy, O., Jayle, M. F.: Compt. Rend. Soc. Biol. (Paris) **151**, 447 (1957).
84. Kellie, A. E., Smith, E. R.: Biochem. J. **66**, 490 (1957).
85. Jayle, M. F., Malassis, D., Pinaud, H.: Acta Endocrinol. **31**, 1 (1959).
86. Staib, W., Teller, W., Scharf, F.: Z. Physiol. Chem. **318**, 163 (1960).
87. Vande Wiele, R. L., Lieberman, S., in: Biological activities of steroids in relation to cancer (G. Pincus and E. P. Volmer, eds.), p. 93. New York: Academic Press 1960.
88. Mauvais-Jarvis, P., Baulieu, E.-E.: J. Clin. Invest. **41**, 1690 (1962).
89. Bitman, J., Cohen, S. L.: J. Biol. Chem. **191**, 351 (1951).

90. Keutmann, E. H., Mason, W. B.: J. Clin. Endocrinol. Metab. **27**, 406 (1967).
91. Heinrichs, W. L., Colás, A., in: Androgens in normal and pathological conditions (A. Vermeulen, ed.), p. 62. New York: Excerpta Medica Foundation 1966.
92. McCullagh, E. P., McCullagh, D. R., Hicken, N. F.: Endocrinology **17**, 49 (1933).
93. a) Mühlbock, O.: Z. Physiol. Chem. **250**, 139 (1937); — b) Lancet **236**, 634 (1939).
94. Migeon, C. J., Plager, J. E.: Recent Progr. Hormone Res. **9**, 235 (1954).
95. Staib, W., Teller, W., Pelzer, H.: Biochim. Biophys. Acta **31**, 591 (1959).
96. Conrad, S. H., Mahesh, V., Herrmann, W. L.: J. Clin. Invest. **40**, 947 (1961).
97. Oertel, G. W., Kaiser, E.: Clin. Chim. Acta **7**, 221 (1962).
98. McKenna, J., Rippon, A. E.: Biochem. J. **95**, 107 (1965).
99. Conrad, S. H., Lindberg, M. C., Herrmann, W. L.: Am. J. Obstet. Gynecol. **91**, 449 (1965).
100. Migeon, C. J., Keller, A. R., Lawrence, B., Shepard, T. H. II: J. Clin. Endocrinol. Metab. **17**, 1051 (1957).
101. Deshpande, H., Bulbrook, R. D.: J. Endocrinol. **36**, 363 (1966).
102. DeMoor, P., Heyns, W., in: Androgens in normal and pathological conditions (A. Vermeulen, ed.), p. 54. New York: Excerpta Medica Foundation 1966.
103. Eik-Nes, K. B., Oertel, G. W., Nimer, R., Tyler, F. H.: J. Clin. Endocrinol. Metab. **19**, 1405 (1959).
104. Lamb, E. J., Dignam, W. J., Pion, R. J., Simmer, H. H.: Acta Endocrinol. **45**, 243 (1964).
105. Migeon, C. J., Keller, A. R., Holmstrom, E. G.: Bull. Johns Hopkins Hosp. **97**, 415 (1955).
106. Colás, A., Heinrichs, W. L., Tatum, H. J.: Steroids **3**, 417 (1964).
107. Simmer, H. H., Easterling, W. E., Jr., Pion, R. J., Dignam, W. J.: Steroids **4**, 125 (1964).
108. Nichols, J., Lescure, O. L., Migeon, C. J.: J. Clin. Endocrinol. Metab. **18**, 444 (1958).
109. Colás, A., Heinrichs, W. L.: Steroids **5**, 753 (1965).
110. Siiteri, P. K., MacDonald, P.: Proc. Intern. Congr. Hormonal Steroids, 2nd., Milan, Italy, May 1966, p. 726 (Pub. 1967).
111. Dirscherl, W., Breuer, H.: Acta Endocrinol. **44**, 403 (1963).
112. Steeno, O., Schirren, C., Heyns, W., DeMoor, P.: J. Clin. Endocrinol. Metab. **26**, 353 (1966).
113. Oertel, G. W.: Biochem. Z. **334**, 431 (1961).
114. — Biochem. Z. **339**, 125 (1963).
115. — Groot, K.: Z. Physiol. Chem. **341**, 1 (1965).
116. — Rindt, W., in: Androgens in normal and pathological conditions (A. Vermeulen, ed.), p. 192. New York: Excerpta Medica Foundation 1966.
117. a) Baulieu, E.-E.: Compt. Rend. **251**, 1421 (1960); — b) J. Clin. Endocrinol. Metab. **22**, 501 (1962).
118. Saez, J. M., Rivarola, M. A., Migeon, C. J.: J. Clin. Endocrinol. Metab. **27**, 615 (1967).
119. Weinheimer, B., Oertel, G. W., Leppla, W., Blaise, H., Bette, L., in: Androgens in normal and pathological conditions (A. Vermeulen, ed.), p. 36. New York: Excerpta Medica Foundation 1966.

120. Rivarola, M. A., Saez, J. M., Migeon, C. J.: J. Clin. Endocrinol. Metab. **27**, 624 (1967).
121. Cohn, G. L., Mulrow, P. J., Dunne, V. C.: J. Clin. Endocrinol. Metab. **23**, 671 (1963).
122. Wallace, E. Z., Lieberman, S.: J. Clin. Endocrinol. Metab. **23**, 90 (1963).
123. Pulkkinen, M. O.: Acta Physiol. Scand. **59**, Suppl. **213**, 124 (1963).
124. Dixon, R., Vincent, V., Kase, N.: Steroids **6**, 757 (1965).
125. Wallace, E. Z., Silberman, N.: J. Biol. Chem. **239**, 2809 (1964).
126. Gallegos, A. J., Berliner, D. L.: J. Clin. Endocrinol. Metab. **27**, 1214 (1967).
127. MacDonald, P. C., Chapdelaine, A., Gonzalez, O., Gurpide, E., Vande Wiele, R. L., Lieberman, S.: J. Clin. Endocrinol. Metab. **25**, 1557 (1965).
128. Chapdelaine, A., MacDonald, P. C., Gonzales, O., Gurpide, E., Vande Wiele, R. L., Lieberman, S.: J. Clin. Endocrinol. Metab. **25**, 1569 (1965).
129. Sandberg, E., Gurpide, E., Lieberman, S.: Biochemistry **3**, 1256 (1964).
130. Baulieu, E.-E.: J. Clin. Endocrinol. Metab. **20**, 900 (1960).
131. Tamm, J., Busch, D., Voigt, K. D.: Acta Endocrinol. **28**, 435 (1958).
132. Wieland, R. G., Levy, R. P., Katz, D., Hirschmann, H.: Biochim. Biophys. Acta **78**, 566 (1963).
133. Kase, N., Conrad, S. H.: Am. J. Obstet. Gynecol. **90**, 125 (1964).
134. O'Malley, B. W., Lipsett, M. B., Jackson, M. A.: J. Clin. Endocrinol. Metab. **27**, 311 (1967).
135. Mahesh, V. B., Greenblatt, R. B.: Abstr. 288, 6th Pan American Congr. Endocrinology, Mexico City, October 1965, p. E. 130.
136. Pion, R. J., Dignam, W. J., Lamb, E. J., Moore, J. G., Frankland, M. V., Simmer, H. H.: Am. J. Obstet. Gynecol. **93**, 1067 (1965).
137. Rivarola, M. A., Saez, J. M., Meyer, W. J., Kenny, F. M., Migeon, C. J.: J. Clin. Endocrinol. Metab. **27**, 371 (1967).
138. Johnsen, S. G.: Acta Endocrinol. **21**, 157 (1956).
139. Landau, R. L., Knowlton, K., Lugibihl, K., Kenyon, A. T.: Endocrinology **48**, 489 (1951). — — — Munson, P. L.: Endocrinology **48**, 506 (1951).
140. Jayle, M. F., Scholler, R., Garrone, G., Morel, F., in: La fonction endocrine du testicule, p. 201. Paris, France: Masson & Cie. 1957.
141. — Baulieu, E.-E., Scholler, R., Lisboa, B. P., Savoie, J. C.: Rev. Franç. Études Clin. Biol. **3**, 455 (1958).
142. Huis in't Veld, L. G., Louwerens, B., Reilingh, W.: Acta Endocrinol. **46**, 185 (1964).
143. Clark, A. F., Raeside, J. I., Solomon, S.: Endocrinology **76**, 427 (1965).
144. Baulieu, E.-E., Fabre-Jung, I., Huis in't Veld, L. G.: Endocrinology **81**, 34 (1967).
145. a) Frandsen, V. A., Stakemann, G.: Acta Endocrinol. **38**, 383 (1961); — b) Acta Endocrinol. **43**, 184 (1963).
146. Pulkkinen, M. O.: Acta Physiol. Scand. **52**, Suppl. **180** (1961).
147. Warren, J. C., Timberlake, C. E.: J. Clin. Endocrinol. Metab. **22**, 1148 (1962).
148. — — Obstet. Gynecol. **23**, 689 (1964).
149. Ryan, K. J.: J. Biol. Chem. **234**, 268 (1959).
150. a) Siiteri, P. K., MacDonald, P. C.: Steroids **2**, 713 (1963); — b) J. Clin. Endocrinol. Metab. **26**, 751 (1966).

151. MacDonald, P. C., Siiteri, P. K.: J. Clin. Invest. **44**, 465 (1965).
152. Baulieu, E.-E., Dray, F.: J. Clin. Endocrinol. Metab. **23**, 1298 (1963).
153. Barlow, J. J., Goldstein, D. P., Reid, D. E.: J. Clin. Endocrinol. Metab. **27**, 1028 (1967).
154. Bolté, E., Mancuso, S., Eriksson, G., Wiqvist, N., Diczfalusy, E.: Acta Endocrinol. **45**, 535 (1964).
155. — — — — — Acta Endocrinol. **45**, 560 (1964).
156. — — — — — Acta Endocrinol. **45**, 576 (1964).
157. Lipsett, M. B., Sarfaty, G. A., Wilson, H., Bardin, C. W., Fishman, L. M.: J. Clin. Invest. **45**, 1700 (1966).
158. Dorfman, R. I., Ungar, F.: Metabolism of steroid hormones. New York: Academic Press 1965.
159. Baulieu, E.-E., Robel, P.: Steroids **2**, 111 (1963).
160. Roberts, K. D., Vande Wiele, R. L., Lieberman, S.: J. Biol. Chem. **236**, 2213 (1961).
161. Kellie, A. E.: J. Endocrinol. **22**, 1 (1961).
162. Gurpide, E., MacDonald, P. C., Vande Wiele, R. L., Lieberman, S.: J. Clin. Endocrinol. Metab. **23**, 346 (1963).
163. — — Chapdelaine, A., Vande Wiele, R. L., Lieberman, S.: J. Clin. Endocrinol. Metab. **25**, 1537 (1965).
164. Korenman, S. G., Lipsett, M. B.: J. Clin. Invest. **43**, 2125 (1964).
165. — — Steroids **5**, 509 (1965).
166. Mauvais-Jarvis, P., Baulieu, E.-E.: J. Clin. Endocrinol. Metab. **25**, 1167 (1965).
167. a) Loras, B., Migeon, C. J.: Steroids **7**, 459 (1966). — b) Jeffcoate, S. L., Brooks, R. V., Lui, N. Y., London, D. R., Prunty, F. T. G., Spathis, G. S.: J. Endocrinol. **37**, 401 (1967).
168. Horton, R., Tait, J. F.: J. Clin. Endocrinol. Metab. **27**, 79 (1967).
169. Gurpide, E., Mann, J., Lieberman, S.: J. Clin. Endocrinol. Metab. **23**, 1155 (1963).
170. Lindberg, M. C., McClenaghan, C., Herrmann, W. L.: Am. J. Obstet. Gynecol. **95**, 743 (1966).
171. Slaunwhite, W. R., Jr., Burgett, M. J., Sandberg, A. A.: J. Clin. Endocrinol. Metab. **27**, 663 (1967).
172. Aakvaag, A., Hagen, A. A., Eik-Nes, K. B.: Biochim. Biophys. Acta **86**, 622 (1964).
173. Sandberg, E. C., Jenkins, R. C., Trifon, H. M.: Steroids **8**, 249 (1966).
174. Fahmy, D., Griffiths, K., Turnbull, A. C., Symington, T.: Biochem. J. **104**, 5 P (1967).
175. Kim, M. H., Herrmann, W. L.: J. Clin. Endocrinol. Metab. **28**, 187 (1968).
176. Warren, J. C., French, A. P.: J. Clin. Endocrinol. Metab. **25**, 278 (1965).
177. Howard, E.: Endocrinology **70**, 131 (1962).
178. Marks, P. A., Banks, J.: Proc. Natl. Acad. Sci., U. S. **46**, 447 (1960).
179. Ashworth-Tsutsui, E., Marks, P. A., Reich, P.: J. Biol. Chem. **237**, 3009 (1962).
180. Nielson, M. H., Warren, J. C.: Biochim. Biophys. Acta **97**, 532 (1965).
181. Levy, H. R.: J. Biol. Chem. **238**, 775 (1963).
182. Kowal, J., Forchielli, E., Dorfman, R. I.: Steroids **4**, 77 (1964).

183. Sharma, D. C., Forchielli, E., Dorfman, R. I.: J. Biol. Chem. **238**, 572 (1963).
184. Oelkers, W., Dulce, H. J.: Z. Physiol. Chem. **337**, 150 (1964).
185. Lipsett, M. B., Coffman, G. D., Nixon, W. E.: J. Clin. Endocrinol. Metab. **25**, 993 (1965).
186. Drucker, W. D., Blumberg, J. M., David, R.: Clin. Res. **14**, 279 (1966).
187. Peillon, F., Racadot, J.: Ann. Endocrinol. (Paris) **26**, 312, 419 (1965).
188. Goldman, A. S., Yakovac, W. C., Bongiovanni, A. M.: J. Clin. Endocrinol. Metab. **26**, 14 (1966).
189. Šonka, J., Gregorová, I.: Acta Univ. Carolinae **1**, 1 (1963).
190. Lopez-S., A., Krehl, W. A.: Lancet **2**, 485 (1967).
191. Hendrikx, A., Heyns, W., Steeno, O., DeMoor, P., in: Androgens in normal and pathological conditions (A. Vermeulen, ed.), p. 63. New York: Excerpta Medica Foundation 1966.
192. Bongiovanni, A. M., Eberlein, W. R.: J. Clin. Endocrinol. Metab. **17**, 238 (1957).
193. Puche, R. C., Nes, W. R.: Endocrinology **70**, 857 (1962).
194. Plager, J. E.: J. Clin. Invest. **44**, 1234 (1965).
195. Wang, D. Y., Bulbrook, R. D.: J. Endocrinol. **39**, 405 (1967).
196. Baulieu, E.-E., Corpéchot, C.: Bull. Soc. Chim. Biol. **47**, 443 (1965).
197. Crépy, O., Pasqualini, J. R., Ducret, M. A., Jayle, M. F.: European J. Steroids **1**, 195 (1966).
198. Bolté, E., Wiqvist, N., Diczfalusy, E.: Acta Endocrinol. **52**, 583 (1966).
199. Payne, A. H., Mason, M.: Steroids **6**, 323 (1965).
200. Lieberman, S.: Proc. Intern. Congr. Hormonal Steroids, 2nd, Milan, Italy, May 1966, p. 22 (Pub. 1967).
201. Morato, T., Lemus, A. E., Gual, C.: Steroids **5**, Suppl. **1**, 59 (1965).
202. Sandberg, A. A., Slaunwhite, W. R., Jr.: J. Clin. Invest. **36**, 1266 (1957).
203. MacDonald, P. C., Gonzalez, O., Vande Wiele, R. L., Lieberman, S.: J. Clin. Endocrinol. Metab. **23**, 665 (1963).
204. — Vande Wiele, R. L., Lieberman, S.: J. Clin. Endocrinol. Metab. **22**, 1222 (1962).
205. YoungLai, E., Solomon, S.: Endocrinology **80**, 177 (1967).
206. Hirschmann, H., Courcy, C. de, Levy, R. P., Miller, K. L.: J. Biol. Chem. **235**, PC48 (1960).
207. Sjövall, J., Vihko, R.: Steroids **7**, 447 (1966).
208. Baulieu, E.-E., Corpéchot, C., Emiliozzi, R.: Steroids **2**, 429 (1963).
209. Schneider, J. J., Lewbart, M. L.: Recent Progr. Hormone Res. **15**, 201 (1959).
210. Okada, M., Fukushima, D. K., Gallagher, T. F.: J. Biol. Chem. **234**, 1688 (1959).
211. Stárka, L., Šulcová, J., Šilink, K.: Clin. Chim. Acta **7**, 309 (1962).
212. Kocic, D., Stárka, L.: Abstr. 517, Intern. Congr. Hormonal Steroids, 2nd, Milan, Italy, May 1966, p. 278.
213. Stárka, L., Hampl, R.: Naturwissenschaften **51**, 164 (1964).
214. — Janata, J., Breuer, H., Hampl, R.: European J. Steroids **1**, 37 (1966).
215. — Döllefeld, E., Breuer, H.: Abstr. 223, Intern. Congr. Hormonal Steroids, 2nd, Milan, Italy, May 1966, p. 142.
216. Lieberman, S., Dobriner, K., Hill, B. R., Fieser, L. F., Rhoads, C. P.: J. Biol. Chem. **172**, 263 (1948).

217. Fukushima, D. K., Kemp, A. D., Schneider, R., Stokem, M. B., Gallagher, T. F.: J. Biol. Chem. **210**, 129 (1954).
218. — Gallagher, T. F.: J. Biol. Chem. **229**, 85 (1957).
219. Gallagher, T. F.: J. Clin. Endocrinol. Metab. **18**, 937 (1958).
220. Baulieu, E.-E., Emiliozzi, R., Corpéchot, C.: Experientia **17**, 110 (1961).
221. Guillon, J., Colas, J., Trichereau, R., Delumeau, G., Baulieu, E.-E.: Ann. Endocrinol. (Paris) **22**, 331 (1961).
222. Fotherby, K., Colás, A., Atherden, S. M., Marrian, G. F.: Biochem. J. **66**, 664 (1957).
223. Teich, S., Rogers, J., Lieberman, S., Engel, L. L., Davis, J. W.: J. Am. Chem. Soc. **75**, 2523 (1953).
224. a) Magendantz, H. G., Ryan, K. J.: Federation Proc. **23**, 275 (1964); — b) J. Clin. Endocrinol. Metab. **24**, 1155 (1964).
225. Fishman, J., Brown, J. B., Hellman, L., Zumoff, B., Gallagher, T. F.: J. Biol. Chem. **237**, 1489 (1962).
226. Gurpide, E., Angers, M., Vande Wiele, R. L., Lieberman, S.: J. Clin. Endocrinol. Metab. **22**, 935 (1962).
227. Troen, P.: J. Clin. Endocrinol. Metab. **21**, 895 (1961).
228. Little, B., Shaw, A., Purdy, R.: Acta Endocrinol. **43**, 510 (1963).
229. Cedard, L., Varangot, J., Yannotti, S.: Compt. Rend. **254**, 3896 (1962).
230. Kirschner, M. A., Wiqvist, N., Diczfalusy, E.: Acta Endocrinol. **53**, 584 (1966).
231. Gurpide, E., Schwers, J., Welch, M. T., Vande Wiele, R. L., Lieberman, S.: J. Clin. Endocrinol. Metab. **26**, 1355 (1966).
232. Reynolds, J. W.: Steroids **3**, 77 (1964).
233. Hirschmann, H.: J. Biol. Chem. **150**, 363 (1943).
234. Marrian, G. F., Butler, G. C.: Biochem. J. **38**, 322 (1944).
235. Reynolds, J. W.: Steroids **8**, 719 (1966).
236. Weinmann, S. H., Demoisson, F. L., Baulieu, E.-E., Jayle, M. F.: Compt. Rend. Soc. Biol. (Paris) **151**, 518 (1957).
237. Dell'Acqua, S., Mancuso, S., Wiqvist, N., Ruse, J. L., Solomon, S., Diczfalusy, E.: Acta Endocrinol. **55**, 389 (1967).
238. Corpéchot, C., Baulieu, E.-E.: Compt. Rend. **258**, 3584 (1964).
239. Schwers, J., Govaerts-Videtsky, M., Wiqvist, N., Diczfalusy, E.: Acta Endocrinol. **50**, 597 (1965).
240. Matsui, M., Fukushima, D. K.: Biochemistry **8**, 2997 (1969).

Chapter 6

Formation, Metabolism, and Transport of Estrogen Conjugates

EGON DICZFALUSY and MORTIMER LEVITZ

Contents

I. Introduction

Several outstanding features of estrogen conjugates impart particular interest to their study. Consider first the variety of conjugates. In addition to the usual sulfate and glucuronide forms, estriol, a quantitatively important estrogen, occurs as the double conjugate, estriol-3-sulfate-16-glucuronide [1–6]. In addition, an N-acetylglucosamine conjugate of 17α-estradiol has been isolated from the urine of rabbits [7], and preliminary evidence has been obtained that sulfo-N-acetylglucosaminides of 15α-hydroxyestrone and 15α-hydroxyestradiol are secreted with the bile following the injection of estrone sulfate into human subjects [8]. Another interesting feature is that some estrogen conjugates, particularly those of estriol, may undergo, in sequence, hydrolysis and then conversion to another conjugate a number of times, prior to excretion. Moreover, the steroid moiety of urinary estrogen conjugates displays significant biological activity. This is not generally true for the other steroid hormones. Finally, certain estrogen conjugates, which are barely detectable in the body fluids and tissues in the nonpregnant state, are found in thousand-fold amounts in late pregnancy, and in proportions distinctly different from those characterizing nonpregnant conditions.

This chapter will deal primarily with the biochemistry of the estrogen conjugates as it pertains to man. Observations made in animal experiments will be included only when important principles have been established which may be relevant to our understanding of the conditions existing in man. Coverage of this rapidly expanding field will be restricted to studies in which the estrogen conjugate in question has been isolated and reasonably characterized. However, a number of studies directed toward the identification and quantitation of estrogen conjugates in body fluids in which enzyme hydrolysis or solvolysis is the sole basis for identification will be included, when the data provided are the best available. The major emphasis to be placed on estrogen conjugates in pregnancy is not only a reflection of the comparatively high concentration of estrogen conjugates in tissues and body fluids in pregnancy, but also of the amount of information available.

II. Biochemical Preparation of Estrogen Conjugates

The most precise information on the metabolism of estrogens and their conjugates is obtained in tracer studies. This is why there is an urgent need for labeled conjugates with high specific activities. Such conjugates can be prepared by both chemical and biochemical methods. The latter are better suited for the preparation of estrogen conjugates since microgram amounts are usually sufficient for several studies. Chemical methods for the synthesis of some estrogen conjugates are described in Chapter 1, while the more important biochemical procedures are reviewed here.

A. Glucuronides

Procedures used for the preparation of the estrogen glucuronides are generally patterned after the methods described by Slaunwhite *et al.* [9] for the synthesis of estriol–16–glucuronide [3,17β–dihydroxyestra–1,3,5(10)–trien–16α–yl–β–D–glucopyranosiduronic acid], and those described by Smith and Breuer [10] for the synthesis of estrone glucuronide [17–oxoestra–1,3,5(10)–trien–3–yl–β–D–glucopyranosiduronic acid].

1. Estrone Glucuronide

Incubation at pH 8.0 of 6,7–^{3}H estrone with guinea pig liver homogenate fortified with UDPGA (uridine diphosphate glucuronic acid) yielded about 55% of radioactive water–soluble material, 90% of which was estrone glucuronide, and the remaining 10% was 17β–estradiol–3–glucuronide [17β–hydroxyestra–1,3,5(10)–trien–3–yl–β–D–glucopyranosiduronic acid] [11]. The two glucuronides were separated by countercurrent distribution. Alternatively, estrone glucuronide may be prepared by incubating estrone with rabbit liver microsomes in the presence of UDPGA at the optimal pH of 8.2 [10].

2. 17β–Estradiol–3–glucuronide

This compound may be prepared by the reduction of estrone glucuronide with potassium borohydride (unpublished observations of U. Goebelsmann). However, convenient methods for its rigorous purification have not yet been described.

3. Estriol–16–glucuronide

Incubation of estriol [estra–1,3,5(10)–triene–3,16α,17β–triol] with human liver homogenate at pH 7.0 in the presence of UDPGA resulted in the formation of the 16–glucuronide in excellent yield [9].

4. Estriol–3–glucuronide [16α,17β–dihydroxyestra–1,3,5(10)–trien–3–yl–β–D–glucopyranosiduronic acid]

Substitution of guinea pig liver homogenate for human liver homogenate in the system described by Slaunwhite *et al.* [9], resulted in the formation of the 3–glucuronide in approximately 90% yield [12].

5. *Estriol–17–glucuronide*[3,16α–dihydroxyestra–1,3,5(10)–trien–17β–yl–β–D–glucopyranosiduronic acid]

Incubation of estriol with the homogenate of human small intestinal mucosa produced estriol–16–glucuronide as the major product, accompanied by smaller amounts of estriol–3– and –17–glucuronides [13]. Following cell fractionation, the supernatant contained the three glucuronyl transferases, but the 17β–transferase was most abundant. Treatment of the supernatant with ammonium sulfate (30 to 60% saturation) precipitated a glucuronyl transferase which formed only estriol–17–glucuronide, albeit in small yield [13, 14].

B. Sulfates

In the mid–fifties, several groups of investigators demonstrated that liver homogenates (preferably microsome–free) from rats and rabbits fortified with ATP (adenosine triphosphate) and magnesium (2+), converted estrogens into sulfates [15–17]. Since then a vast amount of literature has accumulated on the sulfurylation of estrogens by mammalian tissues. The subject has been reviewed recently by Döllefeld and Breuer [18]. However, few, if any, of these studies were conducted with the aim of preparing pure estrogen sulfates for *in vivo* and *in vitro* studies. Furthermore, in our opinion, the evidence for the identification of the estrogen sulfates prepared is equivocal. These incubation studies were carried out with homogenates of liver, a tissue known to be the principal site of metabolism of the estrogen moiety. Therefore, requisite for the preparation of pure estrogen sulfates by liver homogenates is either the extensive purification of the end product, or the extensive purification of the sulfurylating enzyme system. An attractive alternative is the substitution of tissues in which aryl sulfurylation is efficient, but in which the metabolism of the estrogen moiety is minimal. Human fetal lung appears to satisfy this requirement [19–21]. Estrone sulfate [3–sulfooxyestra–1,3,5(10)–trien–17–one], 17β–estradiol–3–sulfate [3–sulfooxyestra–1,3,5(10)–trien–17β–ol], and estriol–3–sulfate [3–sulfooxyestra–1,3,5(10)–triene–16α,17β–diol] can be prepared by the use of fetal lung slices. However, even when such preparations are used, purification of the estrogen sulfate obtained is highly desirable.

On the basis of the chromatographic behavior of the end products obtained, it was reported that microsome–free preparations of rat liver homogenates convert 17β–estradiol [estra–1,3,5(10)–triene–3,17β–diol] into 3–sulfate, 17–sulfate, and 3,17–disulfate [22]. A similar preparation was reported to convert 17α–estradiol into its 3–sulfate and 3,17–disulfate [23]. The end products obtained were not identified by more stringent criteria.

C. Double Conjugates

The biochemical preparation of two unique double conjugates of estrogens has been described. Estriol–3–sulfate–16–glucuronide [3 sulfooxy–17β–hydroxyestra–1,3,5(10)–trien–16α–yl–β–D–glucopyranosiduronic acid] was prepared by incubating estriol–16–glucuronide with the microsome-free fraction of guinea pig–liver homogenate in the presence of ATP and Mg^{2+} [24]. The second double conjugate is 17α–estradiol–3–glucuronide–17–β–N–acetylglucosaminide [estra–1,3,5(10)–triene–3,17α–di–yl–3–β–D–glucopyranosiduronic acid–17–2′–acetamido–2′–deoxy–β D–glucopyranoside] [7]. Incubation of 17β–estradiol with rabbit–liver homogenate fortified with UDPGA and uridine diphosphate N-acetylglucosamine resulted in the formation of 17α–estradiol–3–glucuronide–17–β–N–acetyl-glucosaminide [25]. The yield was approximately 10% of the water–soluble products formed. Apparently, glucuronidation at position 3 must precede the transfer of N–acetylglucosamine. No evidence was found for the transfer of N–acetylglucosamine to the 17β–hydroxyl group of 17β–estradiol [25].

III. Estimation in Body Fluids and Tissues

The present state of knowledge on the estrogen conjugates in human body fluids and tissues is fragmentary. Although some 25 estrogens have been isolated from human sources, only 8 conjugates involving three estrogens have been isolated or characterized (Fig. 6–1).

A critical evaluation of the available information on the quantitation of estrogen conjugates indicates that few reliable data are available even with regard to the conjugated forms of the three classical estrogens shown in Fig. 6–1. The reason for this can be traced to the many problems involved in the extraction, separation, purification, and estimation of estrogen conjugates. These difficulties have virtually precluded the investigation of endogenous estrogen conjugates in nonpregnant conditions, in which the concentration of these substances often approaches the lowest limit of sensitivity of the available methods. Therefore, it is not too surprising that most of the available quantitative data were obtained by methods in which nonspecific hydrolytic procedures were used to deduce the nature of the estrogen conjugate originally present. Another major difficulty has been the paucity of suitable preparations of estrogen conjugates to serve as internal standards to correct for methodological losses. Therefore, the validity of much of the data to be presented in this section is in doubt. However, in our opinion they represent the best information that is available.

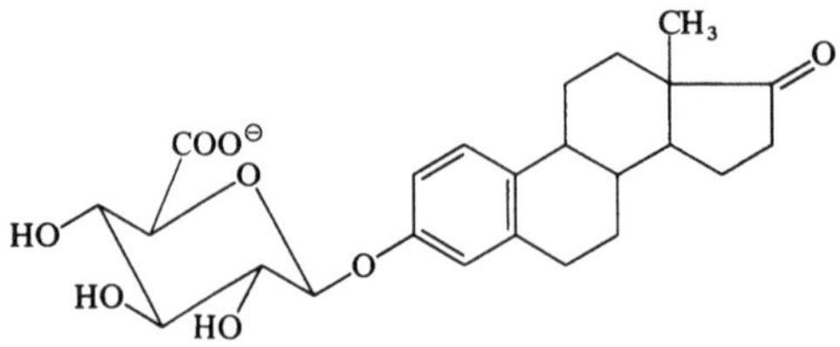

Estrone glucuronide

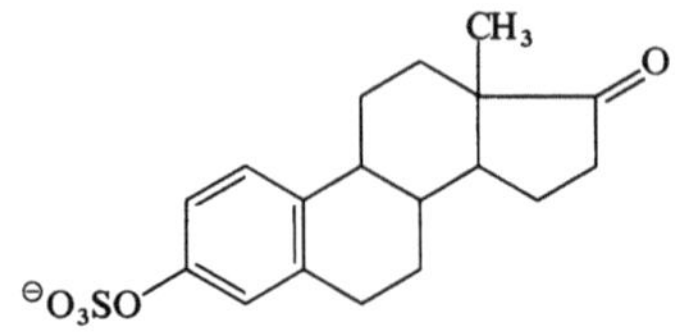

Estrone sulfate

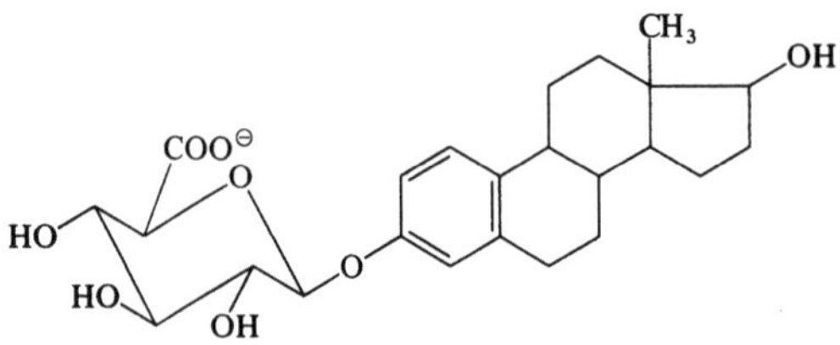

17β–Estradiol–3–glucuronide

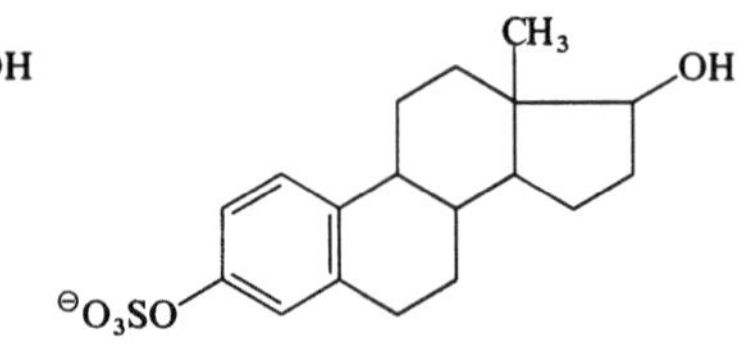

17β–Estradiol–3–sulfate

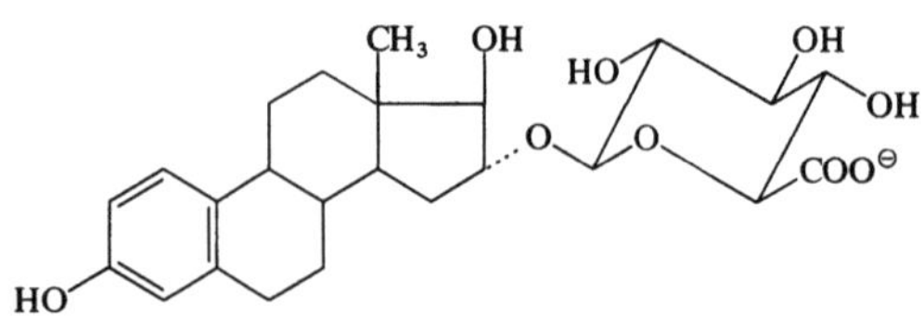

Estriol–16–glucuronide

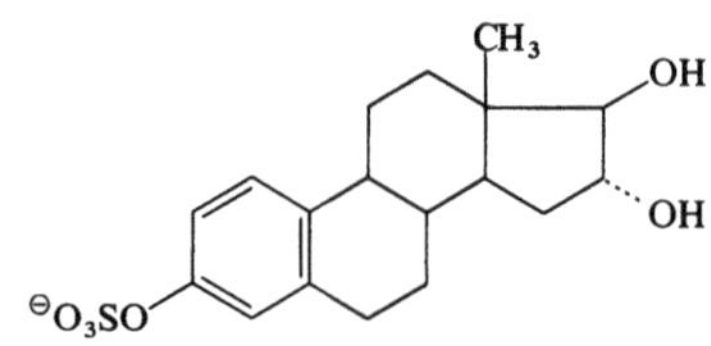

Estriol–3–sulfate

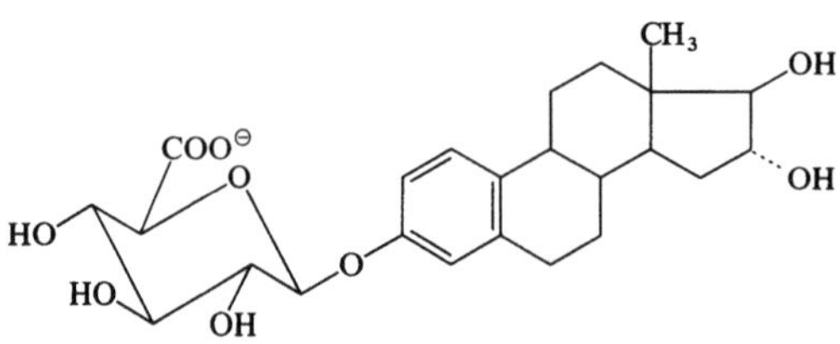

Estriol–3–glucuronide

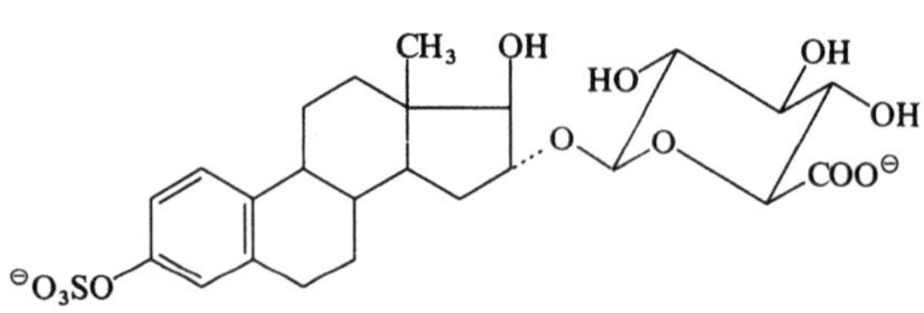

Estriol–3–sulfate–16–glucuronide

Fig. 6–1. Estrogen conjugates isolated from body fluids and tissues and characterized

A. Plasma

Thus far, no reasonable estimates of estrogen conjugates in plasma of nonpregnant subjects have been published. In view of the extremely low concentrations of estrogens in nonpregnant conditions (*e.g.*, ref. [26]), the determination of the individual conjugates with the methods available at present appears to be difficult.

Several studies have been directed toward the determination of estrogen conjugates in pregnancy plasma and cord plasma. Some of these data are presented in Table 6–1. It should be pointed out that the designation "glucuronides" used in the table, represents groups rather than discrete compounds. The sulfate fraction was either analyzed in a mixture of all the conjugates [3, 4], or hydrolyzed by nonspecific methods [27, 28]. In the latter two studies cited, the preliminary fractionation accomplished was inadequate for the separation of individual conjugates.

In 13 instances blood was taken from the newborn by umbilical catherization 2 hours following delivery. Only conjugated estriol was found, in a concentration of 25 μg/ml [29]. The nature of the conjugates was not investigated.

One study has been reported [2] in which extensive fractionation of the estriol conjugates in pools of cord blood was carried out. In this study estriol-3-sulfate, estriol-16-glucuronide, and estriol-3-sulfate-16-glucuronide were identified, and two additional glucuronides were separated, but not identified. In general, the levels were lower than those given in Table 6–1, probably because large pools were worked up and the aim of the study was identification rather than quantitation.

B. Urine

Numerous studies have been conducted in which the structure of urinary estrogens was elucidated following enzyme hydrolysis. Although there is a reasonable assumption that the bulk of the estrogens excreted in pregnancy urine is present in the form of glucuronides, unequivocal proof is lacking [30]. In this section only conjugates that were fractionated prior to enzyme hydrolysis are considered.

Generally speaking, the total amount of the three classical estrogens excreted in the urine in nonpregnant states is considerably less than 100 μg per 24 hr [31]. This may explain why few studies have been directed so far toward the fractionation and quantitation of the forms in which these estrogens are excreted. In contrast, the level of estrogen conjugates in pregnancy urine, particularly toward term, is high. In fact, the first estrogen conjugate, estriol-16(17?)-glucuronide, was isolated from this source as early as 1936 [32]. Using an elegant chemical approach, Hashimoto and

TABLE 6–1. *Levels of estrogen conjugates in plasma in late pregnancy*

Conjugate	Concentration (μg/100 ml)		Mode of hydrolysis	References
	Peripheral blood plasma	Cord blood plasma		
Estrone sulfate	4.1	0.84	Mylase P (arylsulfatase)	3
	6.8		Mylase P or solvolysis	4
	2.9		*Helix pomatia* [b]	27
Estrone glucuronide	0.44	0.35	β–Glucuronidase	3
	0.6		β–Glucuronidase	4
	0.2		*Helix pomatia* [b]	27
17β–Estradiol–3–sulfate	1.6	0.80	Mylase P	3
	0.6		Mylase P or solvolysis	4
	0.2		*Helix pomatia* [b]	27
17β–Estradiol glucuronide [a]	0.69	0.65	β–Glucuronidase	3
	0.5		β–Glucuronidase	4
	0.3		*Helix pomatia* [b]	27
Estriol–3–sulfate	4.4	94	Mylase P	3
	3.1	69	Mylase P	3
	3.0		Mylase P or solvolysis	4
		60	HCl [b]	28
	2.8		*Helix pomatia* [b]	27
Estriol glucuronide [a]	2.9	20	β–Glucuronidase	3
	3.8	22	β–Glucuronidase	3
	6.4		β–Glucuronidase	4
		21	HCl [b]	28
Estriol–3–sulfate–16–glucuronide [a]	4.5		Mylase P, then β–glucuronidase	3
	5.7		Mylase P, then β–glucuronidase	4

[a] Conjugate types in which the positions of attachment were not defined.
[b] Hydrolysis followed separation of conjugate types by solvent partition.

Neeman [33] could find only estriol–16–glucuronide in the estriol–16(17?)–glucuronide fraction of late pregnancy urine previously described by Cohen *et al.* [32]. Despite the fact that estrogen conjugates, particularly glucuronides, are abundant in pregnancy urine (*e.g.*, [30]), systematic studies directed toward their identification and quantitation are few.

Beling [34] achieved a group separation of estrogen conjugates into two peaks (I and II) by filtration on Sephadex® (Pharmacia, Uppsala). It was shown that estrone glucuronide and estriol–3–glucuronide were eluted in peak I. Most of the conjugates studied, including estriol–3–sulfate, were eluted in the second peak, in which estriol–16–glucuronide was the quantitatively most significant constituent. Estriol–3–sulfate–16–glucuronide was not studied. In midpregnancy, the amount of estriol eluted in peak I (presumably estriol–3–glucuronide) was almost as much as that eluted in peak II (mostly estriol–16–glucuronide), whereas, at term, the latter greatly exceeded the former.

Using the same technique, Smith and Kellie [35] obtained evidence for the presence in late pregnancy urine of the following glucuronides: estrone glucuronide, 17β–estradiol–3–glucuronide, estriol–3–glucuronide, estriol–16–glucuronide, 16–epiestriol–3–glucuronide [16β,17β–dihydroxyestra–1,3,5(10)–trien–3–yl–β–D–glucopyranosiduronic acid] and 16α–hydroxyestrone–16–glucuronide [3–hydroxy–17–oxoestra–1,3,5(10)–trien–16α–yl–β–D–glucopyranosiduronic acid].

Slaunwhite *et al.* [36] estimated estriol–16–glucuronide in the urine of two patients from the third to the eighth month of gestation. They reported that the daily excretion rose from 7 to 10 mg/24 hr in the third month, to 37 to 54 mg/24 hr in the eighth month.

In connection with a study designed to investigate the metabolism of estriol–3–sulfate, estriol–16–glucuronide, and estriol–3–glucuronide at midpregnancy, Goebelsmann *et al.* were able to obtain accurate values for the urinary excretion of the corresponding endogenous conjugates [37, 38]. The data presented in Table 6–2 show that the bulk of estriol was eliminated

TABLE 6–2. *Urinary excretion of estriol conjugates by four subjects in the 17th to 20th week of pregnancy*[a]

Subject	Estriol conjugate		
	3–sulfate	16–glucuronide	3–glucuronide
1	0.09	1.9	1.6
2	0.13	3.1	1.5
3	[b]	5.9	3.0
4	[b]	1.3	1.2

Values are in mg/24 hr.

[a] Goebelsmann, U., Eriksson, G., Wiqvist, N., Diczfalusy, E.: Acta Endocrinol. **50**, 273 (1965); Goebelsmann, U., Cooke, I., Wiqvist, N., Diczfalusy, E.: Acta Endocrinol. **52**, 30 (1966).

[b] Not estimated.

both as the 3- and 16-glucuronides, whereas the excretion of the 3-sulfate was limited. The excretion of the double conjugate, estriol-3-sulfate-16-glucuronide, was not studied.

On the basis of enzyme hydrolysis, Smith [39] reported that, during labor, estrone, 17β-estradiol, and estriol are excreted mainly as glucuronides.

C. Bile

The only study of the levels of endogenous estrogen conjugates in human bile appears to be that Adlercreutz [40]. The estrogens in the "sulfate" and "glucuronide" fractions were group-separated by solvent partition, and then were assayed after hydrolyzing the conjugates with a crude enzyme preparation obtained from the gastrointestinal tract of the snail, *Helix pomatia*. This preparation is rich in both arylsulfatase and β-glucuronidase activities. The average values reported are presented in Table 6-3.

D. Amniotic Fluid and Meconium

Virtually all estrogens found in the amniotic fluid thus far are present in the form of conjugated estriol. At midpregnancy, the level is approximately 5 μg/100 ml, which increases to about 73 μg/100 ml at term [28]. Troen *et al.* [2] studied the pattern of estriol conjugates in the amniotic fluid at term. The average values (expressed as μg/100 ml) of 5 individual analyses were: the 3-sulfate, 7.0; the 16-glucuronide, 30; the 3-sulfate-16-glucuronide, 18. Two additional conjugates of estriol were also separated and assayed, but the nature of conjugation was not established in either case. No data are available on the conjugated forms of estrone and 17β-estradiol, except that their concentrations were some 200 times less than that of conjugated estriol [41].

Among all the human sources studied, the highest concentration of estriol (10 mg/100 g) occurred in the meconium. Some 85% of this was conjugated [42, 43]. From this source, estriol-16-glucuronide [44] and estriol-3-sulfate [45] were isolated in approximately equal quantities.

E. Placenta and Fetal Tissues

Endogenous estrogens in placentas and fetal tissues have been separated into unconjugated and conjugated forms of estrone, 17β-estradiol, and estriol, and their levels have been determined. With the exception of the placenta, in which unconjugated estrogens were predominant, in all fetal urines virtually all estrogens were present in a conjugated form [29]. The levels of the conjugates are indicated in Table 6-4. The nature of the conjugated estrogens has not been established. However, the results of

TABLE 6–3. *Estrogen conjugates in human bile* [a]

Clinical condition	Method of collection	Cases studied	Conjugate type [b]					
			"Sulfate"			"Glucuronide"		
			Estrone	17β–Estradiol	Estriol	Estrone	17β–Estradiol	Estriol
Late pregnancy	Duodenal intubation	7	5.8	0.4	4.8	1.2	1.8	367
Cholecystolithiasis (women)	T–tube drainage	9	0.2	0.0	0.2	0.1	0.1	1.3
Cholecystolithiasis (men)	T–tube drainage	8	0.1	0.0	0.1	0.0	0.0	0.5

Values are expressed as μg/100 ml.

[a] Adlercreutz, H.: Acta Endocrinol. **42**, Suppl. **72** (1962).

[b] "Sulfate" and "glucuronide" fractions were separated by solvent partition, hydrolyzed by a *Helix pomatia* enzyme preparation, and the estrogens were estimated by the Brown technique [Brown, J. B.: Biochem. J. **60**, 185 (1955)]. The validity of the partition into sulfates and glucuronides was investigated only with estriol–3–sulfate and estriol–16–glucuronide.

TABLE 6–4. *Concentration of conjugated estrogens in fetal tissues and placenta at the 17th to 21st week of gestation*[a]

Tissue	Conjugated		
	Estrone	17β–Estradiol	Estriol
Liver	0.7	1.5	27.0
Lungs	1.8	1.1	15.0
Brain	0.0	0.0	0.0
Adrenal	0.0	0.0	8.3
Kidney	0.0	0.0	6.5
Small intestines	0.0	0.0	44.0
Large intestines	0.0	0.0	324.0
Placenta	0.0	0.0	1.3
Placenta (at term)	0.0	0.0	12.0

Values are expressed as μg/100 g tissue.

[a] Diczfalusy, E., Cassmer, O., Alonso, C., de Miquel, M.: Recent Progr. Hormone Res. **17**, 147 (1961).

numerous subsequent studies, in which labeled or unlabeled estrogens were introduced into the fetal organism under a variety of conditions, indicate that the pattern of conjugated estrogens present in fetal tissues is characterized by a preponderance of arylsulfates. Small amounts of glucuronides and sulfoglucuronides are also present. Glucuronyl transferase activity appears to be present in the liver and intestinal tract. An extensive discussion of these experiments will be presented in the subsequent section which deals with metabolism of conjugates.

IV. Metabolism of Conjugates

A. Glucuronides

Since the demonstration by Engel *et al.* [46] that human fetal liver slices convert 17β–estradiol into estriol glucuronide, additional *in vitro* studies were reported on the formation of various types of estrogen conjugates. Thus, Dahm and Breuer [13] made the interesting observation that the homogenate of human intestinal mucosa converts estriol to the three possible monoglucuronides. Many of the *in vitro* procedures are suitable for the preparation of specific estrogen conjugates in good yield, and they have been reviewed in the section on Biochemical Preparation. Therefore, this section will concentrate on the metabolism *in vivo*.

1. Nonpregnant Conditions

Early studies by Beer and Gallagher [47] indicated that a substantial part of the urinary metabolites of the administered labeled 17β–estradiol were hydrolyzed by β–glucuronidase. This suggested that glucuronidation is a major pathway of estrogen metabolism in man. Only few studies have been conducted in nonpregnant subjects in which the chemical nature of the glucuronides was firmly established. After the intravenous administration of ^{14}C–estriol–16–glucuronide to a sixteen–year–old patient with sarcoma, Goebelsmann *et al.* [48] found that 85% of the administered radioactivity was excreted in the urine within 96 hours. Virtually all the radioactive material was excreted as estriol–16–glucuronide (81%) and estriol–3–glucuronide (19%). Whereas estriol–16–glucuronide was predominant in the urine excreted during the first day, approximately equal quantities of the 16– and the 3–glucuronide appeared in the urine during the subsequent days. This result is in contrast to the findings of Sandberg and Slaunwhite [49]. They reported that, following intravenous administration to women, estriol–16–glucuronide was excreted mainly, if not entirely, in an unchanged form.

Whereas it has been assumed for a long time that the liver was the major site of estrogen glucuronide formation (*e.g.*, [26]), the experiments in rats by Lehtinen *et al.* [50] suggested that the intestinal mucosa also has the capacity to conjugate estrogens with glucuronic acid. Indeed, when an isolated jejunal loop of a human subject was perfused at laparotomy with estriol, estriol–16(17?)–glucuronide was isolated from the intestinal wall as well as from the effluent venous blood [51]. Another glucuronide, which was present in approximately equal quantities, was also detected but not identified. Its partition characteristics agree with those of estriol 3 glucuronide. In similar experiments in which 17β–estradiol was perfused, estrone glucuronide was identified [52]. The results of an experiment in which ^{14}C–estriol was injected into the duodenum of a subject with a bile fistula, while at the same time ^{3}H–estriol was given intravenously, provided direct proof that estriol–3–glucuronide is an intestinal metabolite in man [53].

2. Pregnancy

a) Metabolism Following Peripheral Administration. Wilson *et al.* [54] injected estriol intravenously into 4 pregnant women 30 minutes prior to the surgical termination of gestation at midpregnancy. At the time of laparotomy, maternal blood was drawn and analyzed for glucuronides. In addition, urine was collected and also analyzed. Less than 1% of the radioactive material recovered from the blood was estriol–16–glucuronide. The pattern of glucuronide excretion in the urine was similar to that found in a nonpregnant subject [49].

TABLE 6–5. *Pattern of urinary estriol glucuronides following the intravenous infusion of labeled estriol–16–glucuronide and estriol–3–glucuronide at the 17th to 21st week of gestation*

Glucuronide administered	Case	Radio-activity recovered (%)	Estriol–16–glucuronide	Estriol–3–glucuronide	References
Estriol–16–glucuronide	1	90 [a]	81 [b]	19 [b]	48
	2	87	77	23	37
	3	83	80	20	37
Estriol–3–glucuronide	4 [c]	85	87	13	38
	5 [c]	78	82	18	38
	4 [c]	91	2.1	97.9	38
	5 [c]	91	2.3	97.7	38

[a] Represent percentage of the administered radioactive material recovered from the urine collected during 24 hours.

[b] Represent percentage of the urinary radioactivity.

[c] In these experiments, estriol–16–glucuronide and estriol–3–glucuronide, labeled differently, were infused simultaneously.

The metabolism of estriol–16–glucuronide administered peripherally at midpregnancy has been studied extensively. The data presented in Table 6–5, which are remarkably consistent, indicate that virtually all radioactive material was recovered from the urine in two forms, estriol–16– and –3–glucuronide. The data of Table 6–5 also indicate that, when estriol–3–glucuronide was infused under identical conditions, only some 2% of the urinary radioactive material was estriol–16–glucuronide, the rest being unchanged. Thus, it may be concluded that in a greater sense estriol–3–glucuronide, rather than estriol–16–glucuronide, is the major end product of estriol metabolism.

b) Metabolism in the Fetus at Midpregnancy. Thirty minutes following the intravenous injection of estriol into the peripheral circulation at midpregnancy, less than 0.5% of the administered dose was found in the amniotic fluid, fetus, and placenta [54]. Therefore, in order to study the metabolism of estrogen conjugates in the conceptus, the substrate must be introduced into the amniotic fluid, or into the intact umbilical circulation at laparotomy. Alternatively, the previable fetus may be perfused with the precursor to be studied. When previable fetuses were perfused with estriol, estriol–16–glucuronide was isolated from the liver, gastrointestinal tract, and perfusate [55, 56], indicating that the fetus is capable of glucuronidating estriol at midpregnancy.

One study has been reported in which labeled estriol-16-glucuronide was injected into the amniotic fluid at midpregnancy 24 hours prior to the termination of gestation [57]. There was a very limited conversion into estriol-3-sulfate by the fetal tissues. The concentration of estriol-3-sulfate was by far highest in the gastrointestinal tract and in the liver, suggesting that these organs were the sites of hydrolysis and subsequent conversion of the 16-glucuronides into the 3-sulfate, and there was no evidence of hydrolysis by the placenta. A major metabolic transformation of the injected estriol-16-glucuronide was its conversion into still more polar conjugates. The nature of these metabolites was not established.

In another study estriol-16-glucuronide was introduced into the intact fetoplacental circulation at laparotomy and the circulation was maintained *in situ* for 15 minutes [5]. Examination of various tissues indicated that the injected glucuronide was converted again to a small extent into estriol-3-sulfate and that the latter was present only in the liver. As in the previous study [57], in all fetal tissues there were considerable quantities of metabolites more polar than estriol-16-glucuronide. About 70% of this fraction was identified as estriol-3-sulfate-16-glucuronide. The remaining 30% was not identified. However, this material was completely hydrolyzed by β-glucuronidase. It is possibly a diglucuronide.

The fetoplacental metabolism of estrone glucuronides, labeled both in the estrogen and glucuronic acid moieties, was investigated in a similar manner [11]. The injected material was converted by the fetus into 17β-estradiol- and estriol-3-glucuronides. The isotopic ratio of the estriol-3-glucuronide was identical to that of the administered estrone glucuronide, indicating that it was converted to estriol-3-glucuronide by the fetus without prior hydrolysis.

One study is available on the formation of glucuronides in a newborn anencephalic monster [58]. Labeled estriol was injected into the umbilical vein. Although no estriol-16-glucuronide was detected in blood drawn at intervals up to 80 minutes, it accounted for approximately 50% of the radioactive material present in the urine.

B. Sulfates

Numerous studies have been conducted in which the estrogen-sulfurylating capacity of various mammalian tissues has been investigated. Recently, most of the relevant information has been extensively reviewed [18]. For details of the incubation and *in vitro* perfusion studies involving human tissues, the reader is referred to several reports [29, 59–64]. It should be pointed out that, except for the first three studies cited [29, 59, 60], the identity of the products of incubation was not firmly established.

Studies on the distribution of sulfatases in tissues derived from nonpregnant adult humans have concentrated on dehydroisoandrosterone sulfate [3β–sulfooxyandrost–5–en–17–one] [65]. Arylsulfatase activity has been detected in the human ovary [66], but systematic studies with other tissues have not been conducted. In pregnancy, arylsulfatase activity is high in the placenta [61, 67, 68] and fetal membranes [69], but is absent from various fetal tissues, with the possible exception of the liver [61]. These conclusions were drawn from the results of *in vitro* experiments. The results of *in vivo* experiments in nonpregnant and pregnant conditions are reviewed.

1. Nonpregnant Conditions

Following the administration of labeled 17β–estradiol to women, estrone sulfate was identified in the plasma [70]. Estrone sulfate is metabolized extensively *in vivo*. Following the administration of estrone–^{35}S–sulfate to patients with advanced breast cancer, the bulk of the radioactive material was excreted in the urine in the form of inorganic ^{35}S–sulfate [71].

The metabolic fate of ^{14}C–labeled estrone sulfate was investigated following both intravenous and oral administration. Metabolism was extensive in both cases. The bulk of the radioactivity in the urine was excreted in the form of β–glucuronidase–hydrolyzable material. Except for the initial two hours of the experiment, the urinary sulfate–like material represented only a small percentage of the excreted radioactive material [72]. The metabolism of estrone sulfate has also been studied in a subject with a bile fistula. 15α–Hydroxyestrone–3–sulfate [15α–hydroxy–3–sulfooxyestra–1,3,5(10)–trien–17–one] was found in the bile and was absent from the urine, while the reverse was true for 16–epiestriol–3–sulfate [3–sulfooxyestra–1,3,5(10)–triene–16β,17β–diol] [73, 8]. In a similar way the fate of estriol–3–sulfate was studied following the intravenous infusion into women with biliary fistulas. Approximately 50% of the urinary radioactivity was present as estriol–16–glucuronide, 22% as estriol–3–sulfate, 19% in the form of estriol–3–sulfate–16–glucuronide. The double conjugate predominated in the bile. No estriol–3–glucuronide was detected [6]. The reason for its absence from the urine and bile in this study will be discussed in a subsequent section dealing with the enterohepatic circulation of estrogens.

2. Pregnancy

a) Metabolism after Peripheral Administration. Thirty minutes following the intravenous administration of labeled estriol to women at midpregnancy, some 15% of the circulating radioactive material was identified as estriol-3-glucuronide [54]. Only small amounts of estriol-3-sulfate were excreted in the urine. In another study, the fate of labeled estriol-3-sulfate infused peripherally was examined. During the infusion period, 26% of the urinary radioactivity was excreted as estriol-3-sulfate, 68% in the form of estriol-16-glucuronide, and 6% as estriol-3-glucuronide. During the postinfusion period, only 2% of the radioactive material was excreted as estriol-3-sulfate, whereas 31% was the 16-glucuronide, and 67% was the 3-glucuronide [37].

b) Metabolism in the Fetus at Midpregnancy. It is established that arylsulfurylation is the major pathway of estrogen conjugation in virtually every fetal tissue. This has been demonstrated in a variety of *in vivo* experiments, including (a) perfusion of or injection into previable fetuses [29, 56, 74]; (b) injection into the amniotic fluid prior to the interruption of pregnancy [29, 75]; and (c) injection into the umbilical circulation of the intact fetal-placental unit [20, 21, 76]. These studies have demonstrated that estriol-3-sulfate is the principal product of fetal estriol metabolism [29, 56, 75]. It was also shown that estriol-3-sulfate is not metabolized further to any appreciable extent [5, 57]. Thus, this conjugate seems to be a fetal, metabolic end product, in contrast to estriol-3-glucuronide, which appears to be an adult end product of estriol metabolism. The metabolism of estrone and 17β-estradiol is more complicated, because, not only the unconjugated, but also the sulfurylated forms are metabolized extensively. Thus, estrone sulfate and 17β-estradiol-3-sulfate were identified following the perfusion of previable fetuses with 17β-estradiol [20, 29]. In experiments in which fetuses were perfused with estrone sulfate, labeled with ^{3}H and ^{35}S, evidence for the direct conversion of the substrate to 17β-estradiol-3-sulfate was obtained [77]. In addition, at least 3 conjugates more polar than 17β-estradiol-3-sulfate were detected, each exhibiting the same isotopic ratio as the injected material. One of the compounds was estriol-3-sulfate. These findings as well as the results obtained in another study [51] indicate the lack of appreciable arylsulfatase activity in the human fetus at midterm. Consequently, the isolation of estriol and 15α-hydroxyestradiol from the conjugated fraction of the liver of previable fetuses perfused with estrone sulfate strongly suggests that these metabolites were formed without prior hydrolysis [78].

In contradistinction to the situation in the fetus, the placenta and the fetal membranes are rich in arylsulfatase activity as demonstrated by numerous *in vivo* studies [5, 20, 21, 29, 54, 57, 75, 76].

It is evident from the foregoing considerations that arylsulfurylation is the predominant mode of conjugation of estrogens by the midterm fetus, while in the adult glucuronidation seems to prevail. Therefore, the pattern of conjugation of labeled estriol administered to an anencephalic newborn monster is of interest. Ten minutes following the injection into the umbilical vein, more than 98% of the circulating radioactive material was in a conjugated form, of which 75% was identified as estriol–3–sulfate [58], as was more than 16% of the radioactive material excreted in the urine.

C. Sulfoglucuronides

Estrogens are the only kinds of steroids reported thus far in which the steroid moiety is conjugated both with sulfuric acid and glucuronic acid. Until the recent report of the presence of 16α–hydroxyestrone–3–sulfate–16–glucuronide [17–oxo–3–sulfooxyestra–1,3,5(10)–trien–16α–yl–β–D–glucopyranosiduronic acid] in late pregnancy urine [35], estriol–3–sulfate–16–glucuronide was the only double conjugate of this type identified. On Sephadex chromatography the double conjugates were eluted in peak I. Estriol–3–sulfate–16–glucuronide is a quantitatively significant constituent in various body fluids and tissues [2–6, 55–57]. The double conjugate is formed *in vitro* when estriol–16–glucuronide is incubated with a microsome–free fraction of guinea pig liver [24]. The sequence of conjugation appears to be specific, since the incubation of estriol–3–sulfate with human liver homogenate under conditions in which estriol is extensively converted into estriol–16–glucuronide produced no detectable estriol–3–sulfate–16–glucuronide.

1. *Metabolism in Nonpregnant Conditions*

In a recent investigation, estriol–3–sulfate labeled with both ^{3}H and ^{35}S was infused intravenously into a patient whose bile was being drained *via* a T–tube in the common duct. The biliary and urinary metabolites were identified. Of the ^{3}H–labeled material in the urine, 19% was the double conjugate, while this conjugate accounted for 75% of the biliary radioactivity. Both the urinary and biliary estriol–3–sulfate–16–glucuronide was virtually devoid of ^{35}S. On the other hand, the estriol–3–sulfate isolated from both sources was rich in ^{35}S [6]. This result is consistent with that of the *in vitro* study [24] and indicates that the pathway of synthesis of estriol–3–sulfate–16–glucuronide is estriol → estriol–16–glucuronide → estriol–3–sulfate–16–glucuronide.

In two other patients with T–tube drainage, labeled estriol–3–sulfate–16–glucuronide was administered *via* the T–tube directly into the duodenum. The T–tube was then clamped for 30 minutes, after which period it was re–opened and the bile was collected. The urinary radioactive material consisted of estriol–16– and –3–glucuronides in approximately equal amounts, while the bile contained estriol–16–glucuronide and estriol–

3–sulfate–16–glucuronide [79]. The metabolism of the double conjugate was also studied in an intact patient [79]. The major urinary metabolite was estriol–16–glucuronide. In addition, some estriol–3–glucuronide and estriol–3–sulfate–16–glucuronide were found.

2. Formation in Pregnancy

In a series of experiments previable fetuses were perfused with labeled estriol [56], and estriol–3–sulfate–16–glucuronide was identified in the extracts of the liver, gastrointestinal tract, and perfusate. In subsequent studies, ^{3}H–labeled estriol–3–sulfate and ^{14}C–labeled estriol–16–glucuronide were administered simultaneously into the intact fetal–placental circulation prior to the interruption of gestation at midpregnancy [5]. Whereas virtually all ^{3}H–labeled material was recovered in the form of the sulfate, up to 50% of the ^{14}C–labeled material recovered from the placenta and various fetal tissues was identified as the 3–sulfate–16–glucuronide. Again, there was no conversion of the 3–sulfate into the double conjugate.

The formation of estriol–3–sulfate–16–glucuronide was also investigated in the study of the anencephalic monster injected with labeled estriol [58]. The radioactive material recovered from the blood and urine, 20 and 25%, respectively, was identified as estriol–3–sulfate–16–glucuronide.

D. Other Double Conjugates

The isolation of 17α–estradiol–3–glucuronide–17–β–N–acetylglucosaminide from the rabbit [7, 25] raises the interesting question whether a similar type of estrogen double conjugate exists in the human. The answer appears to be in the affirmative. Evidence that the sulfo N–acetylglucosaminides of 15α–hydroxyestrone [3,15α–dihydroxyestra–1,3,5(10)–trien–17–one] and 15α–hydroxyestradiol [estra–1,3,5(10)–triene–3,15α,17β–triol] are biliary metabolites of estrone sulfate in man has been presented [8]. In addition, a preliminary report claimed that a diglucuronide of estriol is present in pregnancy urine [80]. This remains to be confirmed.

V. Transport of Estrogen Conjugates

On the basis of equilibrium dialysis experiments, it is believed that estrogen conjugates circulate bound to proteins, particularly albumin and α–globulin [81–84]. The quantitative influence of this phenomenon on the enterohepatic circulation and on the estrogen transport between fetus and mother, which will be discussed in detail in a later section, is not known. In addition, its effect on the urinary excretion of estrogen conjugates is incompletely understood. Estriol glucuronides appear to be secreted by the tubules [39, 85]. The mechanism of excretion of the other conjugates by the kidney is not yet known.

A. Enterohepatic Circulation

The importance of enterohepatic circulation to the metabolism of estrone was indicated by Cantarow *et al.* [86] some 25 years ago. The quantitative significance of this phenomenon in the human was first demonstrated by Sandberg and Slaunwhite [49, 87], and later by Levitz and co-workers [6, 72]. The efficiency of the biliary pathway following the injection of various unconjugated and conjugated estrogens is indicated in Table 6–6.

TABLE 6–6. *Biliary excretion of radioactivity following administration of labeled estrogens to women*

Estrogen administered	Cases	% excreted in bile	References
Estrone	3	48	87
17β–Estradiol	2	65	87
Estriol	5	23	49
Estrone sulfate	1	43	72
Estriol–3–sulfate	2	9	6
Estriol–16–glucuronide	3	0	49

Since it is established that unconjugated estrogens are absent from the bile [6, 49, 72, 87], the data of Table 6–6 could be interpreted as indicating that the efficiency of the biliary pathway is influenced by the structure of the estrogen conjugate. There have been studies directed toward the elucidation of the connection between the structure of estrogen conjugates and their enterohepatic circulation [6, 53, 79]. In Fig. 6–2, an attempt has been made to integrate these results with some other available information into a scheme which reflects our concept of the enterohepatic circulation and general metabolism of estriol. Many of the steps which are presented in the scheme have been deduced from the results of studies on the metabolic fate of estriol–3–sulfate labeled with ^{3}H and ^{35}S. Hence, the ensuing discussion arbitrarily begins with the consideration of blood–borne estriol–3–sulfate, although any form of estriol may be traced in the scheme.

There is convincing evidence that some of the circulating estriol–3–sulfate E_3–3S) is excreted in the urine unchanged as shown in pathway [**A**] [6, 37]. Most of it is taken up by the tissues, a large part by the liver [**B**]. A minor fraction of the estriol–3–sulfate extracted by the liver is secreted into the

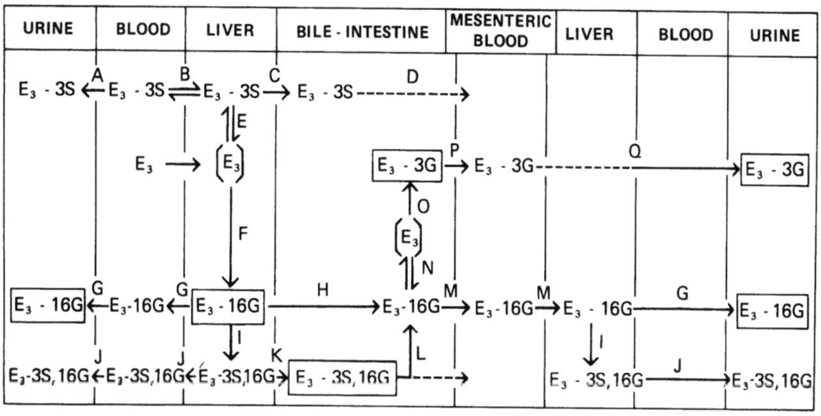

Fig. 6–2. Our present concept of the enterohepatic transport and metabolism of estriol conjugates. ▭ indicates the quantitatively most significant form in the tissue or fluid

intestine *via* the bile [**C**] [6], where its ultimate fate [**D**] remains to be elucidated. Most of the estriol–3–sulfate is converted by the liver [presumably *via* unconjugated estriol (E_3)], [**E**] into estriol–16–glucuronide (E_3–16G), [**F**]. The bulk of the estriol–16–glucuronide is transferred to the blood and then to the urine [**G**] [6, 37]. Only a small part of the estriol–16–glucuronide formed in the liver is secreted into the intestine *via* the bile [**H**] [6]. Estriol–16–glucuronide is converted in significant amounts to estriol–3–sulfate–16–glucuronide in the liver [**I**], from which excretion occurs partly in the urine *via* the blood [**J**] and partly in the bile [**K**]. Indeed, following the intravenous administration of double-labeled estriol–3–sulfate, estriol–3–sulfate–16–glucuronide rich in ^3H but virtually devoid of ^{35}S was the major biliary metabolite. Thus, the conversion of estriol–3–sulfate to the double conjugate appears to occur by the indirect pathway indicated in Fig. 6–2. The metabolism of estriol–3–sulfate–16–glucuronide in the intestine is extensive, and the metabolic events can be deduced only from the study of the urinary metabolites and of the results of *in situ* perfusion experiments as well as incubation studies. With this reservation, the subsequent metabolism is viewed as follows. The estriol–3–sulfate–16–glucuronide in the intestine is converted to estriol–16–glucuronide [**L**], which in part undergoes the enterohepatic pathway [**M**] [79], and in part is converted into estriol–3–glucuronide (E_3–3G) [**O**], probably *via* unconjugated estriol [**N**] [13, 51, 53, 88]. Most of the estriol formed in the intestine would be expected to be converted to estriol–16–glucuronide [**N**] as well as estriol–3–glucuronide [13, 51, 53]. Apparently, estriol–

3–glucuronide, which enters the mesenteric circulation [**P**], is not secreted into the bile [6, 79]; whether any of this conjugate is extracted from the mesenteric blood by the liver remains to be established. It is obvious, however, that it reaches the urine *via* the systemic circulation [**Q**]. Circulating estriol–3–glucuronide is excreted in the urine more rapidly than any other known form of estriol [38].

Gaps exist in our knowledge concerning the extent of hepatic extraction of estriol–16–glucuronide, estriol–3–glucuronide, and estriol–3–sulfate–16–glucuronide. This kind of information may be required to resolve the discrepancy between the results reported by Goebelsmann *et al.* [48] and those of Sandberg and Slaunwhite [49] on the metabolic fate of estriol–16–glucuronide. Sandberg and Slaunwhite found no radioactivity in the bile following the intravenous administration of estriol–16–glucuronide to women, and reported that the conjugate is excreted in the urine unchanged. Goebelsmann *et al.* [48], carrying out similar experiments, did not analyze bile, but observed the conversion of estriol–16–glucuronide to estriol–3–glucuronide in women. If this transformation occurs *via* enterohepatic metabolism, pathway [**G**] between blood and liver would be reversible. However, an alternative pathway for the conversion of the 16– to the 3–glucuronide cannot be excluded. Evidence also points to the reversibility of the transport of estriol–3–sulfate–16–glucuronide between blood and liver [**J**] [79]; however, direct proof is lacking.

It is emphasized that the scheme presented in Fig. 6–2 is valid only for the enterohepatic circulation and metabolism of estriol. Since it is known that, in addition to conjugation, estrone and 17β–estradiol undergo extensive metabolism in the liver and are secreted in large quantities into the bile (*e.g.*, Table 6–6), the enterohepatic metabolism of these compounds must be far more complex than that of estriol.

B. Estrogen Transport between Fetus and Mother

The extent of transfer of estrogen conjugates from the mother to the fetus, which is limited [29, 54, 89], has received relatively little attention. On the other hand, transfer from the fetus to the mother has been studied extensively.

Human placentas have been perfused *in situ* at midpregnancy with labeled estriol, estriol–3–sulfate, and estriol–16–glucuronide [90]. The data summarized in Table 6–7 indicate that unconjugated estriol is transferred across the placenta to the mother at a rate more than 10 times greater than either of the conjugates. The results of other experiments, in which the transport between the fetus and mother was studied at midpregnancy

TABLE 6–7. *Per cent of radioactivity recovered following the in situ perfusion of the placenta at midpregnancy with labeled estriol, estriol–3–sulfate, and estriol–16–glucuronide*[a]

Experiment No.	Perfused compounds	% Radioactivity recovered from			Total recovery
		Urine	Perfusate	Placenta	
1	^{14}C–Estriol	25.7	55.5	8.2	89.4
	^{3}H–Estriol–16–glucuronide	2.6	79.3	18.0	99.9
2	^{14}C–Estriol	32.3	50.7	5.1	88.1
	^{3}H–Estriol–16–glucuronide	0.9	93.9	7.4	102.0
3	^{14}C–Estriol	38.2	47.1	9.1	94.4
	^{3}H–Estriol–3–sulfate	1.7	91.6	3.5	96.8
4	^{14}C–Estriol	22.9	57.1	4.9	84.9
	^{3}H–Estriol–3–sulfate	2.4	90.5	2.7	95.6
5	^{14}C–Estriol–16–glucuronide	0.4	88.5	8.4	97.3
	^{3}H–Estriol–3–sulfate	1.1	90.6	2.1	93.8
6	^{14}C–Estriol–16–glucuronide	0.3	93.1	3.2	96.6
	^{3}H–Estriol–3–sulfate	0.6	96.7	1.1	98.4

In each experiment, the placenta was perfused for a single cycle with 2 forms of estriol, differently labeled, following the delivery of the fetus at laparotomy. The maternal urine perfusate and placenta were analyzed for ^{3}H and ^{14}C.

[a] Levitz, M., Condon, G. P., Dancis, J., Goebelsmann, U., Eriksson, G., Diczfalusy, E.: J. Clin. Endocrinol. Metab. 27, 1723 (1967)

following the injection of the same compounds into the amniotic fluid prior to the interruption of gestation, indicate that the fetal membranes behave similarly to the placenta in this respect [57, 75]. The information available at present on the transport of estriol and its conjugates from the fetus to the mother and on the metabolism of these compounds in the fetus, placenta, and membranes is summarized in Fig. 6–3.

The bulk of the estriol is formed in the placenta predominantly from 16α–hydroxylated neutral and phenolic precursors of fetal origin [21, 91–93]. The bulk of the estriol is transferred mainly to the mother [**A**] [90] and to a lesser extent to the fetus [**B**], where it is converted rapidly, and extensively into estriol–3–sulfate [**C**] [56, 76]. As far as the fetal metabolism is concerned, this reaction is irreversible. Part of the estriol–3–sulfate formed by the fetus is released to the placenta [**D**] where it is converted in part into estriol [**E**] [76], completing the fetal–placental estriol to estriol–

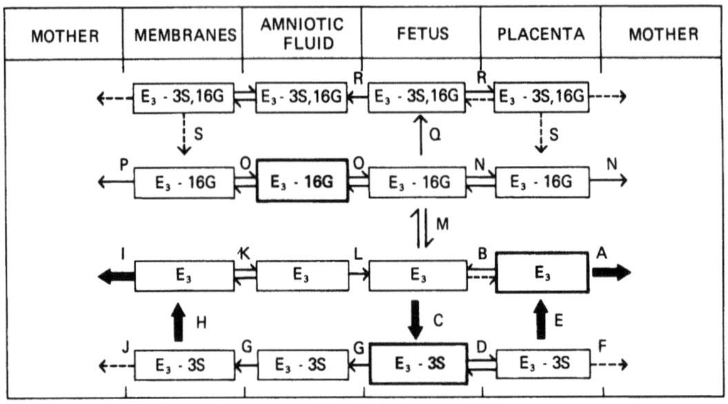

Fig. 6–3. Our present concept of the transport of estriol and its conjugates within the fetal compartments and from the fetal compartments to the mother. Heavy-lined ▭ indicates the quantitatively most significant form in the tissue or fluid. The heavy-lined → indicates the major metabolic or transport pathway

3–sulfate cycle. To what extent, if any, estriol–3–sulfate is transferred as such across the placenta to the mother [**F**] is not known. Some estriol–3–sulfate is excreted by the fetus into the amniotic fluid [**G**] [2]. The fetal membranes hydrolyze estriol–3–sulfate [**H**] [57, 75], and the estriol so formed is transferred to the mother [**I**] and also to the fetus (swallowing) [**L**] [75]. The arylsulfatase activity of the fetal membranes appears to be concentrated in the chorion and absent from the amnion [69]. No information is available on the direct transfer of estriol–3–sulfate across the membranes to the mother [**J**].

Some of the estriol of the fetus is converted to estriol–16–glucuronide [**M**] [56]. This reaction in the fetus is reversible [57]. The estriol–16–glucuronide is in part transferred to the mother across the placenta [**N**], a rather slow process [90]. Although the presence of β–glucuronidase activity in term placental tissue has been reported [94], no evidence of such activity has been found in the *in vivo* studies [5, 11, 57, 90]. It is not surprising therefore, that in experiments in which ^{3}H–labeled estriol–3–sulfate and ^{14}C–labeled estriol–16–glucuronide were perfused simultaneously, the rate of transfer of the ^{3}H-labeled material was higher than that of the ^{14}C–labeled material [90]. The hydrolysis of ^{3}H–estriol–3–sulfate by the placenta produced unconjugated ^{3}H–estriol, which was rapidly transferred across the placenta. This mechanism is not available to the glucuronide.

The estriol–16–glucuronide formed by the fetus is also excreted into the amniotic fluid *via* the fetal kidneys [**O**] [5]. In contradistinction to estriol–3–sulfate, estriol–16–glucuronide is not hydrolyzed by the fetal membranes

[57, 69]. Therefore, the transport across the membranes must occur without hydrolysis [**P**]. Intraamniotically administered estriol-16-glucuronide disappears from the amniotic fluid much slower than estriol-3-sulfate administered simultaneously [57]. These results are in agreement with those obtained in the *in situ* placental perfusion experiments [90] and the same explanation for the difference in behavior between the two conjugates may be offered.

A substantial part of estriol-16-glucuronide is converted by the fetus into estriol-3-sulfate-16-glucuronide [**Q**] [5]. The latter is excreted by the fetus into the amniotic fluid [**R**] [5]. The metabolism and transport of estriol-3-sulfate-16-glucuronide across the placenta and membranes remains to be elucidated. However, it is reasonable to project that the arylsulfatases of the placenta and membranes would convert estriol-3-sulfate-16-glucuronide into estriol-16-glucuronide [**S**].

VI. Biochemical and Pharmacological Effects

A. *In vitro* Experiments

Several reports are available on the influence *in vitro* of estrogen sulfates on various enzyme and transport systems. For instance, it has been reported that estrogen sulfates influence the uptake of uric acid by erythrocytes [95], and the equilibrium distribution of free amino acids in Ehrlich ascites tumor cells [96]. Furthermore, they inhibit a variety of enzyme systems, all of which seem to require pyridoxal phosphate [97, 98]. However, the specificity of estrogen sulfates in inducing these effects remains to be established.

B. *In vivo* Effects

It has been reported that the intravenous administration of estrogen sulfates resulted in increased plasma levels of transcortin (corticosteroid-binding globulin), β-glucuronidase, and ceruloplasmin [99], circulating triglycerides [100], and a nonspecific hyaluronidase inhibitor [101]. It has also been reported that estrogen sulfates effect a decrease in pulmonary diffusing capacity [102]. Several reports have been published on the biological and pharmacological effects of estrogen sulfates. Claims have been made of the beneficial effects on various pathological conditions, including myocardial infarction [103], otorhinolaryngological hemorrhages [104], uterine prolapse [105], secondary hyphema after ocular trauma [106], decreased capillary strength [107], prostatic cancer [108], and mammary cancer [109].

A detailed description and evaluation of these effects is within the province of Chapter 8. The reviewers would merely like to point out that, thus far, all the estrogen conjugates studied by them, and particularly the sulfates, would be hydrolyzed following oral and intravenous administration [6, 37, 72, 76]. It is hardly possible therefore, at present to ascribe a specific effect to any estrogen conjugate.

VII. Concluding Remarks

From a critical appraisal of the literature published during the past 30 years on the metabolism of estrogens, it appears that the main concern of the numerous investigators has been the elucidation of the transformations of the steroid nucleus without apparent consideration of the conjugation processes. It is obvious from the material presented in this chapter that conjugation is an integral part of estrogen metabolism. Estrogen conjugates can be metabolized without prior hydrolysis as is the case in the human fetus. An estrogen molecule may be subjected to a series of different conjugating and hydrolytic steps, occurring alternately, as exemplified by the enterohepatic metabolism. Although many investigators undoubtedly recognized the significance of the role of conjugation in estrogen metabolism, they were seriously handicapped by methodological difficulties due mainly to the unavailability of pure standards. Although some 25 estrogens are known to occur in the human, and many of these estrogens are conjugated in more than one form, only 8 conjugates involving just the three classical estrogens, estrone, 17β–estradiol, and estriol, have been isolated from human sources. Their radioactive forms have become available only within recent years.

Intimately related to the lack of suitable standards is the lack of accurate estimations of estrogen conjugates in human tissues and body fluids. Such standards are indispensable for the evaluation of any procedure. Therefore, most of the quantitative estimates presented in this chapter should be considered tentative.

Most of the information available on the metabolism and transport of estrogen conjugates has been obtained by the use of labeled substrates. Even in these studies definitive information was restricted to the 8 conjugates of the 3 classical estrogens. However, the study of estrogens cannot be restricted to the three classical estrogens. For example, hydroxylation in position 2 instead of position 16 is a quantitatively more important pathway in certain conditions [110], and nothing is known about the role of conjugation in the metabolism of 2–hydroxy– and 2–methoxyestrogens. The same can be said about 15α–, 15β–, 6α–, and 7α–hydroxylated estrogens.

Numerous biochemical, physiological, and pharmacological effects have been ascribed to estrogen conjugates. Much more critical information will be required before the specificity of these effects can be evaluated.

It might even be more difficult at our present state of knowledge to speculate about the raison d'être for estrogen conjugation. An expansion of our knowledge in the field of conjugation alone will certainly not provide the answer, unless it can be related to the chain of events which ultimately result in the manifestation of biological activity.

Finally, to those who feel that we have placed by far too much emphasis on the conditions existing in the human species, we would like to quote Alexander Pope from 1672:

"Know then thyself, presume not God to scan,
The proper study of mankind is man."

Alexander Pope Ep. ii, 1.1

Acknowledgments. The authors were supported by the following grants: USPHS Grant 2RO1 Hd-00462-11A1 from the National Institute of Child Health and Human Development, and by Research Grants from the Ford Foundation; Medical Research Council of Sweden Grant 19 P-703-01 A; and USPHS Grant HD-00413 from the National Institute of Child Health and Human Development; USPHS Grant CA-02071-14 from the National Cancer Institute; Grant P-206H from the American Cancer Society; and USPHS Research Career Development Awardee 5-K3-HD-18, 422-05 from the National Institute of Child Health and Human Development.

VIII. References

1. Shaw, R. F., Katzman, P. A., Doisy, E. A.: Endocrinology **57**, 87 (1955).
2. Troen, P., Nilsson, B., Wiqvist, N., Diczfalusy, E.: Acta Endocrinol. **38**, 361 (1961).
3. Touchstone, J. C., Greene, J. W., Jr., McElroy, R. C., Murawec, T.: Biochemistry **2**, 653 (1963).
4. Smith, O. W., Hagerman, D. D.: J. Clin. Endocrinol. Metab. **25**, 732 (1965).
5. Goebelsmann, U., Eriksson, G., Diczfalusy, E., Levitz, M., Condon, G. P.: Acta Endocrinol. **53**, 391 (1966).
6. Emerman, S., Twombly, G. H., Levitz, M.: J. Clin. Endocrinol. Metab. **27**, 539 (1967).
7. Layne, D. S., Sheth, N. A., Kirdani, R. Y.: J. Biol. Chem. **239**, 3221 (1964).
8. Jirku, H., Levitz, M.: J. Clin. Endocrinol. Metab. **29**, 615 (1969).
9. Slaunwhite, W. R., Jr., Lichtman, M. A., Sandberg, A. A.: J. Clin. Endocrinol. Metab. **24**, 638 (1964).
10. Smith, E. R., Breuer, H.: Biochem. J. **88**, 168 (1963).
11. Zucconi, G., Goebelsmann, U., Wiqvist, N., Diczfalusy, E.: Acta Endocrinol. **56**, 71 (1967).
12. Goebelsmann, U., Diczfalusy, E., Katz, J., Levitz, M.: Steroids **6**, 859 (1965).

13. Dahm, K., Breuer, H.: Z. Klin. Chem. **4**, 153 (1966).
14. — — Biochim. Biophys. Acta **128**, 306 (1966).
15. Segal, H. L.: J. Biol. Chem. **213**, 161 (1955).
16. Schneider, J. J., Lewbart, M. L.: J. Biol. Chem. **222**, 787 (1956).
17. DeMeio, R. H., Lewycka, C., Wizerkaniuk, M., Salciunas, O.: Biochem. J. **68**, 1 (1958).
18. Döllefeld, E., Breuer, H.: Z. Vitamin, Hormon-Fermentforsch. **14**, 193 (1966).
19. Diczfalusy, E., Cassmer, O., Alonso, C., de Miquel, M., Westin, B.: Acta Endocrinol. **37**, 516 (1961).
20. Haynes, R. C., Jr., Mikhail, G., Eriksson, G., Wiqvist, N., Diczfalusy, E.: Acta Endocrinol. **45**, 297 (1964).
21. Schwers, J., Eriksson, G., Diczfalusy, E.: Acta Endocrinol. **49**, 65 (1965).
22. Payne, A. H., Mason, M.: Biochim. Biophys. Acta **71**, 719 (1963).
23. Wengle, B., Boström, H.: Acta Chem. Scand. **17**, 1203 (1963).
24. Levitz, M., Katz, J., Twombly, G. H.: Steroids **6**, 553 (1965).
25. Jirku, H., Layne, D. S.: Biochemistry **4**, 2126 (1965).
26. Diczfalusy, E., Lauritzen, C.: Oestrogene beim Menschen, S. 326. Berlin-Göttingen-Heidelberg: Springer 1961.
27. Adlercreutz, H.: Acta Med. Scand. **175**, Suppl. **412**, 133 (1964).
28. Schwers, J.: Les Oestrogènes au Cours de la Seconde Moitié de la Grossesse. Bruxelles: Editions Arscia S.A. 1965.
29. Diczfalusy, E., Cassmer, O., Alonso, C., de Miquel, M.: Recent Progr. Hormone Res. **17**, 147 (1961).
30. Bugge, S., Nilsen, M., Metcalfe-Gibson, A., Hobkirk, R.: Can. J. Biochem. Physiol. **39**, 1501 (1961).
31. Brown, J. B., Keller, R., Matthew, G. D.: J. Obstet. Gynaecol. Brit. Empire **66**, 177 (1959).
32. Cohen, S. L., Marrian, G. F., Odell, A. D.: Biochem. J. **30**, 2250 (1936).
33. Hashimoto, Y., Neeman, M.: J. Biol. Chem. **238**, 1273 (1963).
34. Beling, C. G.: Acta Endocrinol. **43**, Suppl. **79**, 98 (1963).
35. Smith, E. R., Kellie, A. E.: Biochem. J. **104**, 83 (1967).
36. Slaunwhite, W. R., Jr., Karsay, M. A., Sandberg, A. A.: J. Clin. Endocrinol. Metab. **24**, 263 (1964).
37. Goebelsmann, U., Eriksson, G., Wiqvist, N., Diczfalusy, E.: Acta Endocrinol. **50**, 273 (1965).
38. — Cooke, I., Wiqvist, N., Diczfalusy, E.: Acta Endocrinol. **52**, 30 (1966).
39. Smith, O. W.: Acta Endocrinol. **51**, Suppl. **104**, 3 (1966).
40. Adlercreutz, H.: Acta Endocrinol. **42**, Suppl. **72**, 11 (1962).
41. Diczfalusy, E., Magnusson, A.-M.: Acta Endocrinol. **28**, 169 (1958).
42. Kinsella, R. A., Jr., Francis, F. E., Thayer, S. A., Doisy, E. A.: J. Biol. Chem. **219**, 265 (1956).
43. Diczfalusy, E., Menini, E., Tillinger, K.-G., Westman, A.: Acta Endocrinol. **30**, 539 (1959).
44. Menini, E., Diczfalusy, E.: Endocrinology **67**, 500 (1960).
45. — — Endocrinology **68**, 492 (1961).

46. Engel, L. L., Baggett, B., Halla, M.: Endocrinology **70**, 907 (1962).
47. Beer, C. T., Gallagher, T. F.: J. Biol. Chem. **214**, 335 (1955).
48. Goebelsmann, U., Sjöberg, K., Wiqvist, N., Diczfalusy, E.: Acta Endocrinol. **50**, 261 (1965).
49. Sandberg, A. A., Slaunwhite, W. R., Jr.: J. Clin. Invest. **44**, 694 (1965).
50. Lehtinen, A., Nurmikko, V., Hartiala, K.: Acta Chem. Scand. **12**, 1585 (1958).
51. Diczfalusy, E., Franksson, C., Martinsen, B.: Acta Endocrinol. **38**, 59 (1961).
52. — — Lisboa, B. P., Martinsen, B.: Acta Endocrinol. **40**, 537 (1962).
53. Støa, K. F., Levitz, M.: Acta Endocrinol. **57**, 657 (1968).
54. Wilson, R., Eriksson, G., Diczfalusy, E.: Acta Endocrinol. **46**, 525 (1964).
55. Diczfalusy, E., Cassmer, O., Alonso, C, de Miquel, M.: Acta Endocrinol. **38**, 31 (1961).
56. Mikhail, G., Wiqvist, N., Diczfalusy, E.: Acta Endocrinol. **42**, 519 (1963).
57. Goebelsmann, U., Wiqvist, N., Diczfalusy, E., Levitz, M., Condon, G. P., Dancis, J.: Acta Endocrinol. **52**, 550 (1966).
58. Diczfalusy, E., Barr, M., Lind, J.: Acta Endocrinol. **46**, 511 (1964).
59. Wallace, E. Z., Silberman, N.: J. Biol. Chem. **239**, 2809 (1964).
60. Troen, P., de Miquel, M., Alonso, C.: Biochemistry **5**, 332 (1966).
61. Pulkkinen, M. O.: Acta Physiol. Scand. **59**, Suppl. **213**, 124 (1963).
62. Wengle, B.: Acta Soc. Med. Upsalien. **69**, 105 (1964).
63. Boström, H., Franksson, C., Wengle, B.: Acta Endocrinol. **47**, 633 (1964).
64. — Wengle, B.: Acta Soc. Med. Upsalien. **69**, 41 (1964).
65. Warren, J. C., French, A. P.: J. Clin. Endocrinol. Metab. **25**, 278 (1965).
66. Sandberg, E. C., Jenkins, R. C.: Biochim. Biophys. Acta **113**, 190 (1966).
67. Schwers, J., Rodesch, F.: Ann. Endocrinol. (Paris) **24**, 931 (1963).
68. Alonso, C., Troen, P.: Biochemistry **5**, 337 (1966).
69. Katz, S. R., Dancis, J., Levitz, M.: Endocrinology **76**, 722 (1965).
70. Purdy, R. H., Engel, L. L., Oncley, J. L.: J. Biol. Chem. **236**, 1043 (1961).
71. Lewison, E. F., Levi, J. E., Jones, G. S., Jones, H. W., Jr., Silberstein, H. E.: Cancer **4**, 537 (1951).
72. Twombly, G. H., Levitz, M.: Am. J. Obstet. Gynecol. **80**, 889 (1960).
73. Jirku, H., Hogsander, U., Levitz, M.: Biochim. Biophys. Acta **137**, 588 (1967)
74. Levitz, M., Condon, G. P., Dancis, J.: Endocrinology **68**, 825 (1961).
75. Diczfalusy, E., Tillinger, K.-G., Wiqvist, N., Levitz, M., Condon, G. P., Dancis, J.: J. Clin. Endocrinol. Metab. **23**, 503 (1963).
76. Mikhail, G., Wiqvist, N., Diczfalusy, E.: Acta Endocrinol. **43**, 213 (1963).
77. Emerman, S., Dancis, J., Levitz, M., Wiqvist, N., Diczfalusy, E.: J. Clin. Endocrinol. Metab. **25**, 639 (1965).
78. Schwers, J., Gevaerts-Videtsky, M., Wiqvist, N., Diczfalusy, E.: Acta Endocrinol. **50**, 597 (1965).
79. Levitz, M., Katz, J.: J. Clin. Endocrinol. Metab. **28**, 862 (1968).
80. Felger, C. B., Katzman, P. A.: Federation Proc. **20**, 199 (1961).
81. Antoniades, H. N., McArthur, J. W., Pennell, R. B., Ingersoll, F. M., Ulfelder, H., Oncley, J. L.: Am. J. Physiol. **189**, 455 (1957).
82. Bischoff, F., Stauffer, R. D.: Am. J. Physiol. **191**, 313 (1957).

83. Touchstone, J. C., Chang, E.: Proc. Soc. Exptl. Biol. Med. **112**, 694 (1963).
84. Slaunwhite, W. R., Jr., Rosenthal, H., Sandberg, A. A.: Abstr. 107, 6th Pan-American Congress of Endocrinology, Mexico City, Mexico, October 1965, p. E 53.
85. Brown, C. H., Saffan, B. D., Howard, C. M., Preedy, J. R. K.: J. Clin. Invest. **43**, 295 (1964).
86. Cantarow, A., Rakoff, A. E., Paschkis, K. E., Hansen, L. P., Walking, A. A.: Endocrinology **31**, 515 (1942).
87. Sandberg, A. A., Slaunwhite, W. R., Jr.: J. Clin. Invest. **36**, 1266 (1957).
88. Inoue, N., Slaunwhite, W. R., Jr., Graham, J. B., Sandberg, A. A.: Abstr. 146, 49th Meeting of the Endocrine Soc., Miami, Florida, June 1967, p. 101.
89. Migeon, C. J., Bertrand, J., Gemzell, C. A.: Recent Progr. Hormone Res. **17**, 207 (1961).
90. Levitz, M., Condon, G. P., Dancis, J., Goebelsmann, U., Eriksson, G., Diczfalusy, E.: J. Clin. Endocrinol. Metab. **27**, 1723 (1967).
91. Magendantz, H. G., Ryan, K. J.: J. Clin. Endocrinol. Metab. **24**, 1155 (1964).
92. Colás, A., Heinrichs, W. L., Tatum, H. J.: Steroids **3**, 417 (1964).
93. Bolté, E., Wiqvist, N., Diczfalusy, E.: Acta Endocrinol. **52**, 583 (1966).
94. Fishman, W. H., Anlyan, A. J.: Cancer Res. **7**, 808 (1947).
95. Christensen, H. N., Jones, J. C.: J. Biol. Chem. **236**, 76 (1961).
96. Riggs, T. R., Walker, L. M.: Endocrinology **74**, 483 (1964).
97. Mason, M., Schirch, LaV.: Federation Proc. **20**, 200 (1961).
98. Pulkkinen, M. O., Willman, K.: Steroids **8**, 51 (1966).
99. Musa, B. U., Seal, U. S., Doe, R. P.: J. Clin. Endocrinol. Metab. **25**, 1163 (1965).
100. Robinson, R. W., Lebeau, R. J.: J. Atherosclerosis Res. **5**, 120 (1965).
101. Wayne, L., Glueck, H. I., Brodine, C., Coots, M.: Proc. Soc. Exptl. Biol. Med. **116**, 85 (1964).
102. Pecora, L. J., Putnam, L. R., Baum, G. L.: Am. J. Med. Sci. **246**, 82 (1963).
103. Marmorston, J., Moore, F. J., Kuzma, O. T., Magidson, O., Weiner, J.: Clin. Res. **9**, 62 (1961).
104. del Carril, A. V., de Gastaldo, L., Saporiti, E. M., Marino, A.: Semana Med. (Buenos Aires) **122**, 980 (1963); – Chem. Abstr. **59**, 15827f (1963).
105. Anderson, H. E.: J. Am. Med. Assoc. **168**, 173 (1958).
106. Goldberg, J. L.: Arch. Ophthalmol. **63**, 1001 (1960).
107. Clemetson, C. A. B., Blair, L. M., Reed, D. H.: Am. J. Obstet. Gynecol. **83**, 1261 (1962).
108. Kaufman, J. J., Goodwin, W. E.: J. Am. Geriat. Soc. **4**, 296 (1956).
109. Hertz, R., Young, J. P., Tullner, W. W.: Ciba Found. Colloq. Endocrinol. **1**, 157 (1952).
110. Fishman, J., Hellman, L., Zumoff, B., Gallagher, T. F.: J. Clin. Endocrinol. Metab. **22**, 389 (1962).

Chapter 7

The Isolation and Metabolism of Conjugates of Neutral Steroids from Natural Sources

SAMUEL SOLOMON and BHAGU R. BHAVNANI

Contents

I. Introduction

As was typical for all of the early discoveries in endocrinology, the search for a biologically active male hormone led to the finding by Funk, Harrow, and Lejwa in 1929 of a conjugated urinary steroid [1]. These investigators noted that acidification of male urine to pH 4.0–4.5 and continuous extraction with chloroform led to an enhancement of the biological activity of the extract. This observation was confirmed by Kabak in 1931 [2] and by Adler in 1934 [3] who acidified the urine and hydrolyzed it by boiling. In 1936 Odell and Marrian [4] and Venning and Browne [5] demonstrated that hot acid hydrolysis of a butanol–soluble extract of human pregnancy urine yielded unconjugated pregnanediol. Venning and Browne [5] isolated the C_{19}–conjugate in a crystalline form (m.p. 268–271°), and its C and H analysis was close to that of pregnanediol glucuronide and it gave a positive Tollens Test indicative of glucuronic acid. Pregnanediol glucuronide was regarded as the conjugated water–soluble form in which large amounts of the steroid were excreted in the urine. This was the traditional physiologic role assigned to all steroid conjugates.

The first major publication on the isolation of *neutral* steroid conjugates came from the work of Venning *et al.* [6] who isolated androsterone sulfate from the urine of a man with an interstitial cell tumor of the testes. This discovery was followed by the isolation of dehydroisoandrosterone sulfate from normal male urine by Munson *et al.* [7]. Because relatively sophisticated methods had become available for the chemical manipulation of unconjugated steroids, it was necessary to develop a method for the hydrolysis of steroid conjugates so that the recovery of the steroid moiety would be quantitative and unaffected by the formation of artifacts. The initial success with the use of acid hydrolysis in the cleavage of steroid glucuronides and sulfates was soon overshadowed by the realization that this procedure was at best a balance between hydrolysis and destruction; but what was more detrimental was the formation of numerous artifacts during the use of such procedures. A list of the artifacts formed in the course of isolation of steroids from urine has recently been compiled by Dorfman and Ungar [8]. Enzyme hydrolysis, using various sources of β–glucuronidase, and aryl sulfatase, soon replaced acid hydrolysis, and the development of a solvolytic procedure for the hydrolysis of steroid sulfates by Burstein and Lieberman [9] proved extremely useful and has been widely employed. These newer techniques of hydrolysis greatly accelerated research on steroid conjugates.

A major change in our thinking concerning the role of steroid conjugates came from the demonstration by Baulieu [10, 11] that dehydroisoandrosterone sulfate is secreted by the human adrenal gland. This finding came at a time when Lieberman and collaborators [12] had begun to study the origin of the urinary 17–ketosteroids; and all of these facts taken together indicated that a new role would have to be assigned to steroid conjugates in the origin and metabolism of steroids. From the pioneering work of Drs. Lieberman, Gurpide, Vande Wiele and associates (see Chapter 5) and that of Baulieu and associates [13], the role of steroid conjugates in the formation and metabolism of steroids has evolved to the point that we can now consider these molecules both as intermediates in the formation and metabolism of steroids, and as secretory as well as excretory products.

The scope of this chapter is to review the isolation and metabolism of the conjugates of neutral steroids other than 3β–hydroxy–Δ^5–steroids, which are discussed in Chapter 5. We will not consider isolation and metabolic studies in pathological conditions because this will be covered in Chapter 9. In our review it became apparent that certain criteria had to be set up to narrow the scope of the subject; because every steroid is excreted in the form of a conjugate in human urine, all investigations in this species in some way or another deal with conjugates. After a broad review of the literature we concluded that the isolation and metabolism of conjugates would be considered under three categories: (a) conjugates of steroids isolated and identified in this form (Group I); (b) steroid conjugates which have been purified chromatographically as the conjugate but hydrolyzed by specific methods prior to final characterization (Group II); (c) steroid conjugates isolated in the aqueous soluble form, which have then been hydrolyzed by differential methods and the unconjugated steroid identified (Group III). All other reported investigations dealing with conjugates which did not meet the above criteria were considered to be outside the scope of this review.

II. C_{19}–Steroid Glucuronides

A. Isolation of Testosterone Glucuronide [3–oxoandrost–4–en–17β–yl–β–D–glucopyranosiduronic acid] and Epitestosterone Glucuronide [3–oxoandrost–4–en–17α–yl–β–D–glucopyranosiduronic acid]

In considering the isolation of the glucuronides and sulfates of the C_{19}–steroids, it was apparent that these conjugates would have to be dealt with in a broad group which could then be subdivided; where possible, closely related steroids have been discussed at the same time. For example, in the consideration of the glucuronides of C_{19}–steroids, it is an advantage to discuss testosterone glucuronide and epitestosterone glucuronide at the same time. The isolation of these conjugates is summarized in Table 7–1.

TABLE 7-1. *Isolation of testosterone glucuronide and epitestosterone glucuronide*

Steroid	Source	Means of identi-fication[a]	Refer-ences
Testosterone	Human plasma	I	14
	Human plasma	III	16, 15
	Human urine	III	17, 18
	Human urine after the administration of ^{14}C-testosterone	III	19, 20
	Human intestinal microsomes	I	21
	Human prostate slices	III	22
	Slices of rat liver	I	23
	Slices of rat liver	II	24
	Salmon plasma and testes	III	25, 26
Epitestos-terone	Human urine	III	27
	Human urine after injection of ^{3}H-epitestosterone	III	28
	Human urine after injection of ^{3}H-epitestosterone	III	29

[a] I = Conjugate isolated and identified in this form. II = Conjugate isolated and steroid identified after hydrolysis. III = Conjugate extracted in an aqueous-soluble form and hydrolyzed by differential methods and steroid identified.

The isolation of testosterone glucuronide from human blood was accomplished by Hadd and Rhamy [14]. These authors administered 4-^{14}C-testosterone [17β-hydroxyandrost-4-en-3-one] to an elderly patient with an arrhenoblastoma and isolated the glucuronide as the tri-*O*-acetyl methyl ester by using the isotope dilution technique after the addition of the derivative as carrier, and testosterone as carrier after β-glucuronidase hydrolysis of the conjugate. During their studies on the physiological disposition of testosterone glucuronide in man, Korenman and Lipsett [15, 16] hydrolyzed an aqueous extract of peripheral and hepatic vein plasma with β-glucuronidase and isolated the testosterone released. Schubert and Wehrberger [17] provided the first indication of the presence of testosterone glucuronide in human urine when they hydrolyzed a 30-liter pool of male urine with β-glucuronidase and then isolated testosterone. This observation was confirmed by the work of Camacho and Migeon [18] who were able to identify testosterone following the hydrolysis of human urine with β-glucuronidase. Subsequently, similar observations have been published by a number of investigators who have devised procedures for the measurement of testosterone glucuronide in human urine.

During the course of studying the metabolism of ^{3}H– and ^{14}C–labeled C_{19}–steroids, testosterone glucuronide was isolated from urine and two such examples will be cited. In delineating the "17β–hydroxyl pathway" of testosterone metabolism, Baulieu and Mauvais–Jarvis [19] injected 17α–^{3}H–testosterone and 4–^{14}C–testosterone intravenously into normal human subjects and isolated testosterone containing both labels from the glucuronide fraction of urine after differential hydrolysis, using solvolysis first followed by β–glucuronidase. In a separate study Baulieu and Mauvais–Jarvis [20] injected 4–^{14}C–testosterone and 1,2–^{3}H–androstenedione [androst–4–ene–3,17–dione] intravenously into human subjects and again isolated testosterone from the glucuronide fraction of the urine. These authors concluded that more of the urinary testosterone from the glucuronide fraction was derived from 4–^{14}C–testosterone than from 1,2–^{3}H–androstenedione and that this difference was more pronounced in the male than in the female.

The *in vitro* formation of testosterone glucuronide has also been demonstrated in human intestine and in human prostate. Dahm *et al.* [21] showed that the microsomal fraction of human intestine is capable of forming testosterone glucuronide and to a lesser degree the glucuronides of aryl steroids and ring A–saturated C_{19}–steroids. However, this tissue fraction did not form glucuronides with Δ^{5}–3β–hydroxysteroids. Using slices of human prostate incubated with 4–^{14}C–testosterone, Wotiz *et al.* [22] were able to demonstrate the formation of an aqueous–soluble metabolite which, after hydrolysis with β–glucuronidase, yielded testosterone. The reservation concerning the results, which these authors point to, is that the β–glucuronidase used was not a pure enzyme and, as a result, introduced an element of uncertainty in designating the product formed as testosterone glucuronide.

That rat liver slices are capable of conjugating testosterone with glucuronic acid was demonstrated by Wotiz *et al.* [23]. These authors found that, when the conjugate formed was transformed to the methyl tri–*O*–acetyl derivative, it was identical to methyl[3–oxoandrost–4–en–17β–yl–2′,3′,4′–tri–*O*–acetyl–β–D–glucopyranosid]uronate, thus establishing the structure of testosterone glucuronide. In studying the synthesis of steroid glucuronides in rat liver, Schriefers *et al.* [24] also demonstrated the synthesis of testosterone glucuronide by this tissue. It is of some interest that there is good indirect evidence for the presence of testosterone glucuronide in salmon plasma and testes. This was demonstrated by Grajcer and Idler [25, 26] who hydrolyzed the aqueous extracts of large pools of plasma and testes with β–glucuronidase and identified the testosterone released. The specificity of the hydrolysis was checked with the use of the glucuronidase inhibitor, saccharo-1,4-lactone.

Epitestosterone [17α-hydroxyandrost-4-en-3-one] is excreted in human urine in the form of a conjugate. Korenman *et al.* [27] isolated epitestosterone from the urine of normal males following hydrolysis of the conjugates with β-glucuronidase; and Wilson and Lipsett [28] found that, following the administration of labeled epitestosterone, most of the urinary radioactivity was released after hydrolysis with β-glucuronidase and that almost half of this amount was accounted for by epitestosterone glucuronide. These investigators also found that plasma testosterone is not a precursor of urinary epitestosterone glucuronide and that epitestosterone is not converted to urinary testosterone glucuronide or to androst-16-en-3α-ol. Dray and Ledru [29] injected a mixture of ^{3}H-epitestosterone and ^{14}C-testosterone into normal adult male subjects and found that there was no interconversion of the injected steroids to urinary products isolated following β-glucuronidase hydrolysis of the urinary conjugates.

B. Isolation of Androsterone Glucuronide [17-oxo-5α-androstan-3α-yl-β-D-glucopyranosiduronic acid], Etiocholanolone Glucuronide [17-oxo-5β-androstan-3α-yl-β-D-glucopyranosiduronic acid], and Dehydroisoandrosterone Glucuronide [17-oxoandrost-5-en-3β-yl-β-D-glucopyranosiduronic acid]

The 11-deoxy-17-ketosteroids are composed of the glucuronides and sulfates of androsterone [3α-hydroxy-5α-androstan-17-one], etiocholanolone [3α-hydroxy-5β-androstan-17-one], and dehydroisoandrosterone [3β-hydroxyandrost-5-en-17-one]. In Table 7-2 are summarized the isolations of the glucuronides of the 11-deoxy-17-ketosteroids.

In their early attempt to estimate renal clearance of 17-ketosteroids, Bongiovanni and Eberlein [30] administered large amounts of androsterone by mouth to a normal male and found that, after β-glucuronidase hydrolysis of an aqueous extract of plasma, they could identify the steroid administered. Following the intravenous administration of 7-^{3}H-dehydroisoandrosterone and 4-^{14}C-testosterone to a normal male, urinary androsterone glucuronide and etiocholanolone glucuronide were identified by Siiteri *et al.* [31]. The glucuronides were eluted from a Celite column, hydrolyzed with β-glucuronidase, and the liberated androsterone and etiocholanolone were identified. Hadd and Dorfman [32] isolated androsterone glucuronide from the urine of a subject with an adrenal adenoma, and Schneider and Lewbart [33] isolated the same conjugate following the oral administration of large amounts of androsterone to humans. After the oral administration

TABLE 7–2. *Isolation of androsterone glucuronide, etiocholanolone glucuronide, and dehydroisoandrosterone glucuronide*

Steroid	Source	Means of identification[a]	References
Androsterone	Human plasma	III	30
	Human urine	II	31
	Human urine (adrenal adenoma)	I	32
	Human urine	I	33
	Human urine	III	34, 35
	Human urinc	II	36
	Rat liver slices incubated with testosterone	II	37
	Dog kidney perfusion	II	38
	Human urine after the administration of ^{14}C–testosterone	III	19, 20
	Human urine (children)	III	44
	Human urine after the administration of ACTH	I	39
Etiocholanolone	Human urine after oral administration of dehydroisoandrosterone hemisuccinate	I	33
	Human urine	I	32
	Human urine	III	34
	Human urine	III	35
	Human urine	II	36
	Human urine	II	31
	Human urine after the administration of ^{14}C–testosterone	III	19, 20
	Dog kidney perfusion	II	38
	Human urine (children)	III	44
	Human urine after the administration of ACTH	I	39
Dehydroisoandrosterone	Human urine after the oral administration of dehydroisoandrosterone	I	33
	Human urine	I	31
	Human urine	III	34

[a] I = Conjugate isolated and identified in this form. II = Conjugate isolated and steroid identified after hydrolysis. III = Conjugate extracted in an aqueous–soluble form and hydrolyzed by differential methods and steroid identified.

of large amounts of dehydroisoandrosterone to normal humans, Kirschner *et al.* [34] were able to detect androsterone, etiocholanolone, and small amounts of dehydroisoandrosterone in the β–glucuronidase hydrolyzed fraction of urine extracts. During the course of developing methods for studying the metabolism of C_{19}–steroids, MacDonald *et al.* [35] administered ^3H–dehydroisoandrosterone to normal male subjects and isolated androsterone and etiocholanolone from urine extracts after β–glucuronidase hydrolysis. In these studies the urinary conjugates were first hydrolyzed by solvolysis and then with β–glucuronidase. In a review on the isolation of urinary conjugates performed in his laboratory, Baulieu [36] presented evidence for the isolation of urinary androsterone glucuronide and etiocholanolone glucuronide. This group of investigators separated the urinary glucuronides from the sulfates on alumina or on florisil columns and separated the individual conjugates by column and paper chromatography. Androsterone and etiocholanolone were identified after β–glucuronidase hydrolysis of the purified urinary fractions.

Schriefers *et al.* [37] demonstrated that rat liver slices incubated with labeled testosterone can convert the substrate to androsterone glucuronide. These authors purified the conjugate by chromatography and then identified the steroid after β–glucuronidase hydrolysis. Cohn and Hume [38] injected labeled androsterone and etiocholanolone into the renal artery of the dog and demonstrated the presence of the glucuronide of the two steroids in renal vein plasma. The two glucuronides were purified by electrophoresis and paper chromatography prior to hydrolysis with β–glucuronidase after which unconjugated androsterone and etiocholanolone were identified.

In their studies on the metabolism of testosterone in man, Baulieu and Mauvais–Jarvis [19, 20] isolated androsterone and etiocholanolone from the "glucuronide" fraction of urine after the hydrolysis of conjugates first by solvolysis and then with β–glucuronidase. In these studies mixtures of labeled testosterone and androstenedione were administered to human subjects. Foggitt and Kellie [39] isolated and conclusively identified the glucuronides of androsterone, etiocholanolone, 11β–hydroxyandrosterone [3α,11β–dihydroxy–5α–androstan–17–one], 11β–hydroxyetiocholanolone [3α,11β–dihydroxy–5β–androstan–17–one], and 11–ketoetiocholanolone [3α–hydroxy–5β–androstane–11,17–dione] from human urine after the administration of ACTH [adrenocorticotrophic hormone].

Schneider and Lewbart [33] administered large amounts of dehydroisoandrosterone hemisuccinate [3β–hemisuccinoyloxyandrost–5–en–17–one] to humans by mouth and isolated etiocholanolone glucuronide from the urine. Etiocholanolone glucuronide was subsequently isolated from the urine of a female with an adrenal adenoma [32]. The other isolation studies on etiocholanolone glucuronide cited in Table 7–2 have already been discussed. There are several reports in the literature dealing with the estimation of urinary 17–ketosteroids; however, in most of these studies the individual 17–ketosteroids were not well characterized. Lieberman *et al.* [40], Vande Wiele *et al.* [41], Jailer *et al.* [42], and Lakshmanan and Lieberman [43] modified and developed new techniques for the isolation and purification of individual 17–ketosteroids. In these studies they identified urinary dehydroisoandrosterone, androsterone, etiocholanolone, 11–ketoetiocholanolone, 11–hydroxyandrosterone, and 11–hydroxyetiocholanolone. These steroids were isolated after differential hydrolysis of the urine, followed by pooling of the various extracts; hence, the mode of conjugation of these steroids could not be ascertained with any certainty. Using differential hydrolytic methods where the steroid conjugates in urine are first cleaved by solvolysis and then with β–glucuronidase, Paulsen *et al.* [44] found androsterone and etiocholanolone in the "glucuronide" fraction of children's urine.

Dehydroisoandrosterone glucuronide has been isolated from human urine following the oral ingestion of large amounts of the dehydroisoandrosterone hemisuccinate [33]. Subsequently, dehydroisoandrosterone glucuronide was isolated from normal human urine following the administration of ^{3}H–dehydroisoandrosterone [31]. After the oral administration of dehydroisoandrosterone to normal subjects, Kirschner *et al.* [34] hydrolyzed the urinary conjugates with β–glucuronidase and found small amounts of dehydroisoandrosterone. The excretion of dehydroisoandrosterone in the glucuronide fraction of urine did not increase to any extent even when 1,000 mg/day of dehydroisoandrosterone was given.

C. Isolation of Reduced and Hydroxylated C_{19}–Steroid Glucuronides

In addition to the C_{19}–steroid glucuronides described in Tables 7–1 and 7–2, there are a number of others which have been demonstrated and these are grouped together in Table 7–3. Schriefers *et al.* [37] showed that slices from rat liver incubated with testosterone was capable of forming the glucuronides of 5α–androstane–3β,17β–diol, 5α–androstane–3α,17β–diol, and 5α–dihydrotestosterone [17β–hydroxy–5α–androstan–3–one]. From the analysis of the amount of glucuronic acid released after hydrolysis it would seem that the metabolites formed in rat liver are conjugated as monoglucuronides. Baulieu and Mauvais–Jarvis [19, 20] injected ^{14}C–testosterone and ^{3}H–androstenedione into human subjects and isolated labeled 5α–androstane–3α,17β–diol and 5β–androstane–3α,17β–diol from the glucuronide fraction of urine following hydrolysis of the conjugates with β–glucuronidase. Fukushima *et al.* [45] isolated 3α,11β–dihydroxy-

TABLE 7–3. *Isolation of reduced and hydroxylated C_{19}–steroid glucuronides*

Steroid	Source	Means of identification[a]	References
5α–Androstane–3β,17β–diol 5α–Androstane–3α,17β–diol 17β–Hydroxy–5α–androstan–3–one	Rat liver slices incubated with testosterone	II	37
5β–Androstane–3α,17β–diol 5α–Androstane–3α,17β–diol	Human urine after injection of ^{14}C–testosterone and ^{3}H–androstenedione	III	19, 20
3α,11β–Dihydroxyandrost–4–en–17–one	Human urine after the administration of ^{3}H–11β–hydroxyandrostenedione	III	45
3α,11β–Dihydroxy–5α–androstan–17–one	Human urine after the administration of ACTH	I	39
	Human urine (children)	III	44
3α,11β–Dihydroxy–5β–androstan–17–one	Human urine after the administration of ACTH	I	39
	Human urine (children)	III	44
3α–Hydroxy–5β–androstane–11,17–dione	Human urine after the administration of ACTH	I	39
	Human urine (children)	III	44
3,11β–Dihydroxyandrostane–3,5–dien–17–one	Rat urine after the administration of 11–ketoandrostenedione	II	46
16α–Hydroxyandrostenedione 16α–Hydroxydehydroisoandrosterone 16α–Hydroxyandrosterone 16α–Hydroxyetiocholanolone 3α,16α–Dihydroxyandrost–5–en–17–one Androst–5–ene–3α,16α,17β–triol	Human urine after the administration of 16α–hydroxydehydroisoandrosterone	III	47

[a] See footnote Table 7–1.

androst–4–en–17–one from the glucuronide fraction of human urine following the intravenous administration of ^{3}H–11β–hydroxyandrostenedione [11β–hydroxyandrost–4–ene–3,17–dione]. It was noted by these authors that, during β–glucuronidase hydrolysis, prolonged exposure of the urine at pH 5.0 gave rise to a 1:1 mixture of the 3α– and 3β–epimers, but it was shown that the 3β–epimers were derived from the 3α–hydroxysteroid.

Three 11–oxy–17–ketosteroid conjugates have been isolated from children's urine by Paulsen *et al.* [44]. These authors identified 11β–hydroxyandrosterone, 11β–hydroxyetiocholanolone, 11–ketoetiocholanolone in the glucuronide fraction of urine after differential hydrolysis of the conjugates first by solvolysis followed by β–glucuronidase. Previously the glucuronides of the three 11–oxy–17–ketosteroids were firmly identified after the isolation from human urine following the administration of ACTH [39]. There is a well–documented report in the literature by Wotiz and Fishman [46] on the isolation of the enol–glucuronide of 11β–hydroxyandrostenedione. These investigators administered large amounts of 11–ketoandrostenedione [androst–4–ene–3,11,17–trione] to rats by stomach tube and isolated an aqueous–soluble metabolite from the urine which, on hydrolysis by β–glucuronidase, yielded 11β–hydroxyandrostenedione. To date no confirmatory evidence has been published on enol–glucuronides of 3–keto–4–ene steroids.

A number of metabolites of 16α–hydroxydehydroisoandrosterone [3β,16α–dihydroxyandrost–5–en–17–one] which were isolated from the glucuronide fraction of human urine have been described by YoungLai and Solomon [47]. Following the intravenous administration of ^{3}H–16α–hydroxydehydroisoandrosterone to human subjects, the urinary conjugates were first hydrolyzed by solvolysis and then with β–glucuronidase, and the following steroids were identified in the glucuronide fraction: 16α–hydroxyandrostenedione [16α–hydroxyandrost–4–ene–3,17–dione], 16α–hydroxydehydroisoandrosterone, 16α–hydroxyandrosterone [3α,16α–dihydroxy–5α–androstan–17–one], 16α–hydroxyetiocholanolone [3α,16α–dihydroxy–5β–androstan–17–one], 3α,16α–dihydroxyandrost–5–en–17–one, and androst–5–ene–3α,16α,17β–triol.

III. C_{19}–Steroid Sulfates

A. Isolation of Testosterone Sulfate [17β–sulfooxyandrost–4–en–3–one] and Epitestosterone Sulfate [17α–sulfooxyandrost–4–en–3–one]

While there is no evidence for the glandular secretion of steroid glucuronides, it is known that steroid sulfates can be formed by endocrine tissues, and there is good evidence for the adrenal secretion of dehydroisoandrosterone sulfate [3β–sulfooxyandrost–5–en–17–one]. Because steroid sulfates are intermediates in the metabolism of steroids this class of conjugate has been studied more intensively. Tables 7–4, 7–5, and 7–6 show the sulfates isolated and, Table 7–4 summarizes the isolation of the sulfates of testosterone and epitestosterone.

TABLE 7–4. *Isolation of testosterone sulfate and epitestosterone sulfate*

Steroid	Source	Means of identification[a]	References
Testosterone	Human plasma	II	48
	Human plasma	II	49
	Human urine	II	50
	Human fetal adrenals after injection of ^{3}H–androstenedione and ^{14}C–testosterone into the umbilical vein	I	51
	Human fetal adrenals and gastro–intestinal tract after perfusion of human fetuses with ^{3}H–testosterone and ^{14}C–androstenedione	III	52
	Adrenal tumor slices incubated with ^{14}C–progesterone	III	53
	Normal human testes tissue incubated with ^{14}C–testosterone and ^{3}H–dehydroisoandrosterone sulfate	I	54
Epitestosterone	Human plasma	II	49
	Human plasma	III	29

[a] See footnote Table 7–1.

Testosterone sulfate was identified in human plasma by Saez *et al.* [48] who also devised a method for its measurement in the plasma of normal subjects and patients with gonadal and adrenal disorders. These authors found much higher concentrations of testosterone sulfate in male than in female plasma, and they obtained suggestive evidence for the secretion of this conjugate by abnormal testes, ovaries, and tumors. Dray *et al.* [49] measured testosterone sulfate and epitestosterone sulfate in a pool of plasma from four normal males and found more testosterone sulfate than epitestosterone sulfate. These authors added the ^{3}H– and ^{14}C–labeled sulfates to the plasma as an aid in its isolation and quantitation. Testosterone sulfate has also been identified in the urine of normal male subjects by Tamm *et al.* [50].

The formation of testosterone sulfate by both male and female human fetuses at mid–term has recently been demonstrated by Mancuso *et al.* [51]. These authors injected a mixture of ^{3}H–androstenedione and ^{14}C–testosterone into the umbilical circulation and isolated testosterone sulfate from the aqueous extract of the adrenals. In a subsequent study Benangiano *et al.* [52] perfused male and female previable human fetuses with ^{3}H–testosterone and ^{14}C–androstenedione and isolated a large number of reduced metabolites from the fetal tissues. Following solvolysis of the aqueous extract of the adrenals and the gastrointestinal tract, it was possible to isolate testosterone from both tissues. The *in vitro* formation of testosterone sulfate by slices of a virilizing adrenal tumor incubated with ^{14}C–progesterone [pregn 4–ene–3,20 dione] has been demonstrated by Dixon *et al.* [53]. In addition, Dixon *et al.* [54] incubated a mince of normal adult testes with ^{14}C–testosterone and were able to demonstrate the formation of testosterone sulfate. Testosterone sulfate was also isolated in these latter studies when ^{3}H–dehydroisoandrosterone sulfate was the substrate.

As indicated in Table 7–4, epitestosterone sulfate has been isolated from plasma and urine. Its isolation from plasma by Dray *et al.* [49] has been discussed, and its isolation from human urine by Dray and Ledru [29] was demonstrated following the injection of a mixture of ^{3}H–epitestosterone and ^{14}C–testosterone into normal males.

B. Isolation of Androsterone Sulfate [3α–sulfooxy–5α–androstan–17–one], Isoandrosterone Sulfate [3β–sulfooxy–5α–androstan–17–one], and Etiocholanolone Sulfate [3α–sulfooxy–5β–androstan–17–one]

In Table 7–5 are shown the isolations of the sulfates of androsterone, isoandrosterone [3β–hydroxy–5α–androstan–17–one] and etiocholanolone from human plasma, bile, and urine. Androsterone sulfate was isolated from a 12–liter pool of human blood by Staib *et al.* [55]. These workers prepared a butanol extract of the blood and purified the sulfates of dehydroisoandrosterone and androsterone by high voltage electrophoresis and by paper chromatography. After hydrolysis of the conjugates, they obtained evidence for the presence of unconjugated dehydroisoandrosterone and androsterone. Baulieu [56] analyzed the plasma pools of normal subjects and subjects given ACTH. He found evidence for the presence of dehydroisoandrosterone sulfate, androsterone sulfate, and small amounts of etio-

TABLE 7–5. *Isolation of androsterone sulfate, isoandrosterone sulfate, and etiocholanolone sulfate*

Steroid	Source	Means of identification[a]	References
Androsterone	Human blood	II	55
	Human plasma before and after ACTH	II	56
	Quantitation in human plasma	II	57
	Quantitation in human plasma	II	58
	Quantitation in human plasma	III	59
	Quantitation in human plasma	II	60
	Human pregnancy plasma	II	61
	Human plasma	II	62
	Human plasma after administration of metopirone	II	63
	Human plasma, diurnal variation	II	64
	Human bile	II	65
	Human urine	I	36
Isoandrosterone	Human urine after oral administration of isoandrosterone	I	66
	Human plasma	III	62
	Human plasma	II	64
Etiocholanolone	Human plasma after administration of ACTH	II	56
	Human plasma	III	59
	Human urine	I	67
3β–Hydroxy–5β–androstan–17–one	Human urine after oral administration of the steroid	I	68

[a] See footnote Table 7–1.

cholanolone sulfate in the plasma. The concentration of the three sulfates in blood increased after the administration of ACTH. McKenna and Rippon [57] published a procedure for the determination of the sulfates of dehydroisoandrosterone and androsterone in normal and pathological human plasma. These investigators also reported the presence of etiocholanolone sulfate in the plasma of a subject with idiopathic periodic fever. Conrad *et al.* [58] have also published a procedure for the quantitation of the sulfates of androsterone and dehydroisoandrosterone in human plasma and studied the changes in the concentration of these conjugates in patients treated for virilizing disorders. Oertel and Kaiser [59] also described a procedure for the quantitation of the sulfates of androsterone and dehydroisoandrosterone in human plasma; and more recently, Plager [60] developed a new method to measure these conjugates using ion exchange resins. The range of values for normal males and females of androsterone sulfate and dehydroisoandrosterone sulfate published by these four groups of investigators are approximately the same [57–60].

Sjövall *et al.* [61] analyzed the steroid sulfates in human plasma during pregnancy and presented gas chromatography—mass spectrum evidence for the presence of androsterone in the "monosulfate" fraction. In a separate study, Sjövall and Vihko [62] analyzed the solvolyzable steroids in normal human blood and presented evidence for the presence of the sulfates of dehydroisoandrosterone, androsterone, and isoandrosterone. Following the administration of metopirone [2–methyl–1,2–bis(3–pyridyl)–1–propanone] to human subjects, Jänne *et al.* [63] found that the plasma concentration of the sulfates of androsterone, dehydroisoandrosterone, and isoandrosterone increased. How increased pituitary ACTH secretion causes an increase in the concentration of C_{19}–steroid sulfates in plasma is not at all clear. In their study of the solvolyzable steroids in human plasma, Laatikainen and Vihko [64] obtained suggestive evidence to indicate that there is a diurnal variation in the plasma concentration of the sulfates of dehydroisoandrosterone and androsterone. This is in agreement with an earlier finding of McKenna and Rippon [57] for dehydroisoandrosterone sulfate in human plasma.

There are a number of steroid sulfates in bile, and Laatikainen *et al.* [65] isolated androsterone sulfate from this source. The isolation and identification of androsterone sulfate from human urine has been published by Baulieu [36].

Isoandrosterone sulfate has been isolated and identified in human urine by Baulieu [66] after the oral administration of 500 mg of the unconjugated steroid to a normal eight–year–old subject. This conjugate has also been isolated from human plasma by Laatikainen and Vihko [64] in their study of solvolyzable steroids from this source.

The presence of etiocholanolone sulfate in plasma is not well established. The detection of small amounts of this conjugate in normal plasma and after the administration of ACTH has been reported by Baulieu [56]. Oertel and Kaiser [59] reported data on the quantitation of etiocholanolone sulfate in human plasma. Both of these reports need to be substantiated by unequivocal evidence for the presence of etiocholanolone sulfate in human plasma. The isolation of etiocholanolone sulfate from human urine has been documented by Baulieu [67]. In addition, Baulieu and Emiliozzi [68] isolated 3β–hydroxy–5β–androstan–17–one sulfate [3β–sulfooxy–5β–androstan–17–one] from the urine of a six–year–old child who was given 200 mg of the steroid by mouth. Baulieu *et al.* [69] injected ^{14}C–testosterone and ^{3}H–androstenedione into male and female subjects and isolated etiocholanolone and androsterone from the glucuronide and sulfate fractions of urine after differential hydrolysis. These authors observed a sex difference in the metabolism and subsequent conjugation of these steroids.

TABLE 7–6. *Isolation of reduced and hydroxylated C_{19}–steroid sulfates*

Steroid	Source	Means of identification [a]	References
5α–Androstane–3α,17α–diol	Human bile	II	65
5α–Androstane–3α,17β–diol	Human bile	II	65
11β,17β–Dihydroxyandrost–4–en–3–one	Human fetal adrenals after perfusion of fetuses with ^{3}H–testosterone and ^{14}C–androstenedione	III	70
16α–Hydroxydehydroisoandrosterone	Human urine after administration of ^{3}H–16α–hydroxydehydroisoandrosterone	III	47
16α–Hydroxyandrosterone	Human urine	III	47
16α–Hydroxyetiocholanolone	Human urine	III	47
3α,16α–Dihydroxyandrost–5–en–17–one	Human urine	III	47
Androst–5–ene–3β,16α,17β–triol	Human urine	III	47
Androst–5–ene–3α,16α,17β–triol	Human urine	III	47
Androst–5–ene–3α,16α,17β–triol	Human urine	III	47
5α–Androstane–3α,16α,17β–triol	Human urine	III	47

[a] See footnote Table 7–1.

C. Isolation of Reduced and Hydroxylated C_{19}–Steroid Sulfates

In addition to the C_{19}–steroid sulfates described in Tables 7–4 and 7–5, a number of sulfates of hydroxylated and reduced steroids have been isolated in recent years (Table 7–6). Laatikainen *et al.* [65] chromatographed extracts of human bile on Sephadex LH–20 and obtained a steroid "monosulfate" and "disulfate" fraction. After solvolysis of the "disulfate" fraction, it was possible to tentatively identify 5α–androstane–3α,17α–diol and 5α–androstane–3α,17β–diol by gas chromatography—mass spectroscopy. While studying the metabolism of ^{3}H–testosterone and ^{14}C–androstenedione perfused into the mid–term previable human fetus, Mancuso *et al.* [70] isolated 11β,17β–dihydroxyandrost–4–cn–3–onc following solvolysis of thc aqueous extract of the adrenals. It was mentioned earlier that testosterone sulfate was isolated from the adrenals when the same precursors were perfused into the human fetus [52]. In studying the metabolism of ^{3}H–16α–hydroxydehydroisoandrosterone in humans, YoungLai and Solomon [47] fed the unlabeled steroid and injected the labeled steroid intravenously. From the "sulfate" fraction of urine prepared following solvolysis it was possible to isolate and identify 16α–hydroxydehydroisoandrosterone, 16α–hydroxyandrosterone, 16α–hydroxyetiocholanolone, 3α,16α–dihydroxyandrost–5–en–17–one, androst–5–ene–3β,16α,17β–triol, androst–5–ene–3α,16α,17β–triol, and 5α–androstane–3α,16α,17β–triol.

A new class of nonpolar C_{19}–steroid conjugates, the so–called sulfatidyl–17–ketosteroids and phosphatidyl–17–ketosteroids have been reported by Oertel and collaborators [71–73]. A full discussion of this subject will have to await confirmation of the existence of these conjugates by other investigators.

IV. Metabolism of the Glucuronides and Sulfates of C_{19}–Steroids

An interesting aspect in the study of steroid conjugates is the further metabolism of these compounds without the loss of the glucuronide or sulfate. Such metabolic processes have been referred to as the "direct metabolism" of a conjugate, a term originally used by Dr. E.–E. Baulieu. As is evident in this section, the metabolism of the C_{19}–steroid conjugates has been investigated and a number of interesting observations have emerged. In Table 7–7 is outlined the metabolism of the glucuronides and sulfates of the C_{19}–steroids starting with androsterone glucuronide which was the first glucuronide whose metabolism was well documented.

In order to determine whether a urinary steroid glucuronide can be further metabolized, Siiteri and Lieberman [74] prepared the ^{3}H–labeled

TABLE 7–7. *Metabolism of the glucuronides and sulfates of C_{19}–steroids*

Steroid conjugate	System	Products isolated	References
Androsterone glucuronide	Human *in vivo*	None	74
^{3}H–Testosterone–^{14}C–glucuronide	Human *in vivo*	^{3}H–5β–androstane–3α,17β–diol–^{14}C–glucuronide ^{3}H–Etiocholanolone glucuronide ^{3}H–Etiocholanolone sulfate	75
Testosterone glucuronide	Human *in vivo*	Etiocholanolone 5β–Androstane–3α,17β–diol	76
Dehydroisoandrosterone glucuronide	Human *in vivo*	Dehydroisoandrosterone glucuronide 3β–Hydroxy–5β–androstan–17–one glucuronide Δ^{5}–Androstenediol glucuronide Androstane–3β,17β–diol glucuronide Androsterone Etiocholanolone	13
^{3}H–Dehydroisoandrosterone glucuronide	Human *in vivo*	Androsterone glucuronide Androsterone sulfate Etiocholanolone glucuronide Etiocholanolone sulfate Dehydroisoandrosterone glucuronide Dehydroisoandrosterone sulfate Androst–5–ene–3β,17β–diol glucuronide Androstane–3β,17β–diol glucuronide 5β–Androstane–3β,17β–diol glucuronide	77

TABLE 7–7 (continued)

Steroid conjugate	System	Products isolated	References
^{3}H–Dehydroisoandrosterone–^{14}C–glucuronide	Human *in vivo*	Androst–5–ene–3β,16α,17β–triol 16α–Hydroxydehydroisoandrosterone Androst–5–ene–3β,17β–diol Testosterone Dehydroisoandrosterone 3β–hydroxy–5β–androstan–17–one 3β–Hydroxy–5α–androstan–17–one 3α–Hydroxy–5α–androstan–17–one Androstenedione 5β–Androstane–3,17–dione 5α–Androstane–3,17–dione	78
	Human portal vein and duodenum	Steroid "sulfates" Steroid "glucuronides"	79
^{3}H–Dehydroisoandrosterone–^{14}C–glucuronide and 7–^{3}H–Dehydroisoandrosterone sulfate	Guinea pig duodenum and mesenteric vein	Steroid "sulfates" Steroid "glucuronides"	80
Isoandrosterone glucuronide	Human *in vivo*	Isoandrosterone glucuronide Isoandrosterone sulfate Androsterone glucuronide Androsterone sulfate	13
Androstenedione enol–glucuronide [17–oxoandrosta–3,5–dienyl–β–D–glucopyranosiduronic acid]	Rat liver mince	Unidentified metabolites	81
Testosterone sulfate	Human *in vivo*	Not metabolised	13
	Human placental microsomes	Not aromatised	82
	Rat liver microsomes	17β–hydroxy–5α–androstan–3–one sulfate	83

TABLE 7–7 (continued)

Steroid conjugate	System	Products isolated	References
Androsterone sulfate	Human *in vivo*	Not metabolised	13
	Rabbit liver slices	Androstane–3α,17α–diol sulfate	84
Etiocholanolone sulfate	Human *in vivo*	Not metabolised	13
Isoandrosterone sulfate	Human *in vivo*	Androsterone sulfate Androsterone glucuronide	13

conjugate (randomly labeled) and administered it as a tracer dose to normal humans. The androsterone glucuronide isolated from urine had the same distribution of ^{3}H between the steroid and glucuronic acid as the injected conjugate, and urinary androsterone sulfate did not contain ^{3}H. It was concluded from these studies that androsterone glucuronide was rapidly excreted without undergoing further metabolism. In contrast to androsterone glucuronide there is further metabolism of testosterone glucuronide in man. Robel *et al.* [75] prepared ^{3}H–testosterone and ^{14}C–glucuronide and injected a tracer dose into a normal male. Approximately 20% of the injected steroid was recovered in the urine and 8% was found as urinary ^{3}H–5β–androstane–3α,17β–diol–^{14}C–17–glucuronide. In addition 8% of the injected dose was found as urinary ^{3}H–etiocholanolone glucuronide and 3.5% as urinary ^{3}H–etiocholanolone sulfate. This demonstrated that reduction of ring A can occur in the presence of a 17–glucuronide, and that a portion of the testosterone glucuronide or the 5β–androstane–3α,17β–diol glucuronide was hydrolyzed, allowing the unconjugated steroid to be further metabolized to etiocholanolone, which was then conjugated with sulfuric or with glucuronic acid. In a separate study, Robel *et al.* [76] investigated the metabolic fate of testosterone glucuronide in human subjects and compared its metabolism with that of testosterone. Following the injection of labeled testosterone glucuronide into normal males, only 5β–reduced urinary metabolites were isolated, and more labeled 5β–androstane–3α,17β–diol was found in the urine than labeled 3α–hydroxy–5β–androstan–17–one. When ^{14}C–testosterone was administered along with the ^{3}H–testosterone glucuronide, 5α–reduced metabolites were formed from the former, and more labeled 3α–hydroxy–5β–androstan–17–one was isolated from the urine than 5β–androstane–3α,17β–diol. This is further proof that, during the course of metabolism of testosterone glucuronide, the A–ring is preferentially reduced to yield A/B–*cis* products. Furthermore, these data suggest that the hydrolysis of testosterone glucuronide prior to metabolism is unlikely and that 5β–androstane–3α,17β–diol–17–glucuronide is probably cleaved, with the unconjugated steroid undergoing further metabolism. In this study the authors also calculated that about 5% of the testosterone glucuronide was derived from the injected testosterone.

Baulieu *et al.* [13] went on to study ^{3}H–dehydroisoandrosterone glucuronide and ^{14}C–dehydroisoandrosterone injected into a normal female and found that the conjugate was further metabolized either with the glucuronic acid still at C3 or after its cleavage. Because the glucuronic acid was not labeled in this study, it is difficult to determine the extent of direct metabolism of the conjugate. It is possible that there was a direct reduction of the double bond because of the formation of ^{3}H–3β–hydroxy–5β–androstan–17–one, and a direct reduction of the 17–ketone because of the isolation of Δ^5–androstenediol–3–glucuronide and androstane–3β,17β–diol–3–glucuronide. Further studies by Robel *et al.* [77] on the metabolic fate of dehydroisoandrosterone glucuronide in man are summarized in Table 7–7. These investigators injected ^{3}H–dehydroisoandrosterone glucuronide and ^{14}C–dehydroisoandrosterone into two normal individuals and isolated a number of metabolites from the "sulfate" and "glucuronide" fractions of the urine after differential hydrolysis. From the metabolites isolated it was evident that dehydroisoandrosterone glucuronide can be metabolized directly by hydrogenation at C17 and reduction of the double bond. The authors calculated that dehydroisoandrosterone glucuronide gave 8 times more urinary dehydroisoandrosterone sulfate than dehydroisoandrosterone itself.

A more detailed study of the direct metabolism of dehydroisoandrosterone glucuronide was reported by Knapstein *et al.* [78]. These authors injected ^{3}H–dehydroisoandrosterone–^{14}C–3–glucuronide (^{3}H/^{14}C = 2.02) into a normal female and reported the isolation of eleven steroid glucuronides from the urine (Table 7–7) having the same ^{3}H/^{14}C ratio as the precursors. The authors therefore concluded that the direct conversion of dehydroisoandrosterone–3–glucuronide to metabolites such as androsterone and etiocholanolone glucuronide necessitates the intermediacy of 3,5–dienol–3–glucuronide. The evidence for such dienol–glucuronides is not available at present. It is of interest that the half–life of dehydroisoandrosterone glucuronide in this study was 16.5 minutes, which is a marked contrast to the long half–life of dehydroisoandrosterone sulfate. Knapstein *et al.* [79] injected ^{3}H–dehydroisoandrosterone–^{14}C–glucuronide into the portal vein of normal human subjects and concluded that the C_{19}–steroid "glucuronides" reached the peripheral circulation without hydrolysis, but appeared in bile only after hydrolysis and reconjugation with sulfuric acid or glucuronic acid. When ^{3}H–dehydroisoandrosterone–^{14}C–glucuronide was infused into the duodenum, the conjugate was completely hydrolyzed within the intestinal wall. The interpretation of the data obtained is difficult because the authors did not analyze individual steroid conjugates but rather separated them into groups such as unconjugated steroids, "glucuronides" and "sulfates". In a second study Knapstein *et al.* [80] injected ^{3}H–dehydroisoandrosterone–^{14}C–glucuronide and ^{3}H–dehydroisoandrosterone–^{35}S–sulfate into the duodenum or into a mesenteric vein of guinea pigs in order to study the enterohepatic metabolism of these conjugates. They found that dehydroisoandrosterone sulfate was transported across the intestinal wall and the liver of the guinea pig without hydrolysis. It was also found that dehydroisoandrosterone glucuronide was hydrolyzed in the intestinal wall and was reconjugated therein with sulfuric acid, while in the liver it was secreted into bile without hydrolysis and into the peripheral circulation only after extensive hydrolysis and reconjugation with sulfuric acid. In these studies no individual steroid conjugates or unconjugated steroids were isolated and only group separations are reported, again making it difficult to interpret the results obtained.

Baulieu *et al.* [13] synthesized ^{3}H–isoandrosterone glucuronide [17–oxo–5α–androstan–3β–yl–β–D–glucopyranosiduronic acid] and injected it into a normal male. The most abundant urinary product was isoandrosterone glucuronide, and small amounts of androsterone glucuronide, androsterone sulfate, and isoandrosterone sulfate were also isolated. In contrast to androsterone glucuronide, isoandrosterone glucuronide was hydrolyzed during the course of its metabolism. In a separate study these authors found no labeled isoandrosterone glucuronide in the urine after the administration of ^{3}H–isoandrosterone and it would seem that there is no endogenous production of this conjugate in man.

Wotiz [81] incubated androstenedione and its enol–glucuronide with a rat liver mince and obtained a large number of products which he could not identify. From the characteristics of some of the products formed, the author reached the tentative conclusion that the enol–glucuronide was further metabolized. A definitive identification of the products formed is necessary before definite conclusions can be reached concerning this interesting mode of metabolism of enol–glucuronides.

Baulieu *et al.* [13] prepared ^{14}C–testosterone sulfate and injected it into a normal male. No ^{14}C–androsterone or ^{14}C–etiocholanolone was found in the urinary conjugates, and only 3.5% of the substrate could be recovered. When ^{14}C–testosterone sulfate was administered to a pregnant female, the isolated urinary estrone [3-hydroxyestra-1,3,5(10)-trien-17-one] and 17β-estradiol [estra-1,3,5(10)-triene-3,17β-diol] did not contain any radioactivity, indicating that the conjugate was not hydrolyzed in the placenta. This was borne out by the data obtained by Cheatum *et al.* [82] who incubated testosterone sulfate with a human placental microsome preparation and could not demonstrate the formation of estrogens although testosterone was readily aromatized under the same experimental conditions. In contrast to these findings is the observation of Wu and Mason [83] who demonstrated a further metabolism of testosterone sulfate to the 5α-dihydro derivative by the Δ^4-3-ketosteroid reductase of rat liver microsomes. This is another example of an *in vitro* finding which is difficult to conciliate with other *in vivo* observations. When labeled androsterone sulfate was administered to humans [13] no evidence could be obtained for its further metabolism and only a minute amount of labeled androstane-3α,17β-diol could be detected in the urine. However, when a rabbit-liver preparation was incubated with androsterone sulfate there was a reduction of C 17-ketone to form androstane-3α,17α-diol-3-sulfate [3α-sulfooxy-5α-androstan-17α-ol] [84] demonstrating a direct metabolism *in vitro* in this species. Likewise Baulieu *et al.* [13] found that etiocholanolone sulfate is not further metabolized in man, but that labeled isoandrosterone sulfate when administered to a normal male gives rise to small amounts of androsterone sulfate and androsterone glucuronide excreted into the urine.

A discussion of steroid conjugates would not be complete without touching on the work done by Lieberman and collaborators [85–88] on the *in vivo* metabolism and production of C_{19}-steroids. It is outside the scope of this chapter to discuss production rates based on the specific activities of urinary metabolites, but the authors would like to point to this body of work as having had a great influence on our understanding of the role of steroid conjugates in the metabolism of steroids.

V. C_{21}-Steroid Glucuronides

A. Isolation of the Glucuronides of Reduced and Hydroxylated C_{21}-Steroids

In discussing the C_{21}-steroid glucuronides isolated from natural sources, it is as well to start with the glucuronides of the dihydroxy-pregnanes because this was the first type of neutral steroid isolated as a glucuronide.

In Table 7–8 are shown the glucuronides of the pregnanediols, pregnanetriols, and related steroids isolated mainly from human sources. Pregnan-

TABLE 7–8. *Isolation of the glucuronides of reduced and hydroxylated C_{21}–steroids*

Steroid	Source	Means of identification[a]	References
3α–Hydroxy–5β–pregnan–20–one (pregnanolone)	Female plasma	III	89
5β–Pregnane–3α,20α–diol (pregnanediol)	Human urine	I	5
	Female plasma	III	89
	Human meconium	III	90
	Human fetal liver	I	91
5β–Pregnane–3β,20β–diol	Female plasma	III	89
3α,16α–Dihydroxy–5β–pregnan–20–one	Human fetal liver	III	91
3α–Hydroxy–5β–pregn–16–en–20–one	Human meconium	III	92
Pregnane–3α,17α,20α–triol Pregnane–3α,17α,20β–triol 3α,17α–Dihydroxy–5β–pregnan–20–one	Human urine after injection of ^{14}C–pregnane–3α,17α,20β–triol	III	93
3α,17α–Dihydroxy–5β–pregnan–20–one	Urine of female pseudohermaphrodite	I	94
3–Keto–4–etianic acid	Guinea pig liver slices	I	95

[a] See footnote Table 7–1.

olone [3α–hydroxy–5β–pregnan–20–one] and pregnanediol [5β–pregnane–3α,20α–diol] were detected in the "glucuronide" fraction of plasma of nonpregnant females following the intravenous injection of ^{14}C–progesterone [89]. The first isolation of pregnanediol glucuronide [20α–hydroxy–5β–pregnan–3α–yl–β–D–glucopyranosiduronic acid] from human urine has already been discussed [5]. While pregnanediol exists largely as a sulfate in human meconium, some of it may also be present as a glucuronide [90]. Whether the glucuronide exists in meconium would depend on whether the pH 0.7 continuous ether extraction of the aqueous phase hydrolyzed all of the sulfate present; otherwise, some of it would appear in the glucuronide fraction. Pregnanediol glucuronide has also been isolated from human fetal liver after the injection of ^{14}C–progesterone into the umbilical circulation or following perfusion of the mid–term fetus with the labeled hormone [91].

In addition to pregnanolone and pregnanediol, the "glucuronide" fraction of nonpregnant human plasma also contains 5β–pregnane–3β,20β–diol [89]. Following the injection of ^{14}C–16α–hydroxyprogesterone [16α–hydroxypregn–4–ene–3,20–dione] into the umbilical circulation with the fetus *in utero*, it was possible to isolate 3α,16α–dihydroxy–5β–pregnan–20–one from the fetal liver after β–glucuronidase hydrolysis of the aqueous tissue fraction [91]. It is of interest that 3α–hydroxy–5β–pregn–16–en–20–one was isolated from the glucuronide fraction of human meconium by Francis *et al.* [92]. It is possible that a C16–dehydration could have taken place which was mediated by bacterial enzymes and not artifactually, as Francis *et al.* [92] checked their chemical manipulations with a 16α–hydroxysteroid. The presence of these 16α–hydroxylated glucuronides in fetal tissues and meconium results from the fact that 16α–hydroxyprogesterone is one of the most abundant metabolites of progesterone formed in the fetal adrenals [91]. Following the injection of ^{14}C–pregnane–3α,17α,20β–triol into a male subject it was possible to isolate pregnane–3α,17α,20α–triol, pregnane–3α,17α,20β–triol, and 3α,17α–dihydroxy–5β–pregnan–20–one from the glucuronide fraction of urine [93]. These urinary products were virtually absent from the sulfate fraction of urine after differential hydrolysis of the conjugates. It is of interest that some twenty years ago Mason and Strickler isolated 3α,17α–dihydroxy–5β–pregnan–20–one glucuronide [17α–hydroxy–20–oxo–5β–pregnan–3α–yl–β–D–glucopyranosiduronic acid] from the urine of a female pseudohermaphrodite [94].

There is a glucuronide that has been formed *in vitro*, which is of some interest in that it is an ester glucuronide of a steroid; namely, that of 3–keto–4–etianic acid [95] (see also Chapter 1). This conjugate was described by Schneider [95] following the incubation of deoxycorticosterone [21–hydroxypregn–4–ene–3,20–dione] with slices of rabbit liver.

B. Isolation of the Glucuronides of Corticosteroids

The isolation of the glucuronides of the corticosteroids has not been extensively investigated and only a limited amount of data is available which is summarized in Table 7–9. There are considerably greater amounts of unconjugated cortisol [11β,17α,21–trihydroxypregn–4–ene–3,20–dione] in human plasma than in the conjugated form. Oertel [96] has provided indirect evidence for the presence of cortisol glucuronide [11β,17α–dihydroxy–3,20–dioxopregn–4–en–21–yl–β–D–glucopyranosiduronic acid] in human plasma following the administration of ACTH. The largest amount of this conjugate seems to be present in the hepatic vein when compared with the femoral vein and the adrenal vein concentrations. Following the

TABLE 7–9. *Isolation of the glucuronides of corticosteroids*

Steroid	Source	Means of identi-fication[a]	Refer-ences
Cortisol [11β,17α,21–trihydroxy-pregn–4–ene–3,20–dione]	Human plasma after the administration of ACTH	III	96
	Human urine after the administration of ACTH	III	97
Dihydrocortisol [11β,17α,21–trihydroxy–5β–pregnane–3,20–dione]	Human urine	III	98
Cortisone [17α,21–dihydroxypregn–4–ene–3,11,20–trione]	Human urine after the administration of ACTH	III	97
Corticosterone [11β,21–dihydroxypregn–4–ene–3,20–dione]	Human plasma after the administration of ACTH	III	96
β–Cortolone [3α,17α,20β,21–tetra-hydroxy–5β–pregnan–11–one]	Human urine after the oral administration of 2 g of tetrahydrocortisone	II	99
	Human urine after the oral administration of 2 g of β–cortolone	II	33

[a] See footnote Table 7–1.

administration of ACTH to normal human subjects, cortisol glucuronide and cortisone glucuronide [17α–hydroxy–3,11,20–trioxopregn–4–en–21–yl–β–D–glucopyranosiduronic acid] have also been found in the urine [97]. Kornel [98] found dihydrocortisol [11β,17α,21–trihydroxy–5β–pregnane–3,20–dione] in the glucuronide fraction of normal human urine following sequential hydrolysis of the conjugates, and corticosterone glucuronide [11β–hydroxy–3,20–dioxopregn–4–en–21–yl–β–D–glucopyranosiduronic acid] has been detected in human urine following the administration of ACTH [96]. After the oral administration of 2 g of tetrahydrocortisol [3α,11β,17α,21–tetrahydroxy–5β–pregnan–20–one] to a male, Schneider *et al.* [99] isolated β–cortolone glucuronide from the urine; and this conjugate was also isolated from human urine by Schneider and Lewbart [33] following the administration of 2 g of the unconjugated steroid.

C. Isolation of Tetrahydrocortisol Glucuronide [11β,17α,21-trihydroxy-20-oxo-5β-pregnan-3α-yl-β-D-glucopyranosiduronic acid] and Tetrahydrocortisone Glucuronide [17α,21-dihydroxy-11,20-dioxo-5β-pregnan-3α-yl-β-D-glucopyranosiduronic acid]

A large number of isolations of the glucuronides of tetrahydrocortisol and tetrahydrocortisone [3α,17α,21-trihydroxy-5β-pregnane-11,20-dione] have been published and these are summarized in Table 7-10. Cohn and Bondy [100] developed a method for the measurement of the glucuronides of tetrahydrocortisol and tetrahydrocortisone in plasma. The mean plasma values and standard deviations for the concentrations of these two glucuronides was 7.96 ± 2.76 μg/100 ml for tetrahydrocortisol glucuronide and 4.45 ± 1.29 μg/100 ml for tetrahydrocortisone glucuronide. Tetrahydrocortisol glucuronide has also been found in human plasma after the administration of ACTH by Oertel [96] who noted a high concentration of this conjugate in hepatic vein blood which was greater than that found in adrenal vein or in femoral vein blood. Kornel [98] detected tetrahydrocortisol and tetrahydrocortisone in the solvolyzable fraction of human urine, and tetrahydrocortisol glucuronide has also been detected in human urine after the administration of ACTH [97]. Foggitt and Kellie published a comprehensive study on the isolation and identification of the glucuronides of C_{19}- and C_{21}-steroids [39]. These authors isolated and identified tetrahydrocortisol glucuronide and tetrahydrocortisone glucuronide in human urine after the administration of ACTH. It is noteworthy that the glucuronides of tetrahydrocortisol and tetrahydrocortisone have also been detected in the bile of cattle [101]. Pasqualini [97] reported the presence of allotetrahydrocortisol glucuronide in human urine following the administration of ACTH, and this conjugate was also isolated from human urine by Kornel [98].

Tetrahydrocortisone glucuronide has been found in human blood [102] following the oral administration of large quantities of the unconjugated steroid. This conjugate has been firmly identified in urine after feeding 2 g of the unconjugated steroid to a normal male [99]. In a different type of feeding experiment tetrahydrocortisone glucuronide was isolated from urine after the administration of large amounts of tetrahydrocortisone and β-cortolone [3α,17α,20β,21-tetrahydroxy-5β-pregnan-11-one] [33]. Human newborn urine also contains tetrahydrocortisone glucuronide [103], although the major portion of the conjugates in such a urine are excreted as sulfates. It is of some interest that evidence is available for the formation of tetrahydrocortisone glucuronide *in vitro*. Isselbacher and Axelrod [104] incubated tetrahydrocortisone with microsomes of guinea pig liver and

TABLE 7–10. *Isolation of tetrahydrocortisol glucuronide and tetrahydrocortisone glucuronide*

Steroid	Source	Means of identification[a]	References
Tetrahydrocortisol [3α,11β,17α,21-tetrahydroxy-5β-pregnan-20-one]	Human plasma	II	100
	Human plasma after the administration of ACTH	III	96
	Human urine	III	98
	Human urine after the administration of ACTH	III	97
	Human urine after the administration of ACTH	I	39
	Bovine bile	III	101
Allotetrahydrocortisol [3α,11β,17α,21-tetrahydroxy-5α-pregnan-20-one]	Human urine after the administration of ACTH	III	98
	Human urine	III	97
Tetrahydrocortisone [3α,17α,21-trihydroxy-5β-pregnane-11,20-dione]	Human plasma	II	100
	Human plasma after the feeding of large amounts of tetrahydrocortisone	III	102
	Human plasma after the administration of ACTH	III	98
	Human urine after the administration of ACTH	I	39
	Human urine	III	97
	Human urine after the oral administration of 2 g of tetrahydrocortisone	I	99
	Human urine after the oral administration of tetrahydrocortisone and β-cortolone	II	33
	Human newborn urine	II	103
	Microsomes from livers of guinea pigs	III	104
	Bovine bile	III	101

[a] See footnote Table 7–1.

obtained evidence for the formation of the glucuronide. Similar data were obtained when cortisol was the substrate.

D. Isolation of the Glucuronides of the Tetrahydro-reduced Corticosteroids

During the course of metabolism, the corticosteroids are reduced enzymatically in the A/B rings and the tetrahydro products formed are conjugated to a large extent with glucuronic acid. The glucuronides of tetrahydrocortisol and tetrahydrocortisone were discussed in the previous section and the additional metabolites isolated largely from human urine, are shown in Table 7–11. The glucuronide of tetrahydro-Compound S [3α,17α,21-trihydroxy-5β-pregnan-20-one] has been found in human urine of normal subjects [98], following the administration of ACTH [97], and after the oral administration of 150 mg of Compound S [17α,21-dihydroxypregn-4-ene-3,20-dione] to a normal subject [105]. Allotetrahydro-Compound S [3α,17α,21-trihydroxy-5α-pregnan-20-one] was found in the glucuronide fraction of human urine following the oral administration of 300 mg of 3β,17α,21-trihydroxypregn-5-en-20-one 21-acetate to a normal subject [106]. Following the administration of ACTH to human subjects, tetrahydrocorticosterone [3α,11β,21-trihydroxy-5β-pregnan-20-one] has been isolated and identified as a urinary constituent [39], and again under the same conditions but by indirect means [97]. Evidence has also been presented for the presence of the glucuronide of tetrahydrocorticosterone in human urine following the administration of ^{14}C-corticosterone [11β,21-dihydroxypregn-4-en-3,20-dione] [107], and after the administration to human subjects of ^{3}H corticosterone ^{35}S sulfate [11β hydroxy-21-sulfooxypregn-4-ene-3,20-dione] [108]. Allotetrahydrocorticosterone glucuronide has been identified directly in human urine after the administration of ACTH [39], and also in the same type of study but by indirect means [97]. The presence of this conjugate in human urine has also been indicated after the administration of ^{14}C-corticosterone [107] and again following the administration of ^{3}H-corticosterone-^{35}S-sulfate [108]. Indirect evidence has been presented for the presence of the glucuronide of tetrahydro-11-dehydrocorticosterone [3α,21-dihydroxy-5β-pregnane-11,20-dione] in human urine after the administration of ACTH [97], after the injection of ^{3}H-corticosterone-^{35}S-sulfate [108], and after the injection of ^{14}C-corticosterone [107] to human subjects. There is a single report on the isolation of the glucuronide of tetrahydrodeoxycorticosterone [3α,21-dihydroxy-5β-pregnan-20-one] from human urine of a subject fed 150 mg of deoxycorticosterone [109].

TABLE 7–11. *Isolation of the glucuronides of tetrahydro–reduced corticosteroids*

Steroid	Source	Means of identification[a]	References
Tetrahydro–Compound S [3α,17α,21–trihydroxy–5β–pregnan–20–one]	Human urine	III	98
	Human urine after the administration of ACTH	III	97
	Human urine after the oral administration of Compound S	III	105
Allotetrahydro–Compound S [3α,17α,21–trihydroxy–5α–pregnan–20–one]	Human urine after the administration of 3β,17α,21–trihydroxy–pregn–5–en–20–one	III	106
Tetrahydrocorticosterone [3α,11β,21–trihydroxy–5β–pregnan–20–one]	Human urine after the administration of ACTH	I, III	39, 97
	Human urine after the administration of ^{14}C–corticosterone	III	107
	Human urine after the administration of ^{3}H–corticosterone–^{35}S–sulfate	III	108
Allotetrahydrocorticosterone [3α,11β,21–trihydroxy–5α–pregnan–20–one]	Human urine after the administration of ACTH	I, III	39, 97
	Human urine after injection of ^{14}C–corticosterone	III	107
	Human urine after the administration of ^{3}H–corticosterone–^{35}S–sulfate	III	108
Tetrahydro–11–dehydro-corticosterone [3α,21–dihydroxy–5β–pregnane–11,20–dione]	Human urine after the administration of ACTH	III	97
	Human urine after the administration of ^{3}H–corticosterone–^{35}S–sulfate	III	108
	Human urine after injection of ^{14}C–corticosterone	III	107
Tetrahydrodeoxycorticosterone [3α,21–dihydroxy–5β–pregnan–20–one]	Human urine after the oral administration of deoxycorticosterone	III	109

[a] See footnote Table 7–1.

E. Isolation of the Glucuronides of Aldosterone [11β,21–dihydroxy–3,20–dioxopregn–4–en–18–al] and its Metabolites

Table 7–12 lists the glucuronide conjugates of aldosterone and the metabolites of aldosterone found in tissues and excreted in the urine. As soon as sufficient quantities of aldosterone became available for metabolic studies, it was realized that very little of the aldosterone was excreted as a non–conjugate, but that it was present as a conjugate which, on hydrolysis at pH 1, resulted in the release of unaltered aldosterone (pH 1–labile conjugate) [110]. The excretion of this pH 1–labile conjugate was studied by Ayres *et al.* [111] who called it the 3–oxo–hormone conjugate because the A ring remained intact after pH 1 hydrolysis. It was estimated [111] that about 12% of the aldosterone formed per day is excreted in the form of this conjugate. Underwood and Tait [112] presented good evidence to indicate that the urinary 3–oxo conjugate is an 18–glucuronide of aldosterone. However, these investigators did not determine the exact linkage between the glucuronic acid and the hormone. It will be necessary to isolate large amounts of the aldosterone–18–glucuronide in order to determine the chemical linkage between the sugar and the steroid.

Sandor and Lanthier [113] incubated labeled ^{14}C–aldosterone with human liver and kidney slices and found that the hormone was converted to the 3–oxo conjugate and to aldosterone glucuronide in both tissues, but to tetrahydroaldosterone glucuronide only in the liver slices. There is a single report that a small amount of aldosterone is present in human urine conjugated as a glucuronide [114]. Tetrahydroaldosterone glucuronide was first shown to be present in human urine by Ulick and Lieberman [115] and its structure was elucidated by Ulick *et al.* [116]. In a comprehensive study of the metabolism of aldosterone, Kelly *et al.* [117–120] isolated seven metabolites of the hormone after β–glucuronidase hydrolysis of the urinary conjugates. These metabolites were isolated after the administration of large amounts of aldosterone to human subjects along with ^{3}H–aldosterone. One metabolite has not been completely identified, but the others are 3α,5β–tetrahydroaldosterone [3α,11β,21–trihydroxy–20–oxo–5β–pregnan–18–al], 3α,5α–tetrahydroaldosterone [3α,11β,21–trihydroxy–20–oxo–5α–pregnan–18–al], 3β,5β–tetrahydroaldosterone [3β,11β,21–trihydroxy–20–oxo–5β–pregnan–18–al], 4,5β–dihydroaldosterone [11β,21–dihydroxy–3,20–dioxo–5β–pregnan–18–al], 3α–hydroxy–5β–pregnane–11β, 18S:18S, 20α–diepoxide, 3α,21–dihydroxy–5β–pregnane–11β, 18S:18S, 20α–diepoxide, and 21–deoxytetrahydroaldosterone [3α,11β–dihydroxy–20–oxo–5β–pregnan–18–al].

TABLE 7–12. *Isolation of the glucuronides of aldosterone and its metabolites*

Steroid	Source	Means of identification[a]	References
pH 1–labile conjugate of aldosterone	Human urine	III	110
3–Oxo conjugate of aldosterone	Human urine	III	111
Aldosterone–18–glucuronide	Human urine	III	112
3–Oxo conjugate	Human liver and kidney slices incubated with ^{14}C–aldosterone	III	113
Aldosterone	Human urine	III	114
	Human liver and kidney slices incubated with ^{14}C–aldosterone	III	113
Tetrahydroaldosterone	Human urine	III	115
	Human urine	III	116
	Human liver slices	III	113
3α,5β–Tetrahydroaldosterone	Human urine after the administration of aldosterone	III	118
3α,5α–Tetrahydroaldosterone	Human urine after the administration of aldosterone	III	118
3β,5β–Tetrahydroaldosterone	Human urine after the administration of aldosterone	III	118
4,5β–Dihydroaldosterone	Human urine after the administration of aldosterone	III	120
3α–Hydroxy–5β–pregnane–11β, 18 S: 18 S, 20α–diepoxide	Human urine after the administration of aldosterone	III	117, 119
3α,21–Dihydroxy–5β–pregnane–11β, 18 S: 18 S, 20α–diepoxide	Human urine after the administration of aldosterone	III	117, 119
21–Deoxytetrahydro-aldosterone(3α,5β)	Human urine after the administration of aldosterone	III	120

[a] See footnote Table 7–1.

VI. C_{21}–Steroid Sulfates

A. Isolation of Cortisol Sulfate [11β,17α–dihydroxy–21–sulfooxypregn–4–ene–3,20–dione] and Corticosterone Sulfate

It is only in recent years that C_{21}–steroid sulfates have been investigated. Initially it was the belief of a number of investigators that C_{21}–steroids having a C17–hydroxyl group could not be sulfurylated at C21. Then it was shown that cortisol sulfate was formed in man and this misconception disappeared. At the present time a large number of C_{21}–steroid sulfates have been isolated and these are summarized in the next three tables. In Table 7–13 is shown the isolation of the sulfates of cortisol and corticosterone, both of which have been studied extensively.

As can be seen from Table 7–13, cortisol sulfate has been found in a number of tissues. Following the administration of ACTH, Oertel [96] reported the presence of cortisol sulfate in human plasma. The identification involved the solvolysis of a plasma fraction from a DEAE–Sephadex column and the characterization of the cortisol released. With the use of the solvolytic procedure of hydrolysis, Kornel [98] was able to characterize cortisol sulfate extracted from human urine. Cortisol sulfate was also found in human urine by Pasqualini [97] following the administration of ACTH and after the oral administration of large amounts of cortisol [121]. Cortisol sulfate has also been found in the aqueous extract of human fetal adrenals after the injection of ^{14}C–progesterone into the umbilical circulation with the fetus *in situ* [91], and the conjugate has been isolated from human newborn urine by Drayer and Giroud [103].

Corticosterone sulfate has been isolated from a number of sources. In addition to cortisol sulfate, Oertel [96] also detected corticosterone sulfate in human plasma following the administration of ACTH. This conjugate has also been isolated from human urine following the administration of ACTH [97], and after the oral ingestion of 300 mg of corticosterone by a normal subject [121]. Lebeau and Baulieu [122] injected ^{3}H–corticosterone–^{35}S–sulfate into a normal male and, from the specific activity of the urinary corticosterone sulfate, they were able to calculate a production of 5.5 mg/day. These data suggest that corticosterone sulfate is secreted; but firm evidence will have to be obtained by isolation of the conjugate from adrenal vein blood.

Corticosterone sulfate has been identified in the aqueous extract of human fetal adrenals [91], and following the incubation of adrenal tissue of the premature infant with ^{14}C–progesterone [123], and with a mixture of ^{14}C–progesterone and ^{3}H–pregnenolone [3β–hydroxypregn–5–en–20–

TABLE 7–13. *Isolation of cortisol sulfate and corticosterone sulfate*

Steroid	Source	Means of identification[a]	References
Cortisol [11β,17α,21-trihydroxy-pregn-4-ene-3,20-dione]	Human plasma after the administration of ACTH	III	96
	Human urine	III	98
	Human urine after the administration of ACTH	III	97
	Human urine after the oral administration of cortisol	II	121
	Human fetal adrenals	I	91
	Human urine—newborn	III	103
Corticosterone [11β,21-dihydroxypregn-4-ene-3,20-dione]	Human plasma after the administration of ACTH	III	96
	Human urine after the administration of ACTH	III	97
	Human urine after the oral administration of corticosterone	II	121
	Human urine after injection of ^{3}H-corticosterone-^{35}S-sulfate	II	122
	Human fetal adrenals	I	91
	Human adrenals—premature infant incubated with ^{14}C-progesterone	II	123, 124
	Human urine—newborn	III	103

[a] See footnote Table 7–1.

one] [124]. This conjugate has also been isolated from human newborn urine by Drayer and Giroud [103].

B. Isolation of the Sulfates of Corticosteroids

In Table 7–14 are shown some of the sulfates of the corticosteroids isolated mainly from human tissues. Cortisone sulfate has been detected in adult human urine following the administration of ACTH [97] and from

TABLE 7–14. *Isolation of the sulfates of corticosteroids*

Steroid	Source	Means of identification[a]	References
Cortisone [17α,21–dihydroxypregn–4–ene–3,11,20–trione]	Human urine after administration of ACTH	III	97
	Human newborn urine	III	103
11–Deoxycortisone [17α,21–dihydroxypregn–4–ene–3,20–dione]	Adrenal slices of premature infant incubated with ^{14}C–progesterone	I	123
11–Dehydrocorticosterone [21–hydroxypregn–4–ene–3,11,20–trione]	Human urine after injection of ^{3}H–corticosterone	III	122
	Human urine after administration of ACTH	III	97
	Human fetal adrenals	I	91
11–Deoxycorticosterone [21–hydroxypregn–4–ene–3,20–dione]	Human fetal adrenals	I	91
	Adrenals of premature infant incubated with ^{14}C–progesterone and ^{3}H–pregnenolone	II	124
	Adrenal tissue of premature infant incubated with ^{14}C–progesterone	I	123
	Rabbit liver	III	125
6β–Hydroxycortisol [6β,11β,17α,21–tetrahydroxypregn–4–ene–3,20–dione]	Human urine	III	98

[a] See footnote Table 7–1.

human newborn urine [103]. Following the incubation of ^{14}C–progesterone with adrenal slices from a premature infant, Compound S sulfate [17α–hydroxy–21–sulfooxypregn–4–ene–3,20–dione] was isolated and identified by Klein and Giroud [123]. When ^{3}H–corticosterone was injected into a normal male, it was possible to characterize 11–dehydrocorticosterone sulfate [21–sulfooxypregn–4–ene–3,11,20–trione] as one of the metabolites in the urine. This conjugate was also isolated from normal male urine after the administration of ACTH [97]. In addition, 11–dehydrocorticosterone sulfate has been isolated from human fetal adrenal after the injection of ^{14}C–progesterone into the umbilical circulation with the fetus *in utero* [91]. In this latter study it was also possible to isolate 11–deoxycorticosterone sulfate [21–sulfooxypregn–4–ene–3,20–dione] from the fetal adrenal [91]; and this conjugate has also been isolated following the incubation of adrenal slices from premature infants with ^{14}C–progesterone and ^{3}H–pregnenolone [124], and with ^{14}C–progesterone [123]. Schneider and Lewbart [125] incubated the $110{,}000 \times g$ supernatant fraction of rabbit liver with 11–deoxycorticosterone and obtained indirect evidence for the formation of the sulfate. In addition to the corticosteroid sulfates discussed above, it is noteworthy that Kornel [98] was able to demonstrate the presence of 6β–hydroxycortisol [6β,11β,17α,21–tetrahydroxypregn–4–ene–3,20–dione] in the solvolyzable fraction of human urine.

C. Isolation of the Sulfates of the Tetrahydro reduced Corticosteroids

A number of tetrahydro–reduced corticosteroids have been isolated largely from urine as sulfates and these are shown in Table 7–15. Indirect evidence for the presence of tetrahydrocortisol sulfate [11β,17α,21–trihydroxy–3α–sulfooxy–5β–pregnan–20–one] in human plasma after the administration of ACTH has been provided [96]. In addition, tetrahydrocortisol sulfate has been shown to be a component of the solvolyzable fraction of human urine [98], and in human urine following the administration of ACTH [97]. Tetrahydrocortisone sulfate is a component of the solvolyzable fraction of adult human urine [98] and human newborn urine [103]. Tetrahydro–Compound S sulfate has been characterized in the solvolyzable fraction of human urine [98]. Tetrahydrocorticosterone–21–sulfate [3α,11β–dihydroxy–21–sulfooxy–5β–pregnan–20–one] has been found in human urine after the administration of ACTH [97] and after the injection of ^{3}H–corticosterone–^{35}S–sulfate [108]. Pasqualini and Faggett [126] reported that slices of rat liver incubated with ^{3}H–corticosterone–^{35}S–sulfate can convert this substrate to tetrahydrocorticosterone sulfate thus indicating a direct metabolism of the conjugate *in vitro*. Allotetrahydrocorticosterone sulfate [3α,11β–dihydroxy–21–sulfooxy–5α–pregnan–20–one] has been characterized in human urine following the administration of ^{3}H–corticosterone–^{35}S–sulfate [108]. The formation of 3β,11β,21–trihydroxy–5α–pregnan–20–one sulfate [3β,11β–dihydroxy–21–sulfooxy–5α–pregnan–20–one] by slices of rat liver incubated with ^{3}H–corticosterone–^{35}S–sulfate has also been reported [126] and the same conjugate has been found in rabbit urine following the administration of ^{3}H–corticosterone to the animals [127].

TABLE 7–15. *Isolation of the sulfates of the tetrahydro–reduced corticosteroids*

Steroid	Source	Means of identification[a]	References
Tetrahydrocortisol [3α,11β,17α,21–tetrahydroxy–5β–pregnan–20–one]	Human plasma after administration of ACTH	III	96
	Human urine	III	98
	Human urine after administration of ACTH	III	97
Tetrahydrocortisone [3α,17α,21–trihydroxy–5β–pregnane–11,20–dione]	Human urine	III	98
	Human newborn urine	III	103
Tetrahydro–Compound S [3α,17α,21–trihydroxy–5β–pregnan–20–one]	Human urine	III	98
Tetrahydrocorticosterone [3α,11β,21–trihydroxy–5β–pregnan–20–one]	Human urine after administration of ACTH	III	97
	Human urine after administration of ^{3}H–corticosterone–^{35}S–sulfate	II	108
	Rat liver slices incubated with ^{3}H–corticosterone–^{35}S–sulfate	III	126
Allotetrahydrocorticosterone [3α,11β,21–trihydroxy–5α–pregnan–20–one]	Human urine after administration of ^{3}H–corticosterone–^{35}S–sulfate	II	108
3β,11β,21–Trihydroxy–5α–pregnan–20–one	Rat liver slices incubated with ^{3}H–corticosterone–^{35}S–sulfate	III	126
	Rabbit urine after administration of ^{3}H–corticosterone	III	127

[a] See footnote Table 7–1.

D. Isolation of the Sulfates of Reduced and Hydroxylated Pregnanes

In Table 7–16 are shown a large number of reduced and hydroxylated pregnanes which have been isolated as conjugates, and, as is evident, they have been isolated from many sources. Pregnanolone sulfate [3α–sulfooxy–

TABLE 7–16. *Isolation of the sulfates of reduced and hydroxylated pregnanes*

Steroid	Source	Means of identification[a]	References
3α–Hydroxy–5β–pregnan–20–one (pregnanolone)	Human fetal liver	I	91
	Human bile	II	65
3β–Hydroxy–5α–pregnan–20–one	Human fetal residual tissues	I	91
3β–Hydroxy–5β–pregnan–20–one	Pregnant mare's urine	I	128
3β–Hydroxy–5α–pregnan–20–one	Human pregnancy plasma	II	61
5β–Pregnane–3α,20α–diol (pregnanediol)	Human fetal liver	I	91
	Human meconium	III	90
	Human bile	II	65
5α–Pregnane–3α,20α–diol	Human bile	II	65
5α–Pregnane–3β,20α–diol	Rat liver 105,000 × g supernatant	III	129
5α–Pregnane–3β,20β–diol	Rat liver 105,000 × g supernatant	III	129
5α–Pregnane–3β,20β–diol	Pregnant mare's urine	I	130
5α–Pregnane–3β,20,21–triol	Supernatant fraction of rabbit liver	III	125
5α–Pregnane–3α,20α,21–triol	Human pregnancy plasma	II	131
3α,16α–Dihydroxy–5β–pregnan–20–one	Human fetal liver	III	91
3α,17α,21–Trihydroxy–5α–pregnan–20–one	Human urine after ingestion of 3β,17α,21–trihydroxypregn–5–en–20–one–21–acetate	III	106
Pregn–5–ene–3β,20β–diol	Rat liver 105,000 × g supernatant	III	129
20α–Dihydrocortisol [11β,17α,20α,21–tetrahydroxypregn–4–en–3–one] 20β–Dihydrocortisol [11β,17α,20β,21–tetrahydroxypregn–4–en–3–one]	Human urine after administration of ACTH	III	97

[a] See footnote Table 7–1.

5β–pregnan–20–one] and 3β–hydroxy–5α–pregnan–20–one sulfate [3β–sulfooxy–5α–pregnan–20–one] have been isolated from human fetal liver and residual tissues, respectively, following the injection of ^{14}C–progesterone into the umbilical circulation with the fetus *in situ* [91]. In addition, pregnanolone sulfate has been isolated from human bile [65]. Paterson and Klyne [128] isolated 3β–hydroxy–5α–pregnan–20–one sulfate from pregnant mare's urine, and it has also been tentatively identified in pregnancy plasma using gas chromatography—mass spectroscopy [61]. Pregnanediol sulfate [3α–sulfooxy–5β–pregnan–20α–ol] has been isolated from human fetal liver after the injection of ^{14}C–progesterone into the umbilical circulation; and this conjugate has also been found in the "sulfate" fraction of human meconium [90] and in human bile [65]. It is noteworthy that pregnanediol sulfate was isolated from human fetal tissues and from bile as a monosulfate.

Laatikainen *et al.* [65] have also identified 5α–pregnane–3α,20α–diol sulfate in human bile. Rat liver has an active sulfokinase system, and Wengle and Boström [129] have provided some evidence for the formation of the monosulfate of 5α–pregnane–3β,20α–diol, and both the mono– and disulfate of 5α–pregnane–3β,20β–diol [3β,20β–disulfooxy–5α–pregnane]. At the moment there does not seem to be a relationship between the ability to form mono– or disulfates, and structural requirements of the pregnane–3,20–diols. It is of some interest that the 20–sulfate of 5α–pregnane–3β,20β–diol [20β–sulfooxy–5α–pregnan–3β–ol] has been isolated by Hirschmann and Williams [130]. To our knowledge this is the only 20–sulfate which has been firmly identified.

Schneider and Lewbart [125] obtained indirect evidence for the formation of the sulfate of 5α–pregnane–3β,20,21–triol by incubating the steroid with 110,000 × g supernatant fraction of rabbit liver. This tissue fraction seems to have an active sulfurylating enzyme system for a number of steroids. Another triol, 5α–pregnane–3α,20α,21–triol, was found in the "monosulfate" fraction of human pregnancy plasma [131]. To date it has not been isolated from any other source. Following the injection of ^{14}C–16α–hydroxyprogesterone into the umbilical circulation with the fetus *in utero*, it was possible to isolate the sulfate of 3α,16α–dihydroxy–5β–pregnan–20–one from the fetal liver [91]. Whether this conjugate was a 3– or 16–sulfate was not determined. Pasqualini and Jayle [106] administered 300 mg of 3β,17α,21–trihydroxypregn–5–en–20–one 21–acetate to a normal male and obtained evidence for the excretion in the urine of the sulfate of 3α,17α,21–trihydroxy–5α–pregnan–20–one. Wengle and Boström [129] provided indirect evidence for the sulfation of pregn–5–ene–3β,20β–diol by the 10,500 × g supernatant fraction of rat liver, and, in addition, claimed that

17α–hydroxyprogesterone [17α–hydroxypregn–4–ene–3,20–dione] can also be converted to a sulfate without stating which carbon is involved in the sulfurylation. Pasqualini [97] administered ACTH to normal males and provided evidence for the presence of the sulfates of 20α–dihydrocortisol [11β,17α,20α,21–tetrahydroxypregn–4–en–3–one] and 20β–dihydrocortisol [11β,17α,20β,21–tetrahydroxypregn–4–en–3–one] in the urine.

VII. Metabolism of C_{21}–Steroid Sulfates

In contrast to the C_{19}–steroid sulfates and glucuronides, not much is known about the metabolism of C_{21}–steroid sulfates; and there is no data on the metabolic fate of C_{21}–steroid glucuronides. The small number of metabolic conversions involving C_{21}–steroid sulfates are outlined in Table 7–17. It is probable that the C_{21}–steroid glucuronides are so rapidly cleared into the urine that extensive metabolism in humans cannot occur. As for the sulfates, some limited amount of data is available. In a preliminary note Pasqualini and Faggett [126] reported that, when slices of rabbit liver were incubated with ^{3}H–cortisol sulfate it was possible to isolate unconjugated cortisol, cortisol sulfate, and tetrahydrocortisol sulfate from the incubation medium. When ^{3}H–corticosterone–^{35}S–sulfate was incubated, unconjugated corticosterone and its sulfate were found in the medium, in addition to the sulfates of tetrahydrocorticosterone and 3β–allotetrahydrocorticosterone[3β,11β–dihydroxy–21–sulfooxy–5α–pregnan–20–one][126].

Lebeau and Baulieu [122] injected ^{3}H–corticosterone–^{35}S–sulfate into a normal male and isolated the conjugate from the urine. From the specific activity of the urinary conjugate it was possible to estimate a production rate of 5.5 mg/day for corticosterone sulfate. The authors also found that 2% of the injected sulfate appeared as tetrahydro reduced glucuronides in the urine, indicating that only a small amount of the sulfate was hydrolyzed. Kielmann *et al.* [132] injected ^{3}H–corticosterone sulfate to normal males and found its half–life to be 45 minutes. Approximately 70–80% of the radioactivity in the urine was present in the sulfate fraction in which corticosterone sulfate was the major component. When these authors [132] injected ^{3}H–corticosterone sulfate and ^{14}C–corticosterone into male subjects they found that the glucuronide fraction of the urine had a number of ^{14}C–labeled metabolites which did not contain ^{3}H and that corticosterone sulfate only contained ^{3}H and no ^{14}C. The interpretation of these data was that corticosterone sulfate was a secretion product of normal human adrenals and that the conjugate was metabolized directly without hydrolysis. The production rate of corticosterone sulfate reported in this study was

TABLE 7–17. *Metabolism of C_{21}–steroid sulfates*

Steroid	System	Product	References
^{3}H–Cortisol sulfate	Slices of rat liver	Cortisol, tetrahydrocortisol sulfate, cortisol sulfate	126
^{3}H–Corticosterone–^{35}S–sulfate	Slices of rat liver	Corticosterone sulfate, tetrahydrocorticosterone sulfate, allotetrahydrocorticosterone sulfate	126
^{3}H–Corticosterone–^{35}S–sulfate	Human—*in vivo*	Measurement of production rate	122
^{3}H–Corticosterone sulfate	Human—*in vivo*	Measurement of production rate	132
^{3}H–Corticosterone sulfate and ^{14}C–corticosterone	Human—*in vivo*	Corticosterone sulfate, tetrahydrocorticosterone sulfate, tetrahydro–11–dehydrocorticosterone sulfate	107
3β–Hydroxy–5α–pregnan–20–one sulfate	Slices of rabbit liver	3β–Hydroxy–5α–pregnan–20–one 5α–Pregnane–3β,20α–diol 3β–Hydroxy–5α–pregnan–20–one sulfate 5α Pregnane 3β,20α diol sulfate	133
3α–Hydroxy–5β–pregnan–20–one sulfate	Slices of rabbit liver	3α–Hydroxy–5β–pregnan–20–one sulfate 5β–Hydroxy–3α,20α–diol sulfate	133

530 and 780 μg/day in two subjects and 3.5 mg/day for the unconjugated steroid.

Pasqualini [107] injected ^{3}H–corticosterone sulfate and ^{14}C–corticosterone into 3 men and found a rapid initial elimination of the sulfate into the urine (50–60% in the first 4 hours) which was twice the rate of the unconjugated steroid. From the ^{3}H/^{14}C ratio of the urinary metabolites in the glucuronide and sulfate fractions of the urine, the authors deduced that the sulfate was metabolized directly to products having the 21–sulfate, and that the glucuronides were derived from the metabolism of the unconjugated steroid. These data indicate that corticosterone sulfate is rapidly cleared into the urine and is not metabolically hydrolyzed to any great extent, which agrees with the conclusions of Kielmann *et al.* [132].

In a preliminary report, Crépy *et al.* [133] found that, after the incubation of 3β–hydroxy–5α–pregnan–20–one sulfate and 3α–hydroxy–5β–pregnan–20–one sulfate [3α–sulfooxy–5β–pregnan–20–one], the 3β–hydroxy–5α–steroid was metabolized to yield unconjugated and conjugated (sulfates) products, whereas the 3α–hydroxy–5β–steroid was not hydrolyzed but was reduced to 20α–diol. After incubation, 3α–hydroxy–5α–pregnan–20–one sulfate yielded 3β–hydroxy–5α–pregnan–20–one, 5α–pregnane–3β,20α–diol as well as the corresponding sulfate.

VIII. Summary

In the preparation of this chapter some 900 papers were reviewed and most of them could not be used because, in the investigations reported, the steroid conjugates were hydrolyzed by two or more methods and the resulting products were combined in preparation for chromatographic purification and identification, or the unconjugated steroid was poorly characterized. As is evident from the tables, there are only a small number of class I isolations where the conjugate itself was identified. This is especially true for the steroid glucuronides which are difficult to isolate, and authentic standards for physical–chemical comparison are not readily available. A notable example is the paper by Foggitt and Kellie [39] where the C_{19}– and C_{21}–steroid glucuronides isolated were well identified. There is a great need for the type of investigation exemplified by the paper of Foggitt and Kellie and a few others cited herein.

In viewing the neutral steroid glucuronides and sulfates, it is apparent that they fall into two classes—those that will undergo further metabolism and those compounds whose sole fate is excretion into the urine. Androsterone glucuronide is not further metabolized in man but is rapidly excreted into the urine, whereas testosterone glucuronide undergoes both "direct" metabolism and is hydrolyzed to the unconjugated steroid, which is further metabolized. The glucuronides of dehydroisoandrosterone and isoandrosterone are not totally excreted but also undergo further metabolic conversion, whereas testosterone sulfate and androsterone sulfate seem to be metabolically inert. In contrast to the sulfates of 3β–hydroxy–Δ^5–steroids, the sulfates of the neutral steroids discussed herein do not serve as important metabolic intermediates in the formation of steroids. With the possible exception of testosterone sulfate, it is not known whether any of the other C_{19}-steroid sulfates are secretory products of endocrine glands. No critical study has been reported to demonstrate that C_{19}-steroid glucuronides such as dehydroisoandrosterone glucuronide is or is not a secretory product.

So little is known about the glucuronides of the corticosteroids that it is difficult to make general statements. The reduced C_{21}-steroid glucuronides, such as tetrahydrocortisol glucuronide and pregnanediol glucuronide, have been well characterized and would seem to play the role of excretion products. On the other hand, certain of the C_{21}-steroid sulfates may be considered as secretion products. Corticosterone sulfate seems to be secreted by the adrenal gland but, unlike dehydroisoandrosterone sulfate, it undergoes direct metabolism in humans and rapid excretion in the urine. A large number of corticosteroid sulfates have been isolated from the human fetal adrenal, and these conjugates are formed from progesterone by the human newborn adrenal. It is somewhat surprising that there is such a sparsity of published data on the secretion of other C_{21}-steroid sulfates by the human adult adrenal.

It is clear therefore that, while there has been a great deal of interest shown in steroid conjugates in the last ten years, much remains to be learned. Some of the most intriguing areas of research which bear investigation are the interaction of conjugates with proteins in target tissues, the effect of conjugates on steroid-metabolizing enzyme systems, and the interaction of conjugates with enzymes during the course of their direct metabolism.

Acknowledgments. We would like to acknowledge the assistance of Mrs. K. Begor and Miss Jane Woollcombe in the typing of the manuscript. Experimental data quoted from the authors' laboratory were obtained with the aid of research grants from U. S. Public Health Service (AM-04329) and the Canadian Medical Research Council (MT-1658).

IX. References

1. Funk, C., Harrow, B., Lejwa, A.: Proc. Soc. Exptl. Biol. Med. **26**, 569 (1929).
2. Kabak, J. M.: Endokrinologie **9**, 84 (1931).
3. Adler, A. A.: Nature **133**, 798 (1934).
4. Odell, A. D., Marrian, G. F.: Biochem. J. **30**, 1533 (1936).
5. Venning, E. M., Browne, J. S. L.: Proc. Soc. Exptl. Biol. Med. **34**, 792 (1936).
6. — Hoffman, M. M., Browne, J. S. L.: J. Biol. Chem. **146**, 369 (1942).
7. Munson, P. L., Gallagher, T. F., Koch, F. C.: J. Biol. Chem. **152**. 67 (1944).
8. Dorfman, R. I., Ungar, F.: Metabolism of steroid hormones, p. 28. New York: Academic Press 1965.
9. Burstein, S., Lieberman, S.: J. Biol. Chem. **233**, 331 (1958).

10. Baulieu, E.-E.: Compt. Rend. **251**, 1421 (1960).
11. — J. Clin. Endocrinol. Metab. **22**, 501 (1962).
12. Lieberman, S., Vande Wiele, R., in: Biochemistry of steroids (E. Mosettig, ed.), Proc. IV Internat. Congr. of Biochemistry, Vienna, 1958, p. 153. New York: Pergamon Press 1959.
13. Baulieu, E.-E., Corpéchot, C., Dray, F., Emiliozzi, R., Lebeau, M. C., Mauvais-Jarvis, P., Robel, P.: Recent Progr. Hormone Res. **21**, 411 (1965).
14. Hadd, H. T., Rhamy, R. K.: J. Clin. Endocrinol. Metab. **25**, 876 (1965).
15. Korenman, S. G., Lipsett, M. B.: J. Clin. Invest. **43**, 2125 (1965).
16. — — Steroids **5**, 509 (1965).
17. Schubert, K., Wehrberger, K.: Naturwissenschaften **12**, 281 (1960).
18. Camacho, A. M., Migeon, C. J.: J. Clin. Endocrinol. Metab. **23**, 301 (1963).
19. Baulieu, E.-E., Mauvais-Jarvis, P.: J. Biol. Chem. **239**, 1569 (1964).
20. — — J. Biol. Chem. **239**, 1578 (1964).
21. Dahm, K., Breuer, H., Lindlau, M.: Z. Physiol. Chem. **345**, 139 (1966).
22. Wotiz, H. H., Ziskind, B. S., Lemon, H. M., Gut, M.: Biochim. Biophys. Acta **22**, 266 (1956).
23. — Sie, H. G., Fishman, W. H.: J. Biol. Chem. **232**, 723 (1958).
24. Schriefers, H., Keck, B., Otto, M.: Acta Endocrinol. **50**, 25 (1965).
25. Grajcer, D., Idler, D. R.: Can. J. Biochem. Physiol. **41**, 23 (1963).
26. — — Can. J. Biochem. Physiol. **39**, 1585 (1961).
27. Korenman, S. G., Wilson, H., Lipsett, M. B.: J. Biol. Chem. **239**, 1004 (1964).
28. Wilson, H., Lipsett, M. B.: J. Clin. Endocrinol. Metab. **26**, 902 (1966).
29. Dray, F., Ledru, M. J.: Compt. Rend. **262**, 679 (1966).
30. Bongiovanni, A. M., Eberlein, W. R.: J. Clin. Endocrinol. Metab. **17**, 238 (1957).
31. Siiteri, P. K., Vande Wiele, R. L., Lieberman, S.: J. Clin. Endocrinol. Metab. **23**, 588 (1963).
32. Hadd, H. E., Dorfman, R. I.: J. Biol. Chem. **238**, 907 (1963).
33. Schneider, J. J., Lewbart, M. L.: Recent Progr. Hormone Res. **15**, 201 (1959).
34. Kirschner, M. A., Lipsett, M. B., Wilson, H.: Acta Endocrinol. **43**, 387 (1963).
35. MacDonald, P. C., Gonzalez, O., Vande Wiele, R. L., Lieberman, S.: J. Clin. Endocrinol. Metab. **23**, 665 (1963).
36. Baulieu, E.-E.: Ann. Endocrinol. (Paris) **24**, 801 (1963).
37. Schriefers, H., Kley, H. K., Otto, M.: Z. Physiol. Chem. **341**, 215 (1965).
38. Cohn, G. L., Hume, M.: J. Clin. Invest. **39**, 1584 (1960).
39. Foggitt, F., Kellie, A. E.: Biochem. J. **91**, 209 (1964).
40. Lieberman, S., Mond, B., Smyles, E.: Recent Progr. Hormone Res. **9**, 113 (1954).
41. Vande Wiele, R., Christy, N. P., Lieberman, S., Jailer, J. W.: Proc. Soc. Exptl. Biol. Med. **99**, 520 (1958).
42. Jailer, J. W., Vande Wiele, R., Christy, N. P., Lieberman, S.: J. Clin. Invest. **38**, 357 (1959).
43. Lakshmanan, T. K., Lieberman, S.: Arch. Biochem. Biophys. **53**, 258 (1954).
44. Paulsen, E. P., Sobel, E. H., Shafran, M. S.: J. Clin. Endocrinol. Metab. **26**, 329 (1966).

45. Fukushima, D. K., Dobriner, S., Bradlow, H. L., Zumoff, B., Hellman, L., Gallagher, T. F.: Biochemistry **5**, 1779 (1966).
46. Wotiz, H. H., Fishman, W. H.: Steroids **1**, 211 (1963).
47. YoungLai, E. V., Solomon, S.: Biochemistry **6**, 2040 (1967).
48. Saez, J. M., Saez, S., Migeon, C. J.: Steroids **9**, 1 (1967).
49. Dray, F., Mowszowicz, I., Ledru, M. J.: Steroids **10**, 501 (1967).
50. Tamm, J., Volkwein, U., Voigt, K. D.: Experientia **23**, 299 (1967).
51. Mancuso, S., Benagiano, G., Froysa, B., Diczfalusy, E.: Biochim. Biophys. Acta **144**, 183 (1967).
52. Benagiano, G., Mancuso, S., Mancuso, F. P., Wiqvist, N., Diczfalusy, E.: Acta Endocrinol. **57**, 187 (1968).
53. Dixon, W. R., Phillips, J. G., Kase, M.: Steroids **6**, 81 (1965).
54. — Vincent, V., Kase, M.: Steroids **6**, 757 (1965).
55. Staib, W., Teller, W., Pelzer, H.: Biochim. Biophys. Acta **31**, 591 (1959).
56. Baulieu, E.-E.: J. Clin. Endocrinol. Metab. **20**, 900 (1960).
57. McKenna, J., Rippon, A. E.: Biochem. J. **95**, 107 (1965).
58. Conrad, S. H., Lindberg, M. C., Herrmann, W. L.: Am J. Obstet. Gynecol. **91**, 449 (1965).
59. Oertel, G. W., Kaiser, E.: Clin. Chim. Acta **7**, 221 (1962).
60. Plager, J. E.: J. Clin. Endocrinol. Metab. **26**, 1275 (1966).
61. Sjövall, J., Sjövall, K., Vihko, R.: Steroids **11**, 703 (1968).
62. — Vihko, R.: Steroids **7**, 447 (1966).
63. Jänne, O., Luukkainen, T., Vihko, R.: European J. Steroids **2**, 457 (1967).
64. Laatikainen, T., Vihko, R.: J. Clin. Endocrinol. Metab. **28**, 1356 (1968).
65. — Peltokallio, P., Vihko, R.: Steroids **12**, 407 (1968).
66. Baulieu, E.-E.: Compt. Rend. **248**, 3493 (1959).
67. — Compt. Rend. **250**, 4219 (1960).
68. — Emiliozzi, R.: Compt. Rend. **251**, 3106 (1960).
69. — Robel, P., Mauvais-Jarvis, P.: Compt. Rend. **256**, 1016 (1963).
70. Mancuso, S., Benagiano, G., Dell'Acqua, S., Shapiro, M., Wiqvist, N., Diczfalusy, E.: Acta Endocrinol. **57**, 208 (1968).
71. Oertel, G. W.: Biochem. Z. **334**, 431 (1961).
72. — Kaiser, E.: Biochem. Z. **336**, 10 (1962).
73. — — Bruhl, P.: Biochem. Z. **336**, 154 (1962).
74. Siiteri, P. K., Lieberman, S.: Biochemistry **2**, 1171 (1963).
75. Robel, P., Emiliozzi, R., Baulieu, E.-E.: J. Biol. Chem. **241**, 5879 (1966).
76. — — — J. Biol. Chem. **241**, 20 (1966).
77. — — — J. Clin. Endocrinol. Metab. **27**, 1290 (1967).
78. Knapstein, P. von, Rindt, W., Wendlberger, F., Oertel, G. W.: Z. Physiol. Chem. **348**, 93 (1967).
79. — — — — Z. Physiol. Chem. **348**, 89 (1967).
80. — — Treiber, L., Oertel, G. W.: Z. Physiol. Chem. **346**, 192 (1966).
81. Wotiz, H. H.: Biochim. Biophys. Acta **60**, 28 (1962).
82. Cheatum, S. G., Diebold, J. C., Warren, J. C.: J. Clin. Endocrinol. Metab. **28**, 916 (1968).

83. Wu, H. L. C., Mason, M.: Steroids **5**, 45 (1965).
84. Crépy, O., Lachèse, B., Jayle, M. G.: Compt. Rend. **257**, 2163 (1963).
85. Vande Wiele, R. L., MacDonald, P. C., Gurpide, E., Lieberman, S.: Recent Progr. Hormone Res. **19**, 275 (1963).
86. Gurpide, E., MacDonald, P. C., Chapdelaine, A., Vande Wiele, R. L., Lieberman, S.: J. Clin. Endocrinol. Metab. **25**, 1537 (1965).
87. MacDonald, P. C., Chapdelaine, A., Gonzalez, O., Gurpide, E., Vande Wiele, R. L., Lieberman, S.: J. Clin. Endocrinol. Metab. **25**, 1557 (1965).
88. Chapdelaine, A., MacDonald, P. C., Gonzalez, O., Gurpide, E., Vande Wiele, R. L., Lieberman, S.: J. Clin. Endocrinol. Metab. **25**, 1569 (1965).
89. Thijssen, J. H. H., Zander, J.: Acta Endocrinol. **51**, 563 (1966).
90. Francis, F. E., Kinsella, R. A., Jr.: J. Clin. Endocrinol. Metab. **26**, 128 (1966).
91. Solomon, S., Bird, C. E., Ling, W., Iwamiya, M., Young, P. C. M.: Recent Progr. Hormone Res. **23**, 297 (1967).
92. Francis, F. E., Shen, N. C., Kinsella, R. A., Jr.: Biochemistry **1**, 1184 (1962).
93. Bradlow, H. L., Fukushima, D. K., Koerner, D. R., Hellman, L., Gallagher, T. F.: Steroids **8**, 45 (1966).
94. Mason, H. L., Strickler, H. S.: J. Biol. Chem. **171**, 543 (1947).
95. Schneider, J. J., in: Hormonal steroids—biochemistry, pharmacology and therapeutics, vol. 1 (L. Martini and A. Pecile, eds.), p. 127. New York, N.Y.: Academic Press 1964.
96. Oertel, G. W., in: Structure and metabolism of corticosteroids (J. R. Pasqualini and M. F. Jayle, eds.), p. 65. New York, N.Y.: Academic Press 1964.
97. Pasqualini, J. R.: Bull. Soc. Chim. Biol. **45**, 277 (1963).
98. Kornel, L.: Biochemistry **4**, 444 (1965).
99. Schneider, J. J., Lewbart, M. L., Levitan, P., Lieberman, S.: J. Am. Chem. Soc. **77**, 4184 (1955).
100. Cohn, G. L., Bondy, P. K.: J. Biol. Chem. **234**, 31 (1959).
101. Glick, J. H., Jr.: Endocrinology **60**, 368 (1957).
102. Bongiovanni, A. M., Eberlein, W. R.: Proc. Soc. Exptl. Biol. Med. **89**, 281 (1965).
103. Drayer, N. M., Giroud, C. J. P.: Steroids **5**, 289 (1965).
104. Isselbacher, K. J., Axelrod, J.: J. Am. Chem. Soc. **77**, 1070 (1955).
105. Pasqualini, J. R., Jayle, M. F.: Compt. Rend. **257**, 2345 (1968).
106. — — Biochem. J. **88**, 315 (1963).
107. — J. Clin. Endocrinol. Metab. **27**, 885 (1967).
108. — Bull. Soc. Chim. Biol. **47**, 471 (1965).
109. — Jayle, M. F.: Experientia **17**, 450 (1961).
110. Axelrad, B. J., Cates, J. E., Johnson, B. B., Luetscher, J. A., Jr.: Brit. Med. J. **1**, 196 (1955).
111. Ayres, P. J., Barlow, J., Garrod, O., Kellie, A. E., Tait, S. A. S., Tait, J. F., Walker, G., in: An Internat. Symposium on Aldosterone (A. Muller and C. O'Connor, eds.), p. 73. London: J. & A. Churchill Ltd.; Boston: Little, Brown & Co. 1958.
112. Underwood, R. H., Tait, J. F.: J. Clin. Endocrinol. Metab. **24**, 1110 (1964).
113. Sandor, T., Lanthier, A.: Acta Endocrinol. **39**, 87 (1962).

114. Coppage, W. S. J., Island, D. P., Cooner, A. E., Liddle, G. W.: J. Clin. Invest. **41**, 1672 (1962).
115. Ulick, S., Lieberman, S.: J. Am. Chem. Soc. **79**, 6567 (1957).
116. — Kusch, K., August, J. T.: J. Am. Chem. Soc. **83**, 4482 (1961).
117. Kelly, W. G., Bandi, L., Shoolery, J. N., Lieberman, S.: Biochemistry **1**, 172 (1962).
118. — — Lieberman, S.: Biochemistry **1**, 792 (1962).
119. — — — Biochemistry **2**, 1243 (1963).
120. — — — Biochemistry **2**, 1249 (1963).
121. Pasqualini, J. R., Jayle, M. F.: Biochem. J. **81**, 147 (1961).
122. Lebeau, M. C., Baulieu, E.-E.: Compt. Rend. **258**, 6265 (1964).
123. Klein, G. P., Giroud, C. J. P.: Can. J. Biochem. **44**, 1005 (1966).
124. — — Steroids **9**, 113 (1967).
125. Schneider, J. J., Lewbart, M. L.: J. Biol. Chem. **222**, 787 (1956).
126. Pasqualini, J. R., Faggett, J.: J. Endocrinol. **31**, 85 (1964).
127. — Di Costanzo, P.: Compt. Rend. **261**, 2252 (1965).
128. Paterson, J. Y. F., Klyne, W.: Biochem. J. **43**, 614 (1948).
129. Wengle, B., Boström, H.: Acta Chem. Scand. **17**, 1203 (1963).
130. Hirschmann, H., Williams, J. S.: J. Biol. Chem. **238**, 2305 (1963).
131. Sjövall, J., Sjövall, K.: Steroids **12**, 359 (1968).
132. Kielman, N., Stachenko, J., Giroud, C. J. P.: Steroids **8**, 993 (1966).
133. Crépy, O., Judas, O., Jayle, M. F.: Compt. Rend. Soc. Biol. **160**, 891 (1966).

Chapter 8

Biological Properties of Estrogen Sulfates

F. Herr, C. Revesz, A. J. Manson, and J. B. Jewell

Contents

I. Introduction

A. History

For over thirty-five years, estrogens have been under intense clinical investigation resulting in their acceptance for the treatment of many disorders now recognized as estrogen–deficiency states. Their therapeutic range includes the menopausal syndrome, some menstrual disorders, and several degenerative metabolic processes. In this development of estrogen therapy, estrogen sulfates have played an important role.

However, before estrogen sulfates could assume this role, several important events transpired.

In 1929 Glimm and Wadehn [1] noted that estrogens are excreted in the urine in a water–soluble form that cannot be extracted with the usual organic solvents and suggested that such estrogens be combined with an acid.

In the following year two significant events occurred which were to have far-reaching effects on the development of estrogen sulfates as therapeutic agents. The first was the observation by Zondek [2] that pregnant mare's urine was a rich source of water-soluble estrogens. The second event was the report by Campbell and Collip [3] on the clinical evaluation of a mixture of water-soluble estrogens, shown later to be mostly estriol glucuronide [4–9], obtained from human placenta and pregnancy urine. They reported that the condition of 14 of 18 patients with menopausal symptoms, as well as the condition of patients with amenorrhea, dysmenorrhea, and polymenorrhea, was improved by administering this mixture of estrogens, which, although used as a therapeutic agent (Emmenin®, Ayerst) for over two decades, was later replaced by estrogens that were more potent and more accessible.

The first isolation of a naturally occurring estrogen sulfate was reported in 1938; Schachter and Marrian [10] isolated a water-soluble estrogen in low yield from pregnant mare's urine and identified it as potassium estrone sulfate. In the following year, Butenandt and Hofstetter [11] synthesized estrone sulfate and investigated the estrogenic properties of its sodium, quinidine, and potassium salts. Using castrated mice in the Allen–Doisy assay, low levels of activity were found for these compounds when administered orally or by injection (see Section II–A).

Further development of estrogen sulfates as therapeutic agents appeared to be at a critical stage when Grant *et al.* began their systematic investigations of pregnant mare's urine. These investigations were made known in a series of published reports [12–15] describing a practical method for the isolation of the estrogen conjugates in this urine, together with the relative estrogenic activities of a number of the individual estrogen sulfates. In contrast to the findings of Butenandt and Hofstetter in castrated mice, Grant *et al.* found that the Allen–Doisy test in castrated rats showed the sodium salt of estrone sulfate, as well as the sodium salts of equilin- and 17β-estradiol-3-sulfates to be about twice as active as the corresponding free estrogen when given orally (see Section II–A). On the basis of these findings a preparation of conjugated equine estrogens (Premarin®, Ayerst) was made available for clinical study in 1941.

Chemical investigations into the nature of conjugated equine estrogens has revealed that it is composed mainly of the sodium salts of the 3-sulfate esters of estrone, equilin, and equilenin, and their corresponding 17α-hydroxy derivatives, with the sulfate esters of estrone and equilin, the major components, approximately 45 and 25%, respectively [16].

As a corollary to the investigations of Grant *et al.*, information on the levels of various estrogens present in the mare's urine at various stages of pregnancy was gathered and reported by Gaudry and Glen [17].

The reports of the first clinical trials with conjugated equine estrogens are almost all found in the third volume of the Journal of Clinical Endocrinology, 1943. This volume contains papers by Freed *et al.* [18], Gray [19], Glass and Rosenblum [20], Sevringhaus and St. John [21], and Fried and Hair [22] describing the successful application of conjugated equine estrogens for the treatment of menopausal symptoms in a total of 376 women. Two other papers by Turner *et al.* [23, 24] in the same volume describe the beneficial results obtained with the conjugated estrogens in the treatment of amenorrhea and functional uterine bleeding. It was these reports that established the conjugated equine estrogens as a potent oral estrogen with minimal side effects in man. Today, preparations of conjugated equine estrogens, as well as some individual estrogen sulfates, are marketed throughout the world.

B. The Role of the Bioassay

It has long been known that estrogens produce a wide variety of specific biologic effects on various target organs, such as the gonads, vagina, uterus, oviducts, and anterior pituitary. Their physiological processes—for example, uterine contractions, menstrual cycle, growth, immunity, and distribution of plasma proteins—have been described by Pincus [25] and, more recently, by Diczfalusy and Lauritzen [26], as well as several contemporary textbooks on physiology [27, 28].

Certain effects, which are easy to quantitate and which show a dose–effect relation, can be used for the bioassay of estrogens to measure their concentrations in various tissues and body fluids, especially blood and urine. The value of bioassays, however, is limited by many variables in animals (*e.g.*, species, breed, food) and, today, reliable chemical methods have almost entirely replaced them. However, experimentation in animals and in man is the only way to determine the estrogenic potency of a new chemical substance.

A large amount of data has accumulated from the measurement of the biological activity of estrogen sulfates, reflecting the intensive investigations of this interesting class of compounds. For convenience, these data will be categorized as hormonal and nonhormonal. These designations do not imply a fundamental difference in physiological action, but only separate activities within and outside the general scope of conventional endocrinology.

II. Hormonal Properties of Estrogen Sulfates

A. Animal Data

The hormonal properties of estrogens as well as estrogen conjugates may be assayed by animal experiments in several ways. These methods have been extensively reviewed by Emmens [29] and recently summarized by Loraine and Bell [30]. Hence, only a description of the more relevant biological assays will be presented here.

1. Vaginal Cornification Assay

The main procedure for the biological determination of estrogenic potency has been the Allen–Doisy test, which depends upon the appearance of nucleated or cornified epithelial cells, excluding leucocytes, in the vaginal smears of castrated female mice [31]. This test, modified by many investigators since its introduction in 1923, has been summarized by Hogg and Korman [32] and by Emmens [29]. Because of the many ways in which this assay can be performed (*i.e.*, different routes and spacing of administration, different species of animals, and variations in the scoring of smears) results may vary widely. For example, Pedersen–Bjergaard [33] found a 500–fold difference in estrogenic potency for the same sample of pregnancy urine, depending on the species of animals and method of administration. Consequently, there are difficulties in interpreting such results unless a pure standard is assayed simultaneously.

In 1939, both Pedersen–Bjergaard [33] and Emmens [34] reported on the variables affecting the estimation of estrogenicity. Emmens concluded that a reasonable degree of accuracy can be obtained for the Allen–Doisy test by employing two doses of the standard and two doses of the test material, each group consisting of not less than 20 animals. Multiple doses must be given and vaginal smears must be taken at intervals.

Emmens [35] improved the sensitivity of the Allen–Doisy test, based on an earlier observation by Robson and Adler [36] that estrogens act locally on the vagina without absorption into the general circulation. This method is capable of detecting picogram quantities of estrogens, but it is not suitable to demonstrate a dose–response relationship.

In 1939, Butenandt and Hofstetter [11] published the first report on the biological properties of estrogen sulfates. They prepared the sodium, quinidine and pyridine salts of estrone sulfate [3–sulfooxyestra–1,3,5(10)–trien–17–one], and assayed these salts by the Allen–Doisy test in castrated mice. According to their results, sodium estrone sulfate possessed only 2% of the activity of estrone [3–hydroxyestra–1,3,5(10)–trien–17–one] whether administered orally or by injection. Similar low levels of activity were found for the quinidine and pyridine salts.

The results on the activity of estrogen sulfates reported by Grant *et al.* in a later series of papers [13–15] differ from the findings of Butenandt and Hofstetter. Using castrated rats rather than mice in the Allen–Doisy test, these investigators found the sodium salts of estrone sulfate, 17β–estradiol–3–sulfate [3–sulfooxyestra–1,3,5(10)–trien–17β–ol], and equilin sulfate [3–sulfooxyestra–1,3,5(10),7–tetraen–17–one] to be about twice as active as their corresponding free estrogens when given orally. A like comparison of equilenin sulfate [3–sulfooxyestra–1,3,5(10),6,8–pentaen–17–one] and its sodium salt showed little difference in oral potency. On the other hand, their results indicated that sulfate conjugation significantly reduces the subcutaneous activity of these estrogens. All four of these estrogen sulfates are components of conjugated equine estrogens.

Grant suggested that the increase in oral activity of the conjugated estrogens could best be explained on the basis of greater ease of absorption of the water-soluble form.

The results of Grant and Beall [15] are given in Tables 8–1 and 8–2. The apparent species difference between rats and mice should be noted, and that

TABLE 8–1. *Comparative activity of the estrogen sulfates in the Allen–Doisy test*[a] *(RD 50, as micrograms, total dose)*[b]

Compound	Oral	Subcutaneous
Sodium estrone sulfate	85–110	3–4
Quinidine estrone sulfate	70–90	
Barium estrone sulfate	70	
Sodium 17β–estradiol–3–sulfate	77; 100	5.8; 7.3
Sodium equilin sulfate	78; 87	7.9
Sodium equilenin sulfate	840–1,170	>105
Sodium 17α–estradiol–3–sulfate	1,040	
Quinidine 17β–estradiol disulfate	>180	>7
Sodium estriol trisulfate	>500	
Estrone	158–249	0.7–1.6; Av. 1.5
17β–Estradiol	170; 198	0.1
Equilin	200; 210	1.4
Equilenin	1,000	70

[a] Grant, G. A., Beall, D.: Recent Progr. Hormone Res. **5**, 307 (1950).

[b] Values are given as micrograms estrogen contained in the sulfate. > values, no response at this level. Animal: adult ovariectomized (o. v.) rats.

TABLE 8–2. *Comparison of activity of sodium estrone sulfate and estrone in the Allen–Doisy test*[a] *(RD 50 total dose)*

Method	Administration	Sodium estrone sulfate (μg estrone)[b]	Estrone (μg)
Adult o.v. rats	Oral	85–110	158–249
	Subcutaneous	3–4	1.5
Adult o.v. mice	Oral	3–5	0.3–0.7
	Subcutaneous	0.6–0.7	0.06–0.08
Intact immature rats	Oral	7–8	6.3; 6.4
	Subcutaneous	6.2; 9.7	2.7–3.0

[a] Grant, G. A., Beall, D.: Recent Progr. Hormone Res. **5**, 307 (1950).
[b] μg estrone contained in the estrogen sulfate.

sulfate conjugation of estrone diminishes rather than increases estrogenic activity in the mouse.

A later report by Harmer and Broom [37] compared estrone and conjugated equine estrogens by means of the Allen–Doisy assay in rats. These authors reported that the conjugated equine estrogens were $2^1/_2$ times as active as estrone when given orally, but only about $^1/_4$ as active when the two agents were administered subcutaneously.

Recent evaluations of estrone, equilin, and the sodium salts of their corresponding sulfate conjugates, are shown in Table 8–3. These results indicate that, on a molar basis, conjugation causes potentiation of activity for both steroids.

A report [38] on a comparative study on a nonsteroidal phenolic estrogen and its sulfate conjugate indicates that potassium hexestrol sulfate [3,4-bis-(4′-sulfooxyphenyl)-hexane potassium salt] is less active than hexestrol

TABLE 8–3. *Comparative activity of estrogen sulfates (ED 50, as micrograms, total dose)*[a]

Compound	Oral	Subcutaneous
Sodium estrone sulfate	155 (113)[b]	5 (3.7)
Sodium equilin sulfate	96 (70)	3 (2.2)
Estrone	150	1.6
Equilin	137	1

[a] Spayed albino rats (80–100 g) used.
[b] Wt. of estrogen contained in the estrogen sulfate.

[3,4–bis–(4′–hydroxyphenyl)–hexane] in the Allen–Doisy assay in rats, both by the oral and subcutaneous routes. The decrease of activity caused by conjugation is much more pronounced subcutaneously than orally. Diethylstilbestrol sulfate [3,4–bis–(4′–sulfooxyphenyl)–3–hexene] has been reported to possess only one–half the oral activity of diethylstilbestrol [3,4–bis–(4′–hydroxyphenyl)–3–hexene] in the rat [39].

2. Uterotrophic Assay

This assay uses the weight of the uterus of spayed rats or mice as the end point. In a series of papers between 1951 and 1954, Dorfman *et al.* [40–42] recorded the oral activity of several estrogen sulfates while demonstrating the utility of this method. They administered these estrogen sulfates, in aqueous solution by stomach tube, to 22– to 23–day–old rats, daily for four days, to determine their relative activities. Twenty–four hours after the last oral administration, the rats were autopsied, and body and uterine weights were determined. Pure sodium estrone sulfate (sample A, Table 8–4) and sodium equilin sulfate were found to have essentially the same activity when administered orally (*i.e.*, 100% *vs.* 110%), whereas sodium 17β–estradiol–3–sulfate [3–sulfooxyestra–1,3,5(10)–trien–17β–ol sodium salt] and 17β–dihydroequilin–3–sulfate [3–sulfooxyestra–1,3,5(10),7–tetraen–17β–ol] were found to have 261 and 882% respectively, of the activity of sodium estrone sulfate [42] (Table 8–4).

We evaluated estrone, equilin, and the corresponding sodium salts of the sulfate conjugates by observing the effect of these estrogens on uterine weight of immature intact mice. The results are reported in Table 8–5.

It is interesting to note the apparent species difference between the mouse and rat. In the mouse, sulfate conjugation seems to diminish the oral estrogenicity of estrone and equilin, whereas in the rat (Tables 8–1, 8–2, and 8–3), the oral activity of both estrone and equilin are apparently enhanced by sulfate conjugation.

3. Chick Oviduct Test

In 1953, Dorfman and Dorfman [43] described an assay method in which various estrogens and estrogen sulfates were added to the diet of chickens. Emmens [29] has remarked that this method is surprisingly accurate in spite of the difficulty of controlling individual dosage. The end point of this assay is the oviduct weight. The chickens were fed a diet of chick starting mash to which the estrogen had been added. After 13 days on this diet, the chickens were autopsied, and the body and oviduct weights were determined. The oviduct ratio shown in Table 8–6 is defined by Dorfman as the milligrams of oviduct per 100 g of body weight.

TABLE 8–4. *Rat uterine response to orally administered sodium estrogen sulfates*[a]

Preparation	Total dose (μg)	No. of rats	Mean uterine ratio[b] (±S.E.)
0	0	11	54± 3.7
0	0	13	56± 1.6
Sodium estrone sulfate[c] (sample A)	2.5	17	77± 5.7
	5.0	22	108± 9.4
	10.0	32	141± 7.8
	20.0	11	184±12.3
Sodium estrone sulfate[d] (sample B)	4.0	13	67± 2.6
	8.0	12	104± 8.9
Sodium 17β–estradiol–3–sulfate[e]	2.0	12	86± 8.0
	4.0	14	103± 5.4
Sodium equilin sulfate	4.0	10	75± 2.6
	8.0	17	108± 5.2
	16.0	10	144± 3.8
	32.0	9	188± 9.3
Sodium 17β–dihydroequilin–3–sulfate[e]	2.0	9	151± 6.0
	4.0	17	191± 4.6
	8.0	9	196+ 5.8
	16.0	7	216±10.2

[a] Dorfman, R. I., Dorfman, A. S.: Endocrinology **55**, 65 (1954).

[b] The uterine ratio is calculated as one hundred times the uterine weight in mg divided by the body weight in grams.

[c] Pure synthetic sample.

[d] Synthesized from a partially purified estrogenic fraction from pregnant mare's urine.

[e] Reduced according to Dorfman and Dorfman, see footnote a.

TABLE 8–5. *Comparative activity of estrogen sulfates*

Compound	Oral	Subcutaneous[a]
Sodium estrone sulfate	8 (5.9)[b]	2 (1.5)
Sodium equilin sulfate	8 (5.6)	2 (1.5) (0.18)
Estrone	2	0.1
Equilin	4	0.5

[a] Weight in micrograms of the amount of estrogen required to effect a three-fold increase in weight of the mouse uterus over controls.

[b] Weight of the estrogen contained in the estrogen sulfate.

TABLE 8–6. *Chick oviduct response to naturally occurring estrogens and derivatives*[a] *(chicks on diet 13 days, starting 2–3 days after hatching)*

Preparation	Dosage estrogen sulfate (mg/kg)	No. of chicks	Mean oviduct ratio[b] (±S.E.)
0	0	32	16± 0.5
0	0	24	16± 0.5
0	0	37	14± 0.5
Sodium estrone sulfate (pure)	40	30	31± 4.0
	80	30	115± 11.0
	80	16	117± 17.0
	160	16	466± 90.0
Sodium estrone sulfate (crude)	40	30	39± 3.8
	80	27	142± 19.0
Conjugated equine estrogens	40	97	63± 8.0
	80	92	192± 15.0
Sodium 17β–estradiol–3–sulfate	40	14	28± 4.0
	80	29	78± 13.0
	160	14	470±129.0
Estrone	20	25	24± 2.0
	40	26	40± 3.0
	80	24	149± 24.0
	160	27	278± 21.0
17β–Estradiol dipropionate [3,17β–di–propionyloxyestra–1,3,5(10)–triene]	80	23	77± 6.0
	160	24	202± 36.0
Ethynylestradiol [17α–ethynylestra–1,3,5(10)–triene–3,17β–diol]	2.5	33	23± 1.0
	5.0	30	26± 3.4
	10.0	33	28± 1.4
	20.0	29	88± 11.0

[a] Dorfman, R. I., Dorfman, A. S.: Endocrinology **53**, 301 (1953).
[b] mg of oviduct per 100 g of body weight.

According to Dorfman, the comparative activities of the two estrogen sulfates, calculated by setting the potency of estrone at 100%, were 164% for sodium estrone sulfate (calculated on the basis of estrone content) and 163% for sodium 17β–estradiol–3–sulfate (calculated on the basis of 17β–

estradiol [estra-1,3,5(10)-triene-3,17β-diol] content. It can also be seen that the conjugated equine estrogens exhibit a greater oral activity than either of the two pure estrogen conjugates, thus supporting the concept that much of the potency of conjugated equine estrogens is derived from equilin sulfate.

B. Human Data

Estrogen preparations for therapeutic use have usually been standardized by animal or chemical assays. However, their effectiveness in humans can only be assessed therapeutically. Since the introduction of the conjugated equine estrogens in the early 1940's, clinical assessment of new estrogen preparations has been greatly refined, due in part to the availability of the sulfates. The various methods used to assay their hormonal properties are: 1) Vaginal smear and endometrial biopsy; 2) induction of withdrawal bleeding; 3) inhibition of ovulation; 4) inhibition of gonadotrophins; 5) relief of menopausal symptoms.

The human data on the hormonal effects of these preparations, their components, and related estrogen sulfates, are presented below according to the various methods used for evaluation. The individual estrogen sulfates mentioned are naturally derived or products of partial or total synthesis.

1. Vaginal Smear and Endometrial Biopsy

Many investigators [19, 21, 22, 44–46] have used the effects of estrogen sulfates on the vaginal epithelium and the endometrium to determine their estrogenic potency, monitored by examination of the exfoliated vaginal cells or by endometrial biopsy, respectively. Although these two methods are sometimes combined, examination of the vaginal cells is used more extensively, since the vaginal epithelium is more sensitive to estrogens than the endometrium [47], and because of the comparative ease of this examination [47, 48].

The degree of correlation between the smear and clinical improvement of the menopausal syndrome has been a controversial subject from time to time [44a, 45, 46, 49]. Reich *et al.* [44a] stated that the investigator must be fully aware of the endocrine status of each patient since such events as the climacteric or the time of the cycle can have a profound effect on vaginal cytology.

Cornification from basal cells is a response to estrogenic effect which can be estimated by the relative number of cornified cells according to the staining techniques of Papanicolaou [50] or Shorr [51]. When assaying the response to administered estrogen, a series of daily smears is necessary for an accurate estimation. The mucosal lining of the vagina is a most sensitive indicator of ovarian function and endocrine activity. Therefore, vaginal smears serve as an index to estrogenic balances and a guide to dosage regimens. In the normal adult woman, the vaginal smear reveals large precornified and cornified cells with small nuclei. During the climacteric, with diminishing ovarian function, regression of the mucosa begins, becoming more marked with advancing years. The vaginal cytology reflects the degree of atrophy; as the mucosa becomes thinner more basal cells are found in the smears.

Several papers [19, 21, 22] appeared in 1943 which evaluated conjugated equine estrogens—at that time commonly referred to as estrone sulfate—by the smear technique. These were among the first publications to report the successful clinical applications of estrogen sulfates and their comparison with other agents then in use. Gray [19] reported that the conjugated equine estrogens were about equivalent to oral diethylstilbestrol, mg for mg, based on their effect on the vaginal epithelium of menopausal women. Sevringhaus and St. John [21] found that 1.25 mg of conjugated equine estrogens were at least as effective as 0.1 mg and probably comparable to 0.3 and 0.5 mg of diethylstilbestrol.

Finally, in the same year, Fried and Hair [22] published their comprehensive clinical study with conjugated equine estrogens in 95 patients. Subjectively, the effectiveness of the drug was noted by relief of symptoms. Objectively, the vaginal smear was used as an index. All patients responded to estrogen therapy to some degree. The original complaints were completely alleviated in 44 (47%) and satisfactorily relieved in 72 (80%). These authors recommended an initial dosage of conjugated equine estrogens of 1.25 mg (t.i.d.) and, after alleviation of the symptoms, a maintenance dosage of 1.25 mg each day, or every other day. The vaginal smear technique was used to determine when the patient should be changed to a maintenance dose.

With similar methods, Reich *et al.* [44] evaluated piperazine estrone sulfate (Sulestrex®, Abbott Labs.) and noted that in 24 out of 25 women with various menopausal symptoms complete relief was obtained with initial doses of 1.5 to 6.0 mg daily. Maintenance dosage varied from 0.75 to 3.0 mg daily. These authors found that subjective relief paralleled the smear change. Reicher *et al.* [45] compared conjugated equine estrogens and 17β-estradiol by the same methods in castrated women and found that 1.25 mg of the conjugated equine estrogens, or 0.1 mg of 17β-estradiol gave relatively the same degree of relief, 67 and 68%, respectively, from menopausal complaints; however, improvement in smear composition did not necessarily accompany relief of subjective symptoms.

2. Induction of Withdrawal Bleeding

Although the vaginal smear method has revealed much information as to the comparative efficacies of estrogens, the menopausal patient introduces a variable—the waning activity of the ovary [52]. Indeed, many investigators have reported a lack of parallelism between smear evaluations and clinical experience (see B–1). Because of this, other investigators have placed more reliance on assays based on the amount of estrogen necessary to induce bleeding in women with primary or secondary amenorrhea or in castrated women with intact uteri.

In the general application of this method, the amenorrheal or castrated woman is treated with an estrogen for a period of two or three weeks. Bleeding will occur approximately one week after cessation of therapy if the estrogen has been administered in sufficient quantity. This bleeding, termed withdrawal bleeding, is indistinguishable from the menstrual period. Prior to shedding, the endometrium does not show progestational effects. Results are usually given in terms of minimum effective dose. Modifying this method, several investigators follow the estrogen treatment with an injection of progesterone [pregn-4-ene-3,20-dione], usually 100 mg. Doubling the effective dose of the estrogens does not affect the latent period or the duration of bleeding [53].

Among the earliest clinical publications on conjugated equine estrogens is a 1943 report by Turner *et al.* [23]. These authors reported the successful induction of withdrawal bleeding in 16 of 18 women with functional amenorrhea. The dose range was 2.5 to 12.5 mg daily but was constant for the same individual. They concluded that conjugated equine estrogens and diethylstilbestrol have essentially the same potency mg for mg.

In 1944, Allen [54] reported a well-designed clinical study, comparing 14 different estrogens and estrogen preparations. His subjects were eight castrated women with intact uteri. The administration of an adequate dose of estrogen to these patients produced withdrawal bleeding. The results of this study, which encompassed 130 observations, are reproduced in Table 8–7.

In a series of papers [52, 55, 56] representing a definitive study of this method of evaluating estrogens, Bishop *et al.* compared the efficiency of several estrogens to produce uterine bleeding in women with amenorrhea. The relative potencies of several of the estrogens are listed in Table 8–8. Diethylstilbestrol is arbitrarily given a potency value of 100. For comparison, the findings of Harmer and Broom [37] for the estrogens by means of the Allen–Doisy test in ovariectomized adult rats is also included in Table 8–8.

TABLE 8–7. *Quantity of various estrogens per week necessary to induce withdrawal bleeding when given for three weeks in approximately 50% of the trials*[a]

Estrogen	mg per week
Ethynylestradiol (orally)	0.40
17β–Estradiol dipropionate (intramuscularly)	0.86
17β–Estradiol–3–monobenzoate [3–benzoyloxyestra–1,3,5(10)–trien–17β–ol] (intramuscularly)	1.06
17β–Estradiol–3–monopropionate [3–propionyloxyestra–1,3,5(10)–trien–17β–ol] (intramuscularly)	1.75
Diethylstilbestrol (orally)	3.00
Estrone (intramuscularly)	3.00
17β–Estradiol (orally)	13.40
Sodium estrone sulfate[b] (orally)	17.50
Estrone (orally)	31.50
Monomethoxy–diethylstilbestrol [3–hydroxyphenyl–4–methoxyphenyl–3–hexene] (orally)	35.00
Dimethoxy–diethylstilbestrol [3,4–bis–(4′–methoxyphenyl)–3–hexene] (orally)	14.00
Hexestrol	10.50
Sodium estriol–3–glucuronide [16α,17β–dihydroxyestra–1,3,5(10)–trien–3–yl–β–D–glucopyranosiduronic acid sodium salt] (orally)	42,000 oral units
Mixed estrogens[c] (orally)	126,000 int. units

[a] Allen, W. M.: Southern Med. J. **37**, 270 (1944).
[b] Conjugated equine estrogens.
[c] Estrogen fraction from hydrolyzed urine of stallions or pregnant mares.

TABLE 8–8. *Relative potency (compared with diethylstilbestrol) of estrogens administered orally to rats and humans*[a]

Estrogen	Rat[b]	Man	No. of cycles studied in man
Diethylstilbestrol	100	100	29
Diethylstilbestrol sulfate	50	44	15
Estrone	2	5	14
Equilin	—	32	17
Conjugated equine estrogens	5	49	23

[a] Bishop, P. M. F., Richards, N. A., Perry, W. L. M.: Lancet **I**, 818 (1951).
[b] Harmer, G. L. M., Broom, W. A.: Lancet **I**, 850 (1950).

Three items are worth noting in this table: 1) The estrogenic potency of a compound determined in the rat by the Allen–Doisy test does not always parallel the potency of the same compound in humans. 2) Conjugated equine estrogens are more potent in humans than free estrone or free equilin, indicating that sulfate conjugation enhances the activity. 3) Sulfate conjugation of the synthetic estrogen, diethylstilbestrol, decreases its activity. Furthermore, Bishop [56] also reported that the sulfates of two other synthetic estrogens, dienestrol [3,4–bis–(4′–hydroxyphenyl)–2,4–hexadiene], and hexestrol, were much less active in humans than the parent compounds.

In 1952, Kupperman *et al.* [57], dissatisfied with the existing methods of evaluating estrogens in amenorrheal patients, published a statistical method for analyzing clinical data. By weighing the various factors, they derived certain values which they named Amenorrheal Indices (A.I.). In their application of this method, patients with primary or secondary amenorrhea were treated with a daily dose of estrogen over a period of twenty–one days or by an injection of a long–acting estrogen. On the twenty–second day, an intramuscular injection of 100 mg of progesterone was given. The A.I. was derived from the formula:

$$\text{A.I.} = \frac{D}{T} \times E$$

in which D = the number of bleeding days, T = the time of onset of withdrawal bleeding, and E = an estimation of the extent of bleeding using an arbitrary scale from 0 to 5+.

The comparative effects of the various estrogens on the A.I. are illustrated in Table 8–9. The doses of estrogens were twice those usually employed in menopausal patients. Under these conditions, ethynylestradiol [17α–ethynylestra–1,3,5(10)–triene–3,17α–diol] (Estinyl®, Schering) alone or in combination with free 17β–estradiol and conjugated equine estrogens gave the best results.

Kupperman *et al.* also investigated the effect of dosage on the A.I. to check the quantitative accuracy and validity of the method. The results illustrated in Table 8–10 show that when the higher doses were employed, a higher A.I. was obtained. Moreover, the use of one–half the original dose resulted in an A.I. approximately one–half as great. These results supported the validity of his method.

Kupperman *et al.* also applied a statistical analysis to evaluate estrogens for the relief of menopausal symptoms. These results will be discussed in Section 4 below.

TABLE 8–9. *The comparative effect of various estrogens on the amenorrheal indices*[a] *(figure in parentheses = number of trials)*

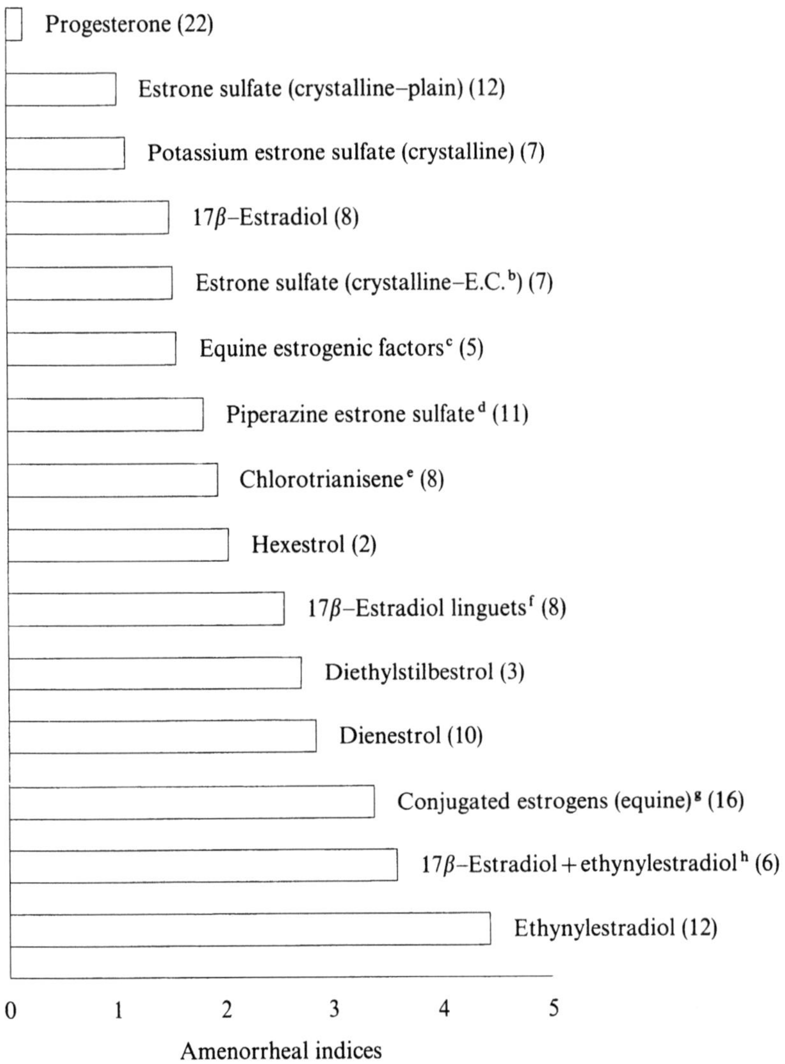

[a] Kupperman, H. S., Blatt, M. H. G., Wiesbader, H., Filler, W.: J. Clin. Endocrinol. Metab. **13**, 688 (1953).

[b] E.C. = enteric coated.

[c] Menagen, Parke Davis.

[d] Sulestrex, Abbott Labs.

[e] TACE, Wm. G. Merrell.

[f] Progynon buccal tablets, Schering.

[g] Premarin, Ayerst, McKenna and Harrison.

[h] Estinyl, Schering.

TABLE 8–10. *The effect of dosage of estrogens on the amenorrheal indices*[a] *(figure in parentheses = number of trials)*

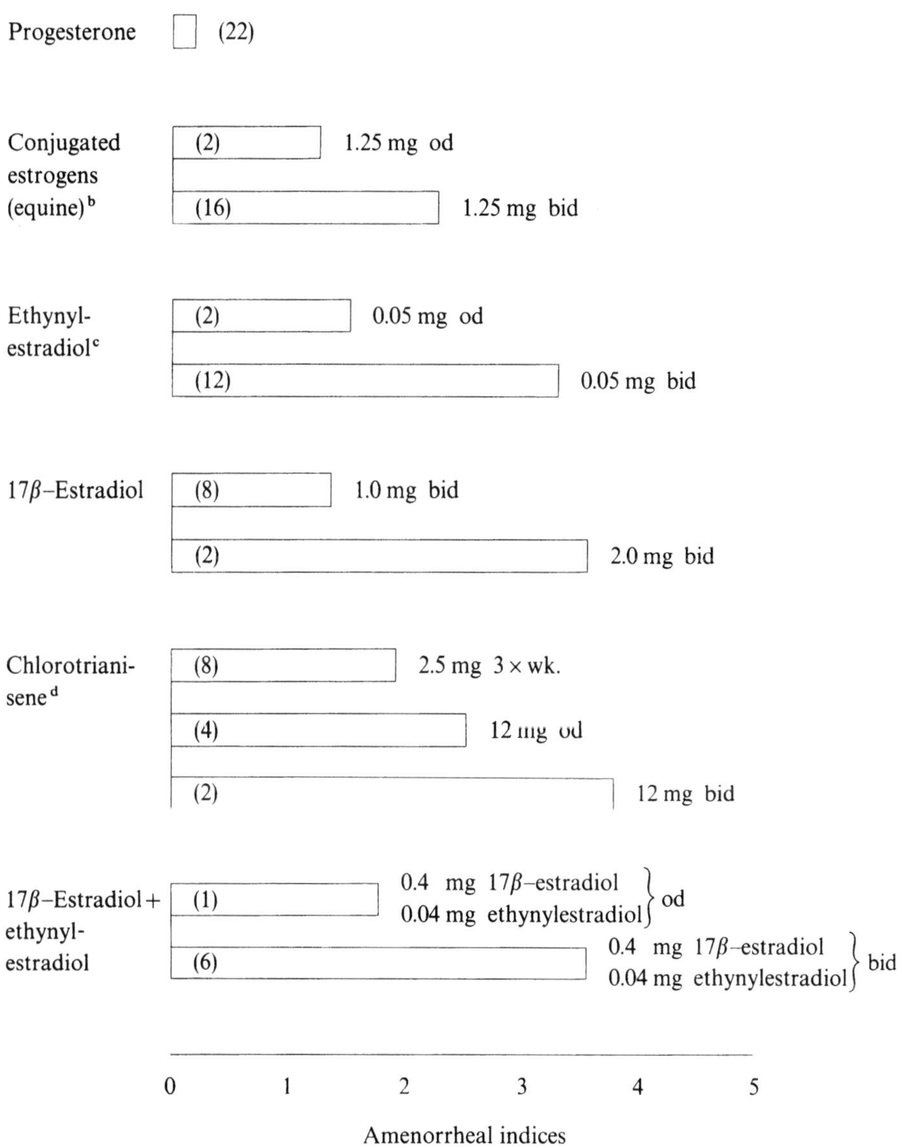

[a] Kupperman, H. S., Blatt, M. H. G., Wiesbader, H., Filler, W.: J. Clin. Endocrinol. Metab. **13**, 688 (1953).

[b] Premarin, Ayerst, McKenna and Harrison.

[c] Estinyl, Schering.

[d] TACE, Wm. S. Merrell.

3. Ovulation Inhibition

Several investigators [58–63] have studied the ability of estrogen sulfates to inhibit ovulation. The exact mechanism of the antiovulatory action of estrogens is still unknown. Preliminary data indicate that this ability is a direct ovarian inhibition to gonadotrophic response as well as a hypothalamic–pituitary suppression of gonadotrophin.

Generally speaking, a correlation exists between the estrogenic and antiovulatory effects of a compound. However, sometimes the antiovulatory effect is less than expected, or even nonexistent (*e.g.*, estriol [estra–1,3,5(10)–triene–3,16α,17β–triol]), presumably because these two effects involve different target organs.

Although a reliable method to determine the time of ovulation has not been found, its occurrence can be quite reliably established by means of urinary pregnanediol [5β–pregnane–3α,20α–diol] assay, basal body temperature, vaginal cytology, changes in the quantity, quality, and "*Spinnbarkeit*" of the cervical mucus or by endometrial biopsy [64].

In 1947, Haus *et al.* [58] from Duke University School of Medicine, and Duke Hospital, reported on a study comparing conjugated equine estrogens with diethylstilbestrol, based on the premise that ovulation is a prerequisite for dysmenorrhea. The effects of these estrogens on basal body temperature and endometrial biopsy in 54 dysmenorrheal patients was recorded for 228 cycles. The investigators noted that all anovulatory cycles were pain–free. Their best results were obtained with a group of patients treated with a daily dose of either 3.75 mg of conjugated equine estrogens or 3.0 mg of diethylstilbestrol from the fifth to the twenty–fifth day of the menstrual cycle for a total of 23 cycles. Eighteen (78.3 %) cycles were totally pain–free. Three years later, in another publication from the same group, Dignam *et al.* [59] found that, with the same dosage schedule, conjugated equine estrogens completely relieved dysmenorrhea in eight patients through 20 cycles. However, ovulation appeared to be suppressed in only one–half of the cycles at this dose. Similar results were obtained by Blewett [60] and Greenblatt [61]. Blewett reported that administration of conjugated equine estrogens at a dosage of 7.5 mg daily from the 3rd to the 23rd day of the cycle suppressed ovulation and completely relieved dysmenorrhea in 37 women through 141 menstrual cycles. Furthermore, Greenblatt reported that painless anovulatory menstrual cycles occurred in a woman with membranous dysmenorrhea when she was given 7.5 mg daily of conjugated estrogens for 20–day periods.

Later, Gual *et al.* [62] cited an unpublished report on a trial of 17β–estradiol–3–sulfate. A dosage of 5 mg daily resulted in six ovulations in 53 cycles, as measured by urinary pregnanediol excretion.

TABLE 8–11. *Antiovulatory activity of several estrogens*[a]

Compound	Dose	Cases	Cycles	Ovulatory cycles
Ethynylestradiol	20 μg	10	20	2
	50 μg	20	44	1[b]
Ethynylestradiol-3-methyl ether	20 μg	10	20	2
	80 μg	18	60	1
17β-Estradiol	1 mg	4	11	5
	2 mg	10	18	7
	5 mg	10	24	3
Estriol	5 mg	5	7	6
Conjugated equine estrogens	1.25 mg	10	18	12
	3.75 mg	15	17	1
Diethylstilbestrol	5 mg	6	12	1
Total		128	247	41

[a] Martinez-Manautou, J., Rudel, H. W., in: Ovulation (R. B. Greenblatt, ed.), p. 243. Philadelphia and Toronto: J. B. Lippincott Co. 1966.

[b] Secretory endometrium one patient on 12th cycle.

Martinez-Manautou and Rudel [63] also reported a comprehensive study on the antiovulatory activity of several synthetic and natural estrogens in 128 fertile women for a total of 247 cycles. Their results are summarized in Table 8–11. Ovulation was determined by secretory changes in the endometrium and by urinary pregnanediol determinations.

It was found that 3.75 mg of conjugated equine estrogens appears to be an effective dose to control ovulation, although more patients must be studied before an adequate comparison with ethynylestradiol and its corresponding 3-methyl ether can be made.

A correlation between estrogenic and antiovulatory activity as determined by these same authors is illustrated in Table 8–12.

4. Inhibition of Gonadotrophins

Urinary gonadotrophin levels are increased in menopausal and castrated women, but these levels can be lowered if sufficient amount of estrogen is administered [65]. This estrogenic effect, used by some investigators [66–69] to estimate the relative potencies of estrogen sulfates, has shown that the minimum dose of estrogen or estrogen sulfate required to depress urinary gonadotrophins is about twice the amount needed to provide optimum

TABLE 8–12. *Correlation between estrogenic and antiovulatory activities*[a]

Compound	Doses	Estrogenic activity	Anti-ovulatory activity
Ethynylestradiol	20 µg	+ +	+ +
	50 µg	+ + + +	+ + +
Ethynylestradiol–3–methyl ether	20 µg	+ +	+ +
	80 µg	+ + + +	+ + +
17β–Estradiol	1 mg	+	+
	2 mg	+ + to + + +	+
	5 mg	+ + + +	+ +
Estriol	5 mg	+ +	0
Conjugated equine estrogens	1.25 mg	+ + +	0 to +
	3.75 mg	+ + + +	+ + +
Diethylstilbestrol	5 mg	+ + + +	+ + +

[a] Martinez–Manautou, J., Rudel, H. W., in: Ovulation (R. B. Greenblatt, ed.), p. 243. Philadelphia and Toronto: J. B. Lippincott Co. 1966.

relief of menopausal symptoms [66]; also, that small doses of these estrogens increase gonadotrophin output [69].

In practice, urine specimens are bioassayed for total gonadotrophin levels before, during, and after estrogen administration, using either mouse or rat uterine weight as an end point. Determinations of urinary gonadotrophins by immunoassay or chemical assay are not yet sufficiently sensitive to justify replacing the biological by the serological procedure [65].

Tokuyama *et al.* [66] reported in 1955 the relative ability of several estrogen sulfates to decrease urinary gonadotrophin excretion, using the rat uterus test. Their results may be summarized as follows:

Diethylstilbestrol	100
Conjugated equine estrogens	50
Potassium estrone sulfate	25
17β–Estradiol	20
Potassium 17β–estradiol–3–sulfate	20

The number of patients in whom each estrogen compound was studied ranged from eight for potassium 17β–estradiol–3–sulfate to 26 for diethylstilbestrol. These authors recommended this method for the clinical evaluation of estrogenic potency. Although their results are in general agreement with those obtained by investigators using other end points, the similar potency of the sulfates of estrone and 17β–estradiol contrast with the usual finding that the 17β–hydroxy estrogens are significantly more potent than their corresponding 17–keto derivatives.

In 1956, Smith and Albert [67] found diethylstilbestrol more effective than conjugated equine estrogens in comparable doses (0.1, 1, and 10 mg/day) in diminishing urinary gonadotrophins in a castrated woman.

First, at a meeting of the Endocrine Society in 1956, and again later, Howard *et al.* [68] reported the comparative effectiveness of several estrogen sulfates in maintaining low urinary gonadotrophin levels for five weeks in castrated and menopausal women. They found that the lowest effective daily dose was 2.5–10 mg for sodium estrone sulfate, 1.2–2.5 mg for conjugated equine estrogens, and 0.6–1.2 mg for sodium equilin sulfate. In 1960, Rosemberg and Engel [69] described a study in which a daily dose of 2.5 mg of conjugated equine estrogens suppressed urinary gonadotrophin levels, whereas the same dose of estrone sulfate had no effect.

Generally speaking, this method for the determination of estrogenic potency has not been widely applied, no doubt due to the availability of methods which measure the more direct effects of estrogens.

Finally, with the use of bioassay methods, Vorys *et al.* [70] investigated the effects of conjugated equine estrogens (0.625, 1.25, 2.5, and 3.75 mg doses); ethynylestradiol (0.02, 0.05, and 0.1 mg doses); ethynylestradiol–3–methyl ether (mestranol) [17α–ethynyl–3–methoxyestra–1,3,5(10)–trien–17α–ol] (0.05, 0.08, and 0.1 mg doses), and other steroids, upon urinary excretion of follicle–stimulating hormone (FSH) and luteinizing hormone (LH). They stated: "Conjugated equine estrogens appear to be the most capable of stimulating LH within a short period of time. The synthetic estrogens, ethynylestradiol and mestranol, appear to require longer periods of time for this action. These findings suggest that steroids in a water–soluble form may be absorbed more quickly and their effect more pronounced."

These authors emphasized that their results (Table 8–13) are only indicative of the effects of estrogens on FSH and LH, being limited by the accuracy of the assay methods and by the relatively small number of subjects. However, continued efforts in this area of research will help to elucidate the mechanism of action of these drugs. Such knowledge will be most helpful, especially in the regulation of fertility.

5. *Relief of Climacteric and Menopausal Symptoms*

A successful evaluation by this method is dependent on the investigators' ability to analyze the data and make exact comparisons. Relief from symptoms is usually categorized by such designations as excellent, good,

TABLE 8–13. *Effect of estrogens on the excretion of FSH and LH*[a]

1. Conjugated equine estrogens		
FSH:	Large doses	— slight supression
	Low doses	— no effect
LH:	High doses	— some stimulation (delayed)
	Low doses	— stimulated more than high doses (elevated)
2. Ethynylestradiol		
FSH:	High doses	— suppression
	Low doses	— stimulates FSH (early)
LH:	High doses	— little change
	Low doses	— little change
3. Ethynylestradiol-3-methyl ether (mestranol)		
FSH:	High doses	— suppression. Mestranol suppressed more than ethynylestradiol alone
	Low doses	— no stimulation or suppression
LH:	High doses	— stimulation but required longer and had more dramatic effect
	Low doses	— stimulated LH earlier

[a] Vorys, N., Ullery, J. C., Stevens, V.: Am. J. Obstet. Gynecol. **93**, 641 (1965).

fair, and poor. The usual early symptoms of the climacteric are vasomotor complaints, such as vascular hyperemia (*e.g.*, hot flushes), paresthesia (sensations of burning, numbness, pricking, etc.), insomnia, and nervousness. Some or all of these symptoms are considered in the evaluation. Most investigators include drug side effects, such as nausea, abnormal bleeding, breast tenderness, or cramps in considering the over-all response to the drug.

Numerous investigators have used this method to evaluate natural and synthetic estrogen sulfates [18–20, 22, 44, 45, 49, 57, 71–73]. Collectively, these early reports all served to demonstrate the effectiveness of estrogen sulfates for the alleviation of menopausal symptoms. Table 8–14 compares the results of several of these investigators in providing relief of menopausal symptoms in patients treated with conjugated equine estrogens.

These data demonstrate the effectiveness of estrogen therapy with conjugated equine estrogens as well as a dose-response relationship. Similar data have been reported for sodium estrone sulfate [73] and piperazine estrone sulfate [44, 49 b].

In addition, Harding [71, 72] reported that 55 out of 138 patients suffering from hypoovarian symptoms (*e.g.*, menopausal symptoms, secondary

TABLE 8–14. *Per cent of patients obtaining complete relief of menopausal symptoms by treatment with conjugated equine estrogens*

Dose (mg) (md = maintenance dose)	% obtaining relief	% obtaining complete relief	No. of patients[a]	References
0.40 tid	83	26	45	Freed *et al.* [18]
0.81 tid	80	38	47	Freed *et al.* [18]
1.20 tid	86	43	60	Freed *et al.* [18]
1.25 daily × 7 then md = 1.25 every other day	84	72	64	Gray [19]
2.5–3.75 daily then md = 3.75–8.75 weekly	92	82	71	Glass and Rosenblum [20]
1.25 tid, md = 1.25 daily or every other day	80	47	95	Fried and Hair [22]
2.5 daily	90–99	95.8	95	Perloff [49a]
1.25 daily	68	40	38	Reicher *et al.* [45]

[a] Both natural and artificial menopausal patients are included.

amenorrhea, senile vaginitis) obtained complete relief. Most of these patients received a daily dose of 1.25 mg of conjugated equine estrogens.

With this method, many comparisons have been made: Freed *et al.* [18] indicated that the ratio of oral effectiveness of conjugated equine estrogens to diethylstilbestrol is no less than, and probably greater than 1:2, whereas Gray [19] considered these two agents equivalent, mg for mg, on the basis of vaginal smears and the ability of these drugs to relieve menopausal symptoms. Perloff [49a] reported complete relief of menopausal symptoms in 90 to 99% of his patients with daily doses of 1.0 mg of 17β–estradiol (Progynon-DH®, Schering), 0.05 mg ethynylestradiol or 2.5 mg conjugated equine estrogens; on the other hand, Reicher *et al.* [45] reported that 0.1 mg of 17β–estradiol appears to be comparable to 1.25 mg of the conjugated estrogens.

One of the most extensive comparisons of agents for the treatment of the menopause was made by Perloff [49a]. In addition to the study reported above, he also compared 11 agents, including conjugated equine estrogens, based on effectiveness and incidence of nausea and bleeding. His results are illustrated in Table 8–15. He concluded that the conjugated equine estrogens

TABLE 8–15. *Comparison of several estrogens in the treatment of menopause*[a]

Medication	Number treated	Number completely relieved	Per cent completely relieved	Nausea		Bleeding		
				No.	Per cent	Incidence in all patients	Incidence in patients with amenorrhea	Incidence in patients capable of bleeding
17β–Estriol[b]	81	68	83.9	3	3.7	8.7	6.2	9.6
17β–Estradiol–3–benzoate[c]	59	55	93.1	0	0	0	0	0
17β–Estradiol, pellets	45	43	95.6	0	0	0	0	0
Bisdehydrodoisynolic acid	24	4	16.6	3	12.5	8.33	4.16	8.33
3,4–Bis(3–methyl–4–hydroxyphenyl)hexane[d]	64	52	81.5	4	6.25	29.7	21.9	33.9
3,4–Bis(4–hydroxyphenyl)–2,4–hexadiene[e]	98	90	90.8	6	6.1	17.2	7.15	20.7
dl–Alpha–tocopherol acetate[f]	65	17	26.1	1	1.5	1.5	1.5	1.8
Estrone	96	76	79.1	0	0	12.5	7.3	14.1
Ethynylestradiol[g]	138	130	94.2	13	9.4	23.2	14.5	27.6
Methyl bisdehydrodoisynolic acid	35	14	40.0	4	11.4	5.7	5.7	6.9
Conjugated equine estrogens[h]	95	91	95.8	4	4.2	11.6	7.4	12.3

[a] Perloff, W. H.: Am. J. Obstet. Gynecol. **58**, 684 (1949).
[b] Progynon–DH, Schering. [c] Progynon–B^2, Schering.
[d] Meprane, Reed and Carnrick. [e] Dienestrol, White Labs.
[f] Ephynal acetate, Hoffman–La Roche.
[g] Estinyl, Schering.
[h] Premarin, Ayerst, McKenna and Harrison.

appeared to have advantages over many of the oral estrogens since they were 95.8% effective in completely relieving menopausal symptoms and produced less nausea and bleeding than most of the other agents.

Another definitive study on the effectiveness of the estrogen sulfates to relieve menopausal symptoms was reported by Kupperman *et al.* [57]. They derived a menopausal index, using a statistical analysis in which the major complaints (*e.g.*, hot flashes, numbness, insomnia and nervousness) were given a weighted value over what they considered minor complaints. The results of this study are given in Table 8–16. With the double–blind testing technique, they confirmed their results obtained from the comparative study of the induction of bleeding in amenorrheal women (see Table 8–9).

In summary, these data point out the wide variation in the effectiveness reported for the same estrogens. These differences indicate the difficulty encountered by the investigators in judging the comparative efficacy of these estrogens by subjective methods and the necessity of employing simultaneously well–defined objective standards. Another factor is the wide variation of estrogen dosages required by menopausal women with similar symptoms. Despite the inherent difficulties of this method, it has been used extensively and does at times give surprisingly good correlation with other assays. In the final analysis, this bioassay is the most useful for predicting the future acceptance of a new drug.

III. Nonhormonal Properties of Estrogen Sulfates

A great variety of nonhormonal effects of estrogens have been reported and comprehensively summarized by Fisher [74], Blanpin [75], and Kappas and Palmer [76]. It is possible that these properties could be demonstrated for many of the estrogen sulfates; however, only the nonhormonal properties actually reported for estrogen sulfates will be discussed in this section.

The literature contains a large amount of data on the nonhormonal properties of estrogen sulfates. Most of these data deal with those effects which have possible clinical application for the treatment of atherosclerosis and capillary hermorrhages of different origin. Accordingly, these data represent the main body of the following presentation. In addition, several noteworthy miscellaneous effects are discussed.

A. Animal Data

Comparatively speaking, animal experiments demonstrating the nonhormonal effects of conjugated estrogen sulfates are few as a result of the wide clinical acceptance of estrogens for the past 35 years. Nevertheless,

TABLE 8–16. *The effect of estrogenic and nonestrogenic*

Preparation	Dose	No. of patients
Conjugated equine estrogens [b]	1.25 mg daily	156
Piperazine estrone sulfate [c]	1.5 mg daily	55
Potassium estrone sulfate	1.25 mg daily	76
Estrone sulfate (crystalline)	1.25 mg daily	80
Estrone sulfate (enteric coated)	1.25 mg daily	81
Ethynylestradiol [d]	0.05 mg daily	88
17β–Estradiol	2.0 mg daily	33
17β–Estradiol linguets [e]	0.25 mg daily	28
Equine estrogenic factors [f]	10,000 i.u. daily	79
Chlorotrianisene [g]	12 mg daily or 24 mg 3 × wk.	85
Diethylstilbestrol	0.1 mg tid	47
Hexestrol	3 mg daily	42
Dienestrol	0.5 mg daily	71
17β–Estradiol [h] Ethynylestradiol [h]	0.4 mg 0.04 mg } daily	66
17β–Estradiol [i] Ethynylestradiol [i] Methyltestosterone [i]	0.4 mg 0.04 mg } daily 2.5 mg	34
dl–Alpha–tocopherol [j]	50–100 mg daily	69
Phenobarbital	15 mg tid	89
Placebo	1 tab. daily	248

[a] Kupperman, H. S., Blatt, M. H. G., Wiesbader, H., Filler, W.: J. Clin. Endocrinol. Metab. **13**, 688 (1953).

[b] Premarin, Ayerst, McKenna and Harrison.

[c] Sulestrex, Abbott Labs. [d] Estinyl, Schering.

[e] Progynon buccal tablets, Schering.

some animal data are available and merit attention. The following deal mostly with the nonhormonal effects of conjugated equine estrogens, the only estrogen sulfate investigated extensively in animals.

1. *Inhibition of Atherogenesis*

As early as 1939, Master *et al.* [77] noted that coronary atherosclerosis occurred with significantly greater frequency in the human male than in the female. This observation prompted Pick *et al.* [78], and Stamler *et al.* [79]

compounds upon menopausal symptoms[a]

Response to therapy					
Poor		Moderate		Excellent	
No. of patients	Per cent	No. of patients	Per cent	No. of patients	Per cent
11	7.0	41	26.3	104	66.7
6	10.9	8	32.7	31	56.4
9	11.8	25	32.9	42	55.3
17	21.2	21	26.3	42	52.5
15	18.4	27	33.3	39	48.3
7	8.0	23	26.0	58	66.0
4	12.1	11	33.4	18	54.5
3	10.8	11	39.2	14	50.0
21	26.6	24	30.4	34	43.0
13	15.2	22	25.8	50	59.0
19	40.4	14	29.8	14	29.8
15	35.7	12	28.6	15	35.7
23	32.4	25	35.2	23	32.4
6	9.1	22	33.3	38	57.6
4	11.8	6	17.7	24	70.5
53	76.75	5	7.25	11	16.0
51	57.3	17	19.2	21	23.5
165	66.6	32	12.8	51	20.6

[f] Menagen, Parke Davis.
[g] TACE, Wm. S. Merrell.
[h] Amboral, Warner Inst. Therapeutic Res.
[i] Bixone, Warner Inst. Therapeutic Res.
[j] Esorb, Wyeth.

to investigate the effects of estrogen on the atherogenesis of cockerels fed cholesterol [cholest-5-en-3β-ol]. These investigators found that oral administration of conjugated equine estrogens to cockerels at 20 mg/bird/day produced feminization, altered lipid patterns and prevented atherosclerosis in the coronary vessels, but not in the aorta [79]. Similar results were reported for 17β-estradiol-3-benzoate [3-benzoyloxyestra-1,3,5(10)-trien-17β-ol] and estrone at doses of 1 and 10 mg/bird/day, respectively, given parenterally [78]. In addition, their studies showed that these estrogens were able to reverse well-developed, cholesterol-induced plaques in the coronary vessels of chicks [79]. These observations suggested an association between the protection of coronary vessels and lipid patterns in cockerels and, no doubt, encouraged many investigators to ascertain if such an association exists in man (see Section III B–1). However, one must be very cautious in assuming a direct relationship of these data obtained from cockerels to man.

Clarke *et al.* [80] reported that, in their studies, conjugated equine estrogens, added to the diet of chickens in doses of 12.5 to 75 mg/day for five weeks, caused a diphasic dose-dependent response of bile cholesterol levels, which were elevated in birds receiving the lowest dose but decreased steadily with increasing doses of the estrogens. Liver cholesterol levels showed a similar pattern. On the other hand, the bile acid levels showed a diphasic dose-dependent respose that was a reversal of this pattern, with a significant decrease in linoleic acid content in all groups receiving conjugated equine estrogens.

Working on the theory that it would be desirable to have an agent with the lipid-shifting qualities of an estrogen with little or no estrogenicity, Arnold *et al.* [81] screened several estrogens in rats. Among the steroids screened were estrone, conjugated equine estrogens, and estrone methyl ether [3-methoxyestra-1,3,5(10)-trien-17-one]. By using serum cholesterol levels and the uterine weight as parameters, this group determined dissociation ratios for these three estrogens of 50, 20, and >3.2, respectively; the lowest ratio indicated the best separation of hypocholesteremic and estrogenic effects. Several other estrogens were noted to have a much better dissociation of these activities.

Current approaches to the therapy of coronary heart disease with estrogens will be discussed in more detail in the next section on human data.

2. *Alteration of Clotting Factors*

Although estrogens, especially conjugated equine estrogens, have been used extensively in humans for the control of capillary bleeding, few animal experiments are available that throw any light on the possible mechanism by which these estrogens may be effective.

At present it can be stated that, in animals, a species difference probably exists for the response of clotting factors to estrogens. In 1957, Johnson [82] reported that conjugated equine estrogens, when administered intravenously to dogs, produced a sharp rise in plasma accelerator globulin (Factor V) and prothrombin concentration, and a small decrease in antithrombin activity, changes which would tend to enhance the coagulability of the blood. On the other hand, Stamler and Warner [83] had previously observed that injecting cockerels with diethylstilbestrol, dienestrol, or hexestrol led to a rise in heparin-like antithrombic activity. Not quite in keeping with either of these observations, Mandoki and Rubio [84] reported in 1966 that, in the male Wistar rat, conjugated equine estrogens increased blood clotting time by 19 and 30% after one and two days' treatment, respectively, whereas diethylstilbestrol, as well as several other estrogens, had no effect.

In exploring the possible value of estrogens in controlling ocular hemorrhage, Lewis [85], in 1959, severed the superior rectus muscles of rabbits and recorded the bleeding time. Later, the contralateral muscle of the same rabbit was severed after the rabbit had received 20 mg of conjugated equine estrogens intravenously, resulting in a pronounced reduction in bleeding time. In another experiment, he performed 26 iridectomies, 13 with and 13 without conjugated equine estrogens. However, in the latter experiments his results were not as conclusive, but seem to indicate less hemorrhage in the animals receiving the estrogens.

Schmidt-Matthiesen and Poliwoda [86] reported on the effect of conjugated equine estrogens, estriol sodium succinate [3-succinoyloxyestra-1,3,5(10)-triene-16α,17β-diol sodium salt]—a sythetic estrogen—and estrone sulfate upon vascular fragility as measured by the cupping glass test in rabbits. All three agents produced a fragility-reducing effect over controls at a single dose of 0.5 mg/kg, the most intense effect being achieved by the first two agents.

The role of the estrogens in this field will be discussed in more detail in the section on human data.

3. *Miscellaneous Effects in Animals*

While investigating the vascular effect of endotoxin in the isolated rat liver, Nolan [87] noted that the vasoconstrictive response to the administration of bacterial lipopolysaccharide failed to occur if female rather than male rat blood was used as a perfusate. Further studies showed that there was a significant protective effect for both male and female rats when the estrogens were given (4 mg/rat) 30 minutes and one hour before bacterial endotoxin injection. No protection was obtained when the estrogens were given at the same time as the toxin, afterwards, or two or more hours before. These results imply a strong relationship between estrogens and immunologic factors. The author noted the similarity of this effect to the stimulatory effects of the natural estrogens, estrone, 17β-estradiol, and estriol, on the recticulo-endothelial system as noted by Nicol *et al.* [88].

In 1963, Rona and Chappel [89] reported on the protective effect of conjugated equine estrogens against the epiphyseal bone lesions of guinea pigs with experimental scurvy. The scorbutic guinea pig was used as a model to study the effects of these estrogens on pathologic epiphyseal bone formation because (a) these estrogens have been used extensively in humans as a hemostatic agent; (b) some authors (see Section III B–2) had reported that the estrogens act on the ground substance of connective tissue; and (c) the morphologic changes in experimental scurvy of the guinea pig are characterized principally by abnormal formation of the ground substance of connective tissue. These authors found that 1.0 mg conjugated estrogens (orally, three times weekly) produced a protective effect against the epiphyseal bone lesions ordinarily observed in guinea pigs with experimental scurvy. This protective action was also exhibited by 0.5 mg 17β–estradiol–3–benzoate (s.c. three times weekly) [90]. In a later study, Hajdu *et al.* [91] found that these conjugated estrogens, but not a synthetic compound—estriol sodium succinate [3–succinoyloxyestra–1,3,5(10)–triene–16α,17β–diol sodium salt]—protected the guinea pig at a dose of 1 mg administered every other day by intraperitoneal injection. No conclusion was drawn about the relationship of these observations to hemostasis.

Two papers on the effect of conjugated equine estrogens on the circulatory system of sheep are noteworthy. Griess and Marston [92] reported that intravenous conjugated equine estrogens caused a significant uterine vasodilation and increased uterine blood flow during ovine pregnancy, and Ueland and Parer [93] noted that similar adminstration of the estrogens caused significant changes in the cardiovascular dynamics of sheep. The latter authors found that the infusion of 60 and 80 mg of conjugated equine estrogens over a 4–hour period resulted in an increase in cardiac output and heart rate, and a decrease in arterial pressure and peripheral resistance. These effects were most pronounced two hours after the infusion and closely resembled effects observed in pregnant ewes. Infusion of 100 mg of 17β–estradiol increased cardiac output and decreased peripheral resistance, but did not change the heart rate or arterial pressure. The same dose of estriol produced no changes in any of these parameters.

Conjugated equine estrogens also tended to lower the blood pressure of cockerels. In addition, such estrogen–treated birds were immune to the chronic pressor effects of deoxycorticosterone acetate [21–acetoxypregn–4–ene–3,20–dione] and cortisone [17α,21–dihydroxypregn–4–ene–3,11,20–trione] [94].

B. Human Data

In humans, the conjugated equine estrogens are the only representatives of the class, estrogen sulfates, to be investigated extensively for nonhormonal properties, mostly in areas related to atherosclerosis and hemostasis.

1. Inhibition of Atherogenesis

Considerable evidence exists that estrogens and estrogen sulfates may exert a protective effect against atherosclerotic processes. For instance, in 1936, Levy and Boas [95] noted, in several thousand patients, that the incidence of coronary artery disease was 4.9 times more frequent in men than in women. Subsequent studies showed that the incidence of this disease increases in the postmenopausal state [96, 97]. Men with carcinoma of the prostate and women with breast cancer treated with large doses of estrogens showed less arteriosclerotic vascular disease than control subjects [98, 99].

This information suggested that ovarian function may be an etiologic factor in atherosclerosis. During the decade 1949 to 1959, this assumption spurred investigations which demonstrated that a sex difference exists in serum cholesterol lipoprotein patterns [79, 100] and that estrogens, including conjugated equine estrogens (*e.g.*, Ref. [79]), were capable of changing male lipoprotein patterns into female patterns.

This evidence led to long-term clinical investigations of the effects of estrogenic hormones on morbidity and mortality in atherosclerotic heart disease [101–104]. In two separate, five-year studies published by Marmorston *et al.* [103] in 1962 and by Stamler *et al.* [104] in 1963, it was observed that conjugated equine estrogens apparently increased the survival time of male patients with myocardial infarcts.

Marmorston's group studied 432 men who had survived a myocardial infarction. These men were randomly treated with conjugated equine estrogens, two synthetic estrogens—ethynylestradiol (Lynoral®, Organon) and mytatrienediol (Anvene®, Searle), [3-methoxy-16α-methylestra-1,3,5(10) triene 16β,17β diol] and a placebo. The estrogens were administered in small, well-tolerated doses of comparable potencies as determined by mild breast tenderness. The following doses were employed:

	Conjugated equine estrogens	Ethynyl-estradiol	Mytatri-enediol
Initial daily dose (mg)	0.625	0.01–0.10	5–10
Usual maintenance daily dose (mg)	1.25–2.5	0.05–0.15	10–15

At these doses, ethynylestradiol and mytatrienediol significantly lowered the cholesterol–phospholipid ratio, whereas conjugated equine estrogens had no effect. On the other hand, according to the statistical evidence, the conjugated equine estrogens provided a better than 50% over-all protection against a recurring infarction compared with placebo treatment. At the doses used, ethynylestradiol and mytatrienediol had no effect on survival.

Stamler's group conducted a similar five-year study on the effects of the conjugated estrogens on the long-term survival of male patients who had a myocardial infarction. In a comparative study of 275 patients who had received conjugated equine estrogens or a placebo, they found that the mortality, five years after the initial infarction, was reduced about 50% or more for the group receiving the estrogen therapy. These investigators found that doses of 1.25 and 2.5 mg daily failed to influence serum lipid patterns. A dose of 4 mg produced a significant change in most patients, but to assure a definite effect in all men, a daily dose of 10 mg was required. This latter dose also produced gynecomastia, depression of libido and potency in almost all patients. For some unexplained reason, an unexpected number of deaths occurred in the first two months of study among those men initially started on 10 mg of estrogen within three months of their most recent infarction. They concluded that conjugated equine estrogens seemed to be effective in the long-term therapy of myocardial infarction when high doses were avoided initially.

These encouraging results with conjugated equine estrogens were, however, tempered by the report of Oliver and Boyd [101] that ethynylestradiol, at doses which lowered serum cholesterol, did not increase longevity in men with previous myocardial infarctions.

In similar studies with patients with cerebral atherosclerosis, Marmorston's group reported that conjugated equine estrogens reduced the incidence of cerebral infarctions in men who had suffered previous myocardial infarction [105], and that small doses of the conjugated equine estrogens administered to patients with cerebral thrombosis reduced significantly the mortality rate. Incidentally, Evans and Marmorston [106] noted an improved mental functioning as indicated by the Rorschach test with the cerebral thrombosis patients treated with the conjugated equine estrogens. On the other hand, Robinson *et al.* [107] reported that large amounts of diethylstilbestrol (5 to 25 mg/day) or conjugated equine estrogens (5 to 10 mg/day) did not prevent the development of myocardial or cerebral infarctions in 57 men with a previous history of this condition. The average duration of treatment for the patients in this study was 5.4 years.

Also in contrast to Marmorston's report relating to the reduced incidence of cerebral infarctions, the United States Veterans Administration Cooperative Study of Atherosclerosis resulted in a preliminary report on the effects of long-term therapy with conjugated equine estrogens in the prevention of recurring cerebral infarction [108]. In observations in

582 men during an average period of 16.7 months, therapy failed to prevent cerebral infarction or to reduce the mortality rate subsequent to this disorder. Initial treatment consisted of 1.25 mg daily of the estrogens for one year, followed by an increase to 2.5 mg daily thereafter. Moreover, estrogen therapy failed to reduce the incidence of myocardial infarction during this period.

Relative to this discussion is the report of Robinson and Lebeau [109] on an investigation of the concerted effects of conjugated equine estrogens on the serum lipids and the clotting mechanism. In a one-month study with postmenopausal women, 24 receiving 1.25 mg daily and 20 receiving 2.5 mg daily of conjugated equine estrogens showed no statistically significant change in any of the clotting parameters from the pretreatment levels. On the other hand, in contrast to reports by Marmorston [103] and Stamler [104] on male patients, they observed modest but significant lipid changes at these doses, including an increase in serum triglycerides, in 72% of the patients.

In summary, these investigations have raised at least as many questions as they have answered. Increased knowledge about the etiologic factors of coronary and cerebral thrombosis and about the physiological role of conjugated equine estrogens may help to answer some of these questions, but ultimately the results of large-scale clinical investigations, such as those now under way, will be necessary to clarify these issues.

2. Hemostasis

For many years, scientists have been studying the relationship of estrogens and bleeding. In 1922, Vogt [110] suggested that hormones may regulate capillary tone and participate in hemostasis. Five years later, Allen [111] postulated that menstruation was due to a cyclic reduction of estrogenic hormone. These postulations have been supported by evidence that certain periods of a woman's life, such as menses and climacteric, are characterized by both a low level of circulating estrogens and a tendency toward capillary fragility [112–114] and bruising [115]. Also, during pregnancy, a time of increased levels of circulating estrogens, elevations of clotting values, such as prothrombin concentration [116, 117], factor VII (proconvertin) [118], fibrinogen [119, 120], and platelet adhesiveness [121] were noted.

The earliest reports on the successful use of estrogens to control hemorrhages appeared in 1936 [122, 123]. In 1949, the water-soluble conjugated equine estrogens were made available for intravenous administration and used for the control of functional uterine bleeding [124–130].

Beginning in 1955, a series of papers [85, 131–148] reported that the intravenous administration of conjugated equine estrogens was effective in controlling epistaxis [131, 135, 138], hyphemia [137, 145], and reduced bleeding following such surgical procedures as tonsillectomy [131, 136, 144] and prostatectomy [141–143]. In contrast, Cooner and Burros [149], in a double blind study, could not demonstrate any hemostatic effect of intravenous conjugated estrogens in prostatectomy. Borchgrevink *et al.* [150], as well, failed to show an effect on nine clotting functions (*e.g.*, whole–blood clotting time, one–stage prothrombin test, prothrombin, factor V, platelet count) in eight normal individuals, nor could they demonstrate, in a group of 71 patients, a difference in blood loss between estrogen–treated patients or those given a placebo. In another double–blind study in 51 normal human subjects, Wayne *et al.* [151] measured several clotting functions, but the only significant change noted was a slight rise in factor V following the administration of 40 mg of the conjugated equine estrogens. However, in a preliminary study which included several subjects with known bleeding disorders, these investigators observed that the estrogens shortened bleeding time and produced a mean increase of 8.27% in factor V. Their studies also showed an inhibition of intracutaneously injected hyaluronidase and an increased inhibition of hyaluronidase activity in the serum following intravenous administration of the estrogens. This would indicate that estrogens favor the formation of ground substance by inhibiting the enzyme presumed to destroy it.

Other investigators [109, 152, 153] also reported that human coagulation values were not altered appreciably by the parenteral administration of 20 mg of conjugated equine estrogens.

Many investigators sought other means to explain the claimed hemostatic action of the conjugated equine estrogens. For example, Stefanini [154], in 1961, noted that the administration of conjugated equine estrogens reversed the abnormal platelet and megakaryocyte morphology and lessened bleeding tendencies in some women with thrombocytopenic purpura. He concluded that most of the intravascular effects of estrogens on hemostasis were not an effect on the prothrombin time or fibrinolysis, but primarily an effect on platelets.

Villadolid *et al.* [155] measured the effect of conjugated equine estrogens on fibrinolysis in normal subjects and patients with uterine bleeding. They reported a correlation between the cessation of bleeding and the inhibition of clot lysis by the estrogens as indicated by an *in vitro* fibrinolysis test.

Since clinical observations have suggested a relationship between estrogen levels and the vascular factors, capillary fragility and capillary permeability [112–115], these two parameters have been used as another approach to explain the hemostatic effect of estrogens.

Using the suction cup method, Clemetson *et al.* [114] reported that the administration of 20 mg of conjugated equine estrogens and estriol sodium succinate were more effective than prednisone [17α,21–dihydroxypregna–1,4–diene–3,11,20–trione] in lowering vascular fragility and raising capillary resistance in 18 out of 26 postmenopausal women who had initially a low capillary strength. Schmidt–Matthiesen and Poliwoda [86] reported that both conjugated equine estrogens and estriol sodium succinate doubled the capillary strength. The latter authors claimed that these capillary vascular effects provided a rational basis for the clinical use of estrogens in the control of various kinds of bleeding.

In still another approach to an explanation for the use of estrogens as hemostatic agents, Schiff and Burn [156, 157] suggested that the ground substance of the connective tissue may play an active role. (This approach has been comprehensively summarized by Rona [158] and Schiff [159]). The ground substance is composed chiefly of mucopolysaccharides and proteins. Using histochemical methods, Schiff and Burn observed that the parenteral administration of conjugated equine estrogens to humans as well as monkeys and hamsters affected the ground substance by increasing the amount of acid mucopolysaccharides and lengthening their polymers. These changes were most marked around the small blood vessels. The authors postulated that these data, together with a shift in the gel–sol equilibrium toward the more solid gel state, could explain the effect of conjugated equine estrogens on vascular hemostasis. Rona [158] pointed out that, although vascular mucopolysaccharide fractions may participate in the coagulation mechanism, some mucopolysaccharide fractions have anticoagulant effects and their exact role in these respects require further investigation.

In summary, the present status of the pharmacological investigation of conjugated equine estrogens does not provide an explanation for the remarkable hemostatic effectiveness of this agent as reported in practice. Because of the myriad of processes involved in hemostasis, many careful studies will have to be done before the significance of the effects of estrogens on hemostasis can be fully understood.

3. *Miscellaneous*

As indicated above, most of the publications on the pharmacology of nonhormonal effects of estrogen sulfates in man have been within the areas of atherosclerosis and hemostasis. The principal publications, which fall outside these areas, may be summarized as follows:

The relationship of estrogens to cancer has been, and will continue to be, a subject of intense investigation. In 1962, Wilson [160] published a critical review of the role of estrogens and progesterone in breast and genital cancer.

He reported statistics from his own study on the incidence of breast and genital cancer in 304 women exposed to high and medium doses of estrogens, mainly conjugated equine estrogens, for an average duration of 718 years. The average age of these women was 50.8 years. No cancer was observed (Table 8–17).

TABLE 8–17. *Incidence of breast and genital cancer in 304 women exposed to high medium to high dosage of exogenous estrogen as indicated by the vaginal smear*[a]

Type	Patient years of exposure	Cases no.
Breast cancer	2,387	0
Genital cancer	1,852[b]	0
Anticipated cancers (breast and genital)	2,387	18[c]

[a] Wilson, R. A.: J. Am. Med. Assoc. **182**, 327 (1962).

[b] Exposure of adnexa, uterus, and cervix was reduced because of 86 total hysterectomies with bilateral–oöphorectomy performed before or during treatment.

[c] Based on the cancer incidence expectancy studies of Mustacchi, P., Gordan, G. S., in: Breast cancer, 2nd Biennial Louisiana Cancer Conference (A. Segaloff, ed.), p. 163. St. Louis, Mo.: C. V. Mosby Co. 1958.

Similarly, Mustacchi and Gordan [161] reported on 120 postmenopausal women treated cyclically with estrogens, principally 1.25 mg daily of conjugated equine estrogens, for a total of 601 patient–years. These authors observed no malignant tumors although, according to their experience with untreated menopausal women, 5 to 6 malignancies were expected. Wilson [160] concluded that the existing data indicate that estrogens are prophylactic to breast and genital cancer to an unknown degree.

In a study involving 944 women, the use of androgens and estrogens was investigated [162] for the palliative treatment of disseminated mammary carcinoma. This study, carried out under the auspices of the Committee on Research of the American Medical Association, indicated that estrogens, limited properly to postmenopausal patients, induced regression in 36% of the patients. Estrogen treatment was preferred to androgen treatment in women past the fourth menopausal year. Conjugated equine estrogens were among the agents used in this study.

Height can be a social or emotional problem for tall girls. For this reason, estrogens have been evaluated for their possible effect in suppressing excessive growth in adolescent girls, since estrogens are known to inhibit the anterior pituitary growth hormone and to fuse epiphyses. A successful application of these ideas was reported by Greenblatt *et al.* [163], who studied the effect of 10–15 mg daily of conjugated equine estrogens on linear growth in 24 adolescent girls. The subjects were divided into subgroups according to age categories. Growth projection was based on two parameters: chronologic and bone age. The results of the two analyses indicated that estrogen–induced inhibition is greater below $12^1/_2$ years than above.

In earlier studies, Goldzieher [164] and Freed [165] reported that conjugated equine estrogens at daily doses of 2.5 to 4.5 mg, respectively, successfully suppressed excessive growth in adolescent girls. On the other hand, Bayley *et al.* [166] noted no statistically significant difference using 1.25 to 15.0 mg of these estrogens given cyclically. This apparent difference in results may be attributed to discrepancies in chronologic and bone age [163].

IV. Conclusion

The estrogen sulfates have biological properties combined with physiochemical properties (*i.e.*, water solubility, ready absorption, etc.) which make these compounds particularly useful pharmacological agents. As a result, they are used extensively for the treatment of those disorders that are benefited by estrogen therapy; for example, the menopausal syndrome, functional uterine bleeding, osteoporosis, senile vaginitis, amenorrhea, and dysmenorrhea. In addition, much attention is now focused on the metabolic effects of the conjugated equine estrogens, against a number of premature degenerative disorders. They are believed to exert a protective metabolic effect on many systems, organs, and tissues of the body, such as cardiovascular, bone, protein and cellular metabolism.

V. References

1. Glimm, E., Wadehn, F.: Biochem. Z. **207**, 361 (1929).
2. Zondek, B.: Klin. Wochschr. **9**, 2285 (1930).
3. Campbell, A. D., Collip, J. B.: Can. Med. Assoc. J. **23**, 633 (1930).
4. Cohen, S. L., Marrian, G. F.: Biochem. J. **30**, 57 (1936).
5. — — Odell, A. D.: Biochem. J. **30**, 2250 (1936).
6. Grant, J. K., Marrian, G. F.: Biochem. J. **43** (1948).
7. — — Biochem. J. **47**, 1 (1950).
8. a) Carpenter, J. G. D., Kellie, A. E.: Biochem. J. **78**, 1 P (1961). — b) Biochem. J. **84**, 303 (1962).

9. Hashimoto, Y., Neeman, M.: J. Biol. Chem. **238**, 1273 (1963).
10. Schachter, B., Marrian, G. F.: J. Biol. Chem. **126**, 663 (1938).
11. Butenandt, A., Hofstetter, H.: Z. Physiol. Chem. **259**, 222 (1939).
12. Grant, G. A., Seemann, C. von: Abstracts, 106th Meeting of the American Chemical Society, Pittsburg, Penn., September 1943, p. 18B.
13. — Souch, J. K.: Abstracts, 106th Meeting of the American Chemical Society, Pittsburg, Penn., September 1943, p. 19B.
14. — Glen, W. L.: J. Am. Chem. Soc. **71**, 2255 (1949).
15. — Beall, D.: Recent Progr. Hormone Res. **5**, 307 (1950).
16. Information supplied by Dr. A. D. Grieve, Control Division of Ayerst, McKenna and Harrison Ltd., Montreal, Canada.
17. Gaudry, R., Glen, W. L.: Ind. Chim. Belge, Suppl. **2**, 435 (1959); Chem. Abstr. **54**, 9030 (1960).
18. Freed, S. C., Eisin, W. M., Greenhill, J. P.: J. Clin. Endocrinol. **3**, 89 (1943).
19. Gray, L. A.: J. Clin. Endocrinol. **3**, 92 (1943).
20. Glass, S. J., Rosenblum, G.: J. Clin. Endocrinol. **3**, 95 (1943).
21. Sevringhaus, E. L., John, R. St.: J. Clin. Endocrinol. **3**, 98 (1943).
22. Fried, P. H., Hair, Q.: J. Clin. Endocrinol. **3**, 512 (1943).
23. Turner, V. H., Davis, C. D., Hamblen, E. C.: J. Clin. Endorcinol. **3**, 453 (1943).
24. — — — J. Clin. Endocrinol. **3**, 455 (1943).
25. Pincus, G., in: The hormones, vol. III (G. Pincus and K. V. Thimann, eds.), p. 665. New York, N. Y.: Academic Press Inc. 1955.
26. Diczfalusy, E., Lauritzen, C.: Oestrogene beim Menschen. Berlin-Göttingen-Heidelberg: Springer 1961.
27. Bard, P.: Medical physiology, 11th ed. St. Louis, Mo.: C. V. Mosby 1961.
28. Best, C. H., Taylor, N. B.: Physiological basis of medical practice, 7th ed. Baltimore, Md.: Williams & Wilkins Co. 1961.
29. Emmens, C. W., in: Methods in hormone research, vol. II (R. I. Dorfman, ed.), p. 59. New York, N. Y.: Academic Press 1962.
30. Loraine, J. A., Bell, E. T.: Hormone assays and their clinical application, 2nd ed., p. 225–301. Baltimore, Md.: Williams & Wilkins Co. 1966.
31. Allen, E., Doisy, E. A.: J. Am. Med. Assoc. **81**, 819 (1923).
32. Hogg, J. A., Korman, J., in: Medicinal chemistry, vol. II (F. F. Blicke and C. M. Suter, eds.), p. 34–217. New York, N. Y.: John Wiley & Sons 1956.
33. Pedersen-Bjergaard, K.: Comparative studies concerning the strengths of oestrogenic substances. London and New York: Oxford Univ. Press 1939.
34. Emmens, C. W.: Med. Research Council (Brit.) Spec. Rept. Ser. 234 (1939).
35. — J. Endocrinol. **2**, 444 (1941).
36. Robson, J. M., Adler, J.: Nature **146**, 60 (1940).
37. Harmer, G. L. M., Broom, W. A.: Lancet **I**, 850 (1950).
38. Revesz, C. (1967) — unpublished.
39. Broom, W. A. (1951), cit. by Bishop, P. M. F., Richards, N. A., and Perry, W. L. M., Lancet **I**, 818 (1951).
40. Rubin, B. L., Dorfman, A. S., Black, L., Dorfman, R. I.: Endocrinology **49**, 429 (1951).

41. Dorfman, R. I.: Physiol. Rev. **34**, 138 (1954).
42. — Dorfman, A. S.: Endocrinology **55**, 65 (1954).
43. — — Endocrinology **53**, 301 (1953).
44. a) Reich, W. J., Nechtow, M. J., Kurzon, A. M., Rubenstein, M. W.: Am. J. Obstet. Gynecol. **62**, 427 (1951). — b) Am. J. Obstet. Gynecol. **64**, 174 (1952).
45. Reicher, N. B., Clark, C. E., Bechtold, E.: N. Y. State J. Med. **51**, 601 (1951).
46. Neustaedter, T.: Am. J. Obstet. Gynecol. **46**, 530 (1943).
47. Neef, J. C. de, in: Clinical endocrine cytology, p. 80. New York, N. Y.: Harper & Row 1965.
48. Rakoff, A. E., in: Ovulation (R. B. Greenblatt, ed.), p. 299. Philadelphia and Toronto: J. B. Lippincott Co. 1966.
49. a) Perloff, W. H.: Am. J. Obstet. Gynecol. **58**, 684 (1949). — b) Am. J. Obstet. Gynecol. **61**, 670 (1951).
50. Papanicolaou, G. N.: Science **95**, 438 (1942).
51. a) Shorr, E.: Science **91**, 579 (1940). — b) Science **94**, 545 (1941).
52. Bishop, P. M. F., Kennedy, G. C., Wynn-Williams, G.: Lancet **II**, 764 (1948).
53. Goldzieher, J. W., Haus, L. W., Hamblen, E. C.: Am. J. Obstet. Gynecol. **54**, 636 (1947).
54. Allen, W. M.: Southern Med. J. **37**, 270 (1944).
55. Bishop, P. M. F., Richards, N. A., Neal Smith, D. J.: Lancet **I**, 848 (1950).
56. — — Perry, W. L. M.: Lancet **I**, 818 (1951).
57. Kupperman, H. S., Blatt, M. H. G., Wiesbader, H., Filler, W.: J. Clin. Endocrinol. Metab. **13**, 688 (1953).
58. Haus, L. W., Goldzieher, J. W., Hamblen, E. C.: Am. J. Obstet. Gynecol. **54**, 820 (1947).
59. Dignam, W. J., Wortham, J. T., Hamblen, E. C.: Am. J. Obstet. Gynecol. **59**, 1124 (1950).
60. Blewett, E. K.: Medical Record and Annals **44**, 182, June (1950).
61. Greenblatt, R. B., in: Office endocrinology, 4th ed., p. 203. Springfield, Illinois: Ch. C. Thomas 1952.
62. Gual, C., Becerra, C., Rice-Wray, E., Goldzieher, J. W.: Am. J. Obstet. Gynecol. **97**, 443 (1967).
63. Martinez-Manautou, J., Rudel, H. W., in: Ovulation (R. B. Greenblatt, ed.), p. 243. Philadelphia and Toronto: J. B. Lippincott Co. 1966.
64. Greenblatt, R. B., ed.: Ovulation. Philadelphia and Toronto: J. B. Lippincott Co. 1966.
65. Emmens, C. W., in: Methods in hormone research, vol. II (R. I. Dorfman, ed.), p. 20. New York, N. Y.: Academic Press 1962.
66. Tokuyama, I., Leach, R. B., Sheinfeld, S., Maddock, W. O.: J. Clin. Endocrinol. Metab. **14**, 509 (1954).
67. Smith, R. A., Albert, A.: Proc. Staff Meetings, Mayo Clinic **30**, 617 (1955).
68. Howard, R. P., Keaty, E. C., Reifenstein, E. C., Jr.: J. Clin. Endocrinol. Metab. **16**, 966 (1956); see also Schneider, R. A., Costiloe, J. P., Howard, R. P., Wolf, S.: J. Clin. Endocrinol. Metab. **18**, 379 (1958).
69. Rosemberg, E., Engel, I.: J. Clin. Endocrinol. Metab. **20**, 1576 (1960).

70. Vorys, N., Ullery, J. C., Stevens, V.: Am. J. Obstet. Gynecol. **93**, 641 (1965).
71. Harding, F. E.: Western J. Surg. Obstet. Gynecol. **52**, 31 (1944).
72. — Am. J. Obstet. Gynecol. **51**, 660 (1946).
73. Birnberg, C. H., Sherber, D. A., Brandman, L. J.: Am. J. Obstet. Gynecol. **63**, 1358 (1952).
74. Fisher, J. J.: Obstet. Surv. Baltim. **9**, 479 (1954).
75. Blanpin, O.: Prod. Pharm. **16**, 433, 477 (1961).
76. Kappas, A., Palmer, R. H.: Pharmacol. Rev. **15**, 123 (1963).
77. Master, A. M., Dack, S., Jaffe, H. L.: Arch. Internal Med. **64**, 767 (1939).
78. Pick, R., Stamler, J., Rodbard, S., Katz, L. N.: Circulation **6**, 276 (1952).
79. Stamler, J., Pick, R., Katz, L. N.: Ann. N.Y. Acad. Sci. **64**, 596 (1956).
80. Clarke, G. B., Jain, S., Pick, R.: Circulation **32**, Suppl. **2**, II–6 (1965).
81. Arnold, A., Potts, G. O., McAuliff, J. P., Christiansen, R. G., Miller, T. C.: Proc. Soc. Exptl. Biol. Med. **121**, 122 (1966).
82. Johnson, J. F.: Proc. Soc. Exptl. Biol. Med. **94**, 92 (1957).
83. Stamler, F. W., Warner, E. D.: Proc. Soc. Exptl. Biol. Med. **78**, 754 (1951).
84. Mandoki, J. J., Rubio, C.: Pharmacologist **8**, 212 (1966).
85. Lewis, E. L.: Eye, Ear, Nose Throat Monthly **38**, 215 (1959).
86. Schmidt-Matthiesen, H., Poliwoda, H.: Arch. Gynaekol. **200**, 231 (1965).
87. Nolan, J. P.: Nature **213**, 201 (1967).
88. Nicol, T., Vernon-Roberts, B., Quantock, D. C.: J. Endocrinol. **33**, 365 (1965).
89. Rona, G., Chappel, C. I.: Endocrinology **72**, 1 (1963).
90. Silberberg, M., Silberberg, R.: Anat. Record **102**, 141 (1948).
91. Hajdu, A., Chappel, C. I., Rona, G.: Experientia **21**, 466 (1965).
92. Greiss, F. C., Jr., Marston, E. L.: Am. J. Obstet. Gynecol. **93**, 720 (1965).
93. Ueland, K., Parer, J. T.: Am. J. Obstet. Gynecol. **96**, 400 (1966).
94. Stamler, J.: Circulation **10**, 896 (1954).
95. Levy, H., Boas, E. P.: J. Am. Med. Assoc. **107**, 97 (1936).
96. James, T. N., Post, H. W., Smith, F. J.: Ann. Internal Med. **43**, 153 (1955).
97. Weinreb, H. L., German, E., Rosenberg, B.: Ann. Internal Med. **46**, 285 (1957).
98. Rivin, A. U., Dimitroff, S. P.: Circulation **9**. 533 (1954).
99. London, W. T., Rosenberg, S. E., Draper, J. W., Almy, T. P.: Ann. Internal Med. **55**, 63 (1961).
100. a) Eilert, M. L.: Am. Heart J. **38**, 472 (1949). — b) Russ, E. M., Eder, H. A., Barr, D. P.: Am. J. Med. **19**, 4 (1955). — c) Steiner, A., Payson, H., Kendall, F. E.: Circulation **11**, 784 (1955). — d) Oliver, M. F., Boyd, G. S.: Circulation **13**, 82 (1956). — e) Robinson, R. W., Higano, N., Cohen, W. D., Sniffen, R. C., Sherer, J. W., Jr.: Circulation **14**, 365 (1956). — f) Eilert, M. L.: Metabolism **2**, 137 (1953). — g) Cohen, W. D., Higano, N., Robinson, R. W.: Circulation **17**, 1035 (1958). — h) Furman, R. H., Howard, R. P., Norcia, L. N., Keaty, E. C.: Am. J. Med. **24**, 80 (1958). — i) Marmorston, J., Moore, F. J., Magidson, O., Kuzma, O., Lewis, J. J.: Ann. Internal Med. **51**, 972 (1959).
101. Oliver, M. F., Boyd, G. S.: Lancet **II**, 499 (1961).
102. Marmorston, J., Magdison, O., Kuzma, O., Moore, F. J.: J. Am. Med. Assoc. **174**, 241 (1960).

103. Marmorston, J., Moore, F. J., Hopkins, C. E., Kuzma, O. T., Weiner, J.: Proc. Soc. Exptl. Biol. Med. **110**, 400 (1962).
104. Stamler, J., Pick, R., Katz, L. N., Pick, A., Kaplan, B. M., Berkson, D. M., Century, D.: J. Am. Med. Assoc. **183**, 632 (1963).
105. Marmoston, J., Moore, F. J., Kuzma, O. T.: Conference of the California Medical Society, San Francisco, 1962.
106. a) Evans, R. B., Marmorston, J.: Proc. Soc. Exptl. Biol. Med. **113**, 698 (1963); — b) Proc. Soc. Exptl. Biol. Med. **118**, 529 (1965).
107. Robinson, R. W., Higano, N., Cohen, W. D.: J. Chronic Diseases **16**, 155 (1963).
108. Report of the Veterans Administration Cooperative Study of Atherosclerosis: Circulation **33**, Suppl. **2**, II–3 (1966).
109. Robinson, R. W., Lebeau, R. J.: J. Atherosclerosis Res. **5**, 120 (1965).
110. Vogt, E.: Deut. Med. Wochschr. **48**, 1007 (1922).
111. Allen, E.: Contrib. Embryol. (No. 98), **19**, 1 (1927).
112. Brewer, J. I.: Am. J. Obstet. Gynecol. **36**, 597 (1938).
113. Salvatore, C. A.: Surg. Gynecol. Obstet. **95**, 13 (1952).
114. Clemetson, C. A. B., Blair, L., Brown, A. B.: Ann. N. Y. Acad. Sci. **93**, 279 (1962).
115. Ellman, P., Parkes-Weber, F.: Brit. J. Dermatol. **47**, 197 (1935).
116. Olwin, J. H., Allen, E.: Am. J. Obstet. Gynecol. **60**, 1363 (1950).
117. Thordarson, O.: Nature **145**, 305 (1940).
118. Alexander, B., Meyers, L., Kenny, J., Goldstein, R., Gurewich, V., Grinspoon, L.: New Engl. J. Med. **254**, 358 (1956).
119. Ratnoff, O. D., Colopy, J. E., Pritchard, J. A.: J. Lab. Clin. Med. **44**, 408 (1954).
120. Dieckmann, W. J., Wegner, C. R.: Arch. Internal Med. **53**, 353 (1934).
121. Wright, H. P.: J. Pathol. Bacteriol. **54**, 461 (1942).
122. de Silva, E. B.: Illinois Med. J. **69**, 81 (1936).
123. Heinberg, G. J.: Arch. Otolaryngol. **24**, 758 (1936).
124. Abarbanel, A. R.: J. Clin. Endocrinol. **9**, 681 (1949).
125. Greenblatt, R. B., Barfield, W. E.: J. Clin. Endocrinol. **11**, 821 (1951).
126. Goldzieher, J. W., Goldzieher, M. A.: Disorders of menstrual functions in endocrine treatment in general practice, p. 303. New York, N. Y.: Springer 1953.
127. Reich, W. J., Rubenstein, M. W., Nechtow, M. J., Reich, J. B.: J. Am. Med. Assoc. **152**, 1679 (1953).
128. Greenblatt, R. B.: Med. Ann. District Columbia **23**, 187 (1954).
129. Randall, L. M.: J. Louisiana Med. Soc. **110**, 160 (1958).
130. Benson, R. C.: Northwest. Med. **58**, 31 (1959).
131. Menger, H. C.: J. Am. Med. Assoc. **159**, 546 (1955).
132. Roman, M. V., Elizondo, P. M.: El Medico **6**, 52 (1956).
133. Whittington, P. B.: J. Am. Dental Assoc. **53**, 595 (1956).
134. Ende, M., Jacobson, P.: J. Am. Geriat. Soc. **5**, 676 (1957).
135. Rigual, R.: Digest Ophth. Otolaryngol. **20**, 3 (1957).
136. Servoss, H., Shapiro, F.: Digest Ophth. Otolaryngol. **20**, 10 (1957).
137. Rigg, J. D.: Digest Ophth. Otolaryngol. **20**, 28 (1957).
138. Jacobson, P.: Virginia Med. Monthly **84**, 396 (1957).
139. Woodruff, G. H.: A. M. A. Arch. Otolaryngol. **67**, 435 (1958).

140. Woodruff, G. H.: Virginia Med. Monthly **85**, 548 (1958).
141. Bleeker, J. J.: Svenska Läkartidn. **56**, 521 (1959).
142. Lally, J. F., Reed, W. A.: Med. Times **87**, 188 (1959).
143. Bobelis, C. K.: Illinois Med. J. **117**, 231 (1960).
144. Fox, S. L.: Eye Ear Nose Throat Monthly **39**, 251 (1960).
145. Goldberg, J. L.: A. M. A. Arch. Ophth. **63**, 1001 (1960).
146. Popper, J.: Dis. Chest **37**, 659 (1960).
147. a) Roberts, H. J.: J. Am. Geriat. Soc. **9**, 184 (1961). — b) J. Am. Geriat. Soc. **9**, 655 (1961).
148. Narang, R. K.: Am. Rev. Resp. Dis. **85**, 436 (1962).
149. Cooner, W. H., Burros, H. M.: J. Urol. **83**, 64 (1960).
150. Borchgrevink, C. F., Andersen, R., Hall, J., Hatteland, K., Ursin-Holm, A.: Brit. Med. J. **II**, 1645 (1960).
151. Wayne, L., Glueck, H. I., Brodine, C., Coots, M.: Proc. Soc. Exptl. Biol. Med. **116**, 85 (1964).
152. Kudish, H. G., Rapaport, S. I.: J. Urol. **83**, 730 (1960).
153. Kaye, B. M., Graziano, J. L., Reaney, B. V.: Am. J. Obstet. Gynecol. **82**, 1221 (1961).
154. Stefanini, M., in: Hemostasis and the effect of estrogens. New York: Excerpta Medica Foundation 1962.
155. Villadolid, L. S., Romero, R. C., Egay, L. M., Millora, A., Negre, M., Abundo, C., Ongioco, G., Llamas, S., Santos, A. A.: Acta Med. Philippina **2**, Ser. 2, No. 2, 82 (1965).
156. Schiff, M., Burn, H. F.: Arch. Otolaryngol. **73**, 43 (1961).
157. — — Laryngoscope **71**, 765 (1961).
158. Rona, G.: Am. J. Obstet. Gynecol. **87**, 434 (1963).
159. Schiff, M., in: Hormones and connective tissue (G. Asboe-Hansen, ed.), p. 282. Copenhagen: Munksgaard (Baltimore, Md.: Williams & Wilkins Co.) 1966.
160. Wilson, R. A.: J. Am. Med. Assoc. **182**, 327 (1962).
161. Mustacchi, P., Gordan, G. S., in: Breast cancer, 2nd Biennial Louisiana Cancer Conference (A. Segaloff, ed.), p. 163. St. Louis, Mo.: C. V. Mosby Co. 1958.
162. Report to the Council of the Committee on Research of the American Medical Association: J. Am. Med. Assoc. **172**, 1271 (1960).
163. Greenblatt, R. B., McDonough, P. G., Mahesh, V. B.: J. Clin. Endocrinol. Metab. **26**, 1185 (1966).
164. Goldzieher, M. A.: J. Clin. Endocrinol. Metab. **16**, 249 (1956).
165. Freed, S. C.: J. Am. Med. Assoc. **166**, 1322 (1958).
166. Bayley, N., Gordan, G. S., Bayer, L. M., Goldberg, M. B., Storment, A.: J. Clin. Endocrinol. Metab. **22**, 1127 (1962).

Chapter 9

Clinical Aspects of Steroid Conjugation

ALFRED M. BONGIOVANNI and ROBERT M. COHN

Contents

I. Introduction

The steroidal hormones are synthesized in several specialized tissues of man and undergo structural alterations in their course of metabolism, sometimes in their organs of origin, and more often in other systems. Reduction, further oxidation, degradation, and conjugation are among the changes which occur through the agency of specific enzymes. Teleologically, these transactions have been regarded as implicit in the concept of humoral regulation, leading to inactivation and disposition of active substances in accordance with basic homeostatic controlling mechanisms. Apart from the justification of such a rationalization, it remains that the steroidal hormones are in large measure further metabolized and conjugated, transported and excreted in a manner somewhat dependent upon the form of the conjugate.

The disposition and conjugation of steroids varies among species. This discussion will confine itself to a consideration of metabolism in man. The usual normal conjugation of steroids will be reviewed with attention to variations associated with age, maturation, and disorders. Most of the studies in man depend upon the indirect identification of steroidal conjugates by differential hydrolysis of suitable extracts of human tissues and fluids. Few examples of the isolation of steroidal conjugates and their identification, intact, are presently available. The matter of conjugation of steroids by the human fetus and the fetal–placental unit will not be included in any detail.

For more than forty years it has been recognized that steroidal hormones were present in human urine as water–soluble, inactive compounds. Only minute quantities of unaltered lipid–soluble, biologically active steroids were normally detected. The metabolites of active precursors were rendered extractable by unpolar solvents after treatment with acid and heat. In 1942 Venning *et al.* [1] isolated androsterone sulfate as a relatively pure conjugated steroid from the urine of a man with an interstitial cell tumor of the testis. In 1944, Munson *et al.* [2] recovered sodium dehydroisoandrosterone sulfate from normal male urine. The detection of pregnanediol in the urine of pregnant women, and its isolation and characterization as a glucuronide by Odell and Marrian [3], Venning and Browne [4], and Heard *et al.* [5] clearly established the conjugation of steroids with glucuronic acid as a major excretory pathway. Subsequently various mammalian and bacterial preparations of β–glucuronidase were found to release large amounts of various steroids in human urine [6–9]. It appears that the major conjugates of steroidal hormones in man are the sulfates and glucuronides. There is evidence for the presence of phosphates, particularly in human plasma. There are probably other, as yet unidentified, conjugates of the steroids in human tissues and fluids.

This presentation will consider the steroidal conjugates and pertinent metabolites in man according to structural and physiologic characteristics: aldosterone; steroids with 19 carbon atoms lacking a side chain at C17; corticosteroids (compounds with 21 carbon atoms and a 2–carbon side chain at C17, as represented by cortisol); estrogens; steroids with 21 carbon atoms related to progesterone with a C21 methyl grouping.

II. Aldosterone [11β,21–dihydroxy–3,20–dioxopregn–4–en–18–al]

A. Normal Conjugation

By isotopic dilution techniques, the normal daily secretory rate of aldosterone is approximately 100 μg per square meter body surface. In human urine, three major components and their relationships have been well studied. A small amount of free, readily extractable aldosterone is present. The so–called "3–oxo conjugate" is readily hydrolyzed at pH 1, releasing unchanged aldosterone. Underwood and Tait [10], Pasqualini [11], and Pasqualini *et al.* [12] have shown this compound to be the 18–glucuronide of the hemiacetal of aldosterone. The major component is hydrolyzed by β–glucuronidase, and, according to Ulick and Lieberman, it is the tetrahydro derivative of aldosterone with hydroxyl groups at C3, 18, and 21 [13], and was derived from administered labeled aldosterone itself.

Flood *et al.* [14] administered labeled aldosterone and cortisol [11β,17α,-21–trihydroxypregn–4–ene–3,20–dione] to normal subjects. They found that over 90% of the dose of aldosterone was excreted in the urine within 24 hours, that 56% was released by β–glucuronidase (*vs.* 64% for cortisol), and 12% at pH 1, room temperature (*vs.* only 2% for cortisol). Thus the 3–oxo conjugate is an important urinary metabolite of aldosterone. According to New *et al.* [15], the percentage of an administered dose of labeled aldosterone appearing in the urine of normal human subjects as the free steroid, 3–oxo conjugate, and tetrahydroaldosterone [3α,11β,21–trihydroxy–20–oxo–5β–pregnan–18–al] is 0.1 to 0.3, 5 to 12, and 20 to 40%, respectively. These results are similar to the earlier ones of Peterson [16], in which studies 80 to 90% of infused labeled aldosterone was recovered in the urine within 3 days. Pasqualini and Jayle [17] measured aldosterone in human plasma after the administration of isotopically labeled steroid and found the greatest part of the aldosterone itself, free, but also detected tetrahydroaldosterone, which was five times greater than the free, in the glucuronide fraction. All reference to the 3–oxo conjugate and tetrahydroaldosterone indicates studies based on differential hydrolysis by acid or β–glucuronidase and suitable chromatographic identification of the free products.

Coppage *et al.* [18] administered labeled aldosterone to human volunteers orally and intravenously. The rate of appearance of the 3–oxo conjugate and tetrahydroaldosterone in the urine was the same by both routes. Bledsoe *et al.* [19] indicated that a considerably higher percentage was converted to the 3–oxo conjugate following intravenous as compared with oral administration, suggesting an important extrahepatic conversion to this conjugate and pointing to a possible source of error in estimating secretion rates.

Several investigators have studied the normal renal clearance of the several forms of aldosterone [20]. Frimpter *et al.* [21] observed the clearance from plasma as compared with inulin. Free aldosterone was cleared in man at a rate of 3 to 6% of inulin, the 3–oxo conjugate at 250 to 300%, and tetrahydroaldosterone glucuronide, 70 to 100%. After large doses, the free aldosterone was excreted at a rate of 22% of the amount filtered, and clearance of the 3–oxo conjugate diminished upon the administration of *p*–aminohippuric acid. Thus it appears that 94 to 97% of free aldosterone, but only a small quantity of tetrahydroaldosterone glucuronide, is reabsorbed by the renal tubules, whereas the 3–oxo conjugate is probably secreted. Cheville *et al.* [22] found that the free aldosterone of arterial blood of man was extracted by the kidney at the same rate as the glomerular filtration rate, but that only 5% was excreted. Since the "missing" aldosterone did not reappear later in the experiment, it was believed that the kidney can remove and possibly convert the free aldosterone to the 3–oxo conjugate. Gfeller and Siegenthaler [23] reported results much the same as above. Furthermore, Deck and Siegenthaler [24], employing the stop–flow technique in dogs, demonstrated that the 3–oxo conjugate is formed by the proximal tubular cells: hence, the use of this metabolite as an indication of secretion rate must be interpreted with caution in the presence of disturbed renal function.

B. Pregnancy and Infancy

Venning and Dyrenfurth [25] found that, although the urinary free aldosterone rose very little during pregnancy, the two conjugates increased markedly and fell soon after delivery. They also found a greater amount of free aldosterone in toxemia of pregnancy. On the other hand, Jones *et al.* [26] reported an increase of the 3–oxo conjugate during pregnancy with a diminution of the conjugates released by β–glucuronidase in urine. Bougas *et al.* [27] reported similar observations with a rise of the 3–oxo conjugate to urinary levels approximating 50% of the secretion rate and a diminution of ring A–reduced metabolites.

New *et al.* [15] reported results in children for the three components similar to those of adults, both in amount and relationships, when compared on the basis of body surface. However, the free urinary aldosterone was relatively greater in young infants.

C. Disease

Hurter and Nabarro [28] found the proportion of labeled aldosterone administered to subjects with hepatic cirrhosis and excreted in the urine as tetrahydroaldosterone glucuronide, much reduced, with an increase of the pH 1-hydrolyzable steroid: *i.e.*, the 3-oxo conjugate. In a single subject with terminal hepatic failure, Ofstad *et al.* [29] also noted an elevation of urinary aldosterone released by hydrolysis at pH 1, and a marked reduction of the fraction released subsequently by β-glucuronidase hydrolysis. Bledsoe *et al.* [19] described similar results, and in subjects with portacaval shunts they failed to observe the relatively greater conversion to the 3-oxo conjugate of aldosterone administered intravenously as compared with the oral route. Coppage *et al.* [18] suggested that these findings point to the reduction of ring A of aldosterone as the rate-limiting step in liver disease. They found that labeled aldosterone administered to subjects with hepatic failure was removed from the plasma at a slower rate than normal, and that both the free aldosterone and the 3-oxo conjugate were elevated in the urine.

Cheville *et al.* [30] studied the disposition of infused aldosterone in twenty-one subjects with circulatory failure. There seemed to be a reduced hepatic clearance of aldosterone with an increased concentration of free aldosterone for any constant rate of secretion. They presented evidence to show that the 3-oxo conjugate is formed both in the liver and the kidney; most of this product formed by the kidney was rapidly excreted. There was a higher concentration of the 3-oxo conjugate in the renal vein than in the arterial blood. The major hepatic product was tetrahydroaldosterone glucuronide.

In a patient with progressive renal failure, Muller and Manning [31] noted a decrease in the total urinary radioactivity following the intravenous administration of labeled aldosterone. There was a diminution of all three components of aldosterone, and these closely paralleled the creatinine clearance, indicating that the limiting factor was the glomerular filtration rate. In a single case of primary aldosteronism, Ulick *et al.* [32] found an increased secretion rate and high urinary level of tetrahydroaldosterone glucuronide, but the amount of the 3-oxo conjugate was in the normal range. Thus, the percentage of conjugates may vary under some special conditions.

III. C_{19}–Steroids

A. Normal Urinary Conjugates

Among the earliest attempts to extract the conjugated 17–ketosteroids from human urine were those of Jayle and Crépy [33]. Extraction with *n*–butanol at pH 11 followed by hydrolysis with β–glucuronidase yielded androsterone [3α–hydroxy–5α–androstan–17–one] and etiocholanolone [3α–hydroxy–5β–androstan–17–one]. The sulfate fraction consisted mainly of dehydroisoandrosterone [3β–hydroxyandrost–5–en–17–one]. They indicated that the neutral 3α–hydroxy steroids were excreted mostly as glucuronides, and the 3β–hydroxy steroids as sulfates. Kellie and Wade [34] extracted and chromatographed 17–ketosteroid conjugates from human urine, and subsequently hydrolyzed the components for identification. The sulfate fraction accounted for 11 to 37% of the total 17–ketosteroids. Among the glucuronides were androsterone, etiocholanolone, 3α–hydroxy–5β–androstane–11,17–dione, 3α–hydroxy–5α–androstane–11,17–dione, and the 11β–hydroxy compounds of etiocholanolone, androsterone, and dehydroisoandrosterone. The sulfates contained dehydroisoandrosterone, androsterone, and etiocholanolone. Malassis *et al.* [35] indicated that almost all dehydroisoandrosterone in human urine is sulfoconjugated. Pelzer *et al.* [36] identified the glucuronides of androsterone, etiocholanolone, and 3α–hydroxy–5β–androstane–11,17–dione in normal human urine. Bush and Gale [37] also showed that dehydroisoandrosterone appears to be present entirely as a sulfate in human urine, and that androsterone and etiocholanolone occur 80 to 95% as glucuronides. They further partially identified the glucuronides of 11β–hydroxyetiocholanolone [3α,11β–dihydroxy–5β–androstan–17–one], 11β–hydroxyandrosterone [3α,11β–dihydroxy–5α–androstan–17–one], and 3α–hydroxy–5β–androstane–11,17–dione. Foggitt and Kellie [38, 39] isolated and identified the glucuronides of androsterone, etiocholanolone, 3α–hydroxy–5β–androstane–11,17–dione, 11β–hydroxyandrosterone, and 11β–hydroxyetiocholanolone. They presented circumstantial evidence that these were present as the 3α–yl–β–D–glucopyranosiduronic acids. Stárka *et al.* [40] isolated 3β,7α–dihydroxyandrost–5–en–17–one from human and rat urine as the monosulfate, with less as the disulfate. Siiteri *et al.* [41] presented indirect but adequate evidence for the presence of dehydroisoandrosterone in urine as the glucuronide. This was isolated after the administration of a tracer dose of the labeled steroid, and the specific activity of the glucuronide was greater than that of the sulfate; thus, it was concluded that the sulfate is derived from other than free endogenous dehydroisoandrosterone alone.

Urinary testosterone glucuronide [3–oxoandrost–4–en–17β–yl–β–D–glucopyranosiduronic acid] has interested a number of investigators, but for the most part, their work has concentrated on establishing a method of reliable measurement of extracts following enzymatic hydrolysis and chromatography. Camacho and Migeon [42] found the daily excretion to be 46 to 106 μg in normal men, and less than 8 in women. They reported [43] an insignificant quantity of free testosterone [17β–hydroxyandrost–4–en–3–one] in the urine following administration of a tracer dose of radioactive testosterone, with most of the radioactivity in the fraction released by hydrolysis with β–glucuronidase, and considerably less in the presumed sulfate fraction. This agrees with the earlier studies of Sandberg and Slaunwhite [44]. Traces of testosterone sulfate [17β–sulfooxyandrost–4–en–3–one] were also isolated from male urine, between 5 to 10 μg daily, by Dessypris *et al.* [45].

1. Administration of Steroids

There have been a number of studies of urinary conjugates following the administration of steroids to human subjects. This discussion will concern those investigations wherein large doses of free steroids were given. When epiandrosterone [3β–hydroxy–5α–androstan–17–one] was administered orally [46], the urinary 17–ketosteroids accounted for 15 to 23% of the dose, and the steroids, extracted as conjugates and subsequently hydrolyzed, proved to consist of epiandrosterone sulfate [3β–sulfooxy–5α–androstan–17–one] and androsterone glucuronide [17–oxo–5α–androstan–3α–yl–β–D–glucopyranosiduronic acid]. Lewbart and Schneider [47] recovered androsterone and dehydroisoandrosterone as the conjugates, after the oral intake of large doses of the hemisuccinate of the latter. Theirs is among the very few studies in man wherein the compounds were isolated and identified as intact conjugates. Crépy *et al.* [48] extracted and chromatographed the urinary conjugates after the administration of dehydroisoandrosterone; they found that 75% was excreted as the sulfate, and the remainder as glucuronides, including androsterone and etiocholanolone. Following the administration of androsterone, the urine contained principally androsterone glucuronide with a small quantity of its sulfate. Schneider and Lewbart [49] isolated androsterone glucuronide with a small amount of epiandrosterone sulfate after the administration of large amounts of androsterone hemisuccinate orally. With larger doses, there was an increase in the sulfate conjugate. They further reported that after large doses of dehydroisoandrosterone, the principal urinary conjugates, separated by countercurrent distribution and identified as such, were androsterone and etiocholanolone glucuronides. In addition, they found 3β,7α–dihydroxyandrost–5–en–17–one sulfate

[7α–hydroxy–3β–sulfooxyandrost–5–en–17–one] and a small amount of dehydroisoandrosterone glucuronide [17–oxoandrost–5–en–3β–yl–β–D–glucopyranosiduronic acid]. Staib *et al.* [50], after oral administration of each of the following compounds, detected dehydroisoandrosterone and isoandrosterone [3β–hydroxy–5α–androstan–17–one] primarily in the sulfate fraction, and androsterone in the glucuronide portion, with a small proportion as the sulfate. Baulieu [51] isolated etiocholanolone sulfate [3α–sulfooxy–5β–androstan–17–one] after its administration to a child. After oral administration of 3β–hydroxy–5β–androstan–17–one, the urine contained its sulfate in addition to the glucuronide and sulfate of etiocholanolone [52].

In a few experiments, "natural" steroid conjugates have been administered to human subjects. Randomly labeled androsterone glucuronide was isolated as such in the urine of 3 subjects tested, with no activity in the sulfate fraction, so that it was regarded as a metabolic end product which is not further altered [53]. Doubly labeled dehydroisoandrosterone sulfate (–^{3}H–^{35}S) [3β–sulfooxyandrost–5–en–17–one] given intraduodenally [54] was excreted in the urine with a higher ^{3}H/^{35}S ratio than the injected compound, pointing to partial hydrolysis and reconjugation. In addition, the sulfate of androst–5–ene–3β,16α,17β–triol, with the double label, was present, suggesting direct hydroxylation and excretion of the administered sulfate. Robel *et al.* [55] found that the administration of testosterone glucuronide yielded primarily 5β–androstane products, whereas free testosterone gave 5α–androstane compounds. With the former there was more 5β–androstanediol [5β–androstane–3α,17β–diol] than etiocholanolone, whereas the reverse was true with the latter. This indicates a different metabolism of the free and conjugate. When the phosphates of androsterone and epiandrosterone were administered, it appeared that they were completely hydrolyzed and excreted free or as other conjugates [56].

B. Plasma Steroids

It has proved extremely difficult to isolate and identify intact conjugates from human plasma, so that most studies ultimately resort to differential hydrolysis for their limited conclusions. Kellie and Smith [57] described the sulfates of dehydroisoandrosterone in human plasma, 12 to 131 μg per 100 ml in males, 5 to 43 in females. Considerably less androsterone sulfate [3α–sulfooxy–5α–androstan–17–one] was found, and the glucuronides of androsterone and etiocholanolone (1.5 to 2.1 μg per 100 ml) were also detected with minute quantities of their 11–oxygenated congeners. Baulieu

[58] measured the sulfates in pooled human plasma with levels of dehydroisoandrosterone, androsterone, and etiocholanolone all very low: 19, 11, and 4.5 μg per 100 ml, respectively. Perhaps the most significant observation of this same author [59] was the isolation and identification of dehydroisoandrosterone sulfate in the urine and plasma of a subject with an adrenal tumor, and the presence of this conjugate in the tumor itself, as well as the adrenal venous blood [60]. No free steroid was found. There seemed to be little question that dehydroisoandrosterone sulfate predominated in the plasma. Results similar to the preceding were obtained by Staib *et al.* [61], and Wieland *et al.* [62]. Etiocholanolone glucuronide [17–oxo–5β–androstan–3α–yl–β–D–glucopyranosiduronic acid] has been detected after stimulation with ACTH [adrenocorticotrophic hormone] [63]. For some years, Oertel and his co–workers have presented evidence for the presence of phosphate esters of dehydroisoandrosterone in plasma [64–70]. This they have described as a lipophilic, nonsolvolyzable conjugate containing 1 mole each of steroid, sulfate, phosphate, and glycerol, and 2 moles of fatty acids. Bush [37] was unable to find any dehydroisoandrosterone associated with phosphate.

A few studies following the administration of stable and radioactive isotopic steroids will be briefly reviewed. Bongiovanni and Eberlein [71, 72] found dehydroisoandrosterone in plasma, presumably as sulfate, and androsterone as glucuronide, after large intravenous doses of dehydroisoandrosterone and androsterone. West *et al.* [73] found conjugates within 20 minutes of intravenous injection of testosterone. Sandberg and Slaunwhite [44] reported that the conjugates exceeded the free levels of testosterone within 15 minutes, and similar results with 11β–hydroxyandrost–4–ene–3,17–dione [74]. The conjugates were not well defined.

C. Renal Clearance

Differences in renal handling of monosulfates and glucuronides appear to be convincing. It has been shown that dehydroisoandrosterone sulfate is cleared at a low rate from the plasma, 6 ml or less per minute, whereas the glucuronides of androsterone and etiocholanolone are higher (70–177 ml per minute) [57, 71, 72]. With high blood levels of androsterone glucuronide, following the administration of the hemisuccinate intravenously, the clearance may fall to lower levels and there is an inverse relationship between the plasma level and renal clearance [72]. This, together with the depression of the clearance by probenecid, suggests tubular secretion of the glucuronides. Furthermore, the sulfates have been found to be more tightly bound to serum proteins than the glucuronides in these studies.

D. Disease

The most interesting disorder related to the conjugation of C_{19}–steroids is "etiocholanolone fever", and probably represents the only recognized syndrome wherein a disturbance in conjugation may itself produce the major symptoms. In 1958, Kappas *et al.* [75] found that the injection of free etiocholanolone caused fever, inflammation, and leukocytosis. Most of the injected steroid was excreted in the urine as the glucuronide. The 3β–hydroxy epimer had no effect. Shortly thereafter, Bondy *et al.* [76] reported the presence of unconjugated etiocholanolone in the plasma of 2 subjects with periodic fever. Further work by this group [76–78] reported normal free levels of dehydroisoandrosterone, etiocholanolone, and androsterone as 1.2 μg per 100 ml or less. During attacks, subjects with periodic fevers had elevations of the free etiocholanolone (4 to 6.5 μg per 100 ml) accompanied by rises in free dehydroisoandrosterone as well. They have suggested several hypotheses: (1) failure of hepatic conjugation of etiocholanolone; (2) diminution of 5α– and rise of 5β–reductase, causing increased production of etiocholanolone at the expense of androsterone; (3) peripheral hydrolysis of conjugated etiocholanolone or its *de novo* formation. Schenker *et al.* [79] have made similar observations, and have found no evidence for any abnormality of the glucuronide–conjugating mechanism or of the excretion of steroids during attacks.

In advanced liver disease the urinary total neutral 17–ketosteroids have been reported by many to be lower than normal. In addition, there is a small but significant increase in the proportion excreted unconjugated [80–83]. One group found that a smaller portion of a dose of testosterone propionate [17β–propionyloxyandrost–4–en–3–one] was excreted in the total neutral 17–ketosteroids in liver disease (10 to 41%) than in normal subjects (13 to 60%) with a greater level of free steroids in the former [81]. Although in one study [80] there appeared to be a particular defect in the excretion of glucuronides, in another [83] there was diminution in both the sulfate and glucuronide fractions. After intravenous testosterone, there was less conjugated steroid in the plasma of subjects with liver disease, but the disappearance rate was not abnormal [73]. In a similar study, the glucuronides of 17–ketosteroids were less after testosterone in a group of women with metastatic mammary carcinoma with hepatic involvement, than in normal individuals or those without liver damage [84]. In the specific disorder associated with a defect of glucuronyl transferase (Crigler–Najjar syndrome), the administration of several steroids produced variable results [85]. Only 27% of a dose of tetrahydrocortisone [3α,17α,21–trihydroxy–5β–pregnane–11,20–dione] was excreted among the urinary glucuronides, *vs.* control, 49%. However, after testosterone, etiocholanolone, and androsterone, the results were not different, suggesting more than one glucuronyl transferase system involved in the formation of glucuronides.

Subjects with endocrine tumors have naturally been the fruitful subject for isolation of steroid conjugates. Several studies of adrenocortical tumors have revealed the following [60, 86–92]: as would be expected, dehydroisoandrosterone sulfate has been repeatedly recovered from urine, and, in addition, the glucuronides of dehydroisoandrosterone, androsterone, and etiocholanolone. The latter two glucuronides have been isolated by countercurrent distribution in one study [92]. As mentioned earlier, dehydroisoandrosterone sulfate was found in peripheral and adrenal venous plasma as well as 3β-hydroxyandrost-5-ene-7,17-dione sulfate [3β-sulfooxyandrost-5-ene-7,17-dione]. Etiocholanolone and androsterone sulfate have been found in both peripheral and adrenal blood, but these are higher in the former in contrast to dehydroisoandrosterone sulfate. Dehydroisoandrosterone sulfate has routinely been isolated from the tumor itself, but glucuronides have not been reported. The tumor may contain free androstenedione [androst-4-ene-3,17-dione] and 11β-hydroxyandrost-4-ene-3,17-dione. Following administration of labeled testosterone to a subject with arrhenoblastoma, the following were isolated from the plasma [93]: testosterone, etiocholanolone, and androsterone glucuronides. Bénard *et al.* [94] reported higher than normal levels of 17-ketosteroids, free and unconjugated, in the plasma of women with carcinoma of the breast, uterus, or ovaries, but there was no differentiation as to possible liver involvement; the total plasma neutral 17-ketosteroids were also elevated in these subjects.

Many studies of hirsutism in women reveal the presence of conjugates similar to those for adrenal carcinoma, although in smaller amount. It has been shown that the sulfates of several 17-ketosteroids were elevated in the urine of patients with the Stein-Leventhal syndrome, whereas the glucuronides were slightly, if at all, raised [95]. Weissman [96] suggested that there is more androsterone sulfate, based on differential hydrolysis, in the urine of women with hirsutism. Many of the observations in the vague assortment of virilizing conditions in women are uncertain and difficult to interpret. Conrad *et al.* [97] measured dehydroisoandrosterone sulfate and androsterone sulfate in human plasma. The former, in males, was 227 ± 56 μg per 100 ml, and 137 ± 52 in females; the latter was the same in both sexes: 48 ± 23. In women with hirsutism of marked degree, these sulfates were considerably elevated, whereas in endocrine disorders with little or no hirsutism the levels were normal. Perhaps the plasma steroid sulfates serve as readily available pools for androgen formation.

Camacho and Migeon [42] reported elevations of urinary testosterone glucuronide of 4.8 to 86.4 μg/day in congenital adrenal hyperplasia in children, and 695 in a single case of virilizing hepatoma.

IV. C_{21}–α–Ketolic Steroids

A number of studies of compounds of this type and their metabolites in clinical medicine have been conducted. Very often group reactions have been employed for the assay of these conjugated steroids; *i.e.*, detection of reducing substances, phenylhydrazine reaction for 17,21–dihydroxy–20–ketosteroids. Most of the studies of 17–hydroxycorticoids to be reviewed have been based on the latter reaction.

A. Urinary Compounds

It has been shown that the major portion of urinary α–ketolic steroids in man is extractable after incubation with β–glucuronidase [98, 99]. The principal compound has been identified as tetrahydrocortisone, and the total daily excretion measured as α–ketolic steroids released was 2.0 to 7.4 mg [99, 100]. These levels rose after stimulation with adrenocorticotropin or the administration of cortisone [17α,21–dihydroxypregn–4–ene–3,11,20–trione] or cortisol. The urinary steroids measured as 17,21–dihydroxy–20–ketosteroids in human urine before and after hydrolysis with β–glucuronidase averaged 0.37 and 6.6 mg/day, respectively [101]. Kornel [102], employing differential extraction for free and conjugated compounds of this type, found a ratio of 1:24, free:conjugated. Thus it appears that the major metabolites of cortisol are excreted as conjugates and principally as glucuronides. Such compounds have sometimes been isolated and identified intact as described later. Other metabolites identified following hydrolysis of urine with β–glucuronidase are as follows: 5β–pregnane–3α,11β,17α,20α,21–pentol (and its 20β epimer), 11–oxo–5β–pregnane–3α,17α,20α,21–tetrol (and its 20β epimer), tetrahydrocortisone, allotetrahydrocorticosterone [3α,11β,21–trihydroxy–5α–pregnan–20–one], tetrahydrocortisol [3α,11β,17α,21–tetrahydroxy–5β–pregnan–20–one], cortisone, cortisol, allotetrahydrocortisol [3α,11β,17α,21–tetrahydroxy–5α–pregnan–20–one], tetrahydro–Compound S [3α,17α,21–trihydroxy–5β–pregnan–20–one] [39, 103–105]. The evidence suggests that unreduced steroids, cortisone and cortisol, are conjugated at C21 [105] with glucosiduronic acid. Human urine contains free cortisone and cortisone glucuronide [17α–hydroxy–3,11,20–trioxo-pregn–4–en–21–yl–β–D–glucopyranosiduronic acid] in quantities of 57 and 73 μg per day; and free cortisol and cortisol glucuronide [11β,17α–dihydroxy-3,20-dioxopregn-4-en-21-yl-β-D-glucopyranosiduronic acid], 19 and 25 μg. It appears that reduced steroids occur as the 3α-yl-β-D-glucuronides, although in some cases these compounds will be conjugated at C21 as well; *e.g.*, tetrahydrocorticosterone and tetrahydrocortisol [39].

Pasqualini and Jayle [106] appear to have identified the 21-sulfates of cortisol and corticosterone [11β,21-dihydroxypregn-4-ene-3,20-dione] in human urine, and these unreduced steroids appeared to occur primarily in this conjugated form. Kornel [107] claimed to have detected, in human urine after solvolysis, highly polar steroids, 17,21-dihydroxy-20-ketosteroids, and that these constituted 15 to 33% of the total. Chromatography revealed, after solvolysis, steroids with the properties of tetrahydrocortisol, cortisol, tetrahydrocortisone, tetrahydro-Compound S, and 6β-hydroxycortisol [6β,11β,17α,21-tetrahydroxypregn-4-ene-3,20-dione]. Finally, it appears that the major unconjugated α-ketolic steroid in normal human urine is 6β-hydroxycortisol [108].

B. Plasma Steroidal Conjugates

In 1952, Savard *et al.* [109] described the presence, in the dialysate of dog blood, of water-soluble corticoids, which were hydrolyzed with β-glucuronidase. Bongiovanni [110] described the presence of phenylhydrazine-reacting corticoids in human plasma, which had been released by β-glucuronidase. The levels of this moiety were approximately equal to the free fraction under normal conditions [111]. Further study indicated that acid hydrolysis released only 30% of the quantity obtained by enzyme hydrolysis [112]. Various other workers found that human plasma contained 17-hydroxycorticoids, which were conjugated, and which were, for the most part, hydrolyzed by mammalian β-glucuronidase [113-117]. Some investigators [114, 117, 118] found that the conjugates may, under some circumstances, exceed the free forms. In man, the free 17-hydroxycorticosteroid in serum was primarily cortisol, whereas indirect evidence suggested that the conjugated fraction contained several reduced compounds, but also some unreduced but conjugated steroids [119]. Brown *et al.* [120] reported the free levels, measured as phenylhydrazine-reacting steroids, as 12.5 ± 0.5 μg per 100 ml, and the glucuronidase-released levels as 10.4 ± 0.6. There was a diurnal pattern for both, which was similar, but the conjugated fraction lagged by 2 to 4 hours.

Others have reported similar levels for free and β-glucuronidase-released 17-hydroxycorticosteroids in human plasma: free, 11 to 15 μg per 100 ml, and conjugated, 12 to 20 μg per 100 ml [121, 122]. By reasonably specific isotopic means, Cohn and Bondy [123] measured tetrahydrocortisone and tetrahydrocortisol in the β–glucuronidase–released fraction of human plasma and found levels of 8.0 and 4.5 μg per 100 ml, respectively.

Klein *et al.* [124] reported acid–hydrolyzable 17–hydroxycorticoids in human plasma. Kornel [107, 125, 126] described steroids of this type as conjugates released by acid solvolysis, after prior β–glucuronidase treatment. He assumed that these were sulfates of C_{21}–steroids, but in smaller amount than the glucuronides; chromatography suggested the presence of tetrahydrocortisone, tetrahydrocortisol, and allotetrahydrocortisol in this fraction. Bègue [127] identified pregnenolone [3β–hydroxypregn–5–en–20–one] as the sulfate in human plasma.

C. Renal Clearance

The few studies of the renal clearance of C_{21}–conjugates are limited to observations concerning glucuronides. The behavior is similar to that of androsterone glucuronide, suggesting tubular secretion, although glomerular filtration and reabsorption may occur [72, 128, 129]. Studies following the administration and measurement of tetrahydrocortisone in man revealed renal clearances of 71 to 119 ml/minute for the glucuronide as judged by differential hydrolysis [112].

D. Administration of ACTH and Steroids

Various maneuvers have been employed as means of measuring the rate of steroid metabolism, including conjugation, in man, and attempts to produce high levels have been pursued in the hope of obtaining sufficient material for the complete characterization of intact conjugates. Nonetheless, this last effort has rarely been successful in human studies, so that many conclusions must be based on indirect evidence.

There are numerous studies [116, 120, 130] attesting to the rise in conjugated C_{21}–steroids in the plasma and urine of man after the administration of ACTH or of steroids themselves. Nelson and Calver [131] recovered 50% of an orally administered dose of cortisone in the glucuronide fraction of urine as 17,21–dihydroxy–ketonic material. Schneider *et al.* [132] isolated and definitively characterized tetrahydrocortisone–3–glucuronide [17α,21–dihydroxy–11,20–dioxo–5β–pregnan–3α–yl–β–D–glucopyranosiduronic acid] from human urine after the oral administration of tetrahydrocortisone. They also isolated the glucuronide of β–cortolone [3α,17α,20β,21–tetrahydroxy–5β–pregnan–11–one] in this experiment, and following the administration of β–cortolone [112]. Fogitt and Kellie [38, 39] isolated the glucuronides of the following steroids from the urine of subjects who had received ACTH: tetrahydrocortisone, tetrahydrocortisol, and tetrahydro–

Compound S. Pasqualini and Jayle [133] isolated the 21-glucuronide of tetrahydro-Compound S from urine after the oral administration of 17,21-dihydroxypregn-4-ene-3,20-dione. After the administration of cortisol or ACTH, the steroids released in human plasma by β-glucuronidase have been partially identified as tetrahydrocortisone and tetrahydrocortisol by various workers [112, 134, 135].

Following the intravenous infusion of cortisol, the free level in human blood falls exponentially after the first 30 minutes as the conjugates released by β-glucuronidase rise to their maximal level at 2 hours [136, 137]. Ninety per cent of the radioactivity of the dose of isotopic cortisol administered was excreted in the urine within 72 hours, and 60% was released by hydrolysis with β-glucuronidase. When deoxycorticosterone [21-hydroxypregn-4-ene-3,20-dione] was infused into man, the glucuronides in the blood reached their peak within 15 to 30 minutes, and fell rapidly thereafter [138]. Migeon *et al.* [139] observed that, after the administration of labeled corticosterone, the rate of appearance of glucuronides in blood was more rapid than with cortisol. However, the urinary metabolites released by β-glucuronidase were considerably less than for cortisol; namely, about 30%.

In a lengthy series of publications, Pasqualini has indicated the presence of cortisol, corticosterone, and 11-dehydrocorticosterone-21-sulfate [21-sulfooxypregn-4-ene-3,11,20-trione] in urine following the administration of ACTH or steroid to human subjects [140–145]. In these studies, the reduced derivatives were present mainly as the glucuronides. Lebeau and Baulieu [146] have presented evidence for the secretion of corticosterone sulfate as such. Following the administration of doubly labeled corticosterone-21-sulfate [11β-hydroxy-21-sulfooxypregn-4-ene-3,20-dione], the isolation of the reduced sulfate derivative in urine revealed a $^{3}H:^{35}S$ ratio identical to that of the injected substance, suggesting direct metabolism of the conjugate in man [147]. After corticosterone, the 6β-OH derivative has also been isolated from urine [148]; about 50% was free and 50% conjugated as a sulfate. Pasqualini and Jayle [149] suggested the presence of 3β,21-dihydroxypregn-5-en-20-one in human urine following ACTH. They also described this steroid in blood and urine as the disulfate, and 3β,17α,21-trihydroxypregn-5-en-20-one as the 3β-sulfate [17α,21-dihydroxy-3β-sulfooxypregn-5-en-20-one] following the administration of each [150]. The renal excretion of the disulfate was more rapid than the monosulfate.

E. Disease

Much attention has been focused on liver disease and steroid conjugation. Studies include observations of baseline values in blood and urine, and rates of disposition of administered steroids. In the latter area, there is some confusion since the α,β–unsaturated steroids require both reduction and conjugation for their disposition, so that some results based entirely upon disappearance rates from plasma are unclear.

Bongiovanni and Eisenmenger [80] noted the presence of large amounts of biologically active (gluconeogenic) steroids in the urine of subjects with cirrhosis. Much more unconjugated cortisol (456 μg/day) was present in the urine of subjects with liver disease than in normal individuals (80 μg) [151], and the total 17–hydroxycorticosteroids measured after hydrolysis with β–glucuronidase were lower in liver disease [152]. The conjugated 17–hydroxycorticosteroids in human plasma were also depressed in liver disease, 5 μg/100 ml, *vs.* 14 in normal subjects. There was a smaller response to ACTH on the part of conjugated plasma steroids [121, 153]: however, similar observations have been made in many chronic wasting diseases [154].

Brown *et al.* [155] infused labeled cortisol into subjects with liver disease and noted a disappearance rate slower than normal, which correlated well with impairment of bromsulphalein excretion; however, the disappearance rate of tetrahydrocortisone was unimpaired, although the recovery of the conjugated urinary metabolites was lower than normal. This would suggest that the principal difficulty is in reduction rather than conjugation. Peterson *et al.* [136] also administered cortisol and found a normal half–life of 70 to 90 minutes, whereas in liver disease, this was 160 to 800 minutes and with more unconjugated urinary steroidal metabolites present. Further studies [156, 157] are in agreement with these, suggesting impairment of the TPNH [reduced triphosphopyridine nucleotide]–dependent reductase, and a diminished rate of cortisol synthesis in liver disease as a homeostatic compensation for the slow disposition. However, some observations continue to point to some degree of disturbance in conjugation of steroids in liver disease [158–160] so that it is possible that two mechanisms are at fault, and these may vary depending upon the precise nature of the hepatic disorder. This matter is not yet well defined. Whereas less tetrahydrocortisone is present in the urine of cirrhotic individuals, the amount of cortol [5β–pregnane–3α,11β,17α,20α,21–pentol] and cortolones [3α,17α,20α(and 20β)-21–tetrahydroxy–5β–pregnan–11–one] is the same as normal, and suggests a different locus for the formation of the two categories of steroid [159]. The majority of studies have concerned the conjugation of certain C_{21}–steroids with glucuronic acid, and little has been learned about possible disturbances of sulfate conjugation in liver disease.

Several studies of the specific hepatic deficiency of glucuronyl transferase with nonhemolytic jaundice have been made, but results were variable. Peterson and Schmid [161] found normal disposition of administered cortisol as concerns disappearance rate from the blood, but the glucuronides appearing in the urine were markedly reduced to about half normal levels. The disappearance rate of tetrahydrocortisol was slower than normal. It was suggested that hepatic glucuronide formation was impaired, but that there remained some undefined ability to form water-soluble metabolites. These curious observations tend to obscure the conclusions that might be drawn from studies in other types of liver disease, as discussed above. Drucker *et al.* [162] reported similar results with administration of tetrahydrocortisone, and found the defect in conjugation of this steroid less pronounced than with other substrates, *e.g.*, menthol and bilirubin, in this disease. Further studies [163] with the microsomal fraction from liver of the Gunn rat (hereditary glucuronyl transferase deficiency) revealed unimpaired glucuronic acid conjugation with steroids *vs.* bilirubin, so that distinctly different enzymes were postulated. In Gilbert's disease (intermittent jaundice of unknown etiology) there was considerably less glucuronide formation after the administration of cortisol than normal [164].

In chronic renal disease, several observers have reported that the blood has normal levels of free 17-hydroxycorticosteroids, but greatly elevated levels of the conjugates released by β-glucuronidase [117, 153, 165–168]. The elevation correlates inversely with the creatinine clearance rate, and directly with the blood urea nitrogen level. The infusion of steroids leads to very high levels of conjugates in the blood. However, the total and the conjugated urinary 17-hydroxycorticoids are diminished. These findings apparently result from normal conjugation with glucuronic acid, but renal failure to clear the metabolites.

In Cushing's syndrome, the plasma levels of conjugated 17-hydroxycorticosteroids are often high, and may exceed the free levels greatly [111, 118]. However, there may be large amounts of free steroids in the urine, and much of this appears to be 6β-hydroxycortisol [108, 169]. In Addison's disease, the administration of tetrahydrocortisol reveals rates of conjugation which are entirely normal, so that prolonged absence of this steroidal substrate does not appear to affect conjugation [162]. In rheumatoid arthritis and lupus erythematosus, the disposition and conjugation of administered cortisol is entirely normal, as is the secretion rate [170]. In sarcoidosis, there is some compromise of conjugation and a large proportion of the urinary metabolites of cortisol are unconjugated [171].

Rates of conjugation of cortisol and its metabolites with glucuronic acid are more rapid in thyrotoxicosis, and diminished in hypothyroidism [172, 173]. The plasma-free 17-hydroxycorticosteroids rise to higher levels after ACTH in hypothyroidism, and there is a smaller than normal increment in glucuronic acid conjugates. With the administration of tetrahydrocortisone, there is a rapid disappearance rate in thyrotoxicosis but a normal rate in hypothyroidism, so that the difficulty in the latter may be better related to less reductase activity. Nonetheless, these observations suggest a relationship between general metabolism and steroidal conjugation as influenced by thyroid hormone. Both the free and glucuronic acid–conjugated 17–hydroxycorticosteroids in plasma rise with surgical stress, and there is no impairment of conjugation [174].

Kornel [107], and Kornel and Motohashi [175] reported that, in essential hypertension, there was a peculiarity of steroid metabolism leading to an elevation of solvolyzable conjugates of 17–hydroxycorticosteroids, and a diminution of glucuronic acid conjugates. Furthermore, they found that this peculiarly conjugated steroid extracted from the plasma of hypertensive patients was more polar than the comparable component in normal individuals.

F. Pregnancy and Infancy

This discussion will not concern itself with the peculiarities of protein–binding of the free steroids, but will be restricted to a consideration of conjugation alone. Szego [176] reported a diminution of conjugation of cortisol with glucuronic acid in late pregnancy which was attributed to a competition for the same enzyme by the large amounts of estrogens and progestins. When labeled cortisol was infused into pregnant women near term, the disappearance rate was prolonged with a half–life of 140 minutes *vs.* 80 minutes for the nonpregnant state, and the plasma steroid conjugated with glucuronic acid was half, or less, that of controls [177]; the urine contained more unconjugated cortisol and relatively less glucuronide. It was also shown that the cord blood contained one–fifth to one–half the level of both free steroid and the glucuronide fraction of the maternal blood. Other studies confirmed the elevation of the free plasma 17–hydroxycorticoids in late pregnancy, but not of the glucuronic acid conjugates, so that the ratio of free conjugates was markedly augmented [178, 179].

Many observations point to alterations in the predominantly glucuronic acid conjugation of the metabolites of cortisol and related steroids during infancy. Following the infusion of cortisol during the first few days of life, there was a great prolongation of high levels with a half–life of several hours, and this pattern slowly approached adult rates after the first two weeks of life [180–182]. There was a slow conversion to glucuronic acid conjugates, but the disappearance of reduced steroids, such as tetrahydrocortisone, was more rapid, so that there may have been a reduced rate of ring–A reduction as well. The level of conjugated 17–hydroxycorticosteroids, measured after β-glucuronidase hydrolysis, was very low during the first 8 days of life, and this portion did not respond to ACTH administration until after the 3rd day of life [183]. The response of the conjugates did not achieve adult levels until the 42nd day of life [184]. The urine of newborn infants contained relatively more free, and less glucuronic acid-conjugated 17-hydroxycorticosteroids [185–187]. The fraction released by β-glucuronidase increased with each day of advancing age. As may be seen later, other pathways for the disposition of these compounds are utilized in infancy.

Labeled cortisol has been administered to full-term and premature infants. Usually, the excretion of the total radioactivity was slower than in older children, and the percentage of the total dose released by hydrolysis with β-glucuronidase was low; *e.g.*, only 10% *vs.* 50% in later life [188]. This diminution in glucuronide was also noted after the administration of tetrahydrocortisone, 20% of the total dose, suggesting a true decrease in glucuronide formation [188]. Aarskog *et al.* described similar findings which were exaggerated in premature infants, and they noted an inverse relationship between the percentage of the steroid dose excreted as glucuronides and the maximal serum bilirubin level attained [189]. Aarskog also found a greater restriction of glucuronide formation in infants of diabetic mothers [190, 191], which appeared to be related to the greater tendency for severe jaundice in this group.

The newborn human infant, therefore, seems to produce less conjugation of α-ketolic steroids and their metabolites with glucuronic acid. There is now evidence that two other mechanisms are involved. Relatively large amounts of 6β-hydroxycortisol were found in urine of the newborn, and, although some was released after hydrolysis with β-glucuronidase, much of it was free [185, 186, 192]. This was a highly polar metabolite which was relatively soluble in water and hence readily excreted in the urine. There is also evidence for the formation of conjugates, in infancy, other than with glucuronic acid, and there are indications that certain steroids, which may ordinarily be found as glucuronides, occur as sulfates. Urine from full-term and premature human infants contained steroids which were released upon solvolysis, following prior extraction of free and β-glucuronidase-released steroids [193, 194]. There was as much, or more, in this sulfate fraction as in the free or glucuronide fractions, which consisted, in part, of corticosterone, cortisone, and cortisol. This mechanism seems to be an answer to the relative difficulty in reducing ring A and in forming glucuronides. However, the glucuronic acid conjugates may rise as a group, following ACTH or stress in some infants, so that the limitation cannot be regarded as absolute. The isolation of pregnanediol [5β-pregnane-3α,20α-diol] from meconium as the sulfate (85 to 95%) is further evidence for expeditious sulfation of steroids in infancy [195]. There is ample evidence for the presence of abundant steroidal sulfokinases in fetal and newborn tissues [196–200].

Steroids have been shown to affect the conjugation of bilirubin with glucuronic acid in the newborn period. Arias *et al.* [201] attributed severe prolonged hyperbilirubinemia to the presence of 5β–pregnane–3α,21β–diol in breast milk, and this compound has been shown to inhibit bilirubin glucuronyl transferase *in vitro*. The administration of estriol [estra–1,3,5(10)–triene–3,16α,17β–triol], pregnanediol, and cortisone within the first 14 days of life increased the level of unconjugated bilirubin, and the administration of certain of these steroids to the nursing mother also affected the infant [202, 203]. This was interpreted as a competition between these steroids and bilirubin for the enzymatic site, although as noted above, there was some doubt that the same enzyme deals with steroids and bilirubin.

With aging, there is diminution of plasma 17–hydroxycorticosteroids conjugated with glucuronic acid [122, 204, 205].

V. Estrogens

A. Menstrual Cycle and Pregnancy

During the nonpregnant state, the ovary elaborates 17β–estradiol [estra–1,3,5(10)–triene–3,17β–diol] and estrone [3–hydroxyestra–1,3,5(10)–trien–17–one], with estriol representing a further stage in their metabolism. In addition to the ovary, estrogenic synthesis has been shown to occur in the adrenal cortex, Leydig cells of the testis, and the placenta.

From the work of Brown [206], and Brown *et al.* [207] during the last decade, has emerged a substantial segment of our knowledge of the secretion rates of the various estrogens during the menstrual cycle. They demonstrated a parallel increase and decrease in the three major estrogens, their level being lowest in the first week of a normal 28–day cycle, rising to the ovulatory peak at approximately day 13. Subsequently, secretion rates fall and then rise again to a second peak, somewhat lower than the first, termed the luteal maximum. As the cycle draws to an end, estrogen secretion again falls, heralding the onset of menstruation. Estriol levels lag slightly behind those of 17β–estradiol and estrone, strongly supporting the contention that the former represents a metabolic product of the other two.

Studies by Barlow and Logan [208] confirmed the occurrence of increased estriol production during the luteal phase, but analysis of their data led them to the conclusion that much of the estriol arose from pathways independent of estrone and 17β–estradiol. Subsequently, Eren *et al.* [209] compared estrogen metabolism in both halves of the menstrual cycle and demonstrated a change in the pattern of estrogen secretion. They utilized a mathematical model involving a series of regression equations and concluded that, although 17β–estradiol was probably the principal precursor of the three estrogens during the preovulatory phase, it was not during the secretory phase. They found that, whereas the specific activity of the estriol was derived completely from 17β–estradiol in the first half of the cycle, in the second half its specific activity from this radioactive precursor was greatly diminished, suggesting a change in the pattern of secretion.

Although measurements of the amount of estrogen secreted into the blood by the ovary have not been made, estimation of the daily production in the menstruating woman have been performed. Thus, by an isotope dilution technique, Goering and Herrmann [210] demonstrated that the production rate of 17β–estradiol during the premenstrual phase is 20 μg/24 hr, increasing to 150–300 μg/24 hr at the time of ovulation.

Early studies on the quantity of urinary estrogens excreted during pregnancy involved the use of hydrolytic methods, either acid or enzyme [211–220]. Although there was great diversity in the amounts recovered, nonetheless the general pattern that emerged was clear. By the 12th week of gestation estriol levels were well beyond those in the range encountered in the nonpregnant individual, achieving a level of from 1 to 3 mg/24 hr. Subsequently, estrogen output increased rapidly until about the 24th to 28th week; then the rise until term was not so marked. The range of average values reported at term was between 11 and 31 mg urinary estriol/24 hr. The predominance of estriol in the urinary estrogens as the pregnancy progressed thus became striking, so that at the time of delivery, it accounted for 92 to 95% of the total urinary estrogens. In an attempt to circumvent the need for hydrolysis and determine the steroids as conjugates, Slaunwhite *et al.* [221] studied estriol–16–glucuronide [3,17β–dihydroxyestra–1,3,5(10)–trien–16α–yl–β–D–glucopyranosiduronic acid] excretion from the 3rd month of pregnancy to term and found that daily values rose from 7 to 10 mg in the 3rd month, to 37 to 54 mg in the 8th month, values somewhat higher than had previously been reported.

Estriol–3–sulfate [3–sulfooxyestra–1,3,5(10)–triene–16α,17–diol] [222, 223] has been shown to be the predominant conjugate in cord blood accompanied by some glucuronide, and some in the free form. The high levels of the sulfate underscored an extensive fetal sulfurylation of estriol as supported by a similar preponderance of estriol–3–sulfate in human fetuses perfused with estriol [222].

Attempts to identify blood estriol during pregnancy revealed a sulfoglucuronide diconjugate (44%), sulfate (25%), and glucuronide (31%) [223]. Subsequently [224], more estrone and 17β–estradiol were found as conjugates in the maternal, than in the cord blood. Estrone sulfate [3–sulfooxyestra–1,3,5(10)–trien–17–one] (4.1 µg/100 ml) and estriol sulfate (4.4 µg/100 ml) were the major estrogens on the maternal side. Notably, there was more of the free form of estrone and 17β–estradiol than of the glucuronides.

More recently Smith [225] measured the free glucuronide, sulfoglucuronide, and sulfate forms of estriol, estrone, and 17β–estradiol in urine and plasma of five normal pregnant women before and during parturition. Plasma and urine levels were not parallel, there being a considerably higher concentration of estriol glucuronide in the urine than in the plasma. Of the total plasma estrogens, 25 to 80% was estriol, 5 to 60% was estrone, and 5 to 15% was 17β–estradiol. In the urine, estriol represented 75 to 97% of the total, estrone 2 to 21%, and 17β–estradiol 1 to 4%. There were wide ranges in the plasma of the three estrogens, reflecting marked fluctuations in blood levels. The plasma estriol was found in the following forms: 15 to 60% as the glucuronide, 15 to 40% as the sulfate, and 7 to 42% as the sulfoglucuronide; while in the urine, 85 to 97% was the glucuronide, and 2 to 13% made up the rest of the conjugated forms. The major circulating form of estrone was the sulfate, which accounted for 50 to 90%, whereas 70 to 95% of the 17β–estradiol was in the free form. Near the onset of labor, there was a decrease in the blood levels of all the major forms, and during labor the amount of circulating estriol–3–glucuronide [16α,17β–dihydroxyestra–1,3,5(10)–trien–3–yl–β–D–glucopyranosiduronic acid] and estriol–3–sulfate–16–glucuronide [3–sulfooxy–17β–hydroxyestra–1,3,5(10)–trien–16α–yl–β–D–glucopyranosiduronic acid] continued to fall, while the estriol sulfate and free 17β–estradiol increased. Thus, it is striking that 90 to 98% of estrone and 17β–estradiol were excreted as the glucuronide, while only trace amounts of these two hormones existed in the blood as the glucuronide. Comparison of urine to plasma concentrations disclosed that the estrogen glucuronides were secreted by the kidney tubules, while the sulfates and free forms were reabsorbed. Studies on renal clearance of estriol [226] of 205 ml/min, with a concomitant creatinine clearance of 131 ml/min, support this finding. These characteristics of the renal handling of estrogen glucuronides resemble those of other steroid glucuronides noted above [57, 71, 72], and once more suggest marked differences between these conjugates and the sulfates.

Hähnel [227] measured the relative levels of the predominating estrogen glucuronide conjugates in the urine of 20 women, 32 to 40 weeks pregnant, and found approximately 60% as estriol–16–glucuronide, 26% as estriol–

3–glucuronide, 10% as estrone–3–glucuronide [17–oxoestra–1,3,5(10)–trien–3–yl–β–D–glucopyranosiduronic acid], and 2% each as two 17β–estradiol glucuronide-like compounds.

B. Fetoplacental Unit

As underscored in the preceding section, normal human pregnancy represents a hyperestrogenic state which follows an inexorable rise only to terminate abruptly after delivery. Moreover, it is noteworthy that the principal estrogen produced by the nonpregnant woman is 17β–estradiol, while in the pregnant female estriol becomes the predominant estrogen. *In vitro* studies have demonstrated the *de novo* synthesis of ovarian estrogens from acetate and cholesterol [cholest–5–en–3β–ol], but have not been able to prove placental synthesis of estrogens from acetate, cholesterol or even progesterone [pregn–4–ene–3,20–dione] [228]. However, Ryan [229] established as fact the great capacity of placental tissue to convert certain neutral steroids, such as dehydroisoandrosterone, androstenedione, and testosterone, to estrogens. Subsequently, *in vivo* studies by Baulieu and Dray [230] and Siiteri and MacDonald [231] demonstrated the conversion of infused dehydroisoandrosterone sulfate to estrogens by the placenta. Bolté *et al.* [232] then showed that placental perfusion *in situ* with dehydroisoandrosterone sulfate led to the formation of estrone and 17β–estradiol.

The observation by Frandsen and Stakemann [233], that lower excretion of estrogen occurred in women bearing anencephalic fetuses, was of importance in the genesis of the concept of the fetoplacental unit. It is known that such fetuses have an absent fetal zone of the adrenal cortex. This finding, in conjunction with the demonstration of high levels of dehydroisoandrosterone sulfate in the cord blood [234, 235], strongly suggests that precursors arising in the fetus might serve in the synthesis of placental estrogen.

Recent discussions of the fetoplacental unit [199, 236] contend that most of the estrogens and progesterone elaborated during pregnancy are formed by this unit. Although neither component of this unit alone possesses all the enzymatic activities essential for this synthesis, together they are able to accomplish formation of these steroids. Thus, fetal and maternal cholesterol and pregnenolone sulfate [3β–sulfooxypregn–5–en–20–one] are seen as building blocks of progesterone. Still another pathway for metabolism of pregnenolone sulfate involves conversion by the fetal adrenals into dehydroisoandrosterone sulfate, which then travels to the placenta and is metabolized to estrone and 17β–estradiol. Transformation to estriol is not accomplished in the placenta; this step requires return of estrone and 17β–estradiol to the fetus where these compounds undergo 16α–hydroxylation, primarily in the liver.

C. Administration of Estrogens

Early attempts to elucidate the metabolism of the estrogens by administration of labeled precursors were made by Beer and Gallagher [237], who followed the excretion pattern of estrogen metabolites after administration of small doses of estrone and 17β–estradiol. In the two patients whose urine they analyzed, they recovered 77 and 58% of the administered dose. While estrone was the principal metabolite immediately following the injection, with the passage of time estriol surpassed estrone in amount [238]. Sandberg and Slaunwhite [239] investigated estrogen metabolism in nine females, and recovered approximately 80% of the administered dose in the urine. They noted that a significant amount of the radioactivity was associated with the red blood cells. Migeon *et al.* [240] reported the first studies on the distribution of 16–^{14}C–estrone in normal males. These workers' confirmed the earlier finding of rapid disappearance of the administered estrogen from the blood stream, and corroborated the finding of significant amounts of radioactivity in association with the erythrocytes. They also observed increased recovery of estriol with the passage of time, speculating that this might reflect extensive enterohepatic circulation of the estrone, enabling it to undergo additional metabolism.

Wilson *et al.* [241] administered 16–^{14}C–estriol to four pregnant women prior to surgical interruption, and found at least 99% of the urinary radioactivity to be estriol. Recovery of radioactivity from the urine was 41 to 47%. Most of the urinary estriol was estriol–16–glucuronide, with only trace amounts of estriol and estriol–3–sulfate. Within 24 to 48 hours, the amount of polar glucuronides increased at the expense of estriol–16–glucuronide.

Goebelsmann *et al.* [242, 243] investigated the formation of estriol–3–glucuronide in two pregnant women by administering 15–^{3}H–estriol–3–sulfate and 16–^{14}C–estriol–16(17?)–glucuronide as a continuous infusion. Recovery of 85% of the ^{14}C–labeled material, and 73% of the ^{3}H–labeled material was accomplished within 24 hours. The radioactivity of estriol–3–sulfate was excreted more slowly, suggesting a more intensive metabolism. No estriol–3–sulfate was formed from infused estriol–16(17?)–glucuronide, the latter's only metabolite being increasing quantities of a polar glucuronide identified as estriol–3–glucuronide [244]. As much as 25% of the radioactivity administered as estriol–3–sulfate was excreted unchanged during the infusion period. Estriol–3–sulfate, however, was rapidly converted to estriol–16(17?)–glucuronide, and, after discontinuation of the infusion, into increasing quantities of estriol–3–glucuronide. Inasmuch as estriol–3–glucuronide is readily formed from both estriol–

3–sulfate and estriol–16(17?)–glucuronide, one must propose extensive hydrolysis of both estriol–3–sulfate and estriol–16(17?)–glucuronide prior to their conversion to urinary estriol–3–glucuronide. These workers concluded that, at midpregnancy, the majority of the estriol conjugates in the blood consisted of glucuronides rather than sulfates. They also noted that urinary estriol–3–glucuronide quantitatively represented a good deal of the metabolism, not only of circulating estriol, but also of estriol–3–sulfate and estriol–16(17?)–glucuronide.

Crowell *et al.* [245] administered 6,7–^{3}H–estrone to male subjects and found conjugated excretory forms in predominance. Estrone excretion was divided equally between sulfate and glucuronide fractions, but 17β–estradiol and estriol excretion was mainly as the glucuronide. The sulfate conjugates were excreted more rapidly than the corresponding glucuronides. Indeed, estriol glucuronide excretion was not complete after four days. The excretion of estrone sulfate, 17β–estradiol sulfate, and of estrone glucuronide followed that of a single exponential, whereas the excretion of 17β–estradiol glucuronide and estriol glucuronide was more complex, reflecting distribution in two or more anatomical pools, one very likely being the biliary enteric circulation.

D. Biliary Conjugates

Sandberg and Slaunwhite [239] studied the hepatobiliary enteric circulation of intravenously administered ^{14}C–estrone and ^{14}C–17β–estradiol in normal women, and women with biliary fistulae. They found a rapid decrease in the free plasma estrogen fraction after administration of the labeled compound followed shortly thereafter by transfer of the majority of the administered estrogen from the peripheral circulation into the biliary excretory pathway with subsequent excretion in the urine. More of the radioactivity of these estrogens was excreted in the bile than has been seen following injection of any other known steroid to humans. Moreover, almost all the biliary estrogen was in conjugated form. About 50% of the radioactivity appeared in the bile collected through T–tube drainage, while approximately 7% appeared in the stools of the non–fistula patients, bespeaking a considerable reabsorption of estrogen metabolites from the gastrointestinal tract.

Subsequent studies by Adlercreutz [246] on estrogen excretion in human bile revealed that in all groups estrone existed mainly as the sulfate, and 17β–estradiol as the glucuronide. Sulfate excretion preceded that of the corresponding glucuronide. As was expected, the excretion of 17β–estradiol and estrone took place earlier than that of estriol. These experiments suggest a rapid conversion in the liver of 17β–estradiol to estrone and estriol, and of 17β–estradiol–3–sulfate [3–sulfooxyestra–1,3,5(10)–trien–17β–ol] to a multiplicity of other conjugated estrogens, both of the sulfate and glucuronide variety. In several experiments, no free estrogens could be detected in human bile. Following estriol administration, practically all found in the bile was in the glucuronide fraction with the rest in the sulfate fraction. The work [247] corroborated the existence of an enterohepatic circulation and supported its physiologic significance as an excretory pathway for estrogens.

It has been shown that concentrations of the individual estrogen sulfates in blood and bile in late pregnancy are approximately of the same magnitude [248]. Since bile contains 30 times more estriol than blood, a consideration of the total concentration of the three estrogens revealed that, in fact, 39% consisted of sulfates in the circulation and only 3% in the bile. This estriol occurred mainly as polar conjugates, presumably glucuronides and sulfoglucuronides. Thus it appears that the formation of estriol and conjugation with glucuronic acid is essential for biliary excretion. Moreover, these studies suggest that the estrogen sulfates, notably estrone sulfate, played an important role in the intermediary metabolism and transport of the estrogens in the human.

Gas chromatographic analysis [249] of the glucuronide fraction of human, late pregnancy bile has established the presence of estriol, 16–ketoestradiol [3,17β–dihydroxyestra–1,3,5(10)–trien–16–one], 16α–hydroxyestrone[3,16α–dihydroxyestra–1,3,5(10)–trien–17–one], 16,17–epiestriol [estra–1,3,5(10)–triene–3,16β,17α–triol], 16–epiestriol [estra–1,3,5(10)–triene–3,16β,17β–triol], 17β–estradiol, and 17α–estradiol [estra–1,3,5(10)–triene–3,17α–diol]. This report represents the first time that 17α–estradiol had been detected in a human.

Levitz *et al.* [250] administered 15–^{3}H–estriol–^{35}S–3–sulfate to three patients and collected their bile and urine. Virtually all of the ^{3}H, and two–thirds of the ^{35}S were recovered from the urine within 24 hours, the majority in the first 6 hours. However, only 9% of the ^{3}H and 1.5% of the ^{35}S were excreted in the bile in 24 hours. Estriol–3–sulfate–16–glucuronide contained 75% of the biliary ^{3}H and was almost totally devoid of ^{35}S. The urine contained the same conjugates but in different relative amounts; *viz.*: glucuronide—47%; sulfate—22%; and sulfoglucuronide—19%. Some of the infused estriol sulfate was identified in the bile and urine as such, but most of it was hydrolyzed. The major metabolite, estriol glucuronide, was excreted primarily in the urine. Furthermore, significant conversion of estriol glucuronide to estriol–3–sulfate–16–glucuronide occurred with the latter constituting the only conjugate excreted efficiently in the bile, while also undergoing partial excretion in the urine.

E. Metabolism in Fetus, Newborn, Infant, and Child

Studies have demonstrated that the fetus is able to carry out conjugation of estrogen in the lungs, liver, intestines, and kidneys as early as the third to fourth month of gestation [251]. Sulfate formation predominated in the lungs, and glucuronide in the intestines. Conjugates identified included estrone sulfate, 17β–estradiol–3–sulfate, estriol–3–sulfate, and estriol–16(17?)–glucuronide.

It is clear that transconjugation of estriol takes place in the human fetus and newborn since estriol–3–sulfate was found predominantly in the cord blood, and estriol–16(17?)–glucuronide in the amniotic fluid and urine of the newborn [252].

Schwers [253], studying estrogens in plasma in the first days of life, demonstrated an increase in total estrogens during the first six hours followed by a decrease commencing on the second day, which declined to one–tenth of the initial value on the third or fourth day. During the first hours after birth, conjugated estriol constituted 87% of the total, and the amount rose relatively and absolutely, so that after the second day it accounted for 100% of the measured estrogens. It is probable that the sulfate, which at birth measured 86% of the conjugated estriol, decreased progressively, being metabolized into the glucuronide.

Mikhail *et al.* [254] injected 16–^{14}C–estriol into the umbilical vein of four previable fetuses and recovered 29% of the radioactivity in the liver, 18.7% in the lungs, and the remainder in the bones, connective tissue, intestinal tract, blood, heart, skin, and adrenals. Almost 90% of the radioactivity was in the conjugated form, preponderantly as estriol–3–sulfate. The presence of estriol 16(17?) glucuronide, and of a sulfoglucuronide like double conjugate in the extracts of the liver, alimentary tract, and perfused blood was established. These findings strongly invite the conclusion that sulfurylation is the principal pathway of fetal estriol metabolism, although the fetus is capable of conjugating estriol early in gestation.

It is of note that the intermediary metabolism of 17β–estradiol in infants and children differs from that of adults. The metabolism of 17β–estradiol assumes the adult pattern, possibly some time during the second year. Thus in infants and young children, estriol apparently does not represent a final stage in estrogen metabolism (in contradistinction to the adult situation) but constitutes an intermediate stage [255].

F. Disease

The question as to whether or not there is impairment of the conjugation of the estrogens in advanced liver cirrhosis has been investigated for many years, but has not been answered definitively. Nonetheless, there is general agreement that there is an increase in the circulating estrogens, be they free or conjugated. The endocrine disturbances of chronic liver disease have usually been attributed to delayed or incomplete inactivation of estrogens, and this hypothesis has been offered to explain menstrual disturbances, gynecomastia, spider angiomas, testicular atrophy, and palmar erythema.

Glass *et al.* [256] found elevations of free estrogen values, but not of conjugated estrogens in advanced cirrhosis, but subsequently reported [257] increased conjugated estrogens in cirrhotic males after intravenous administration of 17β-estradiol and estrone. They contended that, with increasing liver damage, increasing inability to inactivate or convert the endogenous or exogenous estrogen ensued.

Rupp *et al.* [258] observed high urinary estrogen values in patients with liver disease but were unable to establish a correlation with the presence of gynecomastia. Dohan *et al.* [259] reported increased total urinary estrogens in some patients with liver disease, and suggested that the increased excretion was due in part to decreased excretion through the biliary tract. Increases in the conjugated and unconjugated forms of estriol were greater and more frequent than of estrone or 17β-estradiol. There was a high degree of correlation between excretion of estrogens and sulfobromophthalein retention.

On the other hand, Pincus *et al.* [260] found increased total urinary estrogen in only about one-third of their cases, and no consistent correlation between estrogen excretion and palmar erythema, spiders, or gynecomastia. They proposed that the endocrine changes in hepatic disease were more consistent with pituitary depression independent of hyperestrogenism, and quite likely related to nutritional factors occurring in hepatic disease.

Stimmel [261] found decreased ability to conjugate administered natural estrogen in patients with Laennec's cirrhosis. Cameron [262], in his review of liver disease and steroid hormone metabolism, could find no cogent evidence to support the view that the impaired inactivation of estrogenic hormones was responsible for the endocrine features of liver disease.

Diczfalusy *et al.* [263, 264], studying the capacity of the intestine to perform estrogen conjugation both by duodenal intubation of estriol in a patient with a biliary fistula, and by identification of conjugated estriol from effluent venous blood from perfused segments of small bowel, concluded that estriol conjugation occurred very soon after introduction of the compound. They speculated that the intestinal estrogen-conjugating ability might account for the normal urinary estrogen excretion claimed by a number of workers, in patients with liver disease. Thus, they questioned the validity of measuring urinary estrogens alone in assessing liver-conjugating ability.

More recently, Conti *et al.* [265] found that a high estrogen excretion is frequently, but not uniformly, observed in liver cirrhosis, supporting the concept of impaired intrahepatic inactivation of steroids. On the other hand, Brown *et al.* [266] found that the increased urinary estrogen output was, as a rule, due to increased secretion rates of these hormones rather than to impaired estrogen metabolism by the liver. They noted that the ratio of estriol to estrone and 17β–estradiol in the urine was greater than normal, suggesting that in the diseased liver there may be an increased capacity for metabolizing 17β–estradiol and estrone to estriol.

Preedy *et al.* [267] administered tracer 6,7–^{3}H–estrone to normal and cirrhotic males, and found that the excretion in normal individuals of estrone, estrone sulfate, and estrone glucuronide took the pattern of a single exponential. On the other hand, excretion of 17β–estradiol and estriol glucuronide was much slower and not in single exponential form, so that larger amounts of these conjugates were excreted on the second day than on the first day following injection. This was particularly striking with respect to estriol glucuronide, and was thought to reflect secretion into the biliary enteric circulation. Thus, it is significant that they found that the excretion of estriol conjugates was more rapid in cirrhotic than in normal persons, assuming the pattern of a single exponential. They contended that there is impairment of normal estriol glucuronide secretion into, or transfer out of, the biliary enteric circulation. These workers observed no change in the pattern or rate of excretion of estrone conjugates.

Adlercreutz *et al.* [268] investigated a possible etiologic role for the estrogens in the causation of cholestatic jaundice of pregnancy. This study was undertaken because of the observation that cholestatic jaundice has occurred following the administration of methyl testosterone, and oral contraceptives containing both estrogens and progestins. Moreover, abnormal bromsulphalein retention has occurred after administration of large amounts of natural estrogens, and synthetic estrogens have increased serum bilirubin in young women with chronic liver disease. Since estrogen production and excretion is greatly increased during pregnancy, it appeared possible that abnormalities in estrogen metabolism might be related to recurrent jaundice of pregnancy.

Seven patients aged 22 to 32 with recurrent jaundice of pregnancy were studied. Employing Beling's fractionation procedure with a Sephadex G 25® (Pharmacia, Uppsala) medium column, these investigators found increased amounts of peak II conjugates, and interpreted this as reflecting a shift from 3–glucuronide conjugation to 16– or 17–glucuronide conjugation. They reasoned that, since the 3–glucuronide represents a later stage in estrogen metabolism than the 16– or 17–glucuronide, there might be inhibition of metabolism at the earlier step. They further suggested that impaired enterohepatic circulation of the estrogens might account for this phenomenon, since a block in bile transport would divert the estrogens into the urine, leading to an increase in 16– and 17–glucuronides and sulfates.

They proposed that differences both in the nature and amount of the estrogen metabolites found in cholestatic jaundice of pregnancy and those in normal pregnancy might be partially accounted for by the jaundice and impaired enterohepatic circulation of the estrogens. The fact that several patients, who have had recurrent jaundice of pregnancy, have later had jaundice following intake of normal amounts of contraceptive pills led them to propose that some individuals possess an abnormal sensitivity to the effect of normal amounts of estrogens. Thus, although no evidence was found to suggest a primary defect in estrogen metabolism as the cause of this entity, it does appear that the estrogens have a role in the development of the cholestasis, and may be accompanied by alterations of conjugation.

In a series of patients with hypertrophic pulmonary osteoarthropathy [269], increased urinary estrogen excretion has been demonstrated. However, the excretion of estrogens was not increased in another group of patients with bronchial carcinoma unaccompanied by arthropathy, but accompanied in some instances by clubbing. In patients with osteoarthropathy, conjugation of injected 17β–estradiol, as evaluated by measurements of urinary excretion, was unimpaired. Therefore, the authors proposed that increased production accounted for the elevations of the estrogens.

Gabrilove *et al.* [270] recently reviewed clinical findings in 52 patients with feminizing adrenocortical tumors, including one patient on whom clinical and biochemical studies were made. For the first time, evidence was presented for the *in vitro* biosynthesis of estrogens from progesterone, pregnenolone, and androstenedione by the tumor tissue of this latter patient. Of the 52 patients, chemical studies on the urinary estrogens were available for 19, and in these, significant increases were uniformly found. In addition, in 14, partition studies for estrone, 17β–estradiol, and estriol were performed after initial isolation of the estrogen from its conjugate. All three fractions were elevated, with estriol showing the greatest change and 17β–estradiol the smallest. In the new case, they reported significant amounts of 2–methoxyestrone [3–hydroxy–2–methoxyestra–1,3,5(10)–trien–17–one], 16α–hydroxyestrone. 16–ketoestradiol, and 16–epiestriol, in addition to the three common estrogens in the urine. Although the feminizing adrenogenital syndrome in the male is generally due to a tumor, this is not invariably so, and it may be the result of noncancerous adrenocortical hyperfunction.

Flickinger *et al.* [271], studying free and conjugated estrogens of an ovarian cystadenoma and granulosa cell tumor, although unable to detect estrogenic glucuronides, did find estrone sulfate. Barlow and Goldstein [272] have reported for the first time the conversion of dehydroisoandrosterone sulfate in two patients with choriocarcinoma. 17β–Estradiol production rates were elevated in all patients with choriocarcinoma prior to therapy. MacDonald and Siiteri [273] have also demonstrated the utilization of dehydroisoandrosterone sulfate by neoplastic trophoblastic tissue for the synthesis of 17β–estradiol. They also showed that dehydroisoandrosterone sulfate was converted to estriol *via* an estrone–independent pathway. More recently, Kase and Cohn [274] have reported a case of a woman with primary amenorrhea and hypogonadotropic hypogonadism in which the peripheral conversion of dehydroisoandrosterone sulfate to 17β–estradiol was demonstrated.

VI. Progestins

A. Normal Conjugation

The earliest studies on the identification of progesterone conjugates were those of Venning and Browne [4] who, in 1936, reported the isolation of pregnanediol from pregnancy urine as a water–soluble complex salt, sodium pregnanediol glucuronide. Using a gravimetric method, Browne *et al.* [275] demonstrated the excretion of this conjugate during the luteal phase of the menstrual cycle. Normally, during the menstrual cycle, small amounts of pregnanediol are excreted in the urine during the follicular phase, most probably from adrenal sources [276]. Following ovulation, large amounts of pregnanediol are recovered with a peak excretion between the 20th and 24th days of the cycle, the average values being 5 mg/24 hours with continued excretion in lesser amounts up to menstruation [277].

There is no doubt that the placenta produces massive amounts of progesterone during pregnancy [278]. Early studies by Browne *et al.* [275] showed that, up to the 60th day of pregnancy, values of urinary excretion were 4 to 10 mg per day, or in the range found during the menstrual cycle. The rate of excretion rose between the 80th to 100th day, and by the 150th day achieved a level of 40 mg. Subsequently, the rise continued and reached a peak in the 8th month of 80 mg. Within 24 hours of delivery the compound could not be identified in the urine. More recent studies [279] have confirmed that, in the normal human pregnancy, urinary pregnanediol excretion gradually increases until a plateau is reached at about 32–weeks gestation. Isotope dilution studies have demonstrated that in late pregnancy the daily production of progesterone achieves a level of 250 mg [280].

Sandberg and Slaunwhite [281] studied the metabolism of injected labeled progesterone in nine subjects, four with bile fistulas. In all patients, the amount of radioactivity excreted in the urine approximated 50%, with over 95% as conjugates of progesterone metabolites, 40% of which were glucuronides. The urine of the patients without fistulas contained more glucuronides, whereas those patients with bile fistulas excreted more radioactivity as sulfates. Slightly more than 60% of the radioactivity was extractable in the urine after all hydrolytic processes. The average amount of radioactivity excreted in the feces was 13% in the non–fistula group, and 3% in the bile fistula subjects. Total radioactivity excreted over a period of three to five days in the urine and feces of the non–fistula subjects averaged 65%, whereas that in the urine, bile, and feces of the bile fistula subjects averaged 82%. The data indicate that a considerable portion of the metabolites of progesterone are excreted in the bile, partly reabsorbed from the gut, and the remainder excreted in the feces. Apparently the reabsorbed biliary metabolites are not excreted in the urine. Within 15 minutes after injection of the progesterone, the amount of glucuronide in the plasma was eight times higher than that of the unconjugated steroids. The ^{14}C levels in the sulfate fraction were much lower than in the glucuronide fraction.

Wiest *et al.* [282] administered 4–^{14}C–progesterone to a postmenopausal woman with a biliary fistula, and confirmed that the biliary route constitutes a major mode of excretion for progesterone metabolites in the human, in contrast to cortisol and testosterone, which are eliminated primarily by the urine. In this report, 65% of the radioactivity was found in the bile and 35% in the urine. Both bile and urine contained pregnanediol, pregnanolone [3α–hydroxy–5β–pregnan–20–one], and some unidentified polar material, all of which were released by β–glucuronidase hydrolysis. In bile, as in urine, the major conjugate was pregnanediol, but the amount of pregnanolone was proportionately greater in bile than urine.

Chang *et al.* [283] analyzed urinary and biliary excretory products following injection of 4–^{14}C–progesterone [281] and concluded that hydroxylated metabolites of progesterone were excreted chiefly in the urine, while the corresponding ketonic derivatives were found mostly in the bile.

Pasqualini *et al.* [150], studying the mode of conjugation of steroids administered to a normal individual, found that 21–hydroxypregnenolone [3β,21–dihydroxypregn–5–en–20–one] occurred as the disulfate, and 17α,21–dihydroxypregnenolone [3β,17α,21–trihydroxypregn–5–en–20–one] as the monosulfate, both in the blood and urine. The excretion of the disulfate was faster than that of the monosulfate.

Zander *et al.* [284] perfused an anencephalic newborn with 4–^{14}C–progesterone and recovered 65 to 85% in a conjugated form. The majority of the metabolites were sulfates, and the predominant steroid in the sulfate fraction was pregnanediol. These studies indicate that circulating metabolites are extensively sulfurylated by the human anencephalic newborn, and that the conjugation pattern differs significantly from that of the adult.

Recently, Thijssen and Zander [285] reported on the metabolism of trace amounts of injected 4–^{14}C–progesterone in seven nonpregnant women. This work was undertaken because of the author's concern that much of the information in this field had been derived from experiments in which doses far in excess of physiologic amounts were administered. An equilibrium was reached between the progesterone in blood and tissues 10 to 15 minutes after progesterone administration. Glucuronide concentration in blood increased soon after the intravenous injection, and a maximum was reached in about 15 minutes. At this time the glucuronide concentration had exceeded the progesterone level, both levels being equal about 10 minutes after injection. The concentration of sulfates was still increasing at the end of the 90–minute experimental period, but was always much lower than the levels of the glucuronides. Free metabolites isolated included 20α–hydroxypregn–4–en–3–one, pregnanediol, and pregnanolone; and glucuronides characterized included pregnanediol, 5β–pregnane–3β,20β–diol (the first time that the conversion of progesterone to the 20β–diol derivative had been reported), and pregnanolone. These workers calculated that the half life of progesterone was 25 to 29 minutes, and that of the glucuronides was 152 to 162 minutes.

Pion *et al.* [286] injected 7α–^{3}H–pregnenolone sulfate into the umbilical artery immediately following delivery of an infant by cesarean section, but prior to gross and palpable evidence of placental detachment. They found that almost 50% of the radioactivity recovered was progesterone, and thus showed for the first time that pregnenolone sulfate is a precursor to the placental production of progesterone.

Calvin and Lieberman [287] studied the metabolism of pregnenolone sulfate in a female patient with an adrenal carcinoma and found that it was extensively metabolized to a number of compounds, including pregnenediol–3–sulfate [3β–sulfooxypregn–5–en–20α–ol], dehydroisoandrosterone sulfate, androstenediol–3–sulfate [3β–sulfooxyandrost–5–en–17β–ol], and pregnenetriol–3–sulfate [3β–sulfooxypregn–5–ene–17α,20α–diol]. It is of note that the most important fate of dehydroisoandrosterone sulfate is its conversion to estrogens in pregnant women, this transformation involving the cleavage of the sulfate ester group from the steroid in the placenta and aromatization in that organ of the liberated steroid. These investigators were unable to answer the key question as to whether pregnenolone sulfate is converted to dehydroisoandrosterone sulfate in the human adrenal by reactions using only intact steroid sulfates as intermediates. More recently, Conrad *et al.* [288] have described a double isotope–derivative method for the estimation of pregnenolone sulfate in plasma. They demonstrated a gradient between the fetal and maternal compartments, and underscored the importance of this finding in light of the work of Calvin and Lieberman, in which the extensive metabolism of this compound was demonstrated.

B. Disease

The primary physiologic role of progesterone in pregnancy is the preparation of the endometrium for implantation and maintenance of the pregnancy. The progestational endometrium is endowed with a complement of glycogen for nutrition of the ovum. As a result of progesterone deficiency, the endometrial bed may collapse, with consequent abortion. It is generally accepted that the corpus luteum functions in the secretion of progesterone to peak capacity until the 60th day, and thereafter continues to secrete substantial amounts until the 90th day. On the other hand, the placenta does not begin to secrete significant amounts of progesterone until the 60th day, but by the 90th day has usurped the function of secreting all of the progesterone.

As a consequence of its importance in maintaining the integrity of the endometrium, it has been suggested that lack of normal progesterone production may be a factor in habitual abortion. For example, Jones and Delfs [289] found low secretions of progesterone in a number of pregnancies that resulted in abortion. However, because the multitude of factors that are necessary for the fruition of the pregnancy are poorly understood, and because of the variability of results with exogenous progesterone administration in treatment of habitual abortion, there at present appears to be little to substantiate the clinical impression that progesterone is an essential therapeutic factor in the retention of the conceptus.

Similarly, with the question of toxemia of pregnancy, several workers [290, 291] have implicated lowered secretion of progesterone as a factor in the etiology of the disorder, but the work has not been confirmed [292], nor has this unitary hypothesis been adequate to explain the broad range of disorganization seen.

Perhaps the only bona fide situation in which progesterone figures as a prominent factor is the recently described occurrence of prolonged neonatal hyperbilirubinemia in association with breast feeding. Arias *et al.* [201] isolated 5β–pregnane–3α,20β–diol from the maternal milk and demonstrated its inhibitory effect on *in vitro* glucuronidation of bilirubin. It is evident that, at the present time, there is almost no information regarding the conjugation of progestins and metabolites as might be related to human disease.

VII. Drugs

Some pharmacologic agents have been found to affect the conjugation of steroids. The compound, o,p′–DDD [2,2′–bis(2–chlorophenyl–4–chlorophenyl)–1,1–dichloroethane], has been under study as a suppressant of adrenocortical secretion. The exact mechanism of action is unknown, but certain extra–adrenal effects on steroidal metabolism have been noted. In subjects with Addison's disease on a fixed dose of cortisol, a drop in urinary 17–hydroxycorticoids was noted with this drug [293]. In Cushing's syndrome, the drug caused a fall in urinary reduced and glucuronic acid conjugates of cortisol with a rise of free steroids including 6β–hydroxycortisol [294]. The drug diminished the urinary glucuronides expected after the administration of labeled testosterone without changing the steroidal sulfates [295]. It would appear from these observations that steroidal glucuronyl transferase has been affected, but there is no proof of this, and other liver function studies do not necessarily parallel the effect on steroids.

Some comments have been made earlier with respect to the influence of administered steroid on the conjugation. The rise of free plasma 17–hydroxycorticosteroids after estrogen administration in man is now well recognized [296], but there is also a fall in the urinary excretion of steroidal glucuronides and a prolongation in the decay rate of infused cortisol and tetrahydrocortisone as well. This suggests a true effect on glucuronide formation. It has also been noted that in men, given estrogens, the glucuronic acid–conjugated 17–hydroxycorticosteroids in plasma and urine fall slightly and there is an impairment of the response of these conjugates to ACTH [297]. There is also a rise in urinary 6β–hydroxycortisol [298]. There is a suggestion that 17–methyl–19–nortestosterone [17β–hydroxy–17α–methylestr–4–en–3–one], given orally, compromises the formation of the usually expected steroidal glucuronides [299]. Progestins and some estrogens cause a reduction of urinary tetrahydroaldosterone glucuronide and a relative rise in the 3–oxo conjugate [300]. There are indications that the administration of large doses of ACTH causes a reduction in the glucuronides of steroids in man [129, 301, 302].

Diphenylhydantoin produces an increase of urinary, highly polar, steroidal metabolites, principally 6β–hydroxycortisol, with a fall in reduced glucuronides [303]. This phenomenon was further demonstrated by the administration of labeled cortisol, and the effect was enhanced by ACTH. This drug may interfere with ring A reduction, and TPNH production as well. Similar effects have been noted with triparanol (Mer–29) 1–[(4–diethylaminoethoxy)phenyl]–1–(*p*–tolyl)–2–(*p*–chlorophenyl) ethanol [304].

In rats and guinea pigs, 4–allyl–2–methoxyphenol has been found to inhibit the hepatic glucuronyl transferase [305]. Done *et al.* [306] believed that the administration of salicylates led to a reduction of urinary 17–hydroxycorticosteroid glucuronides, but Stempfel *et al.* [307] found that there was no true alteration. The urinary salicylate glucuronides impaired the hydrolysis of steroidal conjugates by β–glucuronidase, and this could be overcome by large amounts of the enzyme. A most interesting observation has been made by Crigler and Gold [308]. In congenital nonhemolytic jaundice due to a deficiency of glucuronyl transferase, there was produced an increase in urinary steroidal glucuronides by the administration of phenobarbital with testosterone propionate. This combination of drugs seems to have caused the induction or facilitation of an otherwise deficient conjugating enzyme.

VIII. Conclusion

In man, the steroid hormones may be secreted and transported in both free and conjugated form. The eventual disposition of these substances involves conjugation as sulfates, glucuronides, and perhaps other as yet unknown compounds. An alternate route may be further hydroxylation into more polar steroids without actual conjugation. The degree of conjugation and the preferred form varies with age and the physiologic state of the individual; *e.g.*, pregnancy. Diseases of man may be associated with alterations of the conjugating mechanism, generally as phenomena secondary to some other principal fault. Rarely do deviations of the usual steroidal conjugating mechanism, *per se*, produce clinical disorders, as may be the case in etiocholanolone fever. Occasionally, as in hereditary nonhemolytic jaundice, the primary basis of the major clinical disturbances may have a direct influence on the conjugation of steroids as well. New studies will probably indicate additional forms of steroidal conjugates and other variations with human disease. A firmer basis for the measurement and precise identification of steroidal conjugates in clinical investigation is sorely needed.

Acknowledgment. Those aspects of the review dealing with investigations from this laboratory have been supported by Grant HD–00371, U.S.P.H.S.

IX. References

1. Venning, E. H., Hoffman, M. M., Browne, J. S. L.: J. Biol. Chem. **146**, 369 (1942).
2. Munson, P. L., Gallagher, T. F., Koch, F. C.: J. Biol. Chem. **152**, 67 (1944).
3. Odell, A. D., Marrian, G. F.: Biochem. J. **30**, 1533 (1936).
4. Venning, E. H., Browne, J. S. L.: Proc. Soc. Exptl. Biol. Med. **34**, 792 (1936).
5. Heard, R. D. H., Hoffman, M. M., Mack, G. E.: J. Biol. Chem. **155**, 607 (1944).
6. Buehler, H. J., Katzman, P. A., Doisy, P. P., Doisy, E. A.: Proc. Soc. Exptl. Biol. Med. **72**, 297 (1949).
7. Venning, E. H.: J. Clin. Endocrinol. **11**, 769 (1951).
8. Cohen, S. L., Goldfine, M. M.: J. Clin. Endocrinol. Metab. **12**, 916 (1952)
9. Venning, E. H., Dyrenfurth, I., Kazmin, V. E.: Recent Progr. Hormone Res. **8**, 27 (1953).
10. Underwood, R. H., Tait, J. F.: J. Clin. Endocrinol. Metab. **24**, 1110 (1964).
11. Pasqualini, J. R.: Compt. Rend. **259**, 934 (1964).
12. — Uhrich, F., Jayle, M. F.: Biochim. Biophys. Acta **104**, 515 (1965).
13. Ulick, S., Lieberman, S.: J. Am. Chem. Soc. **79**, 6567 (1957).
14. Flood, C., Layne, D. S., Ramcharan, S., Rosipal, E., Tait, J. F., Tait, S. A. S.: Acta Endocrinol. **36**, 237 (1961).
15. New, M. I., Miller, B., Peterson, R. E.: J. Clin. Invest. **45**, 412 (1966).
16. Peterson, R. E.: Recent Progr. Hormone Res. **15**, 231 (1959).
17. Pasqualini, J. R., Jayle, M. F.: Nature **198**, 1095 (1963).
18. Coppage, W. S. Jr., Island, D. P., Cooner, A. E., Liddle, G. W.: J. Clin. Invest. **41**, 1672 (1962).
19. Bledsoe, T., Riondel, A. M., Island, D. P., Liddle, G. W.: Clin. Res. **13**, 70 (1965).
20. Laragh, J. H., Ulick, S., Januszewicz, V., Deming, Q. B., Kelly, W. G., Lieberman, S.: J. Clin. Invest. **39**, 1091 (1960).
21. Frimpter, G. W., Siegenthaler, W., Peterson, R. E.: J. Clin. Invest. **42**, 932 (1963).
22. Cheville, R. A., Camargo, C. A., Dowdy, A. J., Nokes, G. W., Hancock, E. W., Luetscher, J. A.: Clin. Res. **13**, 135 (1965).
23. Gfeller, J., Siegenthaler, W.: Acta Endocrinol. **49**, 510 (1965).
24. Deck, K. A., Siegenthaler, W.: Helv. Med. Acta **32**, 407 (1965).
25. Venning, E. H., Dyrenfurth, I.: J. Clin. Endocrinol. Metab. **16**, 426 (1956).
26. Jones, K. M., Lloyd-Jones, R., Riondel, A., Tait, J. F., Tait, S. A. S., Bulbrook, R. D., Greenwood, F. C.: Acta Endocrinol. **30**, 321 (1959).
27. Bougas, J. A., Flood, C., Little, B., Tait, J. F., Tait, S. A. S., Underwood, R., in: Aldosterone (E.-E. Baulieu, P. Robel, eds.), p. 25. Philadelphia, Pa.: Davis Co. 1964.
28. Hurter, R., Nabarro, J. D. N.: Acta Endocrinol. **33**, 168 (1960).
29. Ofstad, J., Støa, K. F., Emberland, R.: Acta Endocrinol. **39**, 123 (1962).
30. Cheville, R. A., Luetscher, J. A., Hancock, E. W., Dowdy, A. J., Nokes, G. W.: J. Clin. Invest. **45**, 1302 (1966).
31. Muller, A. F., Manning, E. L., in: Structure and metabolism of corticosteroids (J. R. Pasqualini and M. F. Jayle, eds.), p. 147. New York, N.Y.: Academic Press 1964.

32. Ulick, S., Laragh, J. H., Lieberman, S.: Trans. Assoc. Am. Physicians **71**, 225 (1958).
33. Jayle, M. F., Crépy, O.: Ann. Endocrinol. (Paris) **14**, 74 (1953).
34. Kellie, A. E., Wade, A. P.: Acta Endocrinol. **23**, 357 (1956).
35. Malassis, D., Baulieu, E.-E., Weinmann, S., Crépy, O., Jayle, M. F.: Compt. Rend. Soc. Biol. **151**, 447 (1957).
36. Pelzer, H., Staib, W., Ott, D.: Z. Physiol. Chem **312**, 15 (1958).
37. Bush, I. E., Gale, M. M.: Acta Endocrinol. **35**, Suppl. **51**, 1027 (1960).
38. Foggitt, F., Kellie, A. E.: Biochem. J. **76**, 62 P (1960).
39. — — Biochem. J. **91**, 209 (1964).
40. Stárka, L., Šulcová, J., Šilink, K.: Clin. Chim. Acta **7**, 309 (1962).
41. Siiteri, P. K., Vande Wiele, R. L., Lieberman, S.: J. Clin. Endocrinol. Metab. **23**, 588 (1963).
42. Camacho, A. M., Migeon, C. J.: J. Clin. Endocrinol. Metab. **23**, 301 (1963).
43. — — J. Clin. Invest. **43**, 1083 (1964).
44. Sandberg, A. A., Slaunwhite, W. R., Jr.: J. Clin. Invest. **35**, 1331 (1956).
45. Dessypris, A., Drosdowsky, M. A., McNiven, N. L., Dorfman, R. I.: Proc. Soc. Exptl. Biol. Med. **121**, 1128 (1966).
46. Jayle, M. F., Huis in't Veld, L. G., Baulieu, E.-E., Crépy, O.: Acta Endocrinol. **21**, 115 (1956).
47. Lewbart, M. L., Schneider, J. J.: Federation Proc. **15**, 300 (1956).
48. Crépy, O., Jayle, M. F., Meslin, F.: Compt. Rend. Soc. Biol. **151**, 234 (1957).
49. Schneider, J. J., Lewbart, M. L.: Recent Progr. Hormone Res. **15**, 201 (1959).
50. Staib, W., Teller, W., Scharf, F.: Z. Physiol. Chem. **318**, 163 (1960).
51. Baulieu, E.-E.: Compt. Rend. **250**, 4219 (1960).
52. — Emiliozzi, R.: Compt. Rend. **251**, 3106 (1960).
53. Siiteri, P. K., Lieberman, S.: Biochemistry **2**, 1171 (1963).
54. Corpéchot, C., Baulieu, E.-E.: Compt. Rend. **258**, 3584 (1964).
55. Robel, P., Emiliozzi, R., Baulieu, E.-E.: J. Biol. Chem. **241**, 20 (1966).
56. Schubert, K., Hobe, G., Bacigalupo, G.: Endocrinologie **50**, 138 (1966).
57. Kellie, A. E., Smith, E. R.: Biochem. J. **66**, 490 (1957).
58. Baulieu, E.-E.: Rev. Franc. Etudes Clin. Biol. **4**, 928 (1959).
59. — Compt. Rend. **248**, 1441 (1959).
60. — Compt. Rend. **251**, 1421 (1960).
61. Staib, W., Teller, W., Pelzer, H.: Biochim. Biophys. Acta **31**, 591 (1959).
62. Wieland, R. G., Levy, R. P., Katz, D., Hirschmann, H.: Biochim. Biophys. Acta **78**, 566 (1963).
63. Baulieu, E.-E.: Acta Endocrinol. **35**, Suppl. **51**, 1029 (1960).
64. Oertel, G. W., Eik-Nes, K. B.: Acta Endocrinol. **28**, 293 (1958).
65. — — Acta Endocrinol. **35**, Suppl. **51**, 221 (1960).
66. — Kaiser, E., Brühl, P.: Acta Endocrinol. **40**, Suppl. **67**, 85 (1962).
67. — Biochem. Z. **339**, 135 (1963).
68. — Groot, K.: Z. Physiol. Chem. **341**, 1 (1965).
69. — — Brühl, P.: Z. Physiol. Chem. **341**, 10 (1965).
70. — Z. Physiol. Chem. **343**, 276 (1966).

71. Bongiovanni, A. M., Eberlein, W. R.: J. Clin. Endocrinol. Metab. **15**, 855 (1955).
72. — — J. Clin. Endocrinol. Metab. **17**, 238 (1957).
73. West, C. D., Tyler, F. H., Brown, H., Samuels, L. T.: J. Clin. Endocrinol. **11**, 897 (1951).
74. Sandberg, A. A., Slaunwhite, W. R., Jr.: Proc. Soc. Exptl. Biol. Med. **96**, 658 (1957).
75. Kappas, A, Hellman, L., Fukushima, D. K., Gallagher, T. F.: J. Clin. Endocrinol. Metab. **18**, 1043 (1958).
76. Bondy, P. K., Cohn, G. L., Herrmann, W., Crispell, K. R.: Yale J. Biol. Med. **30**, 395 (1958).
77. Cohn, G. L., Bondy, P. K., Castiglione, C.: J. Clin. Invest. **40**, 400 (1961).
78. Bondy, P. K., Cohn, G. L., Gregory, P. B.: Medicine **44**, 249 (1965).
79. Schenker, S., Wilson, H., Spickard, A.: J. Clin. Endocrinol. Metab. **23**, 95 (1963).
80. Bongiovanni, A. M., Eisenmenger, W. J.: J. Clin. Endocrinol. **11**, 152 (1951).
81. Cantarow, A., Paschkis, K. E., Williams, T. L., Havens, W. P., Jr.: Federation Proc. **10**, 23 (1951).
82. Williams, T. L., Cantarow, A., Paschkis, K. E., Havens, W. P., Jr.: Endocrinology **48**, 651 (1951).
83. Birke, G.: Acta Med. Scand. **148**, Suppl. **291**, 53 (1954).
84. Schubert, K., Schröder, H.: Acta Endocrinol. **32**, 23 (1959).
85. Drucker, W. D., Engel, T. R., Meislin, A. G., Jirku, H., Levitz, M. M., David, R. R.: Abstr. 229, 48th Meeting of the Endocrine Society, Chicago, Ill., June 1966, p. 141.
86. Crépy, O., Malassis, D., Meslin, F., Jayle, M. F.: Acta Endocrinol. **26**, 43 (1957).
87. Guillon, J., Colás, J., Trichereau, R., Delumeau, G., Baulieu, E.-E.: Ann. Endocrinol. **22**, 331 (1961).
88. — Emiliozzi, R., Corpéchot, C.: Experientia **17**, 110 (1961).
89. Michaud, G., Benhamou, R.: Compt. Rend. Soc. Biol. **155**, 1003 (1961).
90. — J. Clin. Endocrinol. Metab. **22**, 501 (1962).
91. Guinet, P., Mornex, R., Revol, A., Tourniaire, J., Gallois, R.: Ann. Endocrinol. **26**, 554 (1965).
92. Hadd, H. E., Dorfman, R. I.: J. Biol. Chem. **238**, 907 (1963).
93. — Rhamy, R. K.: J. Clin. Endocrinol. Metab. **25**, 876 (1965).
94. Bénard, H., Bourdin, J. S., Saracino, R., Seeman, A.: Compt. Rend. **258**, 6511 (1964).
95. Mauvais-Jarvis, P., Baulieu, E.-E.: J. Clin. Invest. **41**, 1690 (1962).
96. Weissman, M.: Bull. Res. Council Israel, Sect. E **10**, 210 (1963).
97. Conrad, S. H., Lindberg, M. C., Herrmann, W. L.: Am. J. Obstet. Gynecol. **91**, 449 (1965).
98. Glenn, E. M., Nelson, D. H.: J. Clin. Endocrinol. Metab. **13**, 911 (1953).
99. Baggett, B., Kinsella, R. A., Jr., Doisy, E. A.: J. Biol. Chem. **203**, 1013 (1953).
100. Schneider, J. J.: J. Biol. Chem. **194**, 337 (1952).
101. Silber, R. H., Porter, C. C.: J. Biol. Chem. **210**, 923 (1954).
102. Kornel, L.: Metab., Clin. Exptl. **8**, 432 (1959).
103. Fukushima, D. K., Leeds, N. S., Bradlow, H. L., Kritchevsky, T. H., Stokem, M. B., Gallagher, T. F.: J. Biol. Chem. **212**, 449 (1955).
104. Kellie, A. E., in: Structure and metabolism of corticosteroids (J. R. Pasqualini and M. F. Jayle, eds.), p. 21. New York, N. Y.: Academic Press 1964.

105. Brouillet, J. C., Mattox, V. R.: J. Clin. Endocrinol. Metab. **26**, 453 (1966).
106. Pasqualini, J. R., Jayle, M. F.: Compt. Rend. Soc. Biol. **157**, 96 (1963).
107. Kornel, L.: Biochemistry **4**, 444 (1965).
108. Frantz, A. G., Katz, F. H., Jailer, J. W.: J. Clin. Endocrinol. Metab. **21**, 1290 (1961).
109. Savard, K., Kolff, W. J., Corcoran, A. C.: Endocrinology **50**, 366 (1952).
110. Bongiovanni, A. M.: J. Clin. Endocrinol. Metab. **14**, 341 (1954).
111. — Eberlein, W. R., Grumbach, M. M., VanWyk, J. J., Clayton, G.: Proc. Soc. Exptl. Biol. Med. **87**, 282 (1954).
112. — — Proc. Soc. Exptl. Biol. Med. **89**, 281 (1955).
113. Vliers, M., Demey, L., Heusghem, C., van Cauwenberge, H.: Compt. Rend. Soc. Biol. **149**, 2271 (1955).
114. Reddy, W. J., Haydar, N. A., Laidlaw, J. C.: J. Clin. Endocrinol. Metab. **15**, 891 (1955).
115. Ceresa, F., Cravetto, C. A.: Acta Endocrinol. **29**, 321 (1958).
116. Tamm, J., Busch, D., Voigt, K. D.: Klin. Wochenschr. **36**, 297 (1958).
117. Englert, E., Jr., Brown, H., Willardson, D. G., Wallach, S., Simons, E. L.: J. Clin. Endocrinol. Metab. **18**, 36 (1958).
118. Grumbach, M. M., Bongiovanni, A. M., Eberlein, W. R., van Wyk, J. J., Wilkins, L.: Bull. Johns Hopkins Hosp. **96**, 116 (1955).
119. Weichselbaum, T. E, Margraf, H. W.: J. Clin. Endocrinol. Metab. **17**, 959 (1957).
120. Brown, H., Englert, E., Jr., Wallach, S., Simons, E. L.: J. Clin. Endocrinol. Metab. **17**, 1191 (1957).
121. Karl, H. J.: Klin. Wochenschr. **37**, 1124 (1959).
122. Hochstaedt, B. B., Reichenbach, B.: Gerontol. Clin. **3**, 55 (1961).
123. Cohn, G. L., Bondy, P. K.: J. Biol. Chem. **234**, 31 (1959).
124. Klein, R., Papadatos, C., Fortunato, J., Byers, C.: J. Clin. Endocrinol. Metab. **15**, 215 (1955).
125. Kornel, L.: J. Clin. Endocrinol. Metab. **20**, 1445 (1960).
126. — J. Clin. Endocrinol. Metab. **22**, 1079 (1962).
127. Bègue, J. A.: Compt. Rend. **260**, 3777 (1965).
128. Leboeuf, G., Bongiovanni, A. M.: Abstr. 74, 44th Meeting of the Endocrine Society, Chicago, Ill., June 1962, p. 37.
129. Beck, J. C., Blair, A. J., Dyrenfurth, I., Morgen, R. O., Venning, E. H., in: The human adrenal cortex (A. R. Currie, T. Symington, and J. K. Grant, eds.), p. 432. Edinburgh, Scotland: E. and S. Livingston 1962.
130. Turner, M. D.: Federation Proc. **15**, 190 (1956).
131. Nelson, D. H., Calver, G. W.: J. Clin. Endocrinol. Metab. **13**, 849 (1953).
132. Schneider, J. J., Lewbart, M. L., Levitan, P., Lieberman, S.: J. Am. Chem. Soc. **77**, 4184 (1955).
133. Pasqualini, J. R., Jayle, M. F.: Compt. Rend. **257**, 2345 (1963).
134. Vermeulen, A.: Acta Endocrinol. **22**, 115 (1956).
135. Gold, N. I.: Metab., Clin. Exptl. **8**, 878 (1959).
136. Peterson, R. E., Wyngaarden, J. B., Guerra, S. L., Brodie, B. B., Bunim, J. J.: J. Clin. Invest. **34**, 1779 (1955).

137. Migeon, C. J., Sandberg, A. A., Decker, H. A., Smith, D. F., Paul, A. C., Samuels, L. T.: J. Clin. Endocrinol. Metab. **16**, 1137 (1956).
138. Sandberg, A. A., Chang, E., Slaunwhite, W. R., Jr.: J. Clin. Invest. **45**, 1066 (1966).
139. Migeon, C. J., Sandberg, A. A., Paul, A. C., Samuels, L. T.: J. Clin. Endocrinol. Metab. **16**, 1291 (1956).
140. Pasqualini, J. R.: Compt. Rend. **246**, 2945 (1958).
141. — Compt. Rend. **250**, 3892 (1960).
142. — Compt. Rend. **251**, 1236 (1960).
143. — Acta Physiol. Latinoam. **10**, 135 (1960).
144. — Jayle, M. F.: Biochem. J. **81**, 147 (1961).
145. — Bull. Soc. Chim. Biol. **45**, 277 (1963).
146. Lebeau, M. C., Baulieu, E.-E.: Compt. Rend. **258**, 6265 (1964).
147. Pasqualini, J. R.: Bull. Soc. Chim. Biol. **47**, 471 (1965).
148. — Arch. Biochem. Biophys. **106**, 15 (1964).
149. — Jayle, M. F.: J. Clin. Invest. **41**, 981 (1962).
150. — Dutter, F., Jayle, M. F.: Biochim. Biophys. Acta **69**, 331 (1963).
151. Vermeulen, A., Demeulenaere, L.: Rev. Franc. Etudes Clin. Biol. **1**, 398 (1956).
152. Tucci, J. R., Albacete, R. A., Martin, M. M.: Gastroenterology **50**, 637 (1966).
153. Krusius, F. E., Oka, M., Karkela, A.: Scand. J. Clin. Lab. Invest. **10**, 323 (1958).
154. Shuster, S.: J. Clin. Endocrinol. Metab. **20**, 675 (1960).
155. Brown, H., Willardson, D. G., Samuels, L. T., Tyler, F. H.: J. Clin. Invest. **33**, 1524 (1954).
156. Englert, E., Jr., Brown, H., Wallach, S., Simons, E. L.: J. Clin. Endocrinol. Metab. **17**, 1395 (1957).
157. Peterson, R. E.: J. Clin. Invest. **39**, 320 (1960).
158. Støa, K. F., Thorsen, T., Abrahamsen, A. M.: European J. Steroids **1**, 47 (1966).
159. Zumoff, B., Bradlow, H. L., Hellman, L.: J. Clin. Invest. **45**, 1091 (1966).
160. Tucci, J. R., Martin, M. M.: Gastroenterology **51**, 515 (1966).
161. Peterson, R. E., Schmid, R.: J. Clin. Endocrinol. Metab. **17**, 1485 (1957).
162. Drucker, W. D., Sfikakis, A., Borkowski, A. J., Christy, N. P.: J. Clin. Invest. **43**, 1952 (1964).
163. — Abstr. 49, 47th Meeting of the Endocrine Society, New York, N.Y., June 1965, p. 45.
164. Abrahamsen, A. M., Støa, K. F.: Nord. Med. **75**, 179 (1966).
165. Karl, H. J., Torbica, M.: Klin. Wochenschr. **38**, 110 (1960).
166. Pekkarinen, A., Kasanen, A.: Acta Endocrinol. **38**, 13 (1961).
167. Blair, A. J., Morgen, R. O., Beck, J. C.: Can. J. Biochem. Physiol. **39**, 1617 (1961).
168. Pekkarinen, A., Rauramo, L., Thomasson, B.: Acta Endocrinol. **35**, Suppl. **51**, 667 (1960).
169. Pasqualini, J. R., deGennes, J. L., Jayle, M. F.: J. Clin. Endocrinol. Metab. **23**, 651 (1963).
170. Gray, C. H., Shaw, D. A.: J. Endocrinol. **33**, 33 (1965).
171. — — J. Endocrinol. **33**, 57 (1965).
172. Brown, H., Englert, E., Jr., Wallach, S.: J. Clin. Endocrinol. Metab. **18**, 167 (1958).
173. Uete, T.: Metab., Clin. Exptl. **10**, 1052 (1961).

174. Virtue, R. W., Helmreich, M. L., Gainza, E.: Surgery **41**, 549 (1957).
175. Kornel, L., Motohashi, K.: J. Clin. Endocrinol. Metab. **25**, 904 (1965).
176. Szego, C. M.: Ciba Found. Colloq. Endocrinol. **11**, 307 (1957).
177. Migeon, C. J., Bertrand, J., Wall, P. E.: J. Clin. Invest. **36**, 1350 (1957).
178. Bertrand, J., Gilly, R., Loras, B.: Compt. Rend. Soc. Biol. **155**, 813 (1961).
179. Pekkarinen, A., Rauramo, L., Thomasson, B.: Acta Endocrinol. **42**, Suppl. **75**, 1–24 (1962).
180. Bongiovanni, A. M., Eberlein, W. R., Westphal, M., Boggs, T., Jr.: J. Clin. Endocrinol. Metab. **18**, 1127 (1958).
181. Grumbach, M. M., Ducharme, J. R., Morishima, A.: J. Diseases Children **98**, 672 (1959).
182. Reynolds, J. W., Colle, E., Ulstrom, R. A.: J. Clin. Endocrinol. Metab. **22**, 245 (1962).
183. Bertrand, J., Gilly, R., Loras, B.: Compt. Rend. Soc. Biol. **154**, 2317 (1960).
184. — — — Abstr. 410, Internat. Congr. Hormonal Steroids, Milan, Italy, May 1962, p. 279.
185. Ulstrom, R. A., Colle, E., Burley, J., Gunville, R.: J. Clin. Endocrinol. Metab. **20**, 1066 (1960).
186. Støa, K. F.: Acta Endocrinol. **40**, Suppl. **67**, 29 (1962).
187. Bertrand, J., Loras, B., Cautenet, B.: Acta Endocrinol. **45**, Suppl. **89**, 16 (1964).
188. Migeon, C. J.: J. Pediat. **55**, 280 (1959).
189. Aarskog, D., Støa, K. F., Thorsen, T.: Acta Endocrinol. **45**, 286 (1964).
190. — Cortisol in the newborn infant. Oslo, Norway: Universitetsforlaget 1964.
191. — Acta Paediat. Scand. **54**, Suppl. **158**, 1 (1965).
192. Ulstrom, R. A., Colle, E., Burley, J., Gunville, R.: J. Clin. Endocrinol. Metab. **20**, 1080 (1960).
193. Drayer, N. M., Giroud, G. J. P.: Steroids **5**, 289 (1965).
194. Ducharme, J. R., Leboeuf, G., Sandor, T.: Union Méd. Canada **95**, 1447 (1966).
195. Francis, F. E., Kinsella, R. A., Jr.: J. Clin. Endocrinol. Metab. **26**, 128 (1966).
196. Klein, G. P., Giroud, C. J. P.: Steroids **5**, 765 (1965).
197. — — Browne, J. S. L.: Federation Proc. **24**, 574 (1965).
198. Bird, C. E., Solomon, S., Wiqvist, N., Diczfalusy, E.: Biochim. Biophys. Acta **104**, 623 (1965).
199. Diczfalusy, E.: Federation Proc. **23**, 791 (1964).
200. Villee, C. A., Loring, J. M., Rose, M.: Abstr. 98, 47th Meeting of the Endocrine Society, New York, N.Y., June 1965, p. 69.
201. Arias, I. M., Gartner, L. M., Seifter, S., Furman, M.: J. Clin. Invest. **43**, 2037 (1964).
202. Lauritzen, C.: Acta Endocrinol. **50**, Suppl. **100**, 112 (1965).
203. — Lohmann, W. D.: Z. Kinderheilk. **95**, 143 (1966).
204. Hochstaedt, B. B., Schneebaum, M., Shadel, M.: Gerontol. Clin. **3**, 239 (1961).
205. West, C. D., Brown, H., Simons, E. L., Carter, D. B., Kumagai, L. F., Englert, E., Jr.: J. Clin. Endocrinol. Metab. **21**, 1197 (1961).
206. Brown, J. B.: Lancet **I**, 320 (1955).
207. — Klopper, A., Loraine, J. A.: J. Endocrinol. **17**, 401 (1958).
208. Barlow, J. J., Logan, C. M.: Steroids **7**, 309 (1966).

209. Eren, S., Reynolds, G. H., Turner, M. E., Jr., Schmidt, F. H., Mackay, J. H., Howard, C. M., Preedy, J. R. K.: J. Clin. Endocrinol. Metab. **27**, 1451 (1967).
210. Goering, R. W., Herrmann, W. L.: Clin. Res. **12**, 115 (1964).
211. Venning, E. H.: Obstet. Gynecol. Survey **3**, 661 (1948).
212. Bradshaw, T. E. T., Jessop, W. J. E.: J. Endocrinol. **9**, 427 (1953).
213. Anker, R. M.: J. Clin. Endocrinol. Metab. **15**, 210 (1955).
214. Heusghem, C.: Actualités Pharmacol. **11**, 203 (1958).
215. Zondek, B., Goldberg, S.: J. Obstet. Gynaecol. Brit. Empire **64**, 1 (1957).
216. Brown, J. B.: Lancet **I**, 704 (1956).
217. Smith, O. W., Blackham, N. N.: Acta Endocrinol. **25**, 133 (1957).
218. Hobkirk, R., Blahey, P. R., Alfheim, A., Raeside, J. I., Joron, G. E.: J. Clin. Endocrinol. Metab. **20**, 805 (1960).
219. Frandsen, V. A., Stakemann, G.: Danish Med. Bull. **7**, 95 (1960).
220. Greene, J. W., Jr., Touchstone, J. C., Fields, H.: Obstet. Gynecol. **17**, 349 (1961).
221. Slaunwhite, W. R., Jr., Karsay, M. A., Sandberg, A. A.: J. Clin. Endocrinol. Metab. **24**, 263 (1964).
222. Diczfalusy, E.: Abstr. 90, Internat. Congr. Hormonal Steroids Milan, Italy, May 1962, p. 93.
223. Touchstone, J. C., Greene, J. W., Jr., McElroy, R. C., Murawec, T.: Biochemistry **2**, 653 (1963).
224. — Murawec, T.: Biochemistry **4**, 1612 (1965).
225. Smith, O. W.: Acta Endocrinol. **51**, Suppl. **104**, 3 (1966).
226. Brown, C. H., Saffan, B. D., Howard, C. M., Preedy, J. R. K.: J. Clin. Invest. **43**, 295 (1964).
227. Hähnel, R.: J. Endocrinol. **38**, 417 (1967).
228. Levitz, M., Emerman, S., Dancis, J., in: Sterol synthesis in the perfused placenta (F. Polvani and A. Bompiani, eds.), Baltimore, Md.: Williams & Wilkins Co. 1964.
229. Ryan, K. J.: J. Biol. Chem. **234**, 268 (1959).
230. Baulieu, E.-E., Dray, F.: J. Clin. Endocrinol. Metab. **23**, 1298 (1963).
231. Siiteri, P. K., MacDonald, P. C.: Steroids **2**, 713 (1963).
232. Bolté, E., Mancuso, S., Eriksson, G., Wiqvist, N., Diczfalusy, E.: Acta Endocrinol. **45**, 535 (1964).
233. Frandsen, V. A., Stakemann, G.: Acta Endocrinol. **38**, 383 (1961).
234. Migeon, C. J., Keller, A. R., Holmstrom, E. G.: Bull. Johns Hopkins Hosp. **97**, 415 (1955).
235. Colás, A., Heinrichs, W. L., Tatum, H. J.: Steroids **3**, 417 (1964).
236. Diczfalusy, E.: Proc. Second Intern. Congr. Endocrinology, London, England, 1964, p. 732 (Publ. 1965).
237. Beer, C. T., Gallagher, T. F.: J. Biol. Chem. **214**, 335 (1955).
238. — — J. Biol. Chem. **214**, 351 (1955).
239. Sandberg, A. A., Slaunwhite, W. R., Jr.: J. Clin. Invest. **36**, 1266 (1957).
240. Migeon, C. J., Wall, P. E., Bertrand, J.: J. Clin. Invest. **38**, 619 (1959).
241. Wilson, R., Eriksson, G., Diczfalusy, E.: Acta Endocrinol. **46**, 525 (1964).
242. Goebelsmann, U., Eriksson, G., Wiqvist, N., Diczfalusy, E.: Acta Endocrinol. **50**, 273 (1965).

243. Goebelsmann, U., Cooke, I., Wiqvist, N., Diczfalusy, E.: Acta Endocrinol. **52**, 30 (1966).
244. — Sjöberg, K., Wiqvist, N., Diczfalusy, E.: Acta Endocrinol. **50**, 261 (1965).
245. Crowell, G. C., Howard, C. M., Preedy, J. R. K.: J. Clin. Endocrinol. Metab. **26**, 631 (1966).
246. Adlercreutz, H.: Acta Endocrinol. **40**, Suppl. **67**, 89 (1962).
247. — Acta Endocrinol. **42**, Suppl. **72**, 1 (1963).
248. — Acta Med. Scand. **175**, Suppl. **412**, 133 (1964).
249. — Luukkainen, T.: Biochim. Biophys. Acta **97**, 134 (1965).
250. Levitz, M., Twombly, G. H., Emerman, S.: Abstr. 78, 48th Meeting of the Endocrine Society, Chicago, Ill., June 1966, p. 65.
251. Diczfalusy, E., Cassmer, O., Alonso, C., De Miguel, M.: Recent Progr. Hormone Res. **17**, 147 (1961).
252. Troen, P., Diczfalusy, E.: J. Clin. Invest. **40**, 1085 (1961).
253. Schwers, J.: Abstr. 412, Internat. Congr. Hormonal Steroids, Milan, Italy, May 1962, p. 279.
254. Mikhail, G., Wiqvist, N., Diczfalusy, E.: Acta Endocrinol. **42**, 519 (1963).
255. Barr, M., Diczfalusy, E., Tillinger, K. G.: Acta Endocrinol. **37**, 241 (1961).
256. Glass, S. J., Edmondson, H. A., Soll, S. N.: Endocrinology **27**, 749 (1940).
257. — — — J. Clin. Endocrinol. **4**, 54 (1944).
258. Rupp, J., Cantarow, A., Rakoff, A. E., Paschkis, K. E.: J. Clin. Endocrinol. **11**, 688 (1951).
259. Dohan, F. C., Richardson, E. M., Bluemle, L. W., Jr., Gyorgy, P.: J. Clin. Invest. **31**, 481 (1952).
260. Pincus, I. J., Rakoff, A. E., Cohn, E. M., Tumen, H. J.: Gastroenterology **19**, 735 (1951).
261. Stimmel, B. F.: J. Clin. Endocrinol. Metab. **14**, 764 (1954).
262. Cameron, C. B.: Brit. Med. Bull. **13**, 119 (1957).
263. Diczfalusy, E., Franksson, C., Martinsen, B.: Acta Endocrinol. **35**, Suppl. **51**, 743 (1960).
264. — — — Acta Endocrinol. **38**, 59 (1961).
265. Conti, C., Sciarra, F., Concolino, G., Sorcini, G.: Folia Endocrinol. (Roma) **17**, 736 (1964).
266. Brown, J. B., Crean, G. P., Ginsburg, J.: Gut **5**, 56 (1964).
267. Preedy, J. R. K., Howard, C. M., Eren, S.: Clin. Res. **14**, 88 (1966).
268. Adlercreutz, H., Svanborg, A., Ånberg, Å.: Am. J. Med. **42**, 341 (1967).
269. Ginsburg, J., Brown, J. B.: Lancet **II**, 1274 (1961).
270. Gabrilove, J. L., Sharma, D. C., Wotiz, H. H., Dorfman, R. I.: Medicine **44**, 37 (1965).
271. Flickinger, G. L., Murawec, T., Touchstone, J. C.: J. Clin. Endocrinol. Metab. **25**, 1231 (1965).
272. Barlow, J. J., Goldstein, D. P.: Abstr. 80, 48th Meeting of the Endocrine Society, Chicago, Ill., June 1966, p. 66.
273. MacDonald, P. C., Siiteri, P. K.: Steroids **8**, 589 (1966).
274. Kase, N., Cohn, G. L.: New Engl. J. Med. **276**, 28 (1967).
275. Browne, J. S. L., Henry, J. S., Venning, E. M.: J. Clin. Invest. **16**, 678 (1937).

276. Arcos, M., Gurpide, E., VandeWiele, R. L., Lieberman, S.: J. Clin. Endocrinol. Metab. **24**, 237 (1964).
277. Klopper, A. I.: J. Obstet. Gynaecol. Brit. Empire **64**, 504 (1957).
278. Diczfalusy, E., Troen, P.: Vitamins Hormones **19**, 229 (1961).
279. Shearman, R. P.: J. Obstet. Gynaecol. Brit. Empire **66**, 1 (1959).
280. Pearlman, W. H.: Biochem. J. **67**, 1 (1957).
281. Sandberg, A. A., Slaunwhite, W. R. Jr.: J. Clin. Endocrinol. Metab. **18**, 253 (1958).
282. Wiest, W. G., Fujimoto, G. I., Sandberg, A. A.: J. Clin. Endocrinol. Metab. **18**, 972 (1958).
283. Chang, E., Slaunwhite, W. R., Jr., Sandberg, A. A.: J. Clin. Endocrinol. Metab. **20**, 1568 (1960).
284. Zander, J., Holzman, K., Bengtsson, L. P.: Acta Obstet. Gynecol. Scand. **44**, 204 (1965).
285. Thijssen, J. H. H., Zander, J.: Acta Endocrinol. **51**, 563 (1966).
286. Pion, R. J., Conrad, S. H., Wolf, B. J.: J. Clin. Endocrinol. Metab. **26**, 225 (1966).
287. Calvin, H. I., Lieberman, S.: J. Clin. Endocrinol. Metab. **26**, 402 (1966).
288. Conrad, S. H., Pion, R. J., Kitchin, J. D., III: J. Clin. Endocrinol. Metab. **27**, 114 (1967).
289. Jones, G. E. S., Delfs, E.: J. Am. Med. Assoc. **146**, 1212 (1951).
290. Weil, P. G.: Science **87**, 72 (1938).
291. Smith, G. van S., Smith, O. W.: Am. J. Obstet. Gynecol. **39**, 405 (1940).
292. Cope, C. L.: Lancet **II**, 158 (1940).
293. Bledsoe, T., Island, D. P., Ney, R. L., Liddle, G. W.: Abstr. 41, 45th Meeting of the Endocrine Society, Atlantic City, N. J., June 1963, p. 21.
294. Southren, A. L., Tochimoto, S., Strom, L., Ratuschni, A., Ross, H., Gordon, G.: J. Clin. Endocrinol. Metab. **26**, 268 (1966).
295. Bradlow, H. L., Zumoff, B., Gallagher, T. F., Hellman, L.: Abstr. 246, Internat. Congr. Hormonal Steroids, Milan, Italy, May 1962, p. 153.
296. Wallace, E. Z., Carter, A. C.: J. Clin. Invest. **39**, 601 (1960).
297. Marks, L. J., Friedman, G. R., Duncan, F. J.: J. Lab. Clin. Med. **57**, 47 (1961).
298. Katz, F. H., Lipman, M. M., Frantz, A. G., Jailer, J. W.: J. Clin. Endocrinol. Metab. **22**, 71 (1962).
299. Vermeulen, A., Ferin, J.: Acta Endocrinol. **39**, 22 (1962).
300. Layne, D. S.: Hormonal steroids. Proc. Intern. Congr. Hormonal Steroids, 1st, 1964 (1) 433.
301. Berliner, D. L., Keller, N., Dougherty, T. F.: Endocrinology **68**, 621 (1961).
302. — Dougherty, T. F.: Federation Proc. **19**, 160 (1960).
303. Werk, E. E., Jr., MacGee, J., Sholiton, L. J.: J. Clin. Invest. **43**, 1824 (1964).
304. Linquette, M., Fossati, P., May, J. P., Racadot, A., Hubschmann, B.: Ann. Endocrinol. **26**, 192 (1965).
305. Hartiala, K. J. W., Pulkkinen, M., Ball, P.: Nature **210**, 739 (1966).
306. Done, A. K., Ely, R. S., Kelley, V. C.: Metabolism **4**, 129 (1955).
307. Stempfel, R. S., Jr., Sidbury, J. B., Jr., Migeon, C. J.: J. Clin. Endocrinol. Metab. **20**, 814 (1960).
308. Crigler, J. F., Gold, N. I.: J. Clin. Invest. **45**, 998 (1966).

Indices

AUTHOR INDEX

SUBJECT INDEX

A

B

C

D

E

F

G

H

I

J

K

L

M

N

O

P

R

S

T

U

V

W

Y

Z

MIX
Papier aus verantwortungsvollen Quellen
Paper from responsible sources
FSC® C105338

If you have any concerns about our products,
you can contact us on
ProductSafety@springernature.com

In case Publisher is established outside the EU,
the EU authorized representative is:
Springer Nature Customer Service Center GmbH
Europaplatz 3, 69115 Heidelberg, Germany

Printed by Libri Plureos GmbH
in Hamburg, Germany

MIX
Papier aus verantwortungsvollen Quellen
Paper from responsible sources
FSC® C105338

If you have any concerns about our products,
you can contact us on
ProductSafety@springernature.com

In case Publisher is established outside the EU,
the EU authorized representative is:
Springer Nature Customer Service Center GmbH
Europaplatz 3, 69115 Heidelberg, Germany

Printed by Libri Plureos GmbH
in Hamburg, Germany